The ASQ Certified Six Sigma Green Belt Handbook

Third Edition

The ASQ Certified Six Sigma Green Belt Handbook

Third Edition

Roderick A. Munro, Govind Ramu, and Daniel J. Zrymiak, editors

ASQExcellence
Milwaukee, Wisconsin

Published by ASQExcellence, Milwaukee, WI

Produced and distributed by Quality Press, ASQ, Milwaukee, WI

Publisher's Cataloging-in-Publication data

Names: Munro, Roderick A., editor. | Ramu, Govindarajan, editor. | Zrymiak, Daniel J., editor.
Title: The ASQ certified six sigma green belt handbook , third edition / Roderick A. Munro, Govind Ramu, and Daniel Zrymiak, editors.
Description: Includes bibliographic references and index. | Milwaukee, WI: Quality Press, 2022.
Identifiers: LCCN: 2022941555 | ISBN: 978-1-952236-17-4 (hardcover) | 978-1-63694-029-8 (hardcover) | 978-1-952236-18-1 (epub) | 978-1-63694-031-1 (epub) | 978-1-63694-030-4 (pdf) | 978-1-952236-33-4 (pdf) | Subjects: LCSH Six sigma (Quality control standard)--Handbooks, manuals, etc. | Production management--Handbooks, manuals, etc. | Quality control--Statistical methods--Handbooks, manuals, etc. | BISAC BUSINESS & ECONOMICS / Quality Control | STUDY AIDS / Professional
Classification: LCC TS156 .C4235 2022 | DDC 658.5--dc23

ASQ and ASQExcellence advance individual, organizational, and community excellence worldwide through learning, quality improvement, and knowledge exchange.

Bookstores, wholesalers, schools, libraries, and organizations: Quality Press and ASQExcellence books are available at quantity discounts with bulk purchases for business, trade, or educational uses. For more information, please contact Quality Press at 800-248-1946 or books@asq.org.

To place orders or browse the selection of ASQExcellence and Quality Press titles, visit our website at: http://www.asq.org/quality-press.

28 27 26 GP 7 6

ASQ

Quality Press
600 N. Plankinton Ave.
Milwaukee, WI 53203-2914
Email: books@asq.org
Excellence Through Quality™

ASQE™
Inspire Excellence

In remembrance of

DEBORAH L. HOPEN

1953-2020

The authors wish to dedicate this edition of *The ASQ Certified Six Sigma Green Belt Handbook* to the memory of Deborah L. Hopen. After her untimely passing, some came to call her the "First Lady of Quality." Deb was passionate about all aspects of quality and devoted herself to her passions with energy, warmth, and generosity. She asserted herself to establish her role as a leader and promoter, taking ASQ to peak levels of vitality and relevance. She contributed greatly to the methodology of Six Sigma as well as the "soft" quality skills to the more appropriate term of "support" quality skills.

Although Deb preferred to remain out of the limelight, she nevertheless received accolades. Among the honors and awards presented were:

- ASQ Fellow and ASQ Distinguished Service Medal
- American Productivity and Quality Center's C. Jackson Grayson Quality Pioneer Medal
- IAQ Founders Medal
- IAQ Agnes Žaludová Woman of Quality
- Asia-Pacific Quality Organization Miflora M. Gatchalian Medal for Women in Quality Leadership
- ASQ Quality Management Division Roger Berger Spirit Award
- Frank M. Gryna Award for Excellence in a Written Article on Quality Management
- Simon Collier Quality Leadership Award

Table of Contents

List of Figures and Tables

Preface to *The ASQ Certified Six Sigma Green Belt Handbook*, Third Edition

Welcome to *The ASQ Certified Six Sigma Green Belt Handbook,* Third Edition. This reference manual is designed to help those interested in passing the American Society for Quality's (ASQ) certification exam for Six Sigma Green Belts and others who want a handy reference to the appropriate materials needed to conduct successful Green Belt projects. This book is not intended to be a beginner's Six Sigma Green Belt book, but a reference handbook on running projects for those who are already knowledgeable about process improvement and variation reduction.

The primary layout of the handbook follows the American Society for Quality Body of Knowledge (BoK) for the Certified Six Sigma Green Belt (CSSGB) updated in 2022. The authors were involved with the first edition handbook and have utilized previous edition user comments, numerous Six Sigma practitioners, and their own personal knowledge gained through helping others prepare for exams to bring together a handbook that we hope will be very beneficial to anyone seeking to pass the ASQ or other Green Belt exams. In addition to the primary text, we have added two new appendices and an expanded acronym list.

WHERE ARE YOU IN YOUR CAREER?

As your professional career develops, you may wish to choose to use the tools you have learned in advancing your own career. Some have called this *career AQP.* Please visit www.asq.org to see how ASQ conducts exams to be able to advance your career. An overview of the level of accomplishment for each certification is on the next page.

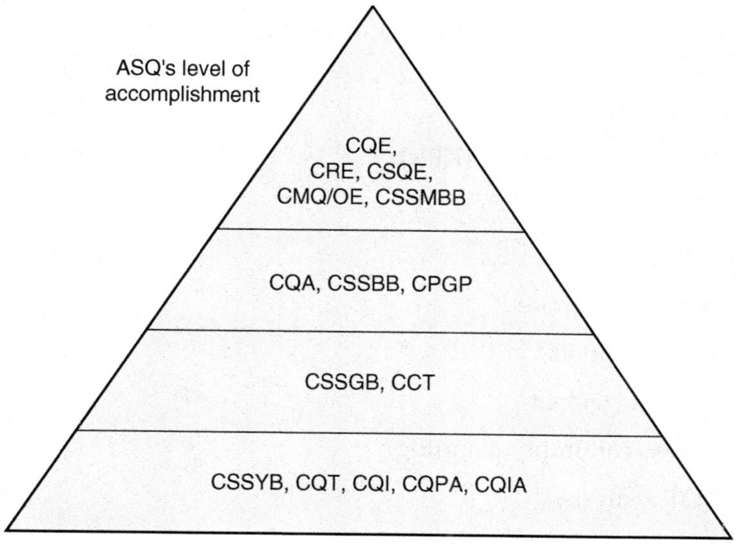

CHANGES TO THE BoK AND THIS HANDBOOK

A detailed cross-matrix of the updated BoK and the previous version can be found in the appendices.

Some of the highlighted sections include: The section on tools used in Six Sigma has been moved to Chapter 7 to align with the new BoK. Several new portions have been added to the third edition to align with the new BoK.

Other major changes to the 2022 CSSGB BoK can be seen in the table below.

Content new to 2022 CSSGB BoK

BoK area	Topic	Subtopic		Description	Bloom's Taxonomy
V	B			Implementation planning	Apply
VI	B	2		Document control	Understand
VI	B	3		Training plans	Apply
VI	B	4		Audits	Remember
VI	B	5		Plan-do-check-act (PDCA)	Apply

You will also find additional content in previously existing sections of the BoK. This includes:

- SMART goals,
- Key process indicators (KPIs),
- Takt time,
- Just-in-time processes,
- Spaghetti diagrams,
- The Kano model,
- Risk management,
- Business continuity planning,
- SWOT analysis,
- RACI charts,
- Data collection plans and quality checks,
- Gap analysis,
- 5 Whys analysis,
- Fault tree analysis,
- And maintaining quality improvements.

Acknowledgments

As in any work of substance, there are a lot of people involved. The authors would like to also thank all these people for their time and support. The reviewers for this third edition worked hard with us to ensure that we covered the ASQ CSSGB BoK well. We also had many individuals assist this ongoing project by helping to develop practice exam questions for a separate book: *The ASQ CSSGB Study Guide*.

We have asked in the past, and continue to, for comments from you, our customers. If you have any comments about this handbook, we welcome to hear from you at books@asq.org.

—Roderick Munro, Govind Ramu, and Daniel Zrymiak

I would like to thank my coauthors Dan and Govind for their ongoing dedication as we strive to continually improve this handbook and help those who are striving to achieve the CSSGB level. I would also like to thank the Quality Press team and ASQE Staff for directions and guidance as we transition our organization in our new structure.

—Roderick Munro

I thank my parents Vasantha and Ramu for instilling in me passion for learning and knowledge sharing, and my wife Anitha and my children Vibha and Vivek for their encouragement, unconditional love, and support.

I want to acknowledge several quality professionals from ASQ and other professional organizations who have provided continual inspiration and guidance in writing and improving this book. Thanks to my fellow authors Rod and Dan for their friendship and support.

—Govind Ramu

I would like to acknowledge my wife Susan, my children Ryan and Brooke, and my extended family members for their personal encouragement. The ongoing and continuous support of my present and past employers and clients, my professional mentors throughout my career, and my peers and partners within the Quality community is most appreciated. Collaborations with prominent and prolific authors and experts like Govind, Rod, and Elizabeth have inspired my efforts and have been among the highlights of my service within the Quality profession.

I must also express gratitude to our community of Six Sigma and Lean Green Belt practitioners who bring these concepts to life within their workplaces and communities. Our Society is stronger through our mutual advocacy and active demonstration of these principles and practices.

—*Daniel Zrymiak*

FIRST EDITION ACKNOWLEDGMENTS

The authors would like to acknowledge the following CSSGB Handbook users who provided inputs and feedback on the errors from the first edition. We have made best effort to include everyone who provided inputs.

Frank Alt
Michael J. Donohoe
Jim Ogren
Arulalan Ram
Eric P. Wayner
Eric J. Lim
Alejandro Ramirez
Sharon Johnson
Jennifer Park Martin
Doreen J. Harris
Chris Jespersen
Timothy Schubach
Mike McWilliams
Veena Nithoo
Sweeny Chhabra
Eric Gemunder
Lester Ramirez Pérez
Joao Felizardo
Scott Lawley
MK
Michelle Gabriel
Dermot Freeman & Associates
Jason Gorman
Greg Crook

SECOND EDITION ACKNOWLEDGMENTS

The authors would like to acknowledge the Quality Class from Davenport College in Grand Rapids, Michigan, with their instructor, Mr. Don Brecken, who provided inputs and feedback on the errors from the second edition. In addition, we would like to thank the Gemba Academy and PQ Systems for their ongoing support of these efforts.

THIRD EDITION ACKNOWLEDGMENTS

The authors would like to thank N. Gomez for a thoughtful detailed review of the section edition handbook that has been used to enhance the third edition. Thanks also to Jim Bossert for the time spent to review the entire manuscript.

Part I

Overview: Six Sigma and the Organization

Each section of this book parallels a section of the ASQ Body of Knowledge (BoK) for the Certified Six Sigma Green Belt exam.

Table 1.0 CSSGB BoK section item allocation.

BoK Section	2014 BoK	2022 BoK	Difference
I. Overview: Six Sigma and the Organization	13	11	-2
II. Define Phase	23	20	-3
III. Measure Phase	23	20	-3
IV. Analyze Phase	15	18	+3
V. Improve Phase	15	16	+1
VI. Control Phase	11	15	+4

Chapter 1

Six Sigma and Organizational Goals

VALUE OF SIX SIGMA

> Recognize why organizations use six sigma, how they apply its philosophy and goals, and the evolution of six sigma from quality leaders such as Juran, Deming, Shewhart, Ishikawa, and others. (Understand)
>
> **Body of Knowledge I.A.1**

Every organization, even not-for-profits, must have a source of income in some form or another to stay in business. If an organization spends more than it takes in over time, then it will be out of business. Thus, the challenge for every organization is to become profitable at whatever it does (even if this involves soliciting contributions) so that it can continue to do what it does. Managers, employees, suppliers, process owners, stakeholders, and customers (internal or external) all have their wants and needs that the business must satisfy in an efficient manner so profit can be achieved. Thus, the first formula that every Six Sigma Green Belt must learn is the calculation for **$** (we sometimes call this S-double bar, and it can be tied to the cost of quality or other financial calculations found in your organization). Without a focus on the financials and the impact that this key performance indicator has on this the bottom line, management will drift off to other random issues that will draw their attention.

Why Use Six Sigma versus Other Methodologies?

After Motorola started promoting their Six Sigma methodology in the late 1980s, there have been many skeptical of its true value. Even Jack Welch of General Electric (GE) initially dismissed the idea of Six Sigma as a passing fad in the early 1990s. However, once GE had a successful launch in one of its divisions, Six Sigma quickly became a driving force in the mid to late 1990s that started spreading across various industries. The Six Sigma buzz, fad, or whatever name it was called, started proving that it was something different, something more than the numerous other business fads that had preceded it.

The real power of Six Sigma is the use of many parts or elements of other methods that have been proven to work, in tandem with managerial focus, to create an organizational network of activities that support the efforts to continually improve on all aspects of the organization, in conjunction with standard accounting practices that demonstrate the impact of continual improvement and variation reduction on the organization's bottom line.

Six Sigma should be a large collection of tools that the organization can bring to bear as appropriate on identified issues to achieve continual improvement across the entire organization. Learning to use these various tools effectively takes time and practice and leads to the distinction of what are called *levels of competence*, or *belts*. Typical titles include White, Yellow, Green, Black, and Master Black Belt (some organizations use fewer or more belts depending on their organizational structure or needs).

How Six Sigma Philosophy and Goals Should Be Applied

With the various successes, there have been even more failures of organizations attempting to implement a Six Sigma methodology. The reasons are many; however, the most common failure is management's lack of commitment to real process improvement. Another leading reason for lack of good payback results is training too many Black Belts in the initial stages of the process, before the organization knows how to deploy the process successfully in the organization or to give those new to the use of process improvement tools time to effectively deploy these applications.

Since many managers often look for the magic bullet, they tend to hire outside consultants to come into the organization and start training Black Belts, who then are expected to conduct projects to save large amounts of cash to the bottom line. The initial waves did typically save a lot of money; however, once the consultants left, there were few internal people who understood the tools at high enough levels to encourage the use of the Six Sigma methodology. Eager to see a return on investment for all the training being done, the consultants are only engaged for short periods, and the managers then expect their internal people to move the process forward.

The truth about any process improvement effort is that it typically takes a person a full two years or more to learn how the tools work and to understand their applications. One experiment conducted in Michigan under a Robert Woods Johnson Grant involved improving performance in practice (doctor's office). Groups in several states had started this process by engaging nurses in those states and teaching them process improvement tools. They did show some success in each program tried. However, the group in Michigan engaged the Automotive Industry Action Group (AIAG), and the top doctors of the U.S. automotive industry got involved by directing that quality engineering and process engineers be used instead of nurses to go into the selected doctors' offices around the state. The AIAG taught the engineers the basics of working in a doctor's office and sent them out to work on processes. Nurses were not excluded, but worked with the quality engineers, who knew how to use the tools.

The results were outstanding and leapfrogged the other seven states involved in the study. One of the head doctors was Joseph Fortuna, MD, who subsequently became the chair of the ASQ Healthcare Division and promoted the use of process engineers in the healthcare field.

What was learned and recognized is that a good understanding of the basic tools should be the first step in setting up a Six Sigma process. This is why some companies today choose to start their process improvement journey by training Green Belts first. Management also needs to learn how these tools work so that they can direct and ask pertinent questions of those running the projects in the organization. The other key is to engage the accounting department very early in the deployment, as they must be able to substantiate the cost savings being claimed or achieved if the Six Sigma methodology is truly going to show the bottom-line (sometime jokingly called the S-double bar) savings and return on investment in the process. A good video demonstrating this was actually created back in the mid-1950s, called *Right First Time* (or sometimes *Right the First Time*).

Thus, management should use the tool of *advanced quality planning* (AQP) to prepare for a Six Sigma deployment to increase the likelihood of success within their organization. In using AQP, managers need to start learning the tools themselves, start engaging the organization using data-driven decision making, and start training Green Belts to work on small projects that can grow in time into larger projects where Black Belts can be trained and utilized more effectively. As the process gains steam within the organization, S-double bar is used (one tool here could be cost of quality) as a focal point for the organization in moving projects through the system to continually improve the processes for customer satisfaction.

The Lead-Up to the Six Sigma Methodology

Over the centuries, managers have tried to find ways to keep their organization in business (sometimes called the *magic bullet*). Many different techniques have been employed over the years to keep customers coming back time and time again. Unfortunately for many organizations, customer wants and needs change over time, leaving the organization with the challenge of finding new and better ways of satisfying those needs and wants. The concept of setting standards of work goes back many centuries and was the foundation of the guilds and crafts trades that developed over the years. During the mid-1800s to early 1900s, separation of work was developed to speed up the process of development and production. Innovators like Frederick W. Taylor and Henry Ford developed ideas and techniques that are still with us today. On the quality side of the production calculation, many techniques have been tried, starting with control charts in the 1920s–1930s by Walter Shewhart.

In the early part of the last century, given the methods of doing business, the *quality control/quality assurance* (QC/QA) specialist was created to ensure that standards were established and maintained so that customers would be satisfied. In many organizations, however, this also created a separation of tasks, and many people in organizations came to think of the responsibility for satisfying customers as only in the hands of the people in the QC/QA groups/departments instead of in the hands of the people who actually did the work of making the product or providing the service. This was especially true in the United States during the decades of the 1950s, 1960s, and 1970s as managers looked for better ways to try to manage all the resources of the organization. Many organizations still struggle with customer satisfaction!

In the mid-1920s a young engineer named Walter Shewhart devised a technique of using graphs to monitor a process to identify whether that process was acting in a predicable manner or if what he termed *special causes* were affecting the process. These charts became known as *quality control charts* (the *p*-chart was the first to be used); however, today we sometimes call them *process behavior charts,* as we want to look at what the process is doing in relation to statistical probabilities. Many other tools and techniques have been developed since then, known by a long list of names. Quality developments over the years are summarized in Table 1.1. (A very good book on the history of quality leading up to and including the Six Sigma process is the book *Fusion Management: Harnessing the Power of Six Sigma, Lean, ISO 9001:2000, Malcom Baldrige, TQM and other Quality Breakthroughs of the Past Century.*[1]) A timeline for the history of quality, lean, and other related events is provided in Appendix N.

Table 1.1 Some approaches to quality over the years.

Quality approach	Approximate time frame	Short description
Quality circles	1979–1981	Quality improvement or self-improvement study groups composed of a small number of employees (10 or fewer) and their supervisor. Quality circles originated in Japan, where they are called "quality control circles."
Statistical process control (SPC)	Mid-1980s	The application of statistical techniques to control a process. Also called "statistical quality control."
ISO 9000	1987–present	A set of international standards on quality management and quality assurance developed to help companies effectively document the quality system elements to be implemented to maintain an efficient quality system. The standards, initially published in 1987, are not specific to any particular industry, product, or service. The standards were developed by the International Organization for Standardization (ISO), a specialized international agency for standardization composed of the national standards bodies of 91 countries. The standards underwent revisions in 2000, 2008, and 2015, and now comprise ISO 9000 (definitions), ISO 9001 (requirements), and ISO 9004 (continuous improvement).
Reengineering	1996–1997	A breakthrough approach involving the restructuring of an entire organization and its processes.

Continued

Table 1.1 Some approaches to quality over the years. *(Continued)*

Quality approach	Approximate time frame	Short description
Benchmarking	1988–1996	An improvement process in which a company measures its performance against that of best-in-class companies, determines how those companies achieved their performance levels, and uses the information to improve its own performance. The subjects that can be benchmarked include strategies, operations, processes, and procedures.
Balanced scorecard	1990s–present	A management concept that helps managers at all levels monitor their results in their key areas.
Baldrige Award criteria	1987–present	An award established by the U.S. Congress in 1987 to raise awareness of quality management and recognize U.S. companies that have implemented successful quality management systems. Two awards may be given annually in each of five categories: manufacturing company, service company, small business, education, and healthcare. The award is named after the late Secretary of Commerce Malcolm Baldrige, a proponent of quality management. The U.S. Commerce Department's National Institute of Standards and Technology manages the award, and ASQ administers it.
Six Sigma	1995–present	As described in Chapter 1.
Lean manufacturing	2000–present	As described in Chapter 2.

Modern Six Sigma

Shortly after the Motorola Company achieved the Malcolm Baldrige National Quality Award (MBNQA) in 1988, they came calling to the Ford Motor Company to try to sell some new radios. The Ford purchasing department had just started a new process called the *supplier quality improvement* (SQI) initiative that was designed to work with external manufacturing suppliers from new design concept to launch of new vehicles. Ford had developed a *planning for quality* process using ASQ's *advanced quality planning* (AQP) and wanted to improve supplier quality delivered to the automotive assembly plants. This effort was instrumental in the development of what is now called *advanced product quality planning* (APQP), used in the automotive industry.

The Motorola sales team presented their newly developed methodology called Six Sigma, which was considered a key to achieving the MBNQA, to a Ford SQI senior quality engineer who was assigned to evaluate the Six Sigma methodology in relation to the Ford Q1 (Ford's top award, which is still available today) and Q-101 (forerunner of the QS-9000 and ISO/TS 16949 and the current IATF 16949) programs. The Ford SQI senior quality engineer liked what he saw except for one particular item: In the early days, a Six Sigma process was described as ±3 standard deviations ($C_p = 1.0$). The Ford requirement was $C_{pk} \geq 1.33$ for ongoing processes and $C_{pk} \geq 1.67$ for startup processes. (Note: Even as late as 1998, Mikel Harry's book *Six Sigma Producibility Analysis and Process Characterization* had a note on page 3-3 stating "Today, some organizations require $1.5 < C_p < 2.0$, or even $C_p < 2.0$.")[2] An interesting aside is that later discussions with Motorola technical individuals did confirm that the salespeople did not understand what they were presenting, and that inside Motorola the 6σ process was actually ±6 standard deviations, but this was well after the initial presentation, and Motorola lost the potential sale to Ford.

Six Sigma is a structured and disciplined process designed to deliver perfect products and services on a consistent basis. It aims at improving the bottom line by finding and eliminating the causes of mistakes and defects/deficiencies in business processes. Today, Six Sigma is associated with process capabilities of $C_{pk} > 2.0$ (some would say $C_p = 2.0$ and $C_{pk} < 1.5$), which are considered world-class performance (this allows for the 1.5 sigma shift factor). Remember that *sigma* is a statistical term that refers to the standard deviation of a process around its mean versus the methodology of problem solving that has been labeled Six Sigma.

A wide range of companies have found that when the Six Sigma philosophy is fully embraced, the enterprise thrives. What is this Six Sigma philosophy? Several definitions have been proposed. The threads common to these definitions are:

- Use of teams that are assigned well-defined projects that have direct impact on the organization's bottom line.
- Training in "statistical thinking" at all levels and providing key people with extensive training in advanced statistics and project management. These key people are designated "Black Belts."
- Emphasis on the DMAIC approach to problem solving: define, measure, analyze, improve, and control.
- A management environment that supports these initiatives as a business strategy.
- Continual effort to reduce variation in all processes within the organization.

Opinions on the definition of Six Sigma can differ:

- *Philosophy.* The philosophical perspective views all work as processes that can be defined, measured, analyzed, improved, and controlled (DMAIC). Processes require inputs and produce outputs. If you control the inputs, you will control the outputs. This is generally expressed as the $y = f(x)$ concept.
- *Set of tools.* Six Sigma as a set of tools includes all the qualitative and quantitative techniques used by the Six Sigma expert to drive process improvement. A few such tools include SPC, control charts, failure mode and effects analysis, and process mapping. There is probably little agreement among Six Sigma professionals as to what constitutes the tool set.

- *Methodology.* This view of Six Sigma recognizes the underlying and rigorous approach known as DMAIC. DMAIC defines the steps a Six Sigma practitioner is expected to follow, starting with identifying the problem and ending with the implementation of long-lasting solutions. While DMAIC is not the only Six Sigma methodology in use, it is certainly the most widely adopted and recognized.
- *Metrics.* In simple terms, Six Sigma quality performance means 3.4 defects per million opportunities (accounting for a 1.5-sigma shift in the mean).

At this point, Six Sigma purists will be quick to say, "You're not just talking about Six Sigma; you're talking about lean, too." Today, the demarcation between Six Sigma and lean has blurred. With greater frequency, we are hearing about terms such as *sigma-lean, LSS,* or *Lean Six Sigma* because process improvement requires aspects of both approaches to attain positive results.

Six Sigma focuses on reducing process variation and enhancing process control, while lean—also known as *lean manufacturing*—drives out waste (non-value-added activities) and promotes work standardization and value stream mapping. Six Sigma practitioners should be well versed in both.

Quality Pioneers

Most of the techniques found in the Six Sigma toolbox have been available for some time thanks to the groundbreaking work of many professionals in the quality sciences. These and many others have contributed to the quality profession. Some of the key contributors include (in alphabetic order):

Subir Chowdhury is one of the new leaders in management thought and is being recognized by many companies and organizations as being on the forefront of customer satisfaction in today's business world. Dr. Chowdhury has written or coauthored a growing number of books on management with other top quality and business leaders. The following themes are found in most of his books:

- Problems can be prevented through continuous improvement—getting it right the first time—and should be the goal of every organization as it designs, develops, and deploys products and services.
- Quality must be the responsibility of every individual in all organizations. The "quality mission" cannot be delegated to one group or individual. It cannot be a "top down" management process. For quality to be robust and sustainable, everyone in the organization must not only accept it, they must believe in it.
- Quality begins at the top. Without the commitment of leadership—and without them demonstrating that commitment in every aspect of their own lives—initiatives will stall or fail over time.
- Everyone has a stake in quality. Not only must quality involve everyone all the time, but in order to achieve robust and sustainable results, everyone must have a stake in its implementation and

continuous improvement through peer reinforcement and other methods.

- Quality is a balance of people power and process power, where "people power" takes into account the role of the quality mind-set—approaching quality with honesty, empathy, and a resistance to compromise. Process power is about solving problems, developing ideas and solutions, and then perfecting those ideas and solutions.
- Improving quality using a cookie-cutter managerial approach does not work. Every organization is unique. Every problem has different issues. Every individual brings different knowledge, skills, and abilities. Therefore, the methods, processes, and procedures used to solve quality issues must be tailored to the specific situation.

One of Dr. Chowdhury's mentors was Philip Crosby. Some of Dr. Chowdhury's books show Crosby's influence, as can been seen in:

- *The Ice Cream Maker.* 2006. Doubleday, Random House.
- *The Power of LEO: The Revolutionary Process for Achieving Extraordinary Results.* 2011. McGraw Hill. LEO is an acronym that stands for *listen* (observe and understand), *enrich* (explore and discover), and *optimize* (improve and perfect).
- *The Difference: When Good Enough Isn't Enough.* 2017. Currency. The difference is summed up for leaders as using STAR in their lives: being Straightforward, being Thoughtful, being Accountable, and having Resolve.

Key contributions:

- Growing number of management books with focus on creating future products for Design for Six Sigma
- Working with top-level management teams to recognize the need for quality
- Thinkers 50 of London named Chowdhury the "quality prophet"

Philip Crosby wrote fourteen books including *Quality Is Free, Quality without Tears, Let's Talk Quality,* and *Leading: The Art of Becoming an Executive.* Crosby, who originated the *zero defects* concept, was an ASQ honorary member and past president. Crosby's fourteen steps to quality improvement as noted in the *Certified Manager of Quality/Organizational Excellence Handbook*[3] are:

1. Make it clear that management is committed to quality.
2. Form quality improvement teams with representatives from each department.
3. Determine how to measure where current and potential quality problems lie.
4. Evaluate the cost of quality and explain its use as a management tool.

5. Raise the quality awareness and personal concern of all employees.
6. Take formal actions to correct problems identified through previous steps.
7. Establish a committee for the zero defects program.
8. Train all employees to actively carry out their part of the quality improvement program.
9. Hold a "zero defects day" to let all employees realize that there has been a change.
10. Encourage individuals to establish improvement goals for themselves and their groups.
11. Encourage employees to communicate to management the obstacles they face in attaining their improvement goals.
12. Recognize and appreciate those who participate.
13. Establish quality councils to communicate on a regular basis.
14. Do it all over again to emphasize that the quality improvement program never ends.

Key contributions:

- Management theory for quality
- Engaged business executives in quality

W. Edwards Deming emphasized the need for changes in management structure and attitudes. He developed a list of "Fourteen Points." As stated in his book *Out of the Crisis*[4] they are:

1. Create constancy of purpose for improvement of product and service.
2. Adopt a new philosophy.
3. Cease dependence on inspection to achieve quality.
4. End the practice of awarding business on the basis of price tag alone; instead, minimize total cost by working with a single supplier.
5. Improve constantly and forever every process for planning, production, and service.
6. Institute training on the job.
7. Adopt and institute leadership.
8. Drive out fear.
9. Break down barriers between staff areas.
10. Eliminate slogans, exhortations, and targets for the workforce.
11. Eliminate numerical quotas for the workforce and numerical goals for management.

12. Remove barriers that rob people of pride of workmanship. Eliminate the annual rating or merit system.
13. Institute a vigorous program of education and self-improvement for everyone.
14. Put everybody in the company to work to accomplish the transformation.

Deming's "Seven Deadly Diseases" include:

1. Lack of constancy of purpose
2. Emphasis on short-term profits
3. Evaluation by performance, merit rating, or annual review of performance
4. Mobility of management
5. Running a company on visible figures alone
6. Excessive medical costs
7. Excessive costs of warranty, fueled by lawyers who work for contingency fees

Deming is known for many other quality processes, which led the Japanese in 1950 to create the Deming Prize (still a very coveted award in Japan for both individuals and companies). It can be argued that the Deming Prize is the foundation on which the U.S. Malcolm Baldrige National Quality Award and similar state and governmental awards are based.

Deming advocated that all managers need to have what he called a *system of profound knowledge*, consisting of four parts:

1. *Appreciation of a system.* Understanding the overall processes involving suppliers, producers, and customers (or recipients) of goods and services (today called the *process approach*).
2. *Knowledge of variation.* The range and causes of variation in quality, and use of statistical sampling in measurements (understanding that variation exists and how to recognize it).
3. *Theory of knowledge.* The concepts explaining knowledge and the limits of what can be known (how to learn). In the ISO Management System process, this is called "organizational knowledge." Other reference materials can be found under the heading of "Human Performance Technology."
4. *Knowledge of psychology.* Concepts of human nature (from the Maslow hierarchy and other literature, and application of the Platinum Rule [Do unto others as they want to have things done for them]).

Key contributions:

- Japan's reconstruction in the 1950s and 1960s; JUSE development of the Deming Prize

- Developments in sampling techniques—applied to census applications
- Management principles: Fourteen Points and Seven Deadly Diseases
- Red bead experiment
- Profound knowledge
- Transformation of American industry (1980s collaboration with Ford Motor Company and Michael Cleary of PQ Systems to teach basic quality principles through community colleges)

Armand Feigenbaum originated the concept of total quality control in his book *Total Quality Control*, published in 1951. In this book Dr. Feigenbaum coined the first use of the term *quality planning*—"The act of planning is thinking out in advance the sequence of actions to accomplish a proposed course of action in doing work to accomplish certain objectives. In order that the planner may communicate his plan to the person or persons expected to execute it, the plan is written out with necessary diagrams, formulas, tables, etc." The book has been translated into many languages, including Japanese, Chinese, French, and Spanish. Feigenbaum is an ASQ honorary member and served as ASQ president for two consecutive terms. He lists three steps to quality:

1. Quality leadership
2. Modern quality technology
3. Organizational commitment

His contributions to the quality body of knowledge include:

- "Total quality control is an effective system for integrating the quality development, quality maintenance, and quality improvement efforts of the various groups in an organization so as to enable production and service at the most economical levels which allow full customer satisfaction."
- The concept of a "hidden" plant—the idea that so much extra work is performed in correcting mistakes that there is effectively a hidden plant within any factory.
- Accountability for quality. Because quality is everybody's job, it may become nobody's job—the idea that quality must be actively managed and have visibility at the highest levels of management.
- The concept of quality costs

Key contributions:

- Quality planning—became AQP
- Quality costs—the hidden factory

Kaoru Ishikawa published four books, is credited with developing the cause-and-effect diagram, and was instrumental in establishing quality circles in Japan. He worked with Deming through the Union of Japanese Scientists and Engineers

(JUSE) and was highly praised by Juran upon his passing. The *Certified Manager of Quality/Organizational Excellence Handbook*[5] summarizes his philosophy with the following points:

1. Quality first—not short-term profit first.
2. Consumer orientation—not producer orientation. Think from the standpoint of the other party.
3. The next process is your customer—breaking down the barrier of sectionalism.
4. Using facts and data to make presentations—utilization of statistical methods.
5. Respect for humanity as a management philosophy—full participatory management.
6. Cross-functional management.

Key contributions:

- Japanese quality circles
- Ishikawa diagram (cause-and-effect diagram, fishbone diagram)
- Developed user-friendly quality control (concepts such as the 5S and other methods to allow operators to know when things are done correctly)
- High focus on internal customers

Joseph M. Juran pursued a varied career in management for over 60 years as an engineer, executive, government administrator, university professor, labor arbitrator, corporate director, and consultant. He developed the Juran trilogy, three managerial processes for use in managing for quality: quality planning, quality control, and quality improvement. Juran wrote hundreds of papers and 12 books, including *Juran's Quality Control Handbook*, *Quality Planning and Analysis* (with F. M. Gryna) and *Juran on Leadership for Quality*. His approach to quality improvement includes the following points:

1. Create awareness of the need and opportunity for improvement.
2. Mandate quality improvement; make it a part of every job description.
3. Create the infrastructure: establish a quality council, select projects for improvement, appoint teams, provide facilitators.
4. Provide training in how to improve quality.
5. Review progress regularly.
6. Give recognition to the winning teams.
7. Propagandize the results.

8. Revise the reward system to enforce the rate of improvement.
9. Maintain momentum by enlarging the business plan to include goals for quality improvement.

The *Juran trilogy* is based on three managerial processes: quality planning, quality control, and quality improvement. Without change, there will be a constant waste; during change there will be increased costs; but after the improvement, margins will be higher, and the increased costs get recouped. Juran founded the Juran Institute in 1979. The Institute is an international training, certification, and consulting company that provides training and consulting services in quality management, lean manufacturing management, and business process management, as well as Six Sigma.

Key contributions:

- Pareto principle—"the vital few and trivial many"
- Management theory for quality
- Juran trilogy

Dorian Shainin started his career in 1936 as an aeronautical engineer and quickly started developing unique solutions to problems. He was mentored by Juran and others and became well known for his unique ability to solve the hardest of problems facing industry and other fields of endeavor. He is credited with saying, "Talk to the parts; they are smarter than the engineers." He was honored with a number of awards in the United States during his career and had a hand in the successful return of Apollo 13 to Earth.

Shainin developed many industrial statistical tools that collectively have become known as the Shainin System for Quality Improvement, or Red "X." Some of the specific tools he developed from his own experience and working with others include the lot plot, reliability service monitoring, pre-control (for control charts), component search, operation search, tolerance parallelogram, overstress testing, B vs. C, paired comparisons, isoplot, variable search, randomized sequencing, resistant limit transform, and rank order ANOVA.

Key contributions:

- Red "X"

Walter Shewhart worked at the Hawthorne plant of Western Electric where he developed and used control charts. He is sometimes referred to as the father of statistical quality control because he brought together the disciplines of statistics, engineering, and economics. He described the basic principles of this new discipline in his book *Economic Control of Quality of Manufactured Product.* He was ASQ's first honorary member.

On a day in May 1924, it is said that Dr. Shewhart presented a little memorandum of about a page in length to his boss (George Edwards). About a third of that page was given over to a simple diagram that we would all recognize today as a schematic control chart. That diagram, and the short text that preceded and followed it, set forth all of the essential principles and considerations that are involved in what we know today as process quality control.

Walter Shewhart was also credited by Dr. Deming as the originator of the plan–do–check–act (PDCA) cycle. This simple tool is the foundation of many problem-solving techniques used today. Deming later updated this to the plan–do–study–act (PDSA) cycle.

Key contributions:

- Father of statistical quality control
- Shewhart cycle—PDCA

D. H. Stamatis has probably published more on the quality profession than any other person. He has developed over 45 volumes relating to quality topics, including an entire series of books on Six Sigma. His *FMEA from Theory to Practice* is considered the foundation work on developing FMEAs for industry. His works are solidly rooted in literature searches, and he has used his skill and the power of the printed word to hone the quality profession.

Key contributions:

- First handbook dedicated to understanding and practical applications of FMEA
- Documented the development of Six Sigma to the present time

Genichi Taguchi was the author or coauthor of six books and received many honors in Japan and the United States for his extensive work in industrial statistics. He taught that any departure from the nominal or target value for a characteristic represents a loss to society. This is the primary function of the *Taguchi loss function*. Instead of long-term focus on specification limits as practiced by many engineering groups, he taught that focusing all efforts on reducing the variation around the target will yield much better results over time and satisfy the customers at much higher levels.

He also popularized the use of fractional factorial designed experiments and stressed the concept of robustness in the Taguchi design of experiments and the use of orthogonal arrays.

Key contributions:

- Taguchi loss function, used to measure financial loss to society resulting from poor quality
- The philosophy of *off-line quality control*, designing products and processes so that they are insensitive ("robust") to parameters outside the design engineer's control
- Innovations in the statistical design of experiments, notably the use of an outer array for factors that are uncontrollable in real life but are systematically varied in the experiment

Processes

A *process* is a series of steps designed to produce products and/or services. A process is often diagrammed with a flowchart depicting inputs, a path that material or information follows, and outputs. An example of a process flowchart is shown in Figure 1.1. Understanding and improving processes is a key part of every Six Sigma project.

The basic strategy of Six Sigma is contained in the acronym DMAIC, which stands for *define, measure, analyze, improve,* and *control.* These steps constitute the cycle used by Six Sigma practitioners to manage problem-solving projects. The individual parts of the DMAIC cycle are explained in subsequent sections, and it is the foundation of the ASQ CSSGB BoK.

Business Systems

A *business system* is designed to implement a process or, more commonly, a set of processes. Business systems make certain that process inputs are in the right place at the right time so that each step of the process has the resources it needs. Perhaps most importantly, a business system must have as its goal the continual improvement of its processes, products, and services. To this end, the business system is responsible for collecting and analyzing data from the processes and other sources that will help in the continual improvement of process outputs. Figure 1.2 illustrates the relationships between systems, processes, subprocesses, and steps.

Process Inputs, Outputs, and Feedback

Figure 1.3 illustrates the application of a feedback loop to help in process control. It is often useful to expand on a process flowchart with more-elaborate diagrams. Various versions of these diagrams are called *process maps, value stream maps,* and so on. Their common feature is an emphasis on inputs and outputs for each process step, the output from one step being the input to the next step. Each step acts as the customer of the previous step and supplier to the next step. The value to the parent enterprise system lies in the quality of these inputs and outputs

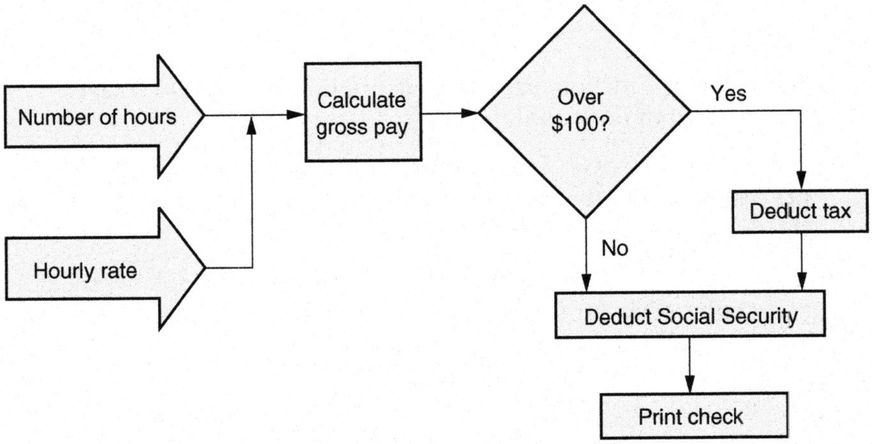

Figure 1.1 Example of a process flowchart.

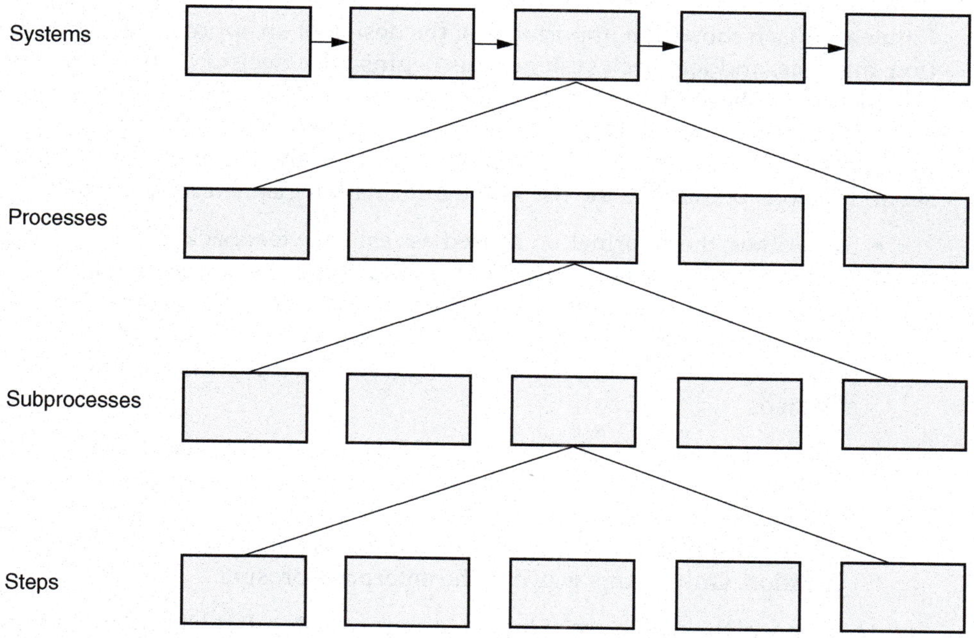

Figure 1.2 Relationships between systems, processes, subprocesses, and steps. Each part of a system can be broken into a series of processes, each of which may have subprocesses. The subprocesses may be further broken into steps.

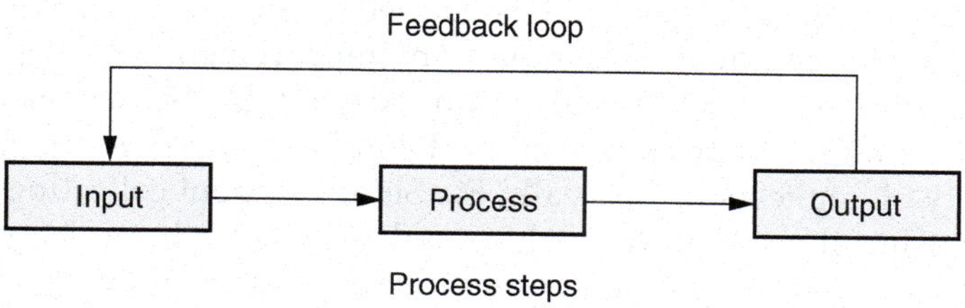

Figure 1.3 A feedback loop.

and the efficiency with which they are managed. There are two ways to look at the method by which efficient use of inputs/resources is implemented to produce quality outputs:

- Some would state that a function of process management is the collection and analysis of data about inputs and outputs, using the information as feedback to the process for adjustment and improvement.
- Another way of thinking about this is that the process should be designed so that data collection, analysis, and feedback for adjustment and improvement are a part of the process itself.

Either approach shows the importance of the design of an appropriate data collection, analysis, and feedback system. This begins with decisions about the points at which data should be collected. The next decisions encompass the measurement systems to be used. Details of measurement system analysis are discussed in Chapter 14. The third set of decisions entails the analysis of the data. The fourth set of decisions pertains to the use of the information gleaned from the data:

- Sometimes, the information is used as real-time feedback to the process, triggering adjustment of inputs. A typical example would involve the use of a control chart. Data are collected and recorded on the chart. The charting process acts as the data analysis tool. The proper use of the chart sometimes suggests that a process input be adjusted.
- Another use for the information would be in the formation of plans for process improvement. If a stable process is found to be incapable, for instance, designed experiments may be required. Any enterprise system must perform process improvement as part of its day-to-day operation. Only in this way can the enterprise prosper.

Figure 1.4 shows the categories of inputs to a process step. It is helpful to list inputs in the various categories and then classify each input as indicated.

Significance of Six Sigma

Six Sigma is just the latest term for the more general concept of continual improvement. Continual improvement can be defined as the use of problem-solving techniques and quick deployment to implement improvements and then using process behavioral studies (Wheeler) to maintain the gains. Six Sigma has been described as a breakthrough system (Juran) and is being used in many organizations today in a variety of applications. Basically, Six Sigma is about collecting data on a process and using those data to analyze and interpret what is happening in that process so that the process can be improved to satisfy the customer (Kano and

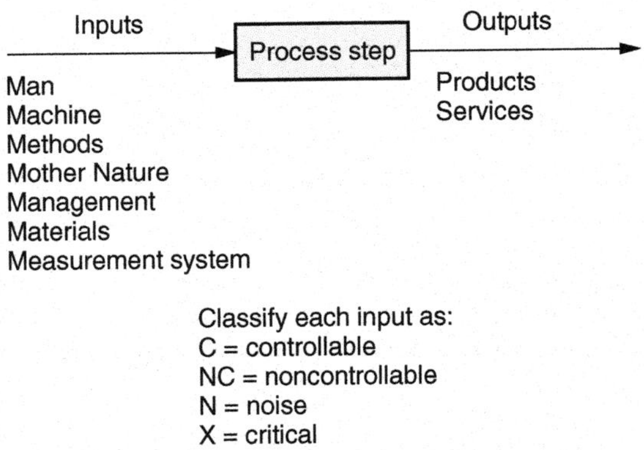

Figure 1.4 Categories of inputs to a process step.

Taguchi). A basic process can be defined as an input, transformation, and output. Six Sigma was first started at Motorola and was then developed more into what we know today at General Electric. By following a prescribed process, the entire organization starts to look at everything that it does in the light of reducing variation and reducing waste, with the result of increasing customer satisfaction. Customers could be anyone from the next person who uses the work we do (internal customer) to the ultimate customer who uses the products or services that our organization produces (external customer). To assist in this process, sometimes the supplier and customer will be added to the basic process definition listed above, creating the SIPOC identification: *suppliers, inputs, process, outputs,* and *customers.* This is used especially to help define the boundaries of what is to be studied.

For some, the idea of improving a process is a waste of time that should not be bothered with ("we are already working the hardest that we can"). But as Juran once said, "Changes creep up on us week by week, a little bit at a time. Over a year or two, there are 50 or 100 of these bits, which amounts to quite a bit. The skills of the men have not necessarily kept pace, and we wake up to the existence of a wide gap."[6] This is one explanation for why accidents and product rejections happen in our shops. If the root cause is actually found for any accident or rejection of product or service, it will usually be traced back to many small changes that occurred either within our own organization or at our supplier.

By using Six Sigma methodologies, we will be able to find those bits of changes and decide which ones should be kept for process improvement and which ones need to be corrected. This process is not meant to be a quick fix (magic bullet) approach. The logical use of the tools over time will save us resources and effort in doing our daily jobs.

A Green Belt's Role

You will find in this process for solving problems a number of tools and methods that you may already be familiar with and a few that may be new to you. You may very well ask, "How is this any different from what we have been doing before?" The direct answer will need to be provided by your organization depending on the various programs that have already been tried. For many of us, this process will be part of an ongoing evolution of how we do our work. One of the main things that you should notice is that upper management will be more involved with your problem-solving efforts and in the everyday problems that are found in your work areas.

During the process, and while using this book, you will be able to reference the Six Sigma model for improvement. It has been shown and demonstrated that by using a model or road map, we can usually accomplish something more quickly than without a guide. Some organizations today use something called the *MAIC model.* They refer to this process as being able to do "magic" without the "garbage" (G) that we find in most operations. Many organizations have added a *define* (D) stage—identifying the process customers—thus making for the DMAIC model.

You may already have control plans, process sheets, standard operating procedures, or any number of other things that you use in your daily work. The use of the Six Sigma model for improvement should not replace anything that you are currently doing, but be used to review daily work to look for areas or methods of improving the process in light of what your customers want and need. Even though we are doing the same things that we might have done before, do our customers still want the same things from us?

We are entering a journey of continual improvement that can involve our work and our lives. Some of us have been on this journey for some time, while others may be just starting. The process involves using what Deming refers to as *profound knowledge*: appreciation for a system, knowledge about variation, theory of knowledge, and psychology. Through the Six Sigma methodology and using the Six Sigma model for improvement, we should see things around us work better and satisfy our customers more.

Potential Tasks

Your organization may already be using something called Six Sigma or some other method (for example, quality operating system [QOS], continuous improvement [CI], total quality management [TQM], process improvement [PI], or some other name). As an operator or owner of a process, you will be asked by your supervisors or management to help implement improvement of the process(es) that you work with. Your challenge will be to look at the process both for the simple improvements that you may already know need to be made (preventive maintenance, cleanliness, parts wearing out, and so on) as well as to assist in measuring certain factors about the process to investigate better ways of performing the process.

You will be asked to use the tools in this book, and maybe others, to study your work and process(es) to look for improvement ideas and to implement those ideas. You may already be familiar with some of these tools, and the challenge will be in how to use them, possibly in new ways, to make the changes that will help your company stay in business in today's fiercely competitive world. We no longer compete only against others within our own country, but against others from countries around the world. How can they do a better job than us, ship the parts that we make, and sell them to our customers faster, better, and cheaper than us? This is the question that should be on your team's mind.

Many of us have found that by using a model or framework we can do things more simply—a picture is worth a thousand words. This is also true when trying to improve processes. Dr. Ishikawa gave us a road map to follow when first looking at a process that needs to be improved. The words may not make much sense right now, but as you work with process improvement (as you work through the DMAIC process), you will come to understand the importance of what is said here:

1. Determine the assurance unit (what is to be measured).
2. Determine the measuring method (how it will be measured).
3. Determine the relative importance of quality characteristics (is this key to our process?).
4. Arrive at a consensus on defects and flaws (does everyone agree on good and bad quality?).
5. Expose latent defects (look at the process over time).

6. Observe quality statistically (use process behavior charting).
7. Distinguish between "quality of design" and "quality of conformance."

After we know what we can change (quality of conformance) versus what we cannot change right now (quality of design—this is left to *design for Six Sigma* [DFSS]), we can start working on our processes. Many operators start out viewing this effort as only more work, but many will find that doing these studies will actually save them a lot of time and grief in the future as things start to improve and machines start to work better. One question to ask yourself now is, how often does your process slow down or stop due to something not working the way it should? Or, is the output ever scrapped by someone down the line (including at your external customers) because something did not happen right at your operation?

Be willing to experiment with the tools and look for ways of applying them to the work and processes to learn as much as you can about how a process operates so that you can modify it as appropriate to give the customers the best output that is possible.

DMAIC Model

The DMAIC model stands for *define, measure, analyze, improve,* and *control* and is very similar to the PDSA or PDCA model that you may already be using.

A key factor in each step is for management to allow the time and resources to accomplish each of the phases to strive for continual improvement. This is one of the driving forces that makes Six Sigma different from other quality improvement programs. The other driving forces include getting everyone in the organization involved, getting the information technology group to assist in supplying data more quickly for everyone, and getting financial data in the form of cost of quality analysis.

Everyone will be asked to get involved with the Six Sigma model and look for continual improvement opportunities in their work areas. Basically, you will do the following in each step:

Define: Identify the issue causing decreased customer satisfaction.

Measure: Collect data from the process.

Analyze: Study the process and data for clues to what is going on.

Improve: Take action, based on the conclusions obtained from the analysis performed in the previous step to change the process for improvement.

Control: Monitor the system to sustain the gains.

A number of tools and methods can be used in each of the steps of the DMAIC model. This is only a quick overview of many of these items. More detailed information can be found in the references, on the Internet, or probably in the quality office of your organization. The DMAIC model uses the following:

Define

Management commitment—PDCA

SIPOC (suppliers, inputs, process, outputs, customers)

Define the problem—five whys and how

Systems thinking

Process identification

Flowchart

Project management

Measure

Management commitment—PDCA

Identify a data collection plan

Measurement systems analysis (MSA)

Collect data—check sheets, histograms, Pareto charts, run charts, scatter diagrams

Identify variability—instability, variation, off-target

Benchmark—start by setting the current baseline for the process

Start cost of quality

Analyze

Management commitment—PDSA

Continual improvement

Preventive maintenance

Cleanliness—5S

Benchmark—continue process

Central limit theorem

Geometric dimensioning and tolerancing (GD&T)

Shop audit

Experiments

Improve

Management commitment—PDSA

Process improvement

Organizational development

Variation reduction

Problem solving

Brainstorm alternatives

Create "should be" flowcharts

Conduct FMEA

Cost of quality

Design of experiments

Control

Management commitment—SDCA

Control plan (CP)—manufacturing controls

End-to-end control plan—administration controls

Dynamic control plan (DCP)

Long-term MSA

Mistake-proofing

Process behavior charts

Update lessons learned

Many will find this process very exciting as they will have the tools and methods to demonstrate the improvements that they are helping the organization to achieve. There have been times in the past when an employee tried in vain to tell a supervisor that something was wrong with a machine or process. Now we have the means to not only tell but show and demonstrate what needs to be done. Following this process creates a road map for continual improvement that once started is a never-ending journey. These tools and methods have proven themselves to be useful everywhere: from shop floors to front offices, from schools to hospitals, and even in churches or at home.

The Six Sigma Road Map

As we prepare for the Six Sigma journey, here is a quick view of the suggested map that we can follow:

1. Recognize that variation exists in everything that we do; standardize your work.
2. Identify what the customer wants and needs. Reduce variation.
3. Use a problem-solving methodology to plan improvements.
4. Follow the DMAIC model to deploy the improvement.
5. Monitor the process using process behavior charts.
6. Update standard operating procedures and lessons learned.
7. Celebrate successes.
8. Start over again for continual improvement—PDSA/SDCA.[7]

Cost–Benefit Analysis (Cost of Quality, Quality Cost, Cost of Poor Quality, Cost of Current Quality)

This is a financial tool that should be used to report how quality levels are being sustained in the target process within an organization. Many things that are worked on throughout the shop can be classified into one of four categories: *prevention costs, appraisal costs, internal failure costs,* or *external failure costs.* However, not all expenses of the company are used, only those that relate in some way to the products or services that are shipped to customers. The real power of this tool is not so much that you use the exact or "right" measures for each expense, but that you look at trends over time to see what you are doing. You want to find out what the *total cost* is to provide your customers with products and services (see Figure 1.5). Traditionally, when cost of quality is first calculated for an organization, a picture such as Figure 1.5 emerges. Part of the reason for this is that many accountants and managers have not been taught about this tool in their formal education, nor does any governmental or professional organization require the reporting of financial data in this format.

On the other hand, organizations that have learned to use the cost–benefit analysis of quality cost, as called for in Six Sigma, are typically very surprised at the amount of waste that is being produced. By focusing on reducing prevention and appraisal costs, initial overall cost may rise; however, failure costs (internal and external) will slowly start to come down. This will not happen overnight and may take years, in stubborn cases, to show improvement as old products work their way out of the customer system. The end goal will be to have total cost of quality lower than when you started the Six Sigma process.

No one should be blamed for the poor results of the first round of measurements. It is important to look at these numbers as a benchmark to measure improvement from. The results of the numbers should be made available to everyone so

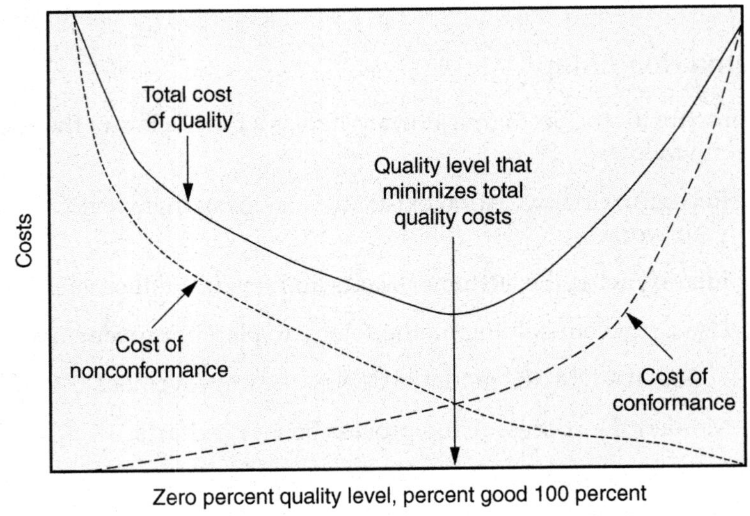

Figure 1.5 Traditional quality cost curves.

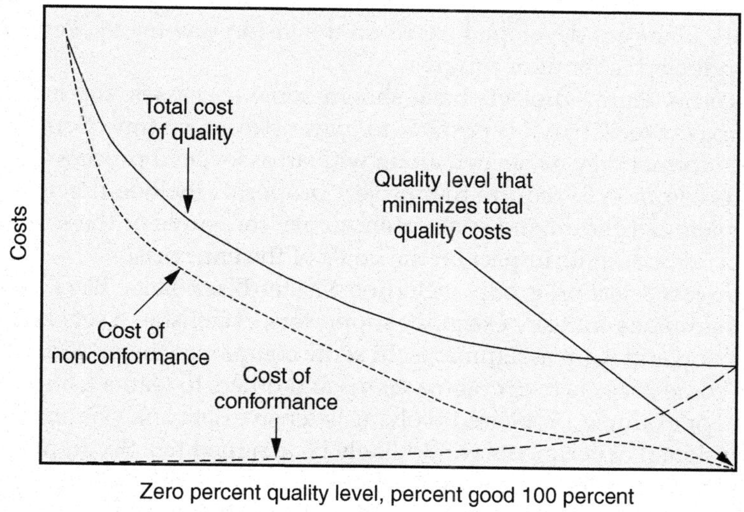

Figure 1.6 Modern quality cost curves.

that ideas can be generated as to what can be done and how. Remember the old management adage: "What gets measured gets done!" Thus, if everyone knows that management is watching the numbers on cost of quality, things should start to improve.

The ultimate goal is to change the overall picture to look like Figure 1.6. As an organization continually improves their products and services, they will see an overall reduction in total cost to manufacture and produce products and services.

ORGANIZATIONAL GOALS AND SIX SIGMA PROJECTS

> Identify the linkages and supports that need to be established between a selected six sigma project and the organization's goals including SMART goals, and describe how process inputs, outputs, and feedback at all levels can influence the organization as a whole. (Understand)
>
> **Body of Knowledge I.A.2**

Linking Projects to Organizational Goals

Organizational goals must be consistent with the long-term strategies of the enterprise. One technique for developing such strategies is called *Hoshin* planning. This is a planning process in which a company develops up to four vision statements that indicate where the company should be in the next five years. Company goals and work plans are developed based on the vision statements. Periodic audits are then conducted to monitor progress.

Once Six Sigma projects have shown some successes, there will usually be more project ideas than it is possible to undertake at one time. Some sort of project proposal format may be needed along with an associated process for project selection. It is common to require that project proposals include precise statements of the problem definition and some preliminary measures of the seriousness of the problem, including its impact on the goals of the enterprise.

A project selection group, including Master Black Belts, Black Belts, organizational champions, and key executive supporters, establishes a set of criteria for project selection and team assignments. In some companies the project selection group assigns some projects to Six Sigma teams and others to teams using other methodologies. For example, problems involving extensive data analysis and improvements using designed experiments would likely be assigned to a Six Sigma team, while a process improvement not involving these techniques might be assigned to a lean manufacturing team. New-product design should follow DFSS guidelines.

The project selection criteria are always a key element to furthering of organizational goals. One key to gauging both the performance and health of an organization and its processes lies with its selection and use of metrics. These are usually converted to financial terms such as return on investment, cost reduction, and increases in sales and/or profit. Other things being approximately equal, the projects with the greatest contributions to the bottom line receive the highest priority.

The formula for expected profit is

$$EP = \Sigma \text{ Profit} \times \text{Probability}$$

A *system* may be thought of as the set of processes that make up an enterprise. When improvements are proposed, it is important to take a systems approach. This means that consideration be given to the effect the proposed changes will have on other processes within the system and therefore on the enterprise as a whole. Operating a system at less than its best is called *suboptimization.* Changes in a system may optimize individual processes but suboptimize the system as a whole.

EXAMPLE

A gambler is considering whether to bet $1.00 on red at a roulette table. If the ball falls into a red cell, the gambler will receive a $1.00 profit. Otherwise, the gambler will lose the $1.00 bet. The wheel has 38 cells, 18 being red.

Analysis: Assuming a fair wheel, the probability of winning is 18/38 ≈ 0.474, and the probability of losing is 20/38 ≈ 0.526. In table form:

Outcome	Profit	Probability	Profit × Probability
Win	$1	.474	$0.474
Loss	–$1	.526	–$0.526
	Expected outcome = –$0.052		

Continued

Continued

In this case the gambler can expect to lose an average of about a nickel (–$0.052) for each $1.00 bet. Risk analysis for real-life problems tends to be less precise primarily because the probabilities are usually not known and must be estimated.

EXAMPLE

A proposed Six Sigma project is aimed at improving quality enough to attract one or two new customers. The project will cost $3M. Previous experience indicates that the probability of getting customer A only is between 60 percent and 70 percent, and the probability of getting customer B only is between 10 percent and 20 percent. The probability of getting both A and B is between 5 percent and 10 percent.

One way to analyze this problem is to make two tables, one for the worst case and the other for the best case, as shown in Table 1.2.

Assuming that the data are correct, the project will improve enterprise profits between $1M and $2.5M.

When estimating the values for these tables, the project team should list the strengths, weaknesses, opportunities, and threats (SWOT) that the proposal implies. A thorough study of this list will help provide the best estimates (see Figure 1.7).

Table 1.2 Risk analysis table.

	Worst case profit			Best case profit		
Outcome		**Probability**	**Profit × Probability**		**Probability**	**Profit × Probability**
A only	$2 M	.60	$1.2 M	$2 M	.70	$1.4 M
B only	$2 M	.10	$0.2 M	$2 M	.20	$0.4 M
A & B	$7 M	.05	$0.35 M	$7 M	.10	$0.7 M
None	–$3 M	.25	–$0.75 M	–$3 M	0	$0 M
	Expected profit = $1 M			Expected profit = $2.5 M		

Strengths: High-quality product Monthly quantity commitment Tooling cost by customer Just-in-time concepts Online interface Product mix	**Weaknesses:** Pricing Union plant High employee turnover Aging equipment—downtime issues
Opportunities: Potential industry leadership More growth Long-term contract	**Threats:** Competition from startups Labor force Union plant Unstable market Unstable labor force

Figure 1.7 A format for SWOT analysis.

Task/objective	S	M	A	R	T
Reduce overtime for clerical staff by 15% by the end of the 3rd quarter	X	X	X	• Training • Process mapping • Office schedule • Electronic filing	12 months
Recruit five nursing assistants for the vaccination program by July 15	X	X	Discuss at December senior management meeting	• Hire tickets • Salary budget • Available talent pool	8 months
Enroll the new quality improvement coordinator into team training for 4th quarter	X	X	X	• Training budget • Appropriate class • Travel funds • Time away from current duties	2 months
Visit three substance program community partners each month between December and July	X	X	X	• Director and health officer schedule • Travel budget • Agency car	5 hrs/ month
Obtain two additional grants totaling $85K for toxic waste cleanup by 2nd quarter next year	X	X	Few funding sources for toxic waste cleanup	• Grant writer • Chemical analysis equipment • Chemist	3 months
	Specific	*Measurable*	*Attainable*	*Resources*	*Time*

Figure 1.8 SMART matrix example.

Examples of suboptimization:

- The resources invested in improving process A might be more profitably invested in process B.
- The throughput rate of a process increases far beyond the ability of the subsequent process to handle it.

A distribution center loads its trucks in a manner that minimizes its work. However, this method requires the receiving organization to expend more time, energy, resources, and dollars unloading the truck. A different loading style/arrangement might be more expensive to the distribution center but would result in significant cost reduction for the entire system.

An addition to the ASQ CSSGB 2022 BoK deals with the use of SMART goals (or objectives), which has become increasingly popular for project management events. SMART is an acronym that has evolved since the early 1980s and most commonly stands for: Specific, Measurable, Attainable (or Achievable), Relevant, and Time (or Time-Bound).

In the early phase of planning for your Six Sigma project, creating a set of SMART goals (see Figure 1.8) will help set boundaries and define steps in the DMAIC process for your team to achieve. These goals help to prevent scope creep and set up targets that can be used in Toll-Gate Reviews (discussed later) to help ensure that the goals have been met. It is a well-known fact that setting measurable goals at the early stages of a project (as well as in your life goals) usually goes a long way in helping to ensure achievement of desired results.

ORGANIZATIONAL DRIVERS AND METRICS

> Recognize key business drivers (profit, market share, customer satisfaction, efficiency, product differentiation, key performance indicators (KPIs)) for all types of organizations. Understand how key metrics and scorecards are developed and how they impact the entire organization. (Understand)
>
> **Body of Knowledge I.A.3**

Key Drivers

All organizations depend heavily on the measurement and analysis of performance. Such measurements should not only derive from business needs and strategy, but they should also provide critical data and information about key processes, outputs, and results. Several types of data and information are needed for performance management. A number of key drivers form the backbone of any business's effort to present performance information to executives and staff. These include customer, product, service, operational, market, competitive, supplier, workforce, cost, financial, governance, and compliance performance. A major consideration in performance improvement and change management involves the selection and use of performance measures or indicators. The measures or indicator that one selects must best represent the factors that lead to improved customer, operational, financial, and ethical performance. A comprehensive set of measures or indicators tied to customer and organizational performance requirements provides a clear basis for aligning all processes with one's organizational goals.

Voice of the Customer (VOC)

One of the key organizational drivers is customer and market knowledge—the ability of an organization to determine the requirements, needs, expectations, and preferences of customers and markets. Also necessary are the relationships with customers and the ability to determine the key factors that lead to customer acquisition, satisfaction, loyalty, and retention, and to business expansion and sustainability. The *voice of the customer* (VOC) is the process for capturing customer-related information. This process is proactive and continuously innovative in order to capture stated, unstated, and anticipated customer requirements, needs, and desires. The goal is to achieve customer loyalty and to build customer relationships, as appropriate. The VOC might include gathering and integrating survey data, focus group findings, Web-based data, warranty data, complaint logs and field reports, and any other data and information that reflect the customer's purchasing and relationship decisions.

Balanced Scorecard

Many business professionals advocate the use of a balanced scorecard type of approach for the selection of project metrics as a method for ensuring that the project meets both customer and business needs. The balanced scorecard approach includes both financial and nonfinancial metrics, as well as lagging and leading measures across four areas or perspectives: financial, customer, internal processes, and employee learning and growth. Lagging measures are those that are measured at the end of an event, while leading measures are measures that help achieve objectives and are measured upstream of the event.

This new approach to strategic management was developed in the early 1990s to help managers monitor results in key areas. The concept was illustrated by Drs. Robert Kaplan and David Norton, who named this system the *balanced scorecard*. Recognizing some of the weaknesses and vagueness of previous management approaches, the balanced scorecard approach provides a clear prescription as to what companies should measure in order to "balance" financial results.

The balanced scorecard is not only a measurement system, but also a management system that enables organizations to focus on their vision and strategy and translate them into actions. It provides feedback on both internal business processes and external outcomes in order to continuously improve strategic performance and results. When fully deployed, the balanced scorecard transforms strategic planning from an academic exercise into the nerve center of the enterprise. Most balanced scorecard metrics are based on brainstorming; however, the brainstorming approach may have limited success in establishing sound metrics that maintain a good balance between lagging and leading measures.

Scoreboard/Dashboard

A *scoreboard*, or *dashboard*, is a visual representation that gives personnel a quick and easy way to view their company's performance in real time. The dashboard should be critical in assisting an employee to predict sales, cash flow, and profit, and gain clarity on the performance and direction of the company. In addition, it should be a critical decision-making tool used in the day-to-day operation of the firm that empowers employees and business owners to make the best decisions for their respective departments that will drive cash flow and profit.

There are three main steps to consider in building an effective dashboard. First, we should know the averages and benchmarks for our industry. Second, we should know what our historical performance has been on these same averages and benchmarks. And third, we have to develop what many call a *balanced scorecard* that comprehensively examines the whole company, not just one or two parts.

Key Performance/Process Indicator (KPI)

Depending on the consultant you talk with, you might get the same definition for three different titles: *key process indicator*, a *key performance indicator*, or a *process performance indicator*. It is hard to distinguish between these three terms. However, for the Green Belt, you will be functioning generally at the levels in an organization that will understand the term *key process indicator*. Thus, a *KPI* is a quantifiable measurement, which is agreed to beforehand, that reflects the critical success factors of a department or group within your organization. They will differ depending on the company.

In nearly all cases of measuring performance of a process, there are usually a lot of things that could be tracked depending on where you are in the process. If you think of any major sporting event, the final score is only one measure of a team's performance. There are many key measures that, when all added up, contribute to what the outcome of the game will be. It's the same in any organization; your biggest challenge may in fact be in trying to sort through all of the data that are being collected to identify the key measures that need to be tracked in order to obtain the desired outcome for the organization. Your management team should have already thought this through and should be able to give direction on what measures you will need to track in your projects.

Chapter 2

Lean Principles in the Organization

LEAN CONCEPTS

> Define and describe lean concepts such as theory of constraints, value chain, flow, takt time, just-in-time (JIT), Gemba, spaghetti diagrams, and perfection. (Apply)
>
> **Body of Knowledge I.B.1**

Lean has been referred to by many names: lean manufacturing, lean office, lean enterprise, lean production, flexible mass production, and others. Toyota is usually credited with creating the concept of lean under their Toyota Production System (TPS) as far back as the 1950s; however, they credit having learned the process from the Ford Motor Company.[1] Three concepts are fundamental to the understanding of lean thinking: value, waste, and the process of creating value without waste. In today's variation of the TPS, some like to identify the *8Ps of lean thinking* as purpose, process, people, pull, prevention, partnering, planet, and perfection.

Essentially, lean is centered on making obvious what adds value by reducing everything else. Some people only look at lean as a set of tools to apply within the organization to eliminate waste (Muda). However, the TPS is much more and involves the developing of a culture in an organization that promotes the continual improvement philosophy and the development of all people in the organization (see Figure 2.1). Far too many systems and management practices in a typical organization prevent operators from doing their best work to satisfy the customers.[2]

One of the more dramatic examples of success using the TPS was the joint venture between Toyota and General Motors at New United Motor Manufacturing, Inc. (NUMMI) in an old GM Fremont, California, plant that operated from 1984 to 2010. The success of the NUMMI plant was demonstrated by literally going from being the worst plant in the GM system to being one of the top plants in less than two years. Part of the initial agreement was that the GM management team would follow the direction of Toyota management and learn the TPS process.

There are a number of books, articles, and web pages that explain the TPS in great detail and are available for more research as your need presents itself. Following are some of the basic tenets of the TPS process.

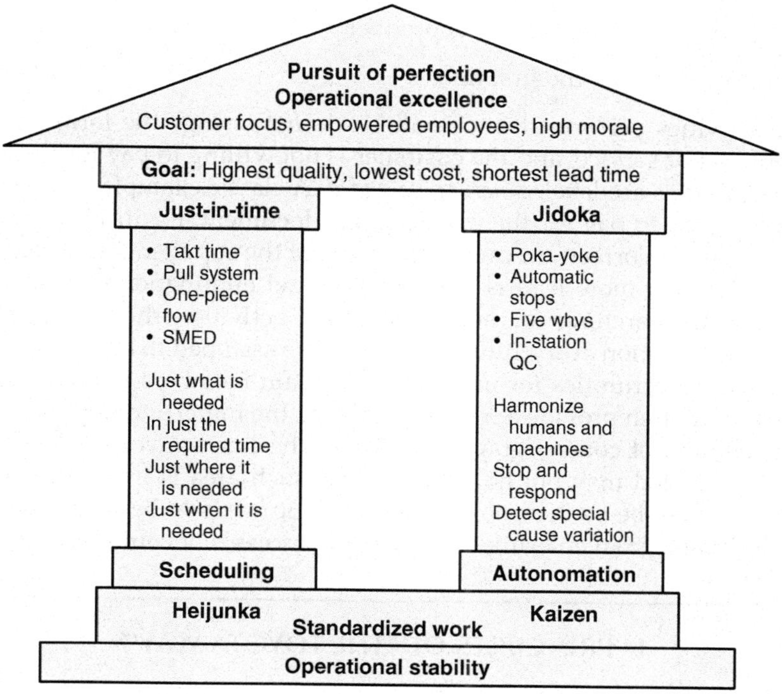

Figure 2.1 TPS house.

Value

The single most important concept that has been brought to awareness in the business community in recent years is *value*. Value is defined by the customer based on their perception of the usefulness and necessity of a given product or service. An excellent industrial video to view on this topic is *Time, the Next Dimension of Quality,* available through CRM Learning at www.crmlearning.com.

While Japanese-made cars and German-made cars are sold in the same markets, some customers prefer Japanese-made for their quality, reliability, resale value, and fuel efficiency. German-made cars can satisfy some of those same expectations and additionally offer a pride of ownership attached to the carmaker. There is a segment of customer that prefers German-made cars for these very reasons. Thus, customers define the value of the product. American carmakers build trucks and vans sturdy enough to handle tough jobs. Some American cars, trucks, and vans are comparable in quality and reliability to the Japanese and German competition. They also have built-in customer loyalty. There is a segment of customer who will buy American-made vehicles for these very reasons.

Once the concept of value is understood, the target cost for the product or service can be determined. According to Womack, this target cost is a mixture of current selling prices of competitors and examination of elimination of waste by lean methods.[3]

Lean experts define a process step as value-added if:

- The customer recognizes the value

- It changes (transforms) the product
- It is done right the first time

Some activities performed in operations do not change the form or function of the product or service, and the customer is not willing to pay for these activities. These activities are labeled *non-value-added*. A classic example is *rework*. The customer expects to pay for the printing of a document, for instance, but does not want to pay for corrections caused by errors of the supplier. A key step in making an organization more lean is the detection and elimination of non-value-added activities. In searching for non-value-added activities, the operative guideline should be "question everything." Steps that are assumed to be necessary are often ripe with opportunities for improvement. Team members not associated with a process will often provide a fresh eye and ask the impertinent questions.

There are, of course, gray areas where the line between valued-added and non-value-added may not be obvious. One such area is inspection and testing. A process may be so incapable that its output needs to be inspected to prevent defective parts from entering downstream processes. It could be argued that this

14 PRINCIPLES OF "THE TOYOTA WAY"[4]

1. Base your management decisions on a long-term philosophy, even at the expense of short-term financial goals.
2. Create a continuous process flow to bring problems to the surface.
3. Use "pull" systems to avoid overproduction.
4. Level out the workload (work like the tortoise, not the hare).
5. Build a culture of stopping to fix problems to get quality right the first time.
6. Standardized tasks and processes are the foundation for continuous improvement and employee empowerment.
7. Use visual controls so no problems are hidden.
8. Use only reliable, thoroughly tested technology that serves your people and process.
9. Grow leaders who thoroughly understand the work, live the philosophy, and teach it to others.
10. Develop exceptional people and teams who follow your company's philosophy.
11. Respect your extended network of partners and suppliers by challenging them and helping them improve.
12. Go and see for yourself to thoroughly understand the situation.
13. Make decisions slowly by consensus, thoroughly considering all options; implement decisions rapidly.
14. Become a learning organization through relentless reflection and continuous improvement.

inspection is a value-added activity because the customer doesn't want defective products. The obvious solution is to work on the process, making it capable and rendering the inspection activity unnecessary. Most authorities would agree that this inspection is non-value-added. On the other hand, a gas furnace manufacturer must fire-test every furnace in order to comply with CSA requirements. Customers are willing to pay for the CSA listing, so this test step is a value-added activity. Studies have shown that an overwhelming percentage of lead time is non-value-added, much of it spent waiting for the next step. Yet, over the years, efforts to decrease lead time have often focused on accelerating value-added functions rather than reducing or eliminating non-value-added functions.

Some of the Top Lean Tools

5S (or 6S or 7S). 5S is a workplace organization method that can help improve the efficiency and management of operations. A process is impacted by its environment, as is the ability of personnel to respond to process change. Improvements in the general state of the work area, including access to hand tools, and so on, are an aid to process control. Especially critical here are the cleanliness, lighting, and general housekeeping status of any area where measurements are conducted since process control data are filtered through the measurement system. Example: A workbench cluttered with tools and accessories wastes the valuable time of skilled workers and causes distraction from work, resulting in poor quality. Similarly, an office table covered with disorganized files and papers can cause clerical errors and delays in processing. 5S is one of the first tools to apply in the path to achieving lean enterprise organizations.

The traditional sequence for 5S is:

Sort. Remove unneeded items. Whether we're in the office or home, we tend to collect items that are very rarely needed or not needed at all. Over a period of time these items accumulate into a mess and make it less efficient to search for needed items, and sometimes even cause safety issues. The first step is sorting through the items as required and cleaning up the work area. Never-used items should be discarded immediately.

Set in order. Arrange the required and rarely required items for ease of accessibility. The items that are required more often, like drawings, instructions, tools, safety goggles, and so on, are placed in designated and marked locations so that they cannot be placed elsewhere. In short, a place for everything and everything in its place. The rarely required items like machine manuals, shop floor layout plans, and so on, can be kept out of the way.

Shine. This involves cleaning the work area and equipment. As simple as this may sound, many quality issues are uncovered through effective cleaning of the work area. Example: Cleaning of the inspection surface plate provides better measurement results, cleaning of the equipment work table provides for better movement, and cleaning of the floor prevents accidents. For some industries, like semiconductor manufacturing, cleanliness is mandatory and is measured in particle count.

Standardize. This involves developing checklists, standards, and work instructions to keep the work area in a clean and orderly condition. Under an ISO Management System umbrella, this can also include reviewing the JHA, JSA, or procedures.

Sustain. This is the most difficult step in 5S. Most organizations are initially successful in the first four steps, but sustaining the efforts and continuing them require support from management and empowerment of employees. Management needs to realize that this time is well spent and be willing to invest in the time. The time invested in 5S improves productivity and overall efficiency, and reduces accidents. Management should also empower the employees by allowing them to take ownership of their work areas.

An article was published by *Quality Progress* in October 2013 called "5S Shakeup: Three Secrets for Sustaining 5S Success" by John Casey.[5] Mr. Casey was a manager at the NUMMI plant and learned directly from Toyota the internal secrets of 5S. The article describes the secrets for typical North American organizations as:

1. Engage management on the cost savings to be achieved with 5S.
2. Establish visible scoreboards that include measures for cleanliness.
3. Once the scoreboards are in place, instead of starting with *sort*, the organization should focus on starting in step 4, *standardize*, so that people know how to change their scores.

Also note that some people add a sixth *S* (safety) and in healthcare (or other heavily regulated industry) a seventh *S* (oversight) (see Figure 2.2).

Andon. A visual feedback system (typically red/yellow/green stacked lights at the work site) that indicates the production status at any given time. It alerts operators and supervisors that assistance may be needed and empowers the employees to stop the process if an issue arises that is not considered good for quality.

With technology improvements, the monitoring of operations is becoming visible from various parts of the operation, and operators are being given ever earlier warnings that something may not be functioning as needed for normal operations.

A3. This tool was originally named after the metric size paper used to publish this reporting tool in Europe. The technique is used to give management a quick overview of key topics/issues of a project on one sheet of paper.[6] This can be used as an overview project template, status report template, or other quick update of the team or management. Examples of these forms are available as part of this handbook's online material.*

Bottlenecks. *See* theory of constraints.

Continuous Flow. Operations where work-in-process smoothly flows through the system with minimal (or no) buffers between steps of the operation. Developing a

*https://asqassets.widencollective.com/portals/8ntnzkee/(H1469)SupplementalFiles-TheCertified SixSigmaGreenBeltHandbook,SecondEdition.

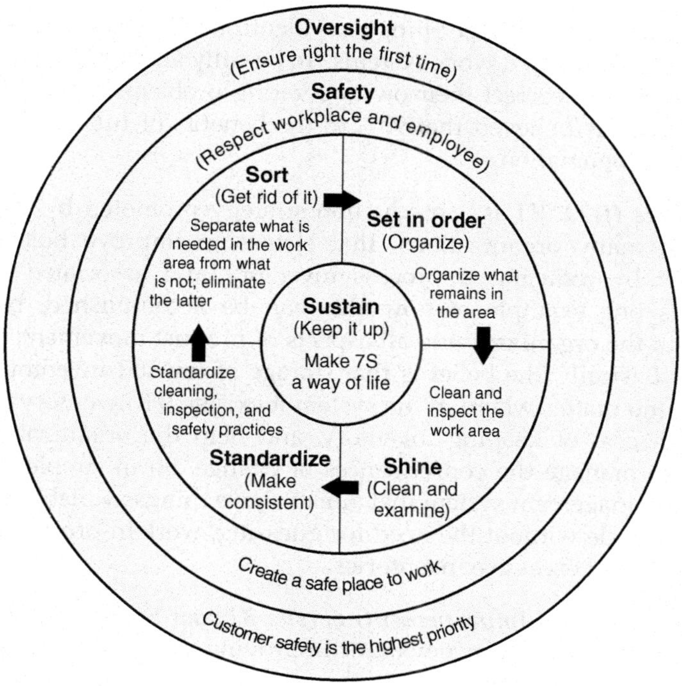

Figure 2.2 7S adaptation (Hirano).

continuous flow eliminates many forms of waste (for example, inventory, waiting time, transport, and overprocessing).

Gemba. The real place! A philosophy that reminds us to get out of our offices and spend time on the operations floor—the place where the real action occurs. In some management circles, this is called "management by walking around" (MBWA). This concept promotes a deeper and more thorough understanding of real-world operational issues by firsthand observation and by talking with employees doing the work. Another common phase for Gemba is "Look and See," and for the green belt, going to where the issue is occurring is very important as the starting point of any improvement project.

Heijunka. A form of production scheduling that purposely produces in much smaller batches by sequencing (mixing) product/service variants within the same process. This tends to reduce lead times (since each product or variant is produced more frequently) and lower inventory levels (since batches are smaller).

Hoshin Kanri. Otherwise known as *policy deployment*, its purpose is to align the goals of the company (strategy), with the plans of middle management (tactics), and the work performed on the operations floor (action). Also ensures that progress toward strategic goals is consistent and thorough and has the benefit of elimination of waste that comes from poor communication and inconsistent direction.

Jidoka (Autonomation). Within the TPS process, the concept is "why have a human do what a machine can do better," especially in the tedious, repetitive jobs that can cause injury over time. Sometimes called "intelligent automation" or "automation with a human touch," Shigeo Shingo has identified 23 stages between purely manual and fully automated work systems. To be fully automated, machines must be able to detect and correct their own operating problems, which is currently not cost-effective. He believed that 90% of the benefits of full automation could be gained by autonomation.

Just-in-Time (JIT). JIT is a production strategy promoted by Toyota, and now applied to many organizations, that strives to improve business return on investment by reducing in-process inventory and associated carrying costs. Kanban is one example of how this can be accomplished, but JIT extends throughout the organization to all aspects of product movement, including from suppliers. Basically, the belief is that storage of unused inventory is a waste of resources (no matter where in the system it exists). JIT inventory systems expose the hidden cost of keeping inventory, and help the organization devise new methods to manage the consequences of change. So in simple terms, JIT is an inventory management system that aims to have things available as needed in the production cycle without the need for guessing, work-in-process (WIP), or other characteristics of excessive inventories.

Kaizen (Continuous Improvement) versus Kaizen Events. *Kaizen* is a Japanese term for change for improvement, or improving processes through small incremental steps. Many people refer to this gradual change as *continual improvement*. Breakthrough improvement (which Juran refers to as big change) is described by another Japanese term, *kaikaku*.

Kaikaku is referred to in North America as a *kaizen event* or *kaizen blitz*. Hence, many practitioners often get confused with the interchangeable usage of kaizen and kaizen event. In lean implementation, kaizen events are used to provide quicker implementation results. Kaizen events are conducted by assembling a cross-functional team for three to five days and reviewing all possible options for improvement in a breakthrough effort. Management support is required for such initiatives. If the employees can't afford taking three to five days to improve a process constraint, then either the problem is unimportant or the organization requires more fundamental cultural adjustment before implementing lean.

Kanban (Pull System). A system is best controlled when material and information flows into and out of the process in a smooth and rational manner. If process inputs arrive before they are needed, unnecessary confusion, inventory, and costs generally occur. If process outputs are not synchronized with downstream processes, delays, disappointed customers, and associated costs may occur. A properly administered kanban system will improve system control by assuring timely movement of products and information. Kanban is implemented using a visual indicator called *kanban cards*. The card indicates the quantity to be replenished once the minimum level is reached.

An empty bin with a kanban card is the signal for production to pull material from the previous step. The kanban quantity is mathematically calculated and fine-tuned during practical implementation. Typically, organizations take a while to perfect kanban. Kanban is a more mature concept. It is important that other fundamentals of lean (5S, standard work, total productive maintenance [TPM], and variation reduction) are put in place before venturing into kanban. If not, frequent equipment failure and unstable or inconsistent processes will defeat the purpose of kanban, resulting in huge kanban sizes to shield against these uncertainties.

Muda. *See* waste

Overall Equipment Effectiveness (OEE). The concept of measuring how effectively a manufacturing operation is utilized was started in the 1960s and has developed into a calculation that multiplies the *availability, performance,* and *quality* of the process to create a percentage of overall effectiveness of the operation ($A \times P \times Q = OEE$). This is one of several measures available to track performance of the operation and is meant to be a benchmark for continual improvement efforts. A perfect 100% would indicate perfect production: manufacturing only good parts, as fast as possible, with no downtime.

PDCA or PDSA. Plan–do–check–act or plan–do–study–act.

Poka-Yoke. *Poka-yoke,* a Japanese term for mistake-proofing or error-proofing, is a method used to prevent errors. There are a number of examples in day-to-day life that use the mistake-proofing concept, such as electrical plugs and sockets that prevent plugging the wrong way, valves that shut once the maximum pressure is reached, fixtures that prevent loading the component in a wrong orientation, and so on. A window envelope is also a mistake-proofing method that allows users to see the letter with the right address sealed in. Similarly, there is detection-type mistake-proofing that alerts a user immediately after an error is made (to prevent further errors). Examples include car alarms that sound when the driver closes the door with the lights on, and an automatic gauging machine that alarms when an oversize or undersize part is produced.

Single Minute Exchange of Die (SMED). The goal of SMED is to provide a rapid and effective way of converting an operating process from running the current product to running the next product. The rapid changeover is key to reducing production lot sizes and thereby improving the flow of the system.

The economic lot size is calculated from the ratio of actual production time to changeover time, which is the time taken to stop production of a product and start production of the same or other product. If changeover takes a long time, then the lost production due to changeover drives up the cost of the actual production itself. The phrase "single minute" does not mean that all changeovers and setups should take only one minute, but that they should take less than 10 minutes (in other words, "single-digit minute").

Standard Work. Basically, *standard work* is a tool that defines the interaction between man and machine in producing a part. It has three components: standard time, standard inventory, and standard sequence. Standard work helps in training new operators and reducing the variation in the process.

The basic idea is to make manufacturing methods and/or service processes consistent. Quality management systems like ISO 9001 provide a basic foundation to lean implementation by incorporating standard work as part of the controlled documentation. Further, by having standard work, equipment, tools, layout, methods, and materials are standardized and thus reduce variation in processes. A detailed process work instruction with all of the above can be a very useful standard work document.

Takt Time. Derived from the German word *taktzeit,* this refers to the baton that an orchestra conductor uses to regulate the speed, beat, or timing at which musicians play. The purpose of takt time is to precisely match production with demand. It provides the heartbeat of a lean production system. Takt time first was used as a production management tool in the German aircraft industry in the 1930s.

Takt time (also referred to as *beat time, rate time,* or *heart beat*) sets the pace of industrial manufacturing lines so that production cycle times can be matched to customer demand rates. Expected customer demand sets the pace at which you need to produce the product to deliver to those customers. Taking the total of customer demand into consideration, the production scheduling department determines what is needed when shipping to the customer. The production operations then set the pace to produce those parts/components/assemblies to match what is needed to ship to the customers.

Theory of Constraints. *Theory of constraints* is a problem-solving methodology that focuses on the weakest link in a chain of processes. Usually, the constraint is the process that is slowest. Flow rate through the system cannot increase unless the rate at the constraint increases. The theory of constraints lists five steps to system improvement:

1. *Identify.* Find the process that limits the effectiveness of the system. If throughput is the concern, then the constraint will often have work-in-process (WIP) awaiting action.
2. *Exploit.* Use kaizen or other methods to improve the rate of the constraining process.
3. *Subordinate.* Adjust (or subordinate) the rates of other processes in the chain to match that of the constraint.
4. *Elevate.* If the system rate needs further improvement, the constraint may require extensive revision (or elevation). This could mean investment in additional equipment or new technology.
5. *Repeat.* If these steps have improved the process to the point where it is no longer the constraint, the system rate can be further improved by repeating these steps with the new constraint.

The strength of the theory of constraints is that it employs a systems approach, emphasizing that improvements to individual processes will not improve the rate of the system unless they improve the constraining process.

Drum–Buffer–Rope (DBR). Goldratt[7] introduced a squad of soldiers walking in single file as an analogy of a string of production processes. As the first soldier moves forward he receives unprocessed material, the fresh ground. Each succeeding soldier performs another process by walking on that same ground. As the last soldier passes over the ground, it becomes finished goods. So, the individual processes are moving over fixed material rather than the other way around. Lead time is the time that it takes for the squad to pass over a certain point. If each soldier moves as fast as he can, the lead time tends to lengthen, with the slower soldiers falling behind and holding up those behind them since passing is not permitted.

The system constraint is the slowest soldier. The ground can't be processed faster than this soldier can move. This soldier sets the drumbeat for the entire system. To avoid lengthening the lead time, a rope connects the lead soldier to the slowest soldier.

Now the squad moves along as a unit with minimum lead time and minimum work-in-process (WIP). If a soldier that is behind the slowest soldier happens to drop his rifle, he'll fall behind a little (especially if the sergeant notices it) but will be able to catch up since he is not the slowest soldier. This is analogous to a minor process problem at one station. If a soldier in front of the slowest soldier drops his rifle, the squad will not have to stop unless the slowest soldier catches up with the one in front of him. So, if the squad has a high tendency to drop their rifles, the rope must be longer. The length of the rope is the size of the buffer. In summary, to avoid long lead times and excess WIP, all system processes should be slowed down (via the rope) to the speed of the slowest process (the drum), with the amount of WIP (or buffer) determined by the dependability of the individual processes. For further explanation of these concepts see Goldratt's *Critical Chain*.

Total Productive Maintenance. If the lean enterprise implementation is to be sustained, the manufacturing or service equipment has to be reliable. In order to have reliable equipment, an organization has to maintain the equipment periodically. Preventive maintenance examples include changing oil at the required frequency, tightening loose parts, and watching for any visible or audible symptoms of failure. A comprehensive maintenance program may need a battery of maintenance technicians. This can be impractical and expensive. Hence, a *total productive maintenance* (TPM) program partners the maintenance technicians and line workers as a team to help each other reduce machine downtime. Management support is required to cross-train line workers to perform simple, basic maintenance and repairs. As the operators are trained to watch for symptoms of common failures, communication reaches maintenance technicians faster, thereby reducing downtime. Mature TPM programs use metrics like overall equipment effectiveness (OEE), which is a product of equipment availability, performance, and quality of output.

Value Stream. *See* value stream mapping later in this chapter.

Visual Factory. Visual factory provides visual identification of the status of material and information throughout the value stream. Examples of visual factory include providing status of material in/out at a raw material warehouse, showing units produced, units to complete order, and total produced by shift or day on a production display board, and indicating machine status with red, yellow, and green lights on the machine. Imagine that we need to find out the current status of a work order for a given customer. Often, this is achieved by talking to line supervisors, referring to logbooks, conducting internal meetings, and so on.

In short, if an employee can walk onto a shop floor and can tell which machines are running, what product is being produced, how many more are to be produced by customer, follow posted safety instructions, and report to management, that is an effective visual workplace.

Waste (Muda)

Some authors list seven or eight categories of waste, or *muda,* as it is referred to in some sources. These lists usually include overproduction, excess motion, waiting, inventory, excess movement of material, defect correction (rework), excess processing, and lost creativity (underutilization of resource skills). The following paragraphs examine the causes and results of each of these wastes.

Overproduction. Defined as making more than is needed or making it earlier or faster than is needed by the next process, the principal symptom of overproduction is excess *work-in-process* (WIP). Companies adopt overproduction for various reasons, including long setup times, unbalanced workload, and a just-in-case philosophy. One company maintains a six-month supply of a particular small part because the machine that produces it is unreliable. In some cases, accounting methods have dictated that machines overproduce to amortize their capital costs. All WIP should be continuously scrutinized for possible reduction or elimination.

Excess motion. This can be caused by poor workplace layout, including awkward positioning of supplies and equipment. This results in ergonomic problems, time wasted searching for or moving supplies or equipment, and often in reduced quality levels. Kaizen events are effectively used to focus a small short-term team on improvements in a particular work area. The team must include personnel with experience in the positions involved, as well as those with similar functions elsewhere. In addition, it is essential to include people with the authority to make decisions. Such teams have made startling changes in two to five days of intense activity.

Waiting. Typically caused by such events as delayed shipments, long setup time, or missing people, waiting results in waste of resources and, perhaps more importantly, demoralization of personnel. Setup time reduction efforts and total productive maintenance are partial answers to this problem. Cross-training of personnel so that they can be effectively moved to other positions is also helpful in some cases. Most important, of course, is carefully planned and executed scheduling.

Inventory. When inventories of raw materials, finished goods, or work-in-process are maintained, costs are incurred for environmental control, record keeping, storage and retrieval, and so on. These functions add no value to the customer. Of course, some inventory may be necessary, but if a competitor finds ways to reduce costs by reducing inventory, business may be lost. One of the most tempting times to let inventory levels rise is when a business cycle is in the economic recovery phase. Instead of increasing inventories based on forecasts, the proper strategy is to synchronize production to increase with actual demand. Similarly, production or administrative functions that use more space or other resources than necessary increase costs without adding value. The common analogy of the sea of inventory, shown in Figure 2.3, illustrates how excess inventory makes it

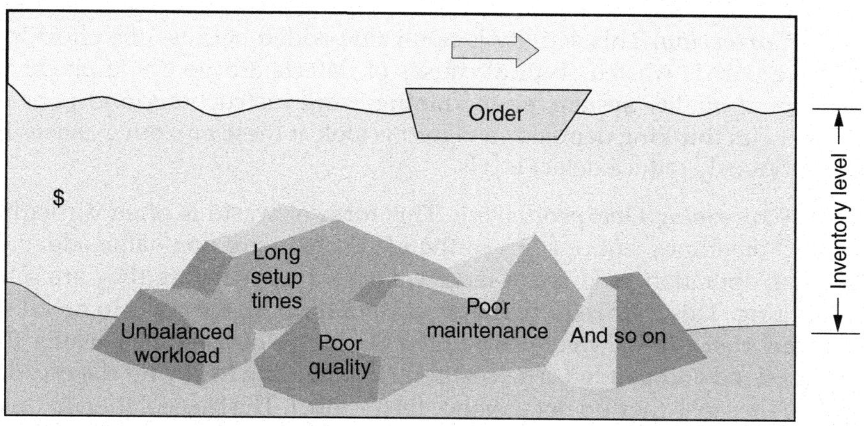

a) The order floats through the system protected from unresolved problems by excess inventory.

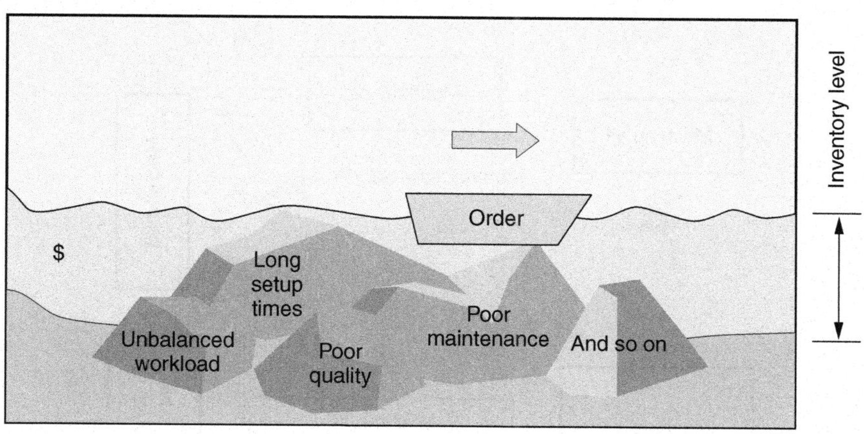

b) When the protective inventory is reduced, problems emerge that must be solved. To reduce cost, we must fix the problems.

Figure 2.3 A sea of inventory often hides unresolved problems.

possible to avoid solving other problems. As the level of inventory is lowered, some problems will rear their ugly heads and need to be solved before further progress is possible.

Excess Movement of Material/Transportation. Large conveyor systems, huge fleets of forklifts, and so on, make production more costly and complex, and often reduce quality through handling and storing. Poor plant layout is usually to blame. Plants with function-oriented departments (all lathes together, all presses together, and so on) require excessive material movement. A better plan is to gather equipment together that is used for one product or product family. This may mean having a manufacturing cell contain several types of equipment requiring personnel with multiple skills. Many companies have had success with cells that form a C shape, as shown in Figure 2.4, because they can be staffed in several ways. If demand for the cell's output is high, six people could be assigned, one per machine. If demand is very low, one person could move from machine to machine, producing parts one at a time.

Defect Correction. This activity is non-value-added because the effort to fix the defective part is wasted. Typical causes of defects are poor equipment maintenance, poor quality system, poor training/work instructions, and poor product design. Lean thinking demands a vigorous look at these and other causes in order to continuously reduce defect levels.

Excess Processing/Overprocessing. This form of waste is often difficult to recognize. Sometimes, entire steps in the value chain are non-value-added. A steel stamping operation produces a large volume of parts before they are scheduled for painting. This may require the practice of dipping the parts in an oil solution to prevent rust as they wait to be painted. As the paint schedule permits, the parts are degreased and painted. The customer is unwilling to pay for the dip/degrease activities because they do not enhance the product. The best solution in this case is to schedule the pre-paint activities so that the parts are painted immediately upon

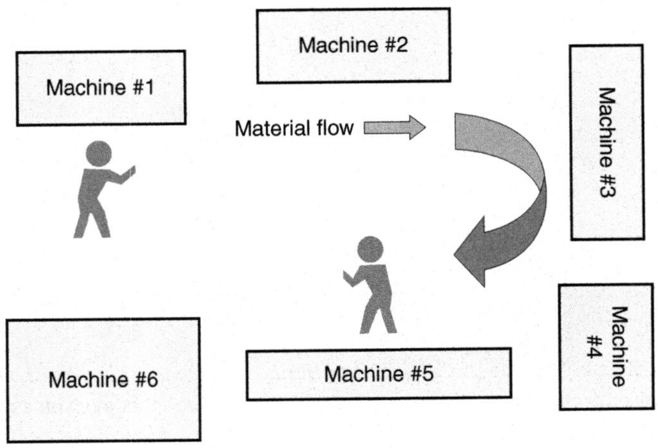

Figure 2.4 C-shaped manufacturing cell.

production. This may require smaller batch sizes and improved communication procedures, among other things.

The purpose of the grinding step that often follows a welding operation is to remove some of the weld imperfections. Improving the welding process may reduce or eliminate the need for grinding. The unnecessary grinding would be classified as excessive processing. Excess processing can occur in the office as well as on the plant floor. Information from customer purchase orders is sometimes entered into a database, and the order itself is filed as a backup hard copy to resolve any later disagreements. A recent study by one company revealed the fact that the hard copies, although they are occasionally pulled from files and initialed, stamped, stapled, and so on, really serve no useful purpose. The company now discards the purchase order once the information has been entered. The processes of filing, storing, and maintaining these records required one-half person performing non-value-added activity.

Additional Forms of Waste

Lost Creativity. This is perhaps the most unfortunate waste. Most manufacturing employees have ideas that would improve processes if implemented. Standard organizational structures sometimes seem designed to suppress such ideas. Union/management divides seem almost impossible to bridge. Lean thinking recognizes the need to involve employees in teams that welcome and reward their input. These teams must be empowered to make changes in an atmosphere that accepts mistakes as learning experiences. The resulting improved morale and reduced personnel turnover impact the bottom line in ways that no accountant has yet calculated. These are the nontangible benefits of lean thinking.

Perfection. The goal of eliminating muda is to strive for perfection. You now understand value-added activities. You also learned about various wastes, both hidden and explicit, in processes. By optimizing value-added activities and eliminating waste, your organization can aim toward achieving "perfection" in lean. This is not a one-time effort. This is a continual learning process.

Employee Demotivation. There is an old saying that hopefully you never have to experience: "Employees don't leave companies, they leave managers." Abraham Maslow created what is called the Hierarchy of Needs to help explain why people behave in business as they do.

The Hierarchy of Needs is usually viewed as a pyramid in this order (first item on top):

- Self-actualization needs: reaching full potential, fulfilling life's purpose
- Esteem needs: achievement, adequacy, confidence, recognition, appreciation
- Belongingness needs: acceptance, intimacy, friendship, trust
- Security needs: stable housing, safe schools, preventive medical care
- Fundamental needs: sleep, food, basic medical care

There can be any number of things at work that demotivate an employee, such as:

- Being apathetic
- Bullies in the workplace
- Culture
- Disorganization
- Failing to address company mission, vision, and values
- Feeling invisible in the workplace
- Going back on their commitments
- Hiring and promoting the wrong people
- Job insecurity
- Lack of clarity
- Lack of communication
- Lack of professional/personal development
- Lack of support
- Lack of trust or micromanagement
- Lack of workplace flexibility
- Letting accomplishments go unrecognized
- Low or unfair pay
- Making a lot of stupid rules
- Organizational direction – do not like the product or service
- Poor management style
- Tolerating poor performance
- Treating everyone equally

Lack of Effective Emergency Response Planning. ISO 22301 deals with Business Continuity, which many companies have failed to do adequately. See Chapter 6 for more information.

Spaghetti Diagrams (Spaghetti Map). A graphic tool intended to show the flow or path that people, products, or paperwork take during the course of normal operations. These can tend to look very messy for processes that have not gone through a process improvement procedure, especially when multiple people are involved in the process. One example that the audit did was to track the patient charts in a doctor's office. The staff had never seen the full extent of the process before (however, the two nurses recognized the paths that they had to run with the charts), and we created a simple flowchart for them to work on improving the process.

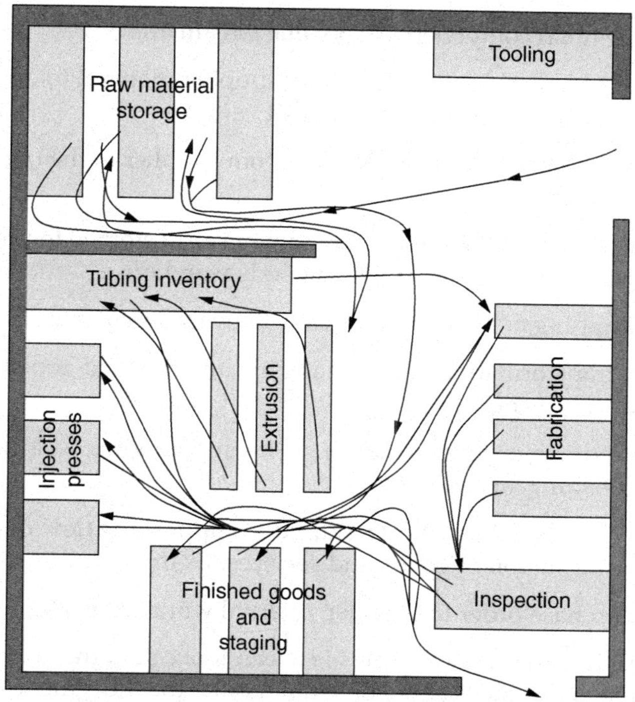

Figure 2.5 Spaghetti diagram showing the flow of materials through an operation.
Source: The Lean Handbook (Milwaukee: ASQ Quality Press, 2012).

VALUE STREAM MAPPING

> Use value-stream mapping to identify value-added processes and steps or processes that produce waste, including excess inventory, unused space, test inspection, rework, transportation, and storage. (Understand)
>
> **Body of Knowledge I.B.2**

Value Stream

A *value stream* is the series of activities that an organization performs, such as order, design, produce, and deliver products and services.[8] A value stream often starts from a supplier's supplier and ends at the customer's customer. Wastes are both explicit and hidden along this value stream.

The three main components of a value stream are:

1. Flow of materials from receipt of supplier material to delivery of finished goods and services to customers. Examples:
   - Raw material shipped weekly from supplier to the organization by truck
   - Movement of material from raw material storage to production process through to finished goods warehouse
   - Shipping of the finished goods to overseas customer via customs
2. The transformation of raw materials into finished goods or inputs into outputs. Example:
   - Production steps like cutting, shaping, forging, welding, polishing, and assembly
3. The flow of information required to support the flow of material and transformation of goods and services. Example:
   - Purchase order to supplier, internal work order, shipping notice

This concept is visually illustrated via a lean tool called the *value stream map.* This map uses simple graphics and icons to illustrate the movement of material, information, inventory, work-in-process, operators, and so on. Value stream mapping is a very powerful tool. The analysis subsequent to value stream mapping, called *value stream analysis,* can help uncover hidden wastes within the organization. An organization that effectively uses lean thinking and applies lean tools to reduce waste throughout the value stream and offer value to their customers is a *lean enterprise* organization.

Achieving a lean enterprise requires a change in attitudes, procedures, processes, and systems. It is necessary to zoom out and look at the flow of information, knowledge, and material throughout the organization. In any organization there are multiple paths through which products, documents, and ideas flow. The process of applying lean thinking to such a path can be divided into the following steps:

1. *Produce a value stream map.* This is also referred to as a *value chain diagram.* This diagram is described in detail by Rother and Shook.[9] It has boxes labeled with each step in the process. Information about timing and inventory is provided near each process box. Some symbols that are used on value stream maps are shown in Figure 2.6. Figure 2.7 shows an example of a value stream map.
2. *Analyze all inventory notes with an eye toward reduction or elimination.* Inventory tends to increase costs because:
   - Storage space may be expensive (rubber awaiting use in a tire factory is stored at 120 °F; wood inventory may need to be humidity-controlled).
   - Quality may deteriorate (rust, spoilage, and so on).

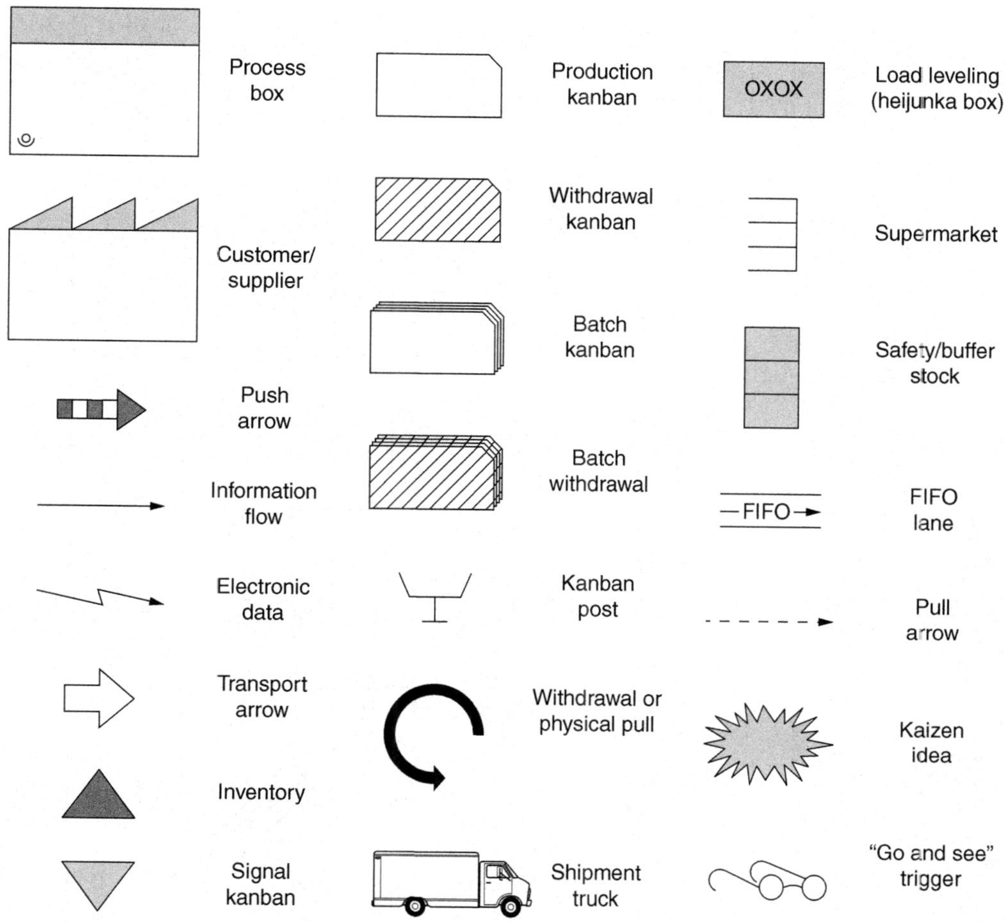

Figure 2.6 Common symbols used in value stream mapping.

- Design changes may be delayed as they work their way through the inventory.
- Money invested in inventory could be used more productively elsewhere.
- Quality problems that are not detected until a later stage in the process will be more expensive to correct if an inventory of defective products has accumulated.

One company refers to its racks of safety stock as the "wall of shame."

3. *Analyze the entire value stream for unneeded steps.* These steps are called non-value-added activities and are discussed in detail earlier in this section.

4. *Determine how the flow is driven.* Strive to move toward value streams in which production decisions are based on the pull of customer demand.

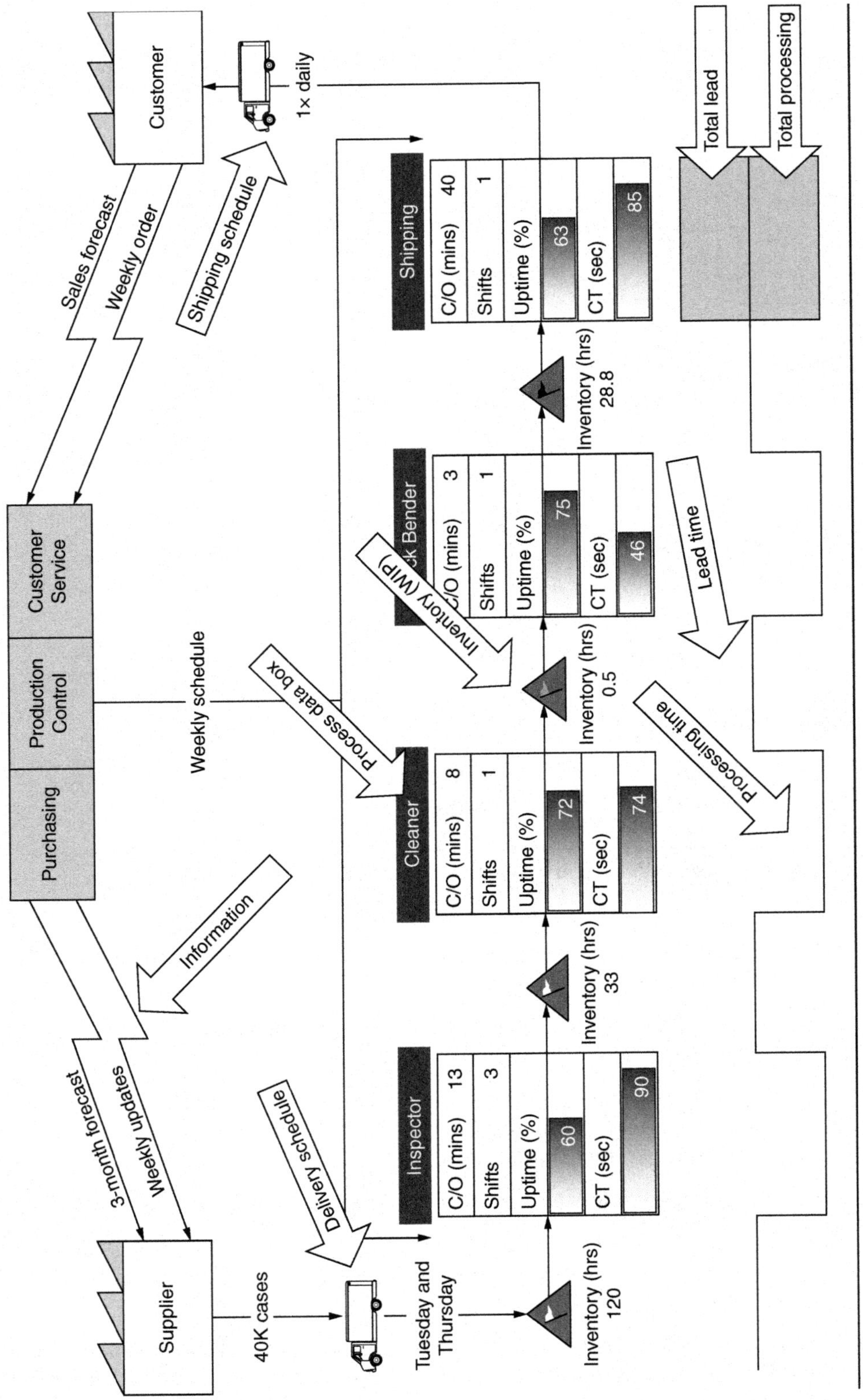

Figure 2.7 Value stream map example.

In a process where pull-based flow has reached perfection, a customer order for an item will trigger the production of all the component parts for that item. These components would arrive, be assembled, and delivered in a time interval that would satisfy the customer. In many situations this ideal has not been reached, and the customer order will be filled from finished goods inventory. The order should still, however, trigger activities back through the value chain that produce a replacement item in finished goods inventory before it is needed by a customer.

5. Extend the value stream map upstream into suppliers' plants. New challenges continue to occur regarding compatibility of communication systems. The flows of information, material, knowledge, and money are all potential targets for lean improvements.

When beginning the process, pick a narrow focus—don't try to boil the ocean, as the saying goes.

Chapter 3

Design for Six Sigma (DFSS) Methodologies

ROAD MAP FOR DFSS

> Distinguish between DMADV (define, measure, analyze, design, verify) and IDOV (identify, design, optimize, verify), and recognize how they align with DMAIC. Describe how these methodologies are used for improving the end product or process during the design (DFSS) phase. Understand how verification and validation are used to compare results against stated goals. (Understand)
>
> **Body of Knowledge I.C.1**

Organizations must extend their design beyond simple functionality and customer wishes to consider fulfilling other attributes and expectations. This holistic approach to design will result in a more stable and robust product that not only reflects customer preferences, but also is capable of being used and applied in the specified environment by the intended user. We typically refer to the use of Six Sigma in the design phase of product development as *design for Six Sigma* (DFSS).

Thus, the definition of DFSS includes:

- DFSS is a business/engineering strategic process that focuses on proactive design quality, rather than reactive design quality.
- DFSS is a systematic process to create produceable designs by reducing and managing variation in order to meet the "customer's" expectations of quality/performance.

DMADV (define, measure, analyze, design, verify) and IDOV (identify, design, optimize, verify) are the most common acronyms used in DFSS (others include DCOV [define, characterize, optimize, verify], ICOV [identify, characterize, optimize, validate], DMEDI [define, measure, explore, develop, implement], IDDOV [identify, define, develop, optimize, verify], and GD[1] [good design, good discussion, good dissection]). These relate to DMAIC and help close the loop on improving the end product/process during the up-front design for Six Sigma phase. When the most common tools are used in the various phases, it is commonly stated that doing IDOV feeds the MAIC of DMAIC. The challenge is that we should be using the various Six Sigma tools on designs for products before they become a reality. As in many Six Sigma applications, scoping projects for DFSS (no matter which acronym you use) can be a challenge. Some issues to keep in mind include:

- Too small a scope—capture enough control factors to achieve robustness.
- Vague scope—no reference to a subsystem.
- Use of separate projects for each symptom.
- Too much time spent on rating systems—why wouldn't we be more robust if we could be?
- DFSS Methodology: Use robust, make robust, keep robust.

One of the central focuses that should be included in a DFSS project is to be very clear about what can happen to your product in the customer's hands. Things that you cannot control in the manufacturing process or how customers use your product are referred to as *noise*. You must allow for a noise strategy to improve a process. If there is a possible way for a customer to misuse your product, the probability is that they will. The root cause failure of DFSS is typically related to the team using too limited control factors (around the noise in the systems). Parts tend to be already designed before teams look for continual improvement opportunities.

Benefits of using DFSS in your organization should include such things as:

- Increased customer satisfaction—measured on the Kano model—see Chapter 5 for more information
- Reduced variation
- Improved robust design
- Decreased warranty costs
- Improved reliability, durability
- Increased market share
- Increased revenue, earnings growth
- Increased production—less production downtime due to defects

Following is a more detailed explanation of the two main DFSS processes in common use today.

IDOV

Woodford[2] refers to IDOV as a four-phase process that consists of *identify, design, optimize,* and *verify.* These four phases parallel the four phases of the traditional Six Sigma improvement methodology, MAIC—*measure, analyze, improve,* and *control.* The similarities can be seen below.

Identify Phase. The *identify* phase tasks link the design to the voice of the customer:

- Identify customer and product requirements
- Establish the business case
- Identify technical requirements, critical to quality (CTQ) variables, and specification limits
- Roles and responsibilities
- Milestones

Design Phase. The *design* phase tasks emphasize CTQ variables and attributes:

- Formulate concept design
- Identify potential risks using failure mode and effects analysis (FMEA)
- For each technical requirement, identify design parameters
- Plan procurement, raw materials, manufacturing, and integration
- Use DOE (design of experiments) and other analysis tools to determine CTQs and their influence on the technical requirements (transfer functions)

Optimize Phase. The *optimize* phase develops detailed design elements to predict performance and optimize design:

- Assess process capabilities to achieve critical design parameters and meet CTQ limits
- Optimize design to minimize sensitivity of CTQs to process parameters
- Design for robust performance and reliability
- Error-proofing
- Establish statistical tolerancing
- Optimize cost through reduced process variation

Validate Phase. The *validate* phase consists of testing and validating the design and recording information for design improvements:

- Prototype test and validation
- Assess performance, failure modes, reliability, and risks
- Design iteration
- Final phase review

DMADV

Breyfogle[3] refers to the approach known as DMADV (define, measure, analyze, design, verify), which he says "is appropriate, instead of the DMAIC approach, when a product or process is not in existence and one needs to be developed. Or the current product/process exists and has been optimized but still doesn't meet customer and/or business needs."

Historically, the redesign process has been found to be a common source of waste that can be reduced by enhancing the original design process. Design for Six Sigma is the process of designing with a particular attribute in mind:

1. ***Define.*** Before beginning a design initiative, the Six Sigma team needs to evaluate and prioritize the primary design objectives for the organization. By targeting the primary priorities, the design efforts will have the most significant impact possible on achieving Six Sigma targets.

2. ***Measure.*** This requires a combination of technical and competitive product management analysis, specifying the design criteria most valued by the industry and customer. In addition, there are expectations imposed by regulators, partners, and other stakeholders.

3. ***Analyze.*** The statistical and investigative approaches used for Six Sigma can identify design priorities with significance and confidence.

4. ***Design.*** Having obtained a clear direction for design objectives, it is incumbent on the Six Sigma team to collaborate with designers to ensure that the final design outputs include the desired attributes. If these are treated as requirements or specifications, the fulfillment of Six Sigma design objectives will be incorporated into the development and testing activities and embedded into the overall solution. Without this approach, the Six Sigma design objectives will have to be an additional layer, which is potentially expensive and wasteful.

Design for Cost (also known as Design to Cost). In most markets, cost has become a major consideration in the design process. This requires a constant search for alternative processes, materials, and methods. People with cost accounting and purchasing backgrounds can assist the design team in this quest.

Design for Manufacturing/Design for Producibility/Design for Assembly. Many companies have found that minor design changes can make the product easier and less costly to produce. Tolerance design can result in savings in machining processes, tooling, and gauging. Designers should be familiar with existing manufacturing equipment and processes and strive to design products that don't require additional capability. Some manufacturers have found that drastic reductions in the number of parts in a product is an effective way to reduce manufacturing costs. As a general rule, the earlier that manufacturing personnel are involved in the design process, the more producible the design.

Design for Test (also known as Design for Testability). In products where in-process testing is critical, designers must make provision for performing tests earlier in the production cycle rather than relying entirely on functional tests of a finished assembly or subassembly.

Design for Maintainability. The ability to perform routine maintenance must be considered in the design process. Products that require long downtimes for diagnosis and repair can cause the user to miss deadlines and alienate customers. Maintainability includes modularity, decoupling, and component standardization.

Design for Robustness. Adequate time must be allowed during the design process to conduct life cycle tests of all parts, subassemblies, and assemblies. Suppliers of purchased parts should be required to document the mean time to failure (MTTF) or mean time between failures (MTBF) for all products they supply. MTTF is used for nonrepairable items and MTBF is used for repairable items. A basic relationship in this area is that between failure rate λ and MTTF or MTBF:

$$\text{MTTF} = 1/\lambda \text{ or MTBF} = 1/\lambda \text{ and, of course,}$$
$$\lambda = 1/\text{MTTF or } \lambda = 1/\text{MTBF}$$

Design for Usability. The quality of a product is determined by *validation,* where it is applied for its prescribed purpose by its intended users in its specified environment. The ability of a user to work comfortably with the product, system, or service to obtain value can be measured and improved.

Design for Extended Functionality. Many products initially designed and intended for a single purpose can have their features applied to extended functionality beyond the initial vision of the designers. Computer software applications are good examples of products that were initially developed for quick mathematical calculation and numerical tracking, but are now preferred tools for graphic design, word processing, and database management.

Design for Efficiency. The product or system must be designed in a way that consumes minimal resources, both to make the product and to use the product or service. This is correlated with design for cost, except the criteria for evaluation are time, resources, and consumption of critical components. Efficiency will have positive effects on long-term cost and reliability.

Design for Performance. Performance refers to the achievement of aggressive benchmarks or breakthroughs on a consistent basis. Terms like "cutting edge" or "latest and greatest" reflect the constant challenge of exceeding once unachievable levels of delivery. Historical examples include the design of aircraft faster than the speed of sound, and the continuous increase in processing power of microchips.

Design for Security. Security is becoming a bigger threat as maladies like computer viruses, identity theft, and product misuse increase in scope and complexity. Security will preserve product integrity and protect the intellectual property and privacy of users and designers.

Design for Scalability. Products or systems deployed for use in a growth market should anticipate expansion or rapid adoption. Without this attribute, quality will be compromised when the product surpasses the threshold of users or scope. An example is the auction website that suddenly has a blank screen during peak periods because it cannot handle the load of 100,000 concurrent users at month-end.

Design for Agility. Many organizations compete on their ability to deliver customized solutions within a short time. This requires a nimble approach to rapid development, a robust architecture or structural foundation, and a ready array of components or vendors who can augment the core product with unique touches.

Table 3.1 Design objectives and outputs traceability matrix.

Design objective	Design output	Status
Extended functionality	Business user can apply software to their operations	Achieved functionality
Maintainability	Modular approach with minimal coupling	Replacement of modules results in quicker diagnosis and maintenance
Efficiency	Point-of-sale transaction system allows sale to be completed within five minutes	Design supports the application of the product to make customers more efficient
Security	Password encryption for user access	Security achieved to prevent unauthorized product use
Compliance to Kyoto standards for emissions	Product did not pass mandated emissions standards	Redesign is required for marketability

An example is a hot tub manufacturer who incorporates the basic hot tub into the style and design of a building or landscape to create a seamless effect.

Design for Compliance. Designers have regulations imposed on them that must be fulfilled in order for the product to be marketed. Compliance requirements can range from achieving specific product performance capabilities to demonstrating that suitable design processes were followed and recorded. If the DFSS initiative is operating in a highly regulated environment, cost-benefit can be derived by the penalties/fines and opportunity costs of noncompliance. An example is designing a process that requires configuration management updates every time the product is changed, to ensure proper and accurate documentation. Example: If the U.S. auto industry does not meet the California emission standards, then cars/trucks cannot be sold there.

5. *Verify.* ISO 9000:2015 clause 3.8.12 defines *verification* as the "confirmation, through the provision of objective evidence, that specified requirements have been fulfilled." This can be completed by the Green Belt or others in the company during the APQP or Toll Gate processes (both described later in this handbook) or independently by outside third-party companies. Verification should be completed prior to the full release of a product or service to the full production phase of making/delivering something.

ISO 9000:2015 clause 3.8.13 defines *validation* as the "confirmation, through the provision of objective evidence, that the requirements for a specific intended use or application have been fulfilled." This can be accomplished through any number of internal audits or the use of the Rice and Munro Evaluation Model (described in Appendix O) for validation of learning outcomes.

BASIC FAILURE MODE AND EFFECTS ANALYSIS (FMEA)

> Use FMEA to evaluate a process or product and determine what might cause it to fail and the effects that failure could have. Identify and use scale criteria, calculate the risk priority number (RPN), and analyze the results. (Analyze)
>
> **Body of Knowledge I.C.2**

The essence of a *failure mode and effects analysis* (FMEA) is the study of risk, and it has grown into multidisciplinary fields of study with many resulting risk methodologies. Risk is about the uncertainty of an event. Another way to say this is that risk is about what might happen, good or bad, that will influence/affect what happens at work, with the product/service, how a customer might use or misuse your product/service, or any other issue that management might identify as a concern.

Risk can also be defined in terms of three aspects: impact (severity), probability (occurrence), and event (detection). *The Risk Management Memory Jogger*[4] lays out a Risk Road Map for ISO 31000:2018 of:

- Plan risk management
- Risk identification tools
- Analyze and evaluate risk
- Plan risk response
- Monitor and control risk

The concepts of what we call failure mode and effects analysis have been around a long time under various names (originally called failure mode, effects, and criticality analysis [FMECA]) and follow the Risk Road Map concepts above. In the past, inventors and product developers thought about possible ways that a product could fail during extreme conditions, handling, and usage. They started to provide countermeasures in the design and manufacturing process to prevent these failure modes. FMEA thus started to evolve informally. A brief history of standards and processes includes (see Appendix N for additional listings):

- The U.S. Military first issues what we now know as FMEA on November 9, 1949—Military P-1629: *Procedures for Performing a Failure Mode, Effects and Criticality Analysis.* This led into the MIL-STD 1629 series of documents.
- NASA's Apollo space program uses RA-006-013-1A: *Procedure for Failure Mode, Effects, and Criticality Analysis (FMECA),* August 1966.
- Enterprise risk management (ERM)—in the early 1970s Gustav Hamilton of Sweden's Statsfoetag proposes the "risk management circle."

- Ford Motor Company starts using FMEA in the late 1970s after the Pinto issue.
- The NASA Challenger disaster on January 28, 1986, exposes a Morton-Thiokol O-ring FMEA in the resulting legal litigation.
- The Committee of Sponsoring Organizations (COSO) is organized in 1985. The full name is Committee of Sponsoring Organizations of the Treadway Commission, and their focus is on the financial aspects of risk management and fraud prevention.
- The Automotive Industry Action Group (AIAG) releases the first Big Three PFMEA in February 1993. (History: February 1993, February 1995, July 2001, fourth edition 2008.)
- SAE International releases SAE J-1739: *Potential Failure Mode and Effects Analysis in Design (Design FMEA), Potential Failure Mode and Effects Analysis in Manufacturing and Assembly Processes (Process FMEA)* 2009-01-15. (History: 1994-07-01, 2000-06-01, 2002-08-02, 2009-01-15.)
- ASQ publishes the first edition of D. H. Stamatis, *Failure Mode and Effect Analysis: FMEA from Theory to Execution* in 1995, second edition June 2003.
- AIAG releases *FMEA for Tooling & Equipment (Machinery FMEA)* November 1, 2001—second edition February 2012
- International Organization for Standardization (ISO) releases ISO 31000: 2018 *Risk management—Principles and guidelines.*
- IATF released a new FMEA Handbook by AIAG & VDA in June 2019.

ISO is also updating the ISO 9001 quality management system (QMS) in 2015 to include what they call "risk-based thinking." ISO 9001 0.5 states in part, "This International Standard makes risk-based thinking more explicit and incorporates it in requirements for the establishment, implementation, maintenance and continual improvement of the quality management system. Organizations can choose to develop a more extensive risk-based approach than is required by this International Standard, and ISO 31000 provides guidelines on formal risk management which can be appropriate in certain organizational contexts."

There are many resources for FMEA and related methodologies. One starting point could be the FMEA Info Centre (www.fmeainfocentre.com). Terms you might consider searching include risk, risk assessment, risk management, enterprise risk management, FMEA, COSO, and many others.

Why Do FMEAs?

The concept of using a risk matrix (see Figure 3.1) to analyze a given situation is not new in today's thinking in many organizations. The automotive industry (through AIAG and SAE International) has been in the forefront of developing the FMEA methodology. The fundamental reason for using an FMEA is to predict the highest likelihood of things that could go wrong (at the concept, design,

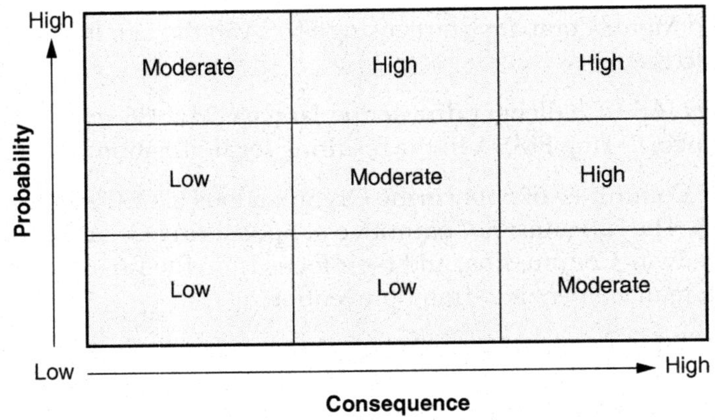

Figure 3.1 Simple risk matrix.

process, machinery, or system level). The old adage of "a stich in time saves nine" very much applies here. If we think through the upcoming process and formally write it out on paper (or in a software system), it gives us a much better chance to predict and prevent occurrences or situations that may cause our organization an unpleasant issue.

The AIAG described FMEA as a systematic group of activities intended to:[5]

- Recognize and evaluate the potential failure of a product/process and the effects of that failure
- Identify actions that could eliminate or reduce the chance of the potential failure occurring
- Document the entire process

Note: The AIAG and VDA FMEA Handbook was meant to replace the fourth edition AIAG FMEA Manual; however, at the time of printing this handbook, the new FMEA reference has not received full endorsement within the automotive community.

The purpose of automotive design FMEA (DFMEA) and process FMEA (PFMEA) is to understand the opportunities for failure and the impact of risks in a product or process design, prioritize the risks, and take actions to eliminate or reduce the impact of these risks. Successful product/process development requires anticipating failure modes and taking actions to eliminate or reduce failure during deployment and life cycle. FMEA is not a one-time event; the product/process design team needs to periodically review and update the failure modes. During the early stages of product/process development, the team identifies the risks based on existing data from similar processes, knowledge, and experience. As the product/process is deployed, new, unforeseen risks and failures may show up. Hence, reviewing the FMEA on a continual basis ensures sustainable success. Carlson[6] includes a number of FMEA checklists in his book that can be found at www.effectivefmeas.com.

FMEA needs to be documented and revision-controlled and should be part of the existing quality management system (QMS). In a well-designed QMS, FMEA is linked to quality function deployment in the design and process "houses of quality," and linked to control plans in the production house of quality. Part of this document control process is needed as the FMEA is considered to be a living document—that is, it should be updated as needed and the revisions controlled to track changes over time.

Another key aspect to the use of the FMEA risk management concept is that FMEA is not just confined to manufacturing applications. FMEA has been successfully used in service/transactional processes, software development, the medical field, and so on.

Once the FMEA is completed, some of the benefits/uses that you should see include:

- Assesses effect on all customers (internal and external)
- Aids in evaluating requirements and alternatives
- Identifies potential design, manufacturing, or assembly cause issues and needs to focus on controls for reducing occurrence and/or increasing detection
- Develops a prioritized list for actions (ongoing as a living document)
- Helps validate the intended design, manufacturing, or assembly process
- Documents the results of the design, manufacturing, or assembly process
- Identifies confirmed special characteristics requiring special controls

Although in principle FMEA is conducted to address the potential failures in product design and process design, FMEA is identified separately as *design FMEA* (DFMEA) and *process FMEA* (PFMEA). (There are also the concept FMEA, machinery FMEA, and system FMEA, which are beyond the scope of this BoK.)

FMEA Forms/Tables

The typical FMEA format is a simple matrix that can be easily duplicated in a simple spreadsheet software program. Specialized software programs are also available to help create the form as your team works through the process. Depending on the industry that you work in, there may be specific formats that are required, so checking with your organization's quality manager or quality engineer might be advisable if you are unfamiliar with this tool. Some common column headers used today include those for design FMEA (see Figure 3.2) and process FMEA (see Figure 3.3).

Steps in Performing FMEA

A team approach has proven to be the most effective method for conducting an FMEA, so it is discussed here. Assemble a cross-functional team with diverse

Potential
Failure Mode and Effects Analysis
(Design FMEA)

____ System
____ Subsystem
____ Component ____________
Model year(s)/Vehicle(s) ____________
Design responsibility ____________
Key date ____________
FMEA number ____________
Page # ________ of ________
Prepared by ____________
FMEA date (orig.) ________ (rev.) ________
Core team ____________

Item / Function	Potential failure mode	Potential effect(s) of failure	S	Class	Potential cause(s)/ mechanism(s) of failure	O	Current design controls • Prevention • Detection	D	RPN	Recommended action(s)	Responsibility and target completion date	Action results				
												Action taken	S	O	D	RPN

or

Item / Function	Potential failure mode	Potential effect(s) of failure	S	Class	Potential cause(s)/ mechanism(s) of failure	O	Current design controls Prevention	Current design controls Detection	D	RPN	Recommended action(s)	Responsibility and target completion date	Action results				
													Action taken	S	O	D	RPN

Figure 3.2 Sample DFMEA headers.

Potential
Failure Mode and Effects Analysis
(Process FMEA)

From ________ Process responsibility ________ FMEA number ________

Model year(s)/Vehicle(s) ________ Key date ________ Page # ____ of ____

Prepared by ________

FMEA date (orig.) ____ (rev.) ____

Core team ________

Item / Function	Potential failure mode	Potential effect(s) of failure	S	Class	Potential cause(s)/ mechanism(s) of failure	O	Current design controls • Prevention • Detection	D	RPN	Recommended action(s)	Responsibility and target completion date	Action results: Action taken	S	O	D	RPN

or

Item / Function	Potential failure mode	Potential effect(s) of failure	S	Class	Potential cause(s)/ mechanism(s) of failure	O	Current design controls Prevention	Current design controls Detection	D	RPN	Recommended action(s)	Responsibility and target completion date	Action results: Action taken	S	O	D	RPN

Figure 3.3 Sample PFMEA headers.

knowledge about the process, product, or service, and customer needs. Functions often included are design, manufacturing, quality, testing, reliability, maintenance, purchasing (and suppliers), shop floor operators, sales, marketing (and customers), and customer service. It is important to have process experts present in design FMEA and design experts in process FMEA. For effective interaction, the team is typically five to seven people. If additional experts are needed to provide inputs on safety, regulatory, or legal issues, they are included in the team as subject matter experts.

Identify the scope of the FMEA. Is it for concept, system, design, process, or service (yet another FMEA type)? What are the boundaries? How detailed should we be? What will be the overall focus of the effort your team is about to work on? A basic flow of the FMEA process is shown in Figure 3.4 and is called the *FMEA flowchart*. This figure shows a basic three-step process for thinking through the various aspects that are needed when working on an FMEA. Figure 3.5 shows how this flowchart fits on a typical FMEA form. As your team works through the various columns of the form, a key item to remember is that there are no absolutes, and if disagreement arises, consider taking a middle ground for the time being (maybe listing a note for further study later on) so that the team does not get bogged down in the overall process. Table 3.2 gives more detail for performing an FMEA.

Severity, Occurrence, and Detection Tables

As the potential failure modes are identified by the cross-functional team, a determination of score needs to be developed for each of the three primary categories of *severity, occurrence,* and *detection.* Just as each word indicates, you will need

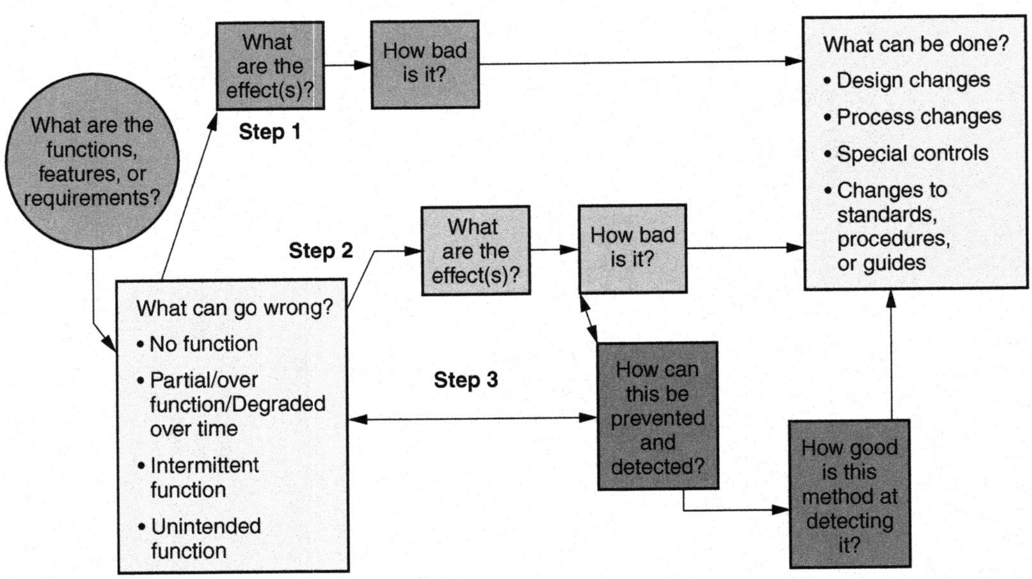

Figure 3.4 FMEA flowchart.

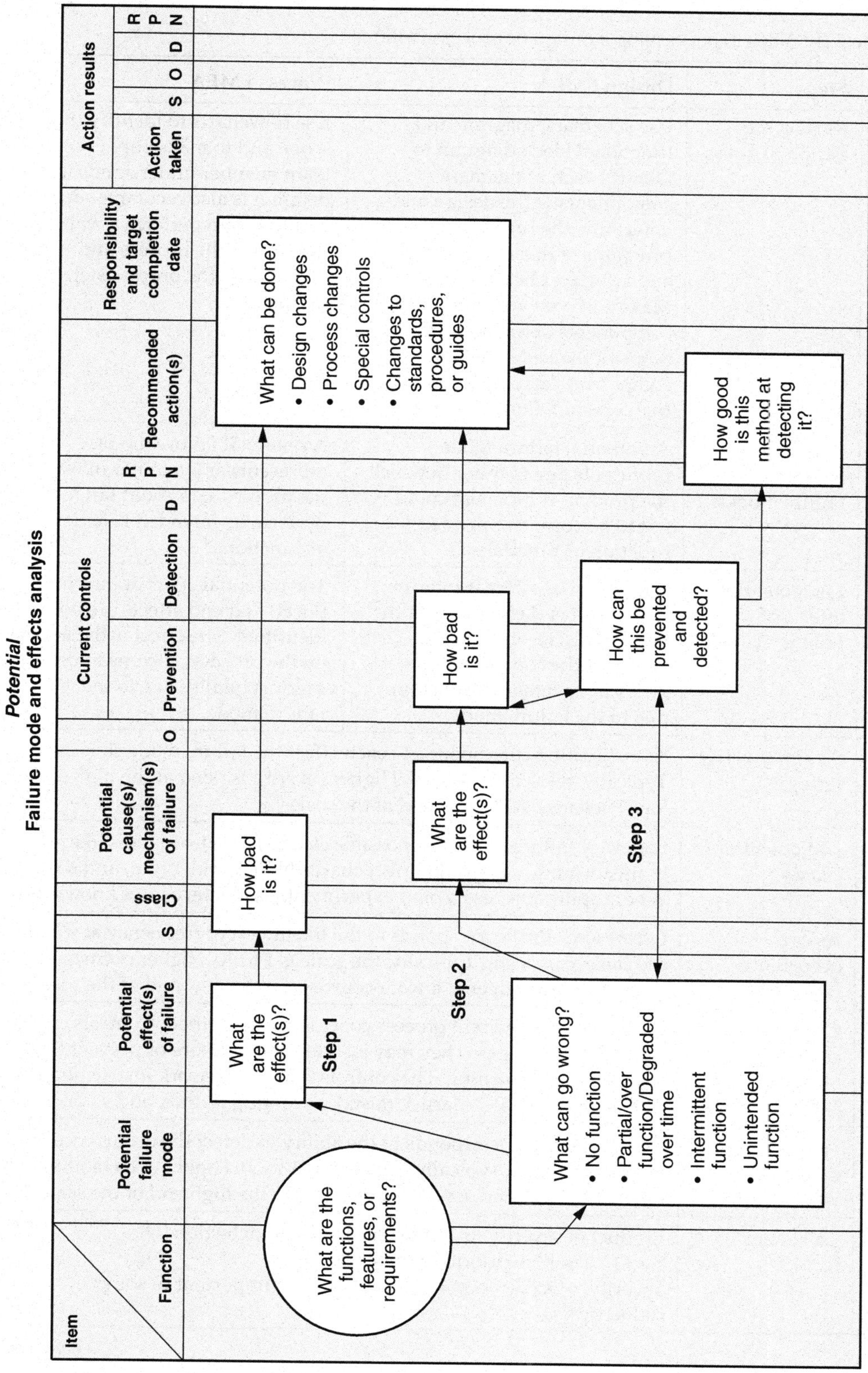

Figure 3.5 FMEA flowchart in line with common FMEA form.

Table 3.2 Steps in performing a design or process FMEA.

	Steps	Design FMEA	Process FMEA
1	Review the design/process	Use schematic diagram and functional block diagram to identify each of the main components of the design and determine the function or functions of those components and interfaces between them. Make sure you are studying all components defined in the scope of the DFMEA. Some components may have more than one function.	Use flowcharts to identify the scope and to make sure every team member understands it in detail. It is also recommended that the team perform a walk-through of the process and understand the process steps firsthand.
2	Brainstorm potential failure modes	A potential failure mode represents any manner in which the product component could fail to perform its intended function or functions.	A potential failure mode represents any manner in which the process step could fail to perform its intended function or functions.
3	List potential effects of failure	The potential effect at interim (local) and end effects are both identified. The effect is the ability of the component to perform its intended function due to the failure mode.	The potential effect at interim (local) and end effects are both identified. The effect is the impact on the process outcome and product quality due to the failure mode.
4	Assign severity rating (S)	*Severity* rating corresponds to each effect the failure mode can cause. Typically, the scale is 1 to 10. Higher severity is rated at the high end, lower severity at the low end of the scale.	
5	List potential causes	For every failure mode, list possible cause(s). Use team tools like brainstorming, cause-and-effect charts, NGT, multivoting, and so on. Where applicable, use a pilot experiment, past data, expert knowledge.	
6	Assign occurrence rating (O)	*Occurrence rating* corresponds to the likelihood or frequency at which the cause can occur. Typically, the scale is 1 to 10. Higher occurrence is rated at the high end, lower occurrence at the low end of the scale.	
7	Current controls	For each cause, current process controls are identified. Controls can be of different types. They may just detect the failure or prevent the failure from happening. The controls range from work instructions to AQL sampling, SPC, alarms, mistake-proofing fixture, and so on.	
8	Assign detection rating (D)	*Detection rating* corresponds to the ability to detect the occurrence of the failure mode. Typically, the scale is 1 to 10. Higher detectability is rated at the low end, lower detectability at the high end of the scale.	
9	Calculate RPN	Product of severity (S), occurrence (O), and detection (D). S × O × D= Risk priority number (RPN). Severity × Occurrence = Criticality is also important in some industries.	

Continued

Table 3.2 Steps in performing a design or process FMEA. *(Continued)*

	Steps	Design FMEA	Process FMEA
10	Develop action plan	Action plan may contain tasks to improve the current controls or reduce the frequency of the occurrence of the cause. In order to reduce the severity, the team may have to think of redesigning the product or process. Assign a realistic completion date and responsibility for tasks.	
11	Take action	This is a step where many FMEAs fall apart due to lack of management support, conflicting priorities, lack of resources, and lack of team leadership. The actions have to be implemented and results should be validated. Building a prototype and testing the action and piloting the process in small scale before mass producing are recommended.	
12	Recalculate the RPN	Bring the team back and objectively recalculate the RPN. Use objective evidence like customer feedback, reliability tests, warranty return rate, yield tracking, and so on, to reassess the score.	
13	Periodically review and update new risks	Carefully evaluate customer feedback, warranty analysis, internal nonconformance reports, ongoing reliability test reports, and so on, to explore new risks and update the FMEA. Keep the FMEA as a living document.	

to evaluate each failure mode, based on the scoring table you are using (sample tables can be found in the J-1739 or PFMEA-4 standards). Many tables used in manufacturing organizations run from 1–10, while some tables can also be scored 1–5. One key here is to have tables that meet the needs and requirements of your specific industry and organization. This could require either reliability engineers or quality engineers working with warranty data and other information from within your organization to create company-specific tables for your needs.

The process involves the team selecting the number from the identified table that most closely indicates the team's estimate. This could include an issue that will cause harm to workers or customers if used improperly, which should yield a severity of 9 or 10 depending on how serious the injury might be.

For occurrence, you generally look at past times that the issue may have occurred, and greater frequency yields higher numbers. Thus, if something happens on a daily basis, it could have a number between 8 and 10. In detection, the numbering usually works in reverse of severity in that the least likelihood of actually detecting something will yield the highest numbers. So, if an operator can obviously see the error every time it might happen, then the score would be a low number such as 1 or 2.

Sample potential failure mode tables are shown for severity (Table 3.3), occurrence (Table 3.4), and detection (Table 3.5).

Table 3.3 Possible severity evaluation criteria.

Effect	Criterion: Severity of effect	Ranking
Hazardous without warning	Very high severity ranking when a potential failure mode affects safe vehicle operation and/or involves noncompliance with government regulation without warning.	10
Hazardous with warning	Very high severity ranking when a potential failure mode affects safe vehicle operation and/or involves noncompliance with government regulation with warning.	9
Very high	Vehicle/item inoperable (loss of primary function).	8
Low	Vehicle/item operable but comfort/convenience item(s) operable at a reduced level of performance. Customer somewhat dissatisfied.	5
Minor	Fit and finish/squeak and rattle item does not conform. Defect noticed by 50% of customers.	3
None	No discernible effect.	1

Table 3.4 Possible occurrence evaluation criteria.

Probability of failure	Possible failure rates	Ranking
Very high: Persistent failure	≥ 100 per thousand vehicles/items	10
	50 per thousand vehicles/items	9
Moderate: Occasional failures	5 per thousand vehicles/items	6
	2 per thousand vehicles/items	5
	1 per thousand vehicles/items	4
Remote: Failure unlikely	≤ 0.01 per thousand vehicles/items	1

Table 3.5 Possible detection evaluation criteria.

Detection	Criterion: Likelihood of detection by design control	Ranking
Absolute uncertainty	Design control will not and/or cannot detect a potential cause/mechanism . . . or there is no design control	10
Moderately high	Moderately high chance the design control will detect potential cause/mechanism and . . .	4
Almost certain	Design control will almost certainly detect a potential cause/mechanism . . .	1

Risk Priority Number

Once the team has identified the three numbers for severity, occurrence, and detection (S-O-D) for a given failure mode, then the three numbers are multiplied together to create what is called the *risk priority number* (RPN) (see Figure 3.6). This is the number that is the starting point for analyzing what failure modes should be addressed first (you will need to review the rules for your industry as to how to apply the RPN, as some groups want the severity number to take higher priority over the overall RPN). Once the entire FMEA is completed, a Pareto diagram can be completed of the various RPN numbers to see which failure modes have the biggest potential for issues in your organization.

Many FMEA forms have a far right-hand column that should be used by the cross-functional team to show when actions have been taken on a particular failure mode to reduce the overall RPN as part of the due diligence of the team to prevent issues before they might occur. Thus, a road map of ongoing work for continual improvement projects can be set up; the FMEA should be considered a living document as the team continually works to prevent issues in the design/process using the FMEA as their guide. Thus, the team should meet periodically to update information in the FMEA and to see where additional efforts might be targeted to improve the overall process.

Do's

- Always provide FMEA training to team members before assignment to an FMEA team.
- Always use the team approach.
- Ask for subject matter expertise if required.
- Talk to your customer about how they intend to use the product.
- Take time as a team to standardize the scales (the scale to be used should be based on the nature of business or the organization versus some generic table that may not apply to your situation). This helps when comparing the overall risks between FMEAs and helps set up a cutoff score.
- Brainstorm all possible failure modes, even if they only happen occasionally.

S	×	O	×	D	=	RPN
2	×	8	×	10	=	160
10	×	5	×	2	=	100
8	×	2	×	9	=	144

Figure 3.6 Example of RPN calculations.

- When two risks have the same overall score, the risk with the higher severity rating is escalated.
- Complete the action and reassess the risks as a team.
- Update the FMEA with any new learned risks.

Don'ts

- Try not to copy the S-O-D scales from another industry or from a different organization. The description of the scale levels and impact may be different.
- Try not to force-fit into a 1 to 10 scale. If there are not many levels of severity, occurrences, or detection in your industry, try a 1 to 5 scale.
- Discourage creating customized scales within the organization unless absolutely essential.
- Don't fight over ratings of small difference, such as between 4 and 5 or 6 and 7. Analyze the impact thoroughly if the team is divided by two or three rating points, for example, 4 and 7.
- Don't get hung up on a numbers game; the objective is to create a reduced-risk product and/or service.
- Don't perform FMEA just to comply with procedures or standards. FMEA is a business risk management tool. It has to be used with commitment to make it work.

Successful FMEA implementation requires leadership and management commitment. Few tools can test the patience of team members as finishing an FMEA; tackle multiple process steps for the product as a team in several meetings. In these cases, split the process into major process blocks and perform FMEA by block. Maintain a good FMEA database. This will significantly reduce the time spent on successive FMEAs.

Once the initial RPN scores are tabulated, the team may decide on a cutoff score. For most organizations, the cutoff score is standardized. The cutoff score of one organization may not be directly applicable to another. Too low a cutoff score can result in spending lots of resources to eliminate or reduce several risks. Too high a cutoff can result in not addressing important risks. Management needs to review the data and agree on a score.

Figures 3.7 through 3.11 show various FMEA implementation tools and examples.

Before taking action					After taking action				
Risk ID	**S**	**O**	**D**	**Initial RPN**	**Risk ID**	**S**	**O**	**D**	**Recalc RPN**
Risk 7	5	8	7	280	Risk 7	5	5	3	75
Risk 1	8	8	4	256	Risk 1	8	4	3	96
Risk 3	5	5	7	175	Risk 3	5	4	4	80
Risk 9	7	5	4	140	Risk 9	7	3	4	84
Risk 5	8	4	4	128	Risk 5	8	3	3	72
Risk 2	7	4	3	84	Risk 2	7	4	3	84
Risk 4	5	5	3	75	Risk 4	5	5	3	75
Risk 6	3	7	2	42	Risk 6	3	7	2	42
Risk 8	5	3	2	30	Risk 8	5	3	2	30
Risk 10	3	3	3	27	Risk 10	3	3	3	27

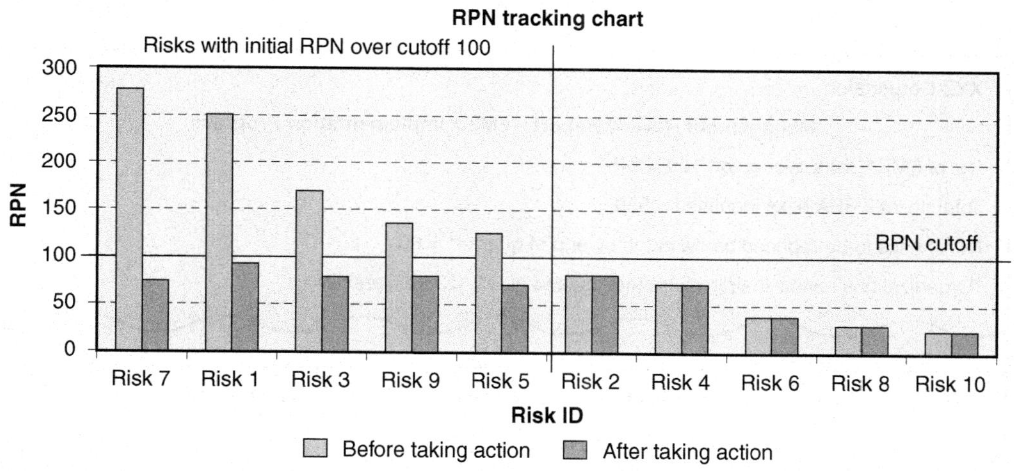

Figure 3.7 Example FMEA reporting and RPN chart.

Note: FMEA is a powerful tool but requires in-depth knowledge to successfully execute. It is recommended that guidance be sought from others who have performed an FMEA prior to or during the application of an FMEA to an improvement project.

Please see the Glossary for some of the fundamental terms used in FMEA such as *failure mode, cause, effect, failure mechanism, severity, occurrence, detection, risk priority number,* and so on.

Note: The automotive industry has developed a new FMEA handbook that does change the concept of the RPN to a risk rating. The concepts and usage of the FMEA remain unchanged in this handbook.

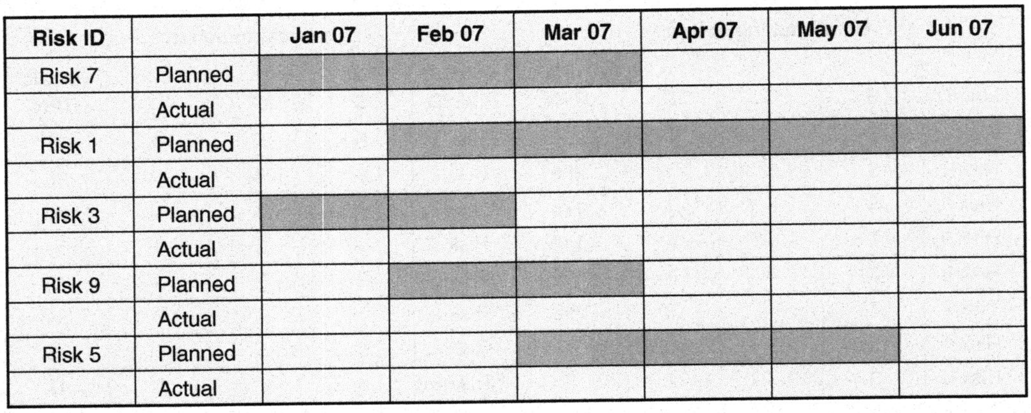

Figure 3.8 A typical risk action plan Gantt chart.

XYZ Corporation

Management Review Report—FMEA Implementation Progress

No. of FMEA risks over cutoff 100 = 245

Total no. of FMEA risks identified = 550

No. of risks to be reduced below cutoff by end of quarter* = 60

*Organizations review their performance by end of Q1, Q2, Q3, and Q4.

Figure 3.9 A typical top-level management review report for FMEA progress.

Function	Potential failure mode	Potential effect(s) of failure	S	Potential cause(s) of failure	O	Current process controls	D	RPN	CRIT	Recommended action(s)	Responsibility and target completion date	Action results					
												Action taken	S	O	D	RPN	CRIT
Dispense amount of cash requested by customer	Does not dispense cash	Customer very dissatisfied	8	Out of cash	5	Internal low-cash alert	5	200	40								
		Incorrect entry to demand deposit system		Machine jams	3	Internal jam alert	10	240	24								
		Discrepancy in cash balancing		Power failure during transaction	2	None	10	160	16								
	Dispenses too much cash	Bank loses money	6	Bills stuck together	2	Loading procedure (riffle ends of stack)	7	84	12								
		Discrepancy in cash balancing		Denomina-tions in wrong trays	3	Two-person visual verification	4	72	18								
	Takes too long to dispense cash	Customer somewhat annoyed	3	Heavy computer network traffic	7	None	10	210	21								
				Power interruption during transaction	2	None	10	60	6								

Figure 3.10 An example of a partially filled-in FMEA.

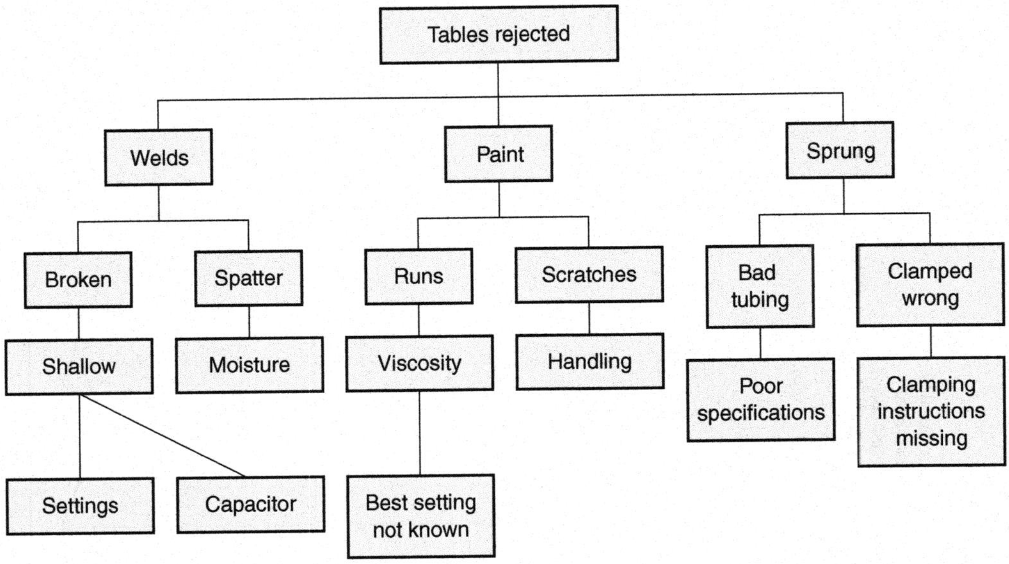

Figure 3.11 This example of a tree diagram is a fault tree (used to study defects and failures).

DESIGN FMEA AND PROCESS FMEA

> Define and distinguish between these two uses of FMEA. (Apply)
>
> **Body of Knowledge I.C.3**

There are any number of uses for the FMEA methodology. Figure 3.12 shows how several of these manufacturing methods (concept, design, process, machinery, and system) can fit and work together. We could also add "management" to this figure, which is found in the ERM systems discussed earlier. All of this together makes up what is referred to in the ISO management system standard series (9001, 14001, 22301, 45001, and 50001) as *risk management* and will be required of your organization if it is to be registered to one of the standards. Note: the ISO now has over 50 management system standards.

For distinguishing between the design and process FMEA, a key item to remember is, what is your end goal in creating the FMEA document? Sequentially, you should consider creating a concept FMEA first, followed by a design FMEA, and then a process FMEA. Many times in a manufacturing setting there is no concept FMEA or design FMEA, so the engineer or manager must work with what they have to create the process FMEA without knowing all aspects of the design intent.

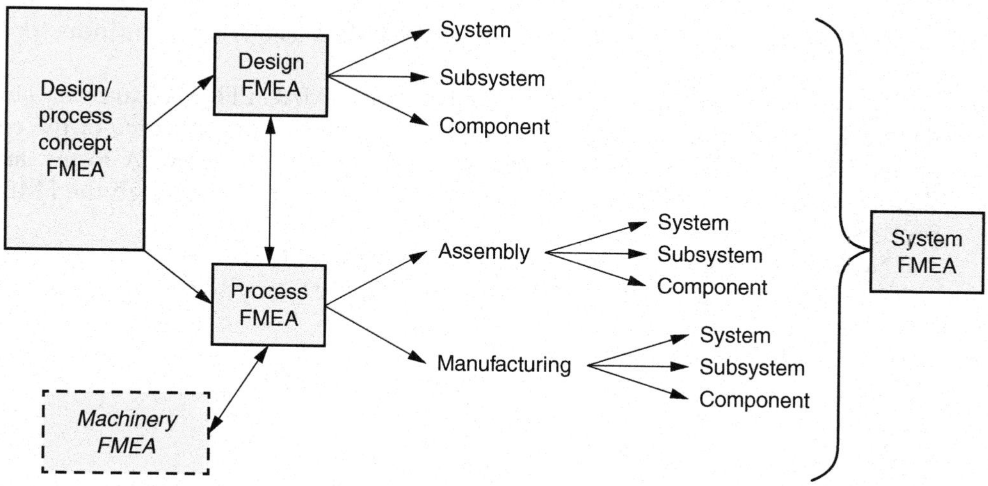

Figure 3.12 Common types of FMEA.

Design FMEA is about the product design and should be used by designers and process engineers to think through possible issues in creating the designs/drawings in certain ways versus other alternatives. If they design something in one particular way versus another, what ramifications could that characteristic have on the part, how it is made, or how it will be used by the customer? If we are designing toys for small children, are there any small components that could come loose and become choking hazards? If we design a tire jack that might slip when used by the vehicle owner with a flat tire, additional injury might be the end result.

As discussed earlier, working through the design FMEA during these early stages before tooling is cut and final processes have been established can save your organization vast amounts of time and money. Design FMEAs can be done at various levels, as seen in Figure 3.12, such as the system level, subsystem level, or component level.

Process FMEA is about the shop floor manufacturing process and should be done at the beginning of developing the manufacturing process layout for either manufacturing or assembly of parts or components. Here, too, we can view manufacturing or assembly at various levels: system level, subsystem level, or component level.

Differences between Design and Process FMEA

If you are reviewing an FMEA to understand the process more, you should first note what the header block says about which type of FMEA you're looking at. The header should clearly indicate the type of FMEA, such as design, process, or some other system-level document. The methods of analyzing and completing the forms are similar, as has been discussed. The difference between design and process FMEA will be in the application of the FMEA as used in your industry or organization.

The general forms used by both SAE J-1739 and AIAG PFMEA-2 are very similar except for the name in the header block and the spacing of some of the columns. Many people use an Excel spreadsheet to create their FMEA forms, and there are also many software packages available for working through the FMEA process.

Note: Most of the discussion around DFMEA and PFMEA was centered around a manufacturing organization and can equally apply to any service organization.

Part II

Define Phase

Part II is an overview of the *define* process for Six Sigma systems. It covers approximately 20 of the 100 questions that will be asked on the ASQ CSSGB Exam.

Where are we? Or, what is the problem? Where do we want to be? How will we get there? How will we know when we are there?

These are critical questions that are asked and answered during the *define* phase. Without understanding these basic tenets, the activity (or project) to resolve the problem or improve performance can flounder aimlessly, wasting needed resources and frustrating personnel to the point of not supporting an improvement culture.

This section will provide:

1. An overview of the define phase that includes process flows and guidelines to help ensure that a project is on track. Here the difference between defining the "improvement project" and defining the "issue or problem" will be discussed. We will also discuss how these two items, "project" and "problem," differ, and how they are similar. Finally, the overview will close out with guidance on tailoring the define phase intensity to the specific activity to be worked, in other words, how to keep this simple.

2. A review of each area of the ASQ Certified Six Sigma Green Belt Body of Knowledge. The goal here is to provide information that can help you successfully pass a certification exam, but more importantly, to ensure that this handbook is a tool that helps you execute improvement projects. In this review, tools that can be used will be discussed at a high level, with more in-depth explanations provided later in the section.

3. A tools section to provide detailed information regarding applicable define phase tools—how they work and how and when to use them. This section also lists additional resources in different media (that is, in print or online). The purpose of listing these additional resources is to ensure a balanced view of the tools and to provide usable templates and resources to execute a Green Belt project.
4. A summary of the items discussed that highlights the most commonly used tools, key resources, critical factors to success, and general process flows for successful execution of a project.

OVERVIEW

The *define* phase of the DMAIC model serves two purposes: to define the project management process for the Green Belt improvement project and to define the problem or issue to be worked on by the Green Belt project team. This overview will outline these two focus areas by detailing basic processes and recommended items for each area, and annotating potential pitfalls to avoid.

As noted, when we execute the define phase, two primary deliverables are the project plan and detailed knowledge of the current state of the problem. The *project plan* outlines several aspects of the project to ensure that the Green Belt project team and key stakeholders understand what needs to be done, what resources (for example, people, financial, tools and equipment, infrastructure) are anticipated, and when things will be completed. The documentation associated with gaining knowledge of the problem and the current state of the process varies widely based on the problem or issue being worked on. The primary goal is to have sufficient detail on what is happening to cause the undesirable performance or nonconformance to keep the project focused through the remaining DMAIC phases.

An organization that has an improvement culture based on a proven methodology, such as DMAIC, will be able to consistently improve performance and eliminate problems. However, if the organization allows itself to become mired in bureaucracy, it could lose this edge and reduce the overall effectiveness and motivation to improve. One method to avoid this pitfall is to ensure that the improvement culture process (based on a methodology like DMAIC) allows for flexibility in the level of project detail and tool selection. For example, some projects may only need a short charter approved by a process owner, champion, Black Belt, or Master Black Belt. Yet others, especially larger projects, may require a more detailed plan, coordination with multiple process owners, intra-phase reviews, and so on.

Keep things simple, adjust based on the process, but stay true to the basic methodology—this should lead to a successful improvement program embraced across the entire organization. One caution: be aware not to focus on sub-area/process optimization to the detriment of the whole process or "system." As noted above, the first section will follow the outline of the ASQ Certified Six Sigma Green Belt Body of Knowledge. Although this makes the most sense for the handbook and as a certification resource, note that in many cases it is better to outline the project management aspects and get stakeholder buy-in prior to spending too many resources on defining the problem and fully understanding the current state of the problem or process.

Key Point: Ensure that the management and money are lined up before starting any project—and check in often.

Key Point: The more time you spend up front in good planning, the higher the probability of a successful project.

Chapter 4

Project Identification

PROJECT SELECTION

> Describe the project selection process and what factors should be considered in deciding whether to use the six sigma DMAIC methodology or another problem-solving process. (Understand)
>
> **Body of Knowledge II.A.1**

When looking for Six Sigma projects, there will usually be more ideas than it is possible to undertake at one time. Some sort of project proposal format may be needed, along with an associated process for project selection. If your organization does not already have a selection process in place, use of some basic tools such as advanced quality planning and quality function deployment may be useful in setting up a selection process.

It is common to require that project proposals include precise statements of the problem definition and some preliminary measures of the seriousness of the problem, including its impact on the goals of the organization. For some managers, these will be the criteria that define which projects to start first based on which ones save the most bottom-line dollars ($).

If available, a project selection group could be made up of Master Black Belts, Black Belts, organizational champions, and key executive supporters. They should establish a set of criteria for project selection and team assignments. In some companies the project selection group assigns some projects to Six Sigma teams and other projects to teams using other methodologies (examples: Operation Excellence, Lean, etc.). For example, problems involving extensive data analysis and improvements using designed experiments would likely be assigned to a Six Sigma team, whereas a process improvement not involving these techniques might be assigned to a lean manufacturing team employing kaizen tools.

The project selection criteria should always have the furthering of organizational goals as a key element. One key to gauging both the performance and the health of an organization and its processes lies with its selection and use of metrics.

These are usually converted to financial terms such as return on investment, cost reduction, and increases in sales and/or profit. Other things being approximately equal, projects with the greatest contributions to the bottom line receive the highest priority.

Project selection typically means that a project is now in suitable form and is ready to be judged against suitable criteria to determine its viability as a Lean Six Sigma project.

PROCESS ELEMENTS

> Define and describe process components and boundaries. Recognize how processes cross various functional areas and the challenges that result for process improvement efforts. (Analyze)
>
> **Body of Knowledge II.A.2**

A process is a step or sequence of steps that uses inputs and produces a product or service as an output. Every process has inputs and outputs. Inputs are traditionally categorized as man (used here to allow for five to eight M's—we do include all people), material, methods/machines, Mother Nature, management, money, and measurement system (as the measurement system may have an impact on the output). Outputs are usually products (hardware, software, systems, and so on) or services. Figure 4.1 depicts the model of a process.

Processes often are made up of smaller subprocesses. For example, a part may be produced through a process that has a machining step. This machining step may be thought of as a process whose steps might include clamping, turning, plunging, facing, and so on. In addition, the plunging step is a process in itself. In a similar manner, a payroll process has subprocesses, which include gathering time clock information, making deductions, and so on. The deduction process

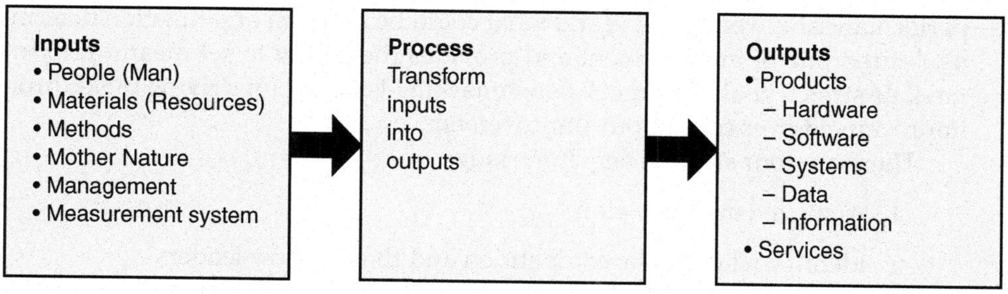

Figure 4.1 Process diagram/model.

itself could be thought of as having subprocesses for tax deductions, insurance, and so on.

When defining a process, it is important to define its start and end points—its boundaries. If, for example, a team is charged with improving a process, they need to know these process boundaries. Cross-functional processes may incur subprocess boundaries defined by the organizational structure, geography, and so on.

BENCHMARKING

> Understand various types of benchmarking, including competitive, collaborative and best practices. (Understand)
>
> **Body of Knowledge II.A.3**

Improving processes and products is often aided by comparing the current state with outstanding processes or products. In some cases, the comparison will be with other divisions of the same company. In other cases, external comparisons are more appropriate. The use of these comparisons is called *benchmarking.* Benchmarking often assists a Six Sigma team in setting targets and finding new ways to achieve them. It is an especially helpful technique when a quality improvement team has run out of new ideas.

Depending on which reference source you review, there are many "types" of benchmarking processes and methodologies utilized today. For the Green Belt, identifying whether you can conduct benchmarking internally and/or externally to your organization will be a starting point.

The information for benchmarking may come from various sources, including publications, professional meetings, university research, customer feedback, site visits, and analysis of competitors' products. A downside of benchmarking within a particular industry is that it tends to put one in second place, at best.

Benchmarking helps teams strive toward excellence while reducing the tendency to feel that locally generated ideas are the only ones worth considering. Moreover, it is important because it provides an organization with the opportunity to see what level of process performance is possible. Seeing a gap in process performance between what is and what could be helps an organization determine its desired rate of improvement and provides the ability to set meaningful intermediate stretch goals or targets. Benchmarking is useful for driving breakthrough improvement over continuous improvement.

There are four steps in benchmarking:

1. Analyze your operation
2. Identify who are the competition and the industry leaders
3. Incorporate the best of the best
4. Determine how your organization can gain superiority

Several types of benchmarking exist, and each has advantages and disadvantages:

- *Internal benchmarking.* Provides easy access to other departments within the same company. However, the search for exceptional performance is often limited by the company's culture, norms, and history.
- *Competitive benchmarking.* Forces an organization to take an external perspective. However, focusing on industry practices may limit opportunities to achieve high levels of performance, particularly if a given industry is not known for its quality.
- *Functional benchmarking.* Compares similar functions, typically outside the organization's industry, and provides ample opportunities to seek out benchmarking partners.
- *Collaborative benchmarking.* Refers to the cooperation between various functions or organizations to achieve benchmarking results. Collaborative benchmarking may permit access to specific benchmarking partners that may not exist with the other types of benchmarking.

One quick example: If you want to benchmark a process that has excellent flow and is capable of taking an order from a customer and servicing that order within minutes (talking about a lean pull system here), then visit your local McDonald's! How is it that they can have a process in place to take orders quickly and provide exactly what is requested, and do it with a group of high school students (in many cases)? If you doubt it—go watch their process.

Some useful tips for planning a formal benchmarking exercise are shown in Table 4.1. Additional information on benchmarking is also included in Chapter 7.

PROCESS INPUTS AND OUTPUTS

> Identify process input and output variables and evaluate their relationships using the supplier, inputs, process, output, customer (SIPOC) model. (Analyze)
>
> **Body of Knowledge II.A.4**

Everything we do can be defined in terms of a process or system. Many of the tools discussed in this book can help identify the causes and effects, the frequency and distribution of measures, the most frequently occurring data, the process flow as it is occurring, and other factors about the process or system. When planning to study a process or system, it is very important to first identify the boundaries to work within. There are many ways to identify the process boundaries; most are not complex and can be implemented easily with common process knowledge and some investigation.

Table 4.1 Tips for benchmarking.

Do . . .	Don't . . .
✓ Select benchmarking projects that are tied to strategic goals/objectives	✗ Benchmark just to say you did it
✓ Benchmark a core process	✗ Expect big paybacks when benchmarking a non-core process
✓ Obtain management commitment	✗ Benchmark without sufficient support
✓ Get the support/involvement of process owners	✗ Leave out the middle managers
✓ Know and clearly map out your own process before attempting to benchmark	✗ Expect to benchmark another's process without a thorough understanding of your own
✓ Identify the important measures of the process	✗ Trust what you can't measure
✓ Allocate adequate resources	✗ Think you can get a big return without some investment of resources
✓ Plenty of research	✗ Forget to research public domain
✓ Limit the number of site visits and the benchmarking team members who participate in visits	✗ Confuse benchmarking with industrial tourism
✓ Research companies/organizations you visit *before* you go	✗ Go on a site visit unprepared
✓ Reciprocate	✗ Ask for information that you would not be willing to share
✓ Debrief benchmarking teams ASAP after each site visit	✗ Delay a debrief more than three days after the site visit
✓ Keep communications flowing up and down the chain of command	✗ Wait until the benchmarking study is complete to get management's thumbs up or thumbs down on progress
✓ Implement the improvements identified by the benchmarking study ASAP	✗ Forget that the primary reason for benchmarking is to implement the best practices
✓ Ask internal/external customer what they think would improve the process	✗ Forget what's important to your customers
✓ Provide guidance/resources/chapter	✗ "Over control" the team

Adapted from: The Department of the Navy Benchmarking Handbook: A Systems View.

The two primary tools used to identify process boundaries are the basic process model identified in Figure 4.1 and the suppliers–inputs–process–outputs–customers (SIPOC) diagram in Figure 4.2. The basic process model provides a quick high-level look at the process but, in some cases, may be too simplistic to provide an improvement team with a clear understanding of the process/problem to work on and the boundaries of where and when to stop.

The SIPOC diagram can be enhanced by also capturing the requirements of the process and customer, as shown in Chapter 7.

Remember, understanding the boundaries of the improvement project and/or process does not prevent outside-of-the-box thinking; it just provides clear guidelines of what to deal with as daily activities and improvement activities are performed. Taking the time to actually list or draw the components of the process will assist in visualizing and being able to find issues quicker than might have otherwise been possible.

Key Point: Many projects flounder due to the lack of clear boundaries for the project.

When identifying the process, it is important to recognize that processes usually affect multiple departments and organizations. Crossing functional areas (departments) and organizations (suppliers, intercompany, teammate company, and so on) can add challenges to an improvement project. The first step in recognizing the challenges is to understand the organizations and functional areas involved with the process. As noted, the SIPOC diagram can help in identifying these organizations and functional areas as process suppliers and customers. The flowchart (especially the "swim-lane" style) and process map are other tools that can be used to help recognize these interactions. Challenges associated with these interactions include, but are not limited to:

- Process ownership (two or more areas may think they own the process and have final decision authority on changes).
- Sharing of information (for example, proprietary issues, hiding poor performance).
- Commonality of measures (for example, finance usually measures things in dollars, production may use defects or productivity, engineering considers productivity and design completion).
- Process knowledge or expertise (that is, manufacturing may not fully understand how the supply chain works or the requirements).

If the challenges identified are significant, it is recommended that the challenges be included as potential risks for the project so associated mitigation activities can be performed. Risk identification and mitigation are discussed in more detail in Chapter 6.

The details of how to use and implement the basic process model diagram and SIPOC diagrams are provided in Chapter 7. SIPOC templates are provided online for both the basic SIPOC diagram and the SIPOC diagram with requirements capture formats.

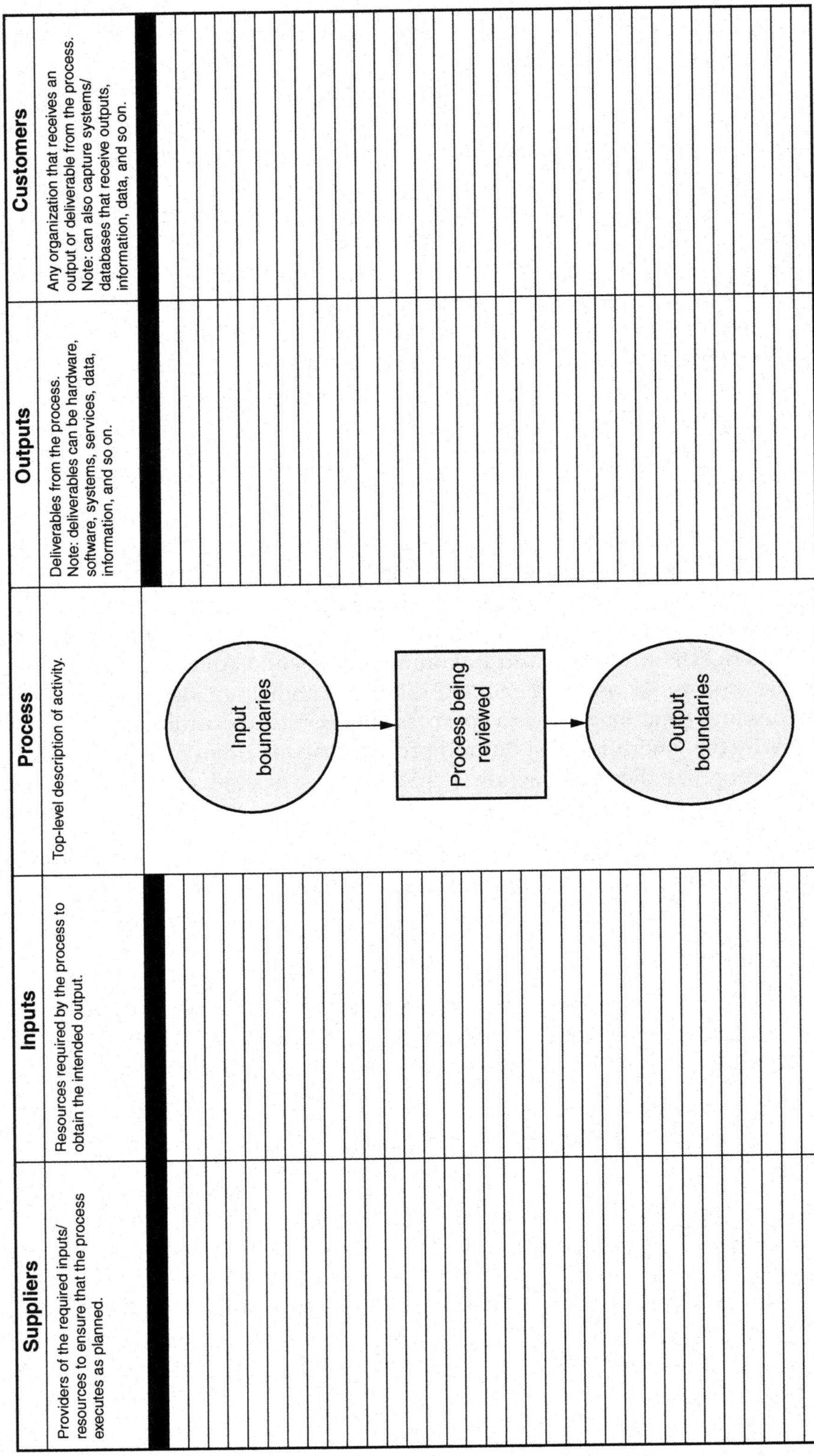

Figure 4.2 Basic SIPOC diagram.

Systems Thinking

According to Dr. E. Rice,[1] the recognition/identification and consideration of all the various individual elements that interrelate with a common purpose toward a whole function of a unit is considered *systems thinking*. Systems thinking is about using tools and methods available to understand what is being done at a specific operation and how that activity affects tasks and products further downstream and how prior tasks and products affect the process being reviewed. Far too often a change made in one part of an operation causes new problems somewhere else—this type of change is effected without applying systems thinking.

Using systems thinking, we strive to understand the process and know that if one factor is changed or influenced in some particular way, then something different might happen downstream in the process. For example, if a supplier ships a material with a higher than normal moisture content, how do processes that use that product need to adjust to ensure that the end product meets requirements? All of this is part of systems thinking, and we are able to use the tools and processes in this book to assist us in reducing variation and satisfying the customer. Note: For those who received training in continuous improvement during the 1980s and early 1990s, please note that in order to actually improve the process, the *process behavior charts* (called *control charts* back then) will sometimes have to go out of control to show an improvement in the system. Thinking that we can have both continuous improvement and control charts that are always in control is *not* systems thinking!

OWNERS AND STAKEHOLDERS

> Identify the process owners and other stakeholders in a project. (Apply)
>
> **Body of Knowledge II.A.5**

Process owners are those who have responsibility for the execution and implementation of a specific process. Process owners are usually formally recognized in this role through related documentation (for example, procedures, work instructions, documentation approval authority), through their job/position description, or through the organization chart.

Stakeholders are those who have a vested interest in the process and/or its products and outputs. Generally, stakeholders of an organization include customers, suppliers, employees, investors, and communities. Stakeholder interest and involvement with the process may change over time depending on economic, contractual, and other influences. Process owners are those that are responsible for the definition, execution, maintenance, and improvement of the process; in some cases, process owners may also be referred to as *subject matter experts*. Personnel involved with process design usually have a specific interest in systems, subprocesses, and individual steps within processes.

The most effective methods of process improvement utilize teams representing process owners *and* all stakeholders because:

- Stakeholders have the best knowledge base about the process.
- Stakeholders tend to have the best ideas for process improvement.
- Stakeholders are often the most aware of unintended consequences of process changes.
- Stakeholders' buy-in is usually necessary to implement real process improvement.

The stakeholders in a process are:

- Process operators and managers from all shifts
- Process customers, internal and external
- Process suppliers, internal and external
- Process design personnel
- Maintenance personnel
- Others impacted in some way by process changes

Chapter 5

Voice of the Customer (VOC)

Understanding the VOC is about knowing what the downstream process will do with your data, service, or products. This section covers a number of tools and techniques to help ensure that what you produce will be useful to your customers (be they the next person in the business flow all the way to the end user). Other terms that you could come across in business include: Voice of the Employee, Voice of the Business, Voice of the Process (VOP), or others.

CUSTOMER IDENTIFICATION

> Identify the internal and external customers of a project, and what effect the project will have on them. (Apply)
>
> **Body of Knowledge II.B.1**

Who is the customer of the project? This question should always be asked up front on any project, and in many cases, a plan is needed to go talk to those individuals (see Customer Data later in this chapter). Customers can be found both internal to the project (individuals who work in the process to be studied) and external to the project (individuals outside of the boundaries of the project), thus the terms *internal customer* and *external customer*. Setting the boundaries of the project will allow for easier identification of who to consider as internal versus external customers. If someone is outside what we might consider as internal or external customers, we sometimes will refer to those individuals as *stakeholders*.

When reviewing the external customer list, please remember to include people who might be outside of your direct organization as stakeholders. These could include the board of directors of your company, organizations who buy the services or products of your company, neighbors to your company's local address(es), and local, state, and federal government agencies. A method for ensuring that you have external customers identified is to look at the social responsibility of your company to the society at large. Your safety manager or lead might be a good source of information on this topic.

It is important to identify and understand a project or process's customers. Depending on the maturity of the product or process, the customers may be known. Even if the customers are known, it is always good practice to identify the customers using some of the tools in this book. Methods used to identify customers include:

- Brainstorming
- SIPOC
- Tracking a product or service to delivery

Internal and external customers should be identified when applicable. Customers can be grouped into segments, with this segmentation driven by customer requirements, and often includes the following categories:

- Internal and external
- Age groups, especially for consumer products
- Geographical location, including climate, language, ethnicity, and so on
- Industry type (for example, construction, agricultural)

Where possible, a listing of customers within a segment should be constructed. When a project team proposes changes of any type, customers—internal and external—must be consulted, or at minimum the customers' concerns must be represented.

CUSTOMER DATA

> Collect feedback from customers using surveys, focus groups, interviews, and various forms of observation. Identify the key elements that make these tools effective. Review data collection questions to eliminate vagueness, ambiguity, and any unintended bias. (Apply)
>
> **Body of Knowledge II.B.2**

A lot of research data are available on how to set up effective communication systems with customers. The primary thing to remember is, how do you know that you really know what the customer needs and/or wants (see Taguchi Loss Function)? The simple answer is to ask them.

It is an interesting phenomenon that many managers and engineers think that they know what the customer wants simply because they are the experts in what they do. This is readily seen when computer software programmers develop new programs or patches without a full testing (should include actual end customer usage) of what was developed. Inevitably, new patches and rework are needed to fix the new software.

Key Point: As the Green Belt for a project, after you get your benchmark process data on the system, talk with the people who are doing the work to get an understanding of what they are doing (see more in the discussion of the Hawthorne Studies) and what might make their jobs easier. Starting here and seeing what can be done to improve the local job can be a key to building support for the project you will be working on. Then start talking with individuals both upstream and downstream of the target project area. What would help them do their work better that might relate to the project that you are working on?

Once identified, the next step is to understand the wants, requirements, and expectations of the customer. One of W. E. Deming's more famous statements is that some of the most important numbers are unknown and unknowable. He was referring to such things as the financial value of customer goodwill. And if a supplier/employee disappoints a customer, the customer will ". . . tell all his friends and some of his enemies . . . " in Deming's words. His point, of course, is that it is easy to underestimate the value of understanding and providing for customers' needs. Without customers we have nothing![1]

There are several tools available for capturing customer data. The most widely used tools include:

- Surveys
- Interviews
- Focus groups
- Taguchi Loss Function
- Quality function deployment (QFD)

Statistically, the most valid procedure for collecting customer data is to randomly select a reasonably large group of customers and obtain complete and accurate data on each one selected. Since this procedure is not possible in many situations, various other methods are employed (talk with your Black Belt or Champion for ideas). Each method usually comprises "statistically valid" procedures in one or more ways.

The data collected should be objective and designed to shed light on customer requirements. It is important to use several independent resources to obtain this information. The results can be played against each other to determine patterns of reinforcement or contradiction of conclusions. After customer data have been collected, the accuracy and consistency of the data should be verified; it is important to resolve conflicts or ambiguous data.

Taking the time to talk with people openly about what you can do to help them (in the context of your project) can lead to good insights about what might be roadblocks in the current system.

CUSTOMER REQUIREMENTS

> Use quality function deployment (QFD), Critical to X (CTX when 'X' can be quality, cost, safety, etc.), Critical to Quality tree (CTQ), and Kano model to translate customer requirements statements into product features, performance measures, or opportunities for improvement. Use weighting methods as needed to amplify the importance and urgency of different kinds of input; telephone call vs. survey response; product complaint vs. expedited service request. (Apply)
>
> **Body of Knowledge II.B.3**

Some of the most important applications of Six Sigma are in the design and redesign of processes and products. The ideal design meets or exceeds customer requirements at the lowest possible cost. Key steps in achieving this ideal are:

1. Link customer requirements to features of products and processes. *Quality function deployment* (QFD) provides a valuable approach to this activity.
2. Design a product and processes that will result in the desired quality (so that the product or process is actually Six Sigma capable from the start—Chowdhury[2]) at the lowest possible cost.
3. Employ design tools that will result in entirely new approaches to problem solving.

QFD (also known as the *house of quality*) provides a process for planning new or redesigned products and services (organizes customer requirements and desires [wants] and allows them to be traced to specifications). The input to the process is the voice of the customer (VOC). The QFD process requires that a team discover the needs and desires of their customer and study the organization's response to these needs and desires. The QFD matrix aids in illustrating the linkage between the VOC and the resulting technical requirements.

A quality function deployment matrix consists of several parts. There is no standard format matrix or key for the symbols, but the example shown in Figure 5.1 is typical. A map of the various parts of Figure 5.1 is shown in Figure 5.2. The matrix is formed by first filling in the customer requirements ①, which are developed from analysis of the voice of the customer (VOC). This section often includes a scale reflecting the importance of the individual entries. The technical requirements are established in response to the customer requirements and placed in area ②. The symbols on the top line in this section indicate whether lower (↓) or higher (↑) is better. A circle indicates that target is better. The relationship area ③ displays the connection between the technical requirements and the customer requirements. Various symbols can be used here. The most common are shown

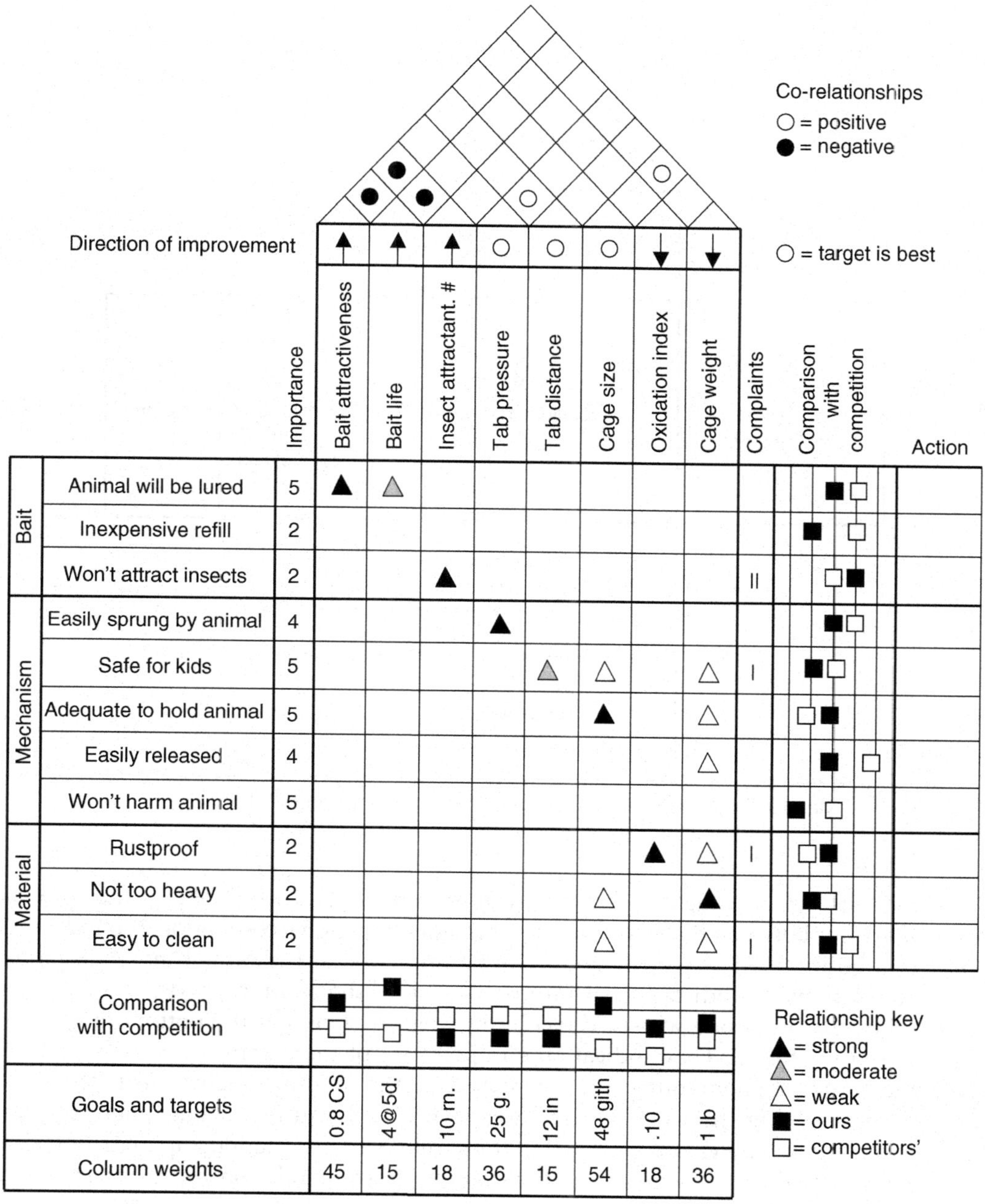

Figure 5.1 Example of a QFD matrix for an animal trap.

in Figure 5.1. Area ④ is not shown on all QFD matrices. It plots comparison with competition of the customer requirements. Area ⑤ provides an index to documentation concerning improvement activities. Area ⑥ is not shown on all QFD matrices. It plots comparison with competition of the technical requirements. Area ⑦

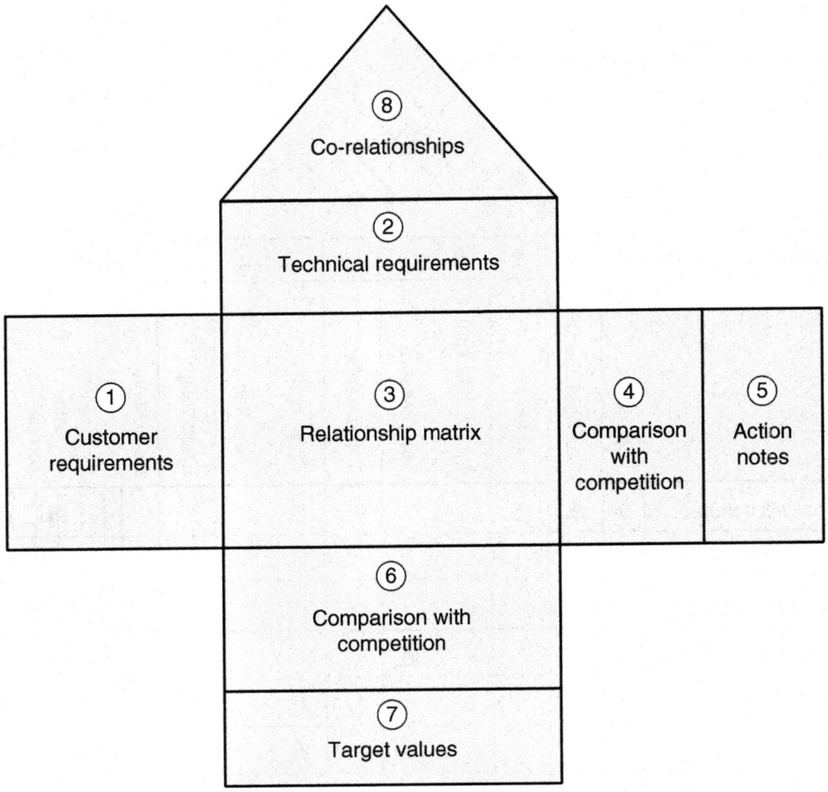

Figure 5.2 Map to the entries for the QFD matrix illustrated in Figure 5.1.

lists the target values for the technical requirements. Area ⑧ shows the co-relationships between the technical requirements. A positive co-relationship indicates that both technical requirements can be improved at the same time. A negative co-relationship indicates that improving one of the technical requirements will make the other one worse. The column weights shown at the bottom of the figure are optional. They indicate the importance of the technical requirements in meeting customer requirements. The value in the column weights row is obtained by multiplying the value in the "importance" column in the customer requirements section by values assigned to the symbols in the relationship matrix. These assigned values are arbitrary; in the example, a strong relationship is assigned a 9, moderate 3, and weak 1.

The completed matrix can provide a database for product development, serve as a basis for planning product or process improvements, and suggest opportunities for new or revised product or process introductions.

The customer requirements section is sometimes called the "what" information, while the technical requirements section is referred to as the "how" area. The basic QFD product-planning matrix can be followed with similar matrices for planning the parts that make up the product and for planning the processes that will produce the parts (see Figure 5.3).

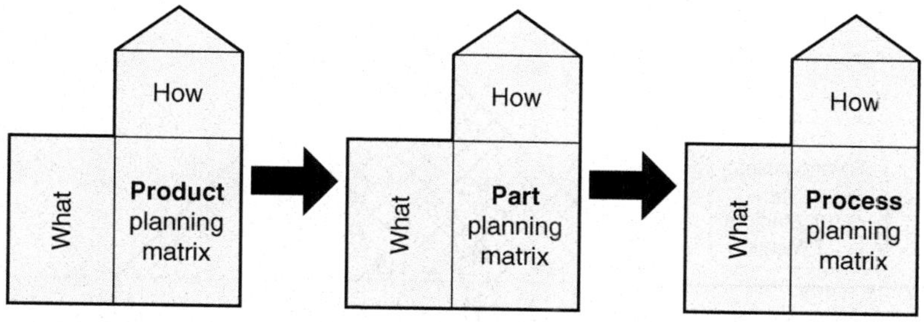

Figure 5.3 Sequence of QFD matrices for product, part, and process planning.

If a matrix has more than 25 customer voice lines, it tends to become unmanageable. In such a situation, a convergent tool such as the affinity diagram (see Chapter 7) may be used to condense the list.

Application

Relationships are tracked and the key links are determined to find out which customer wants are correlated with a specific design approach. These relationships are ranked to determine the targets of opportunity by technical importance, and areas of focus for improvement or innovation.

The completed matrix, as depicted in Figure 5.4, can provide a database for product development, serve as a basis for planning product and/or process improvements, and suggest opportunities for new or revised product and/or process introductions.

QFD can be applied to other processes to improve quality. In a staged environment, the outcomes of one stage are the inputs of the succeeding stages, as shown in Figure 5.5. QFD is a useful technique to entrench the customer requirements into the earliest stages and allow the captured information to cascade to subsequent stages to ensure that there are no gaps to customer satisfaction.

QFD is applied to improve customer satisfaction at acceptable cost. The basic relationship is captured in the input–output matrix (see Figure 5.6). "Whats" can be expressed as required (must have), necessary (expected to have), and optional (nice to have). "Hows" refer to the technical details, and should be defined to address—at minimum—the specific requirements of the customer.

Quality function deployment supports the adoption of customer value within the product. Table 5.1 shows product and service value characteristics.

QFD supports customer-driven quality and the total customer experience by providing an objective and traceable matrix linking customer wants and expectations to technical details and acceptance criteria. This reinforces the robustness of the design, and will more accurately define the terms of product development, delivery, maintenance, and fulfillment.

Critical to Quality Factors

The concept behind Critical to x (CTx), (as defined by T. Kubiak in *The Certified Six Sigma Black Belt Handbook, Second Edition*) where x is a variable, is simply that

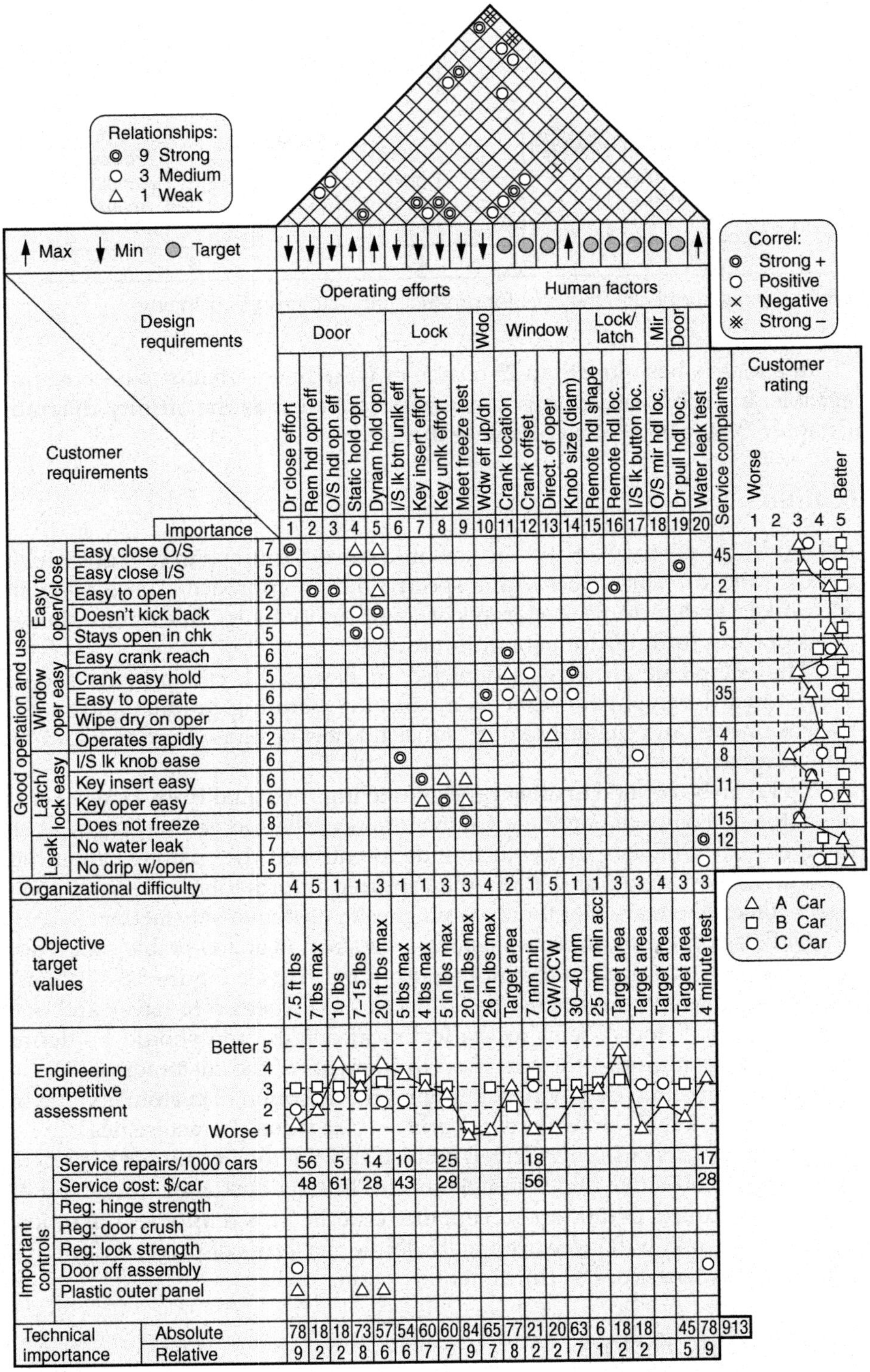

Figure 5.4 Example of a completed QFD matrix.

Source: R. T. Westcott, *The Certified Manager of Quality/Organizational Excellence Handbook*, 3rd ed. (Milwaukee: ASQ Quality Press, 2006): 477.

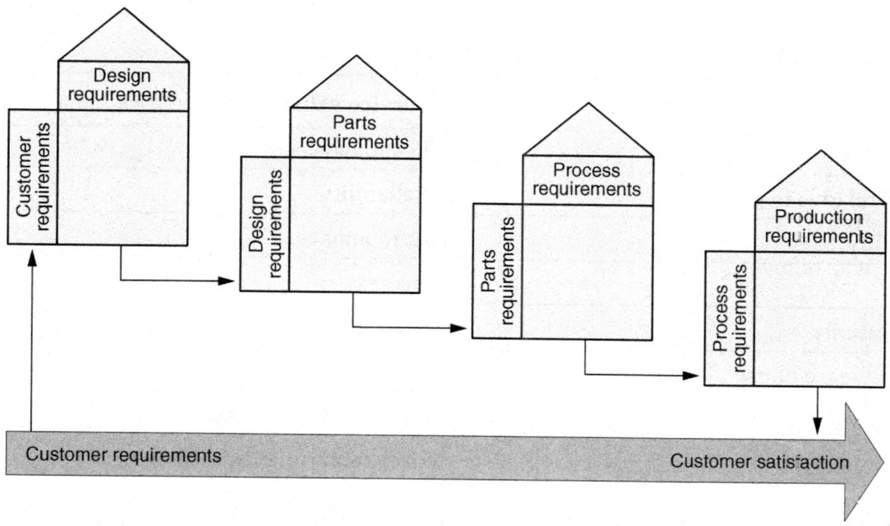

Figure 5.5 QFD stage flow and relationships.

Source: R. T. Westcott, *The Certified Manager of Quality/Organizational Excellence Handbook*, 3rd ed. (Milwaukee: ASQ Quality Press, 2006): 478.

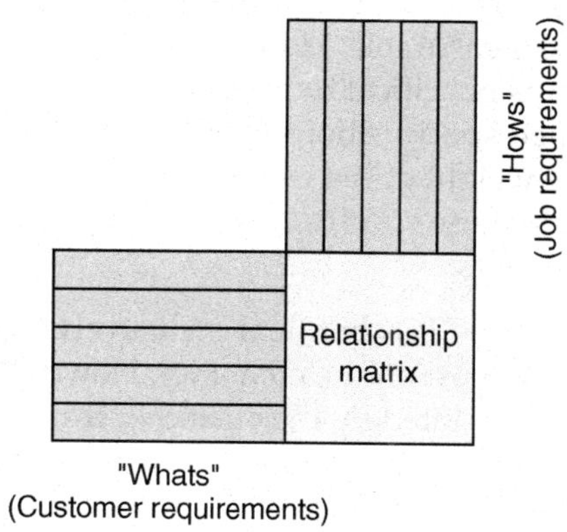

Figure 5.6 QFD input–output requirements matrix.

Source: Connie M. Borror, ed., *The Certified Quality Engineer Handbook*, 3rd ed. (Milwaukee: ASQ Quality Press, 2009).

x is an area or areas of impact on the customer. Critical customer requirements are usually expressed as expectations or needs, but not necessarily in quantifiable terms. Process CTxs represent measurable product or process performance characteristics and act to quantify the critical customer requirements. Let's look at some of the more familiar "xs":

- CTQs may include the physical dimensions Critical-to-Quality (CTQ) of height, width, depth, and weight. They may even include

Table 5.1 QFD value characteristics.

Product value characteristics	Service value characteristics
Performance	Responsiveness
Benefit relative to cost	Reliability
Durability	Competence
Safety	Access
Serviceability	Communication
Usability/ease of use	Credibility/image
Simplicity of design	Confidentiality/security
Functionality	Understanding the customer
Availability	Accuracy/completeness
Performance	Timeliness

the electrical characteristics of a product, such as impedance. They describe the requirements of quality in general terms but lack the specificity to be measurable. However, CTQs can be translated into measurable terms or critical requirements through the inclusion of customer-defined specifications. For example, the customer may specify that the weight of the product be between 15 and 20 pounds. Products outside these specifications will not meet the customer's critical requirements.

- CTCs are similar to CTQs but deal exclusively Critical-to-Cost (CTC) with the impact of cost on the customer. However, CTCs and CTQs may be similar, yet stated by the customer for different reasons. For example, the weight of the product may also serve as a CTC. Again, the customer might require the weight to be between 15 and 20 pounds. However, the specification is necessary to achieve an impact on cost. Products heavier than 20 pounds may require more power consumption, thus increasing the cost.

- CTPs are typically key process input Critical-to-Process (CTP) variables. In terms of the process, we might think of CTPs as the critical *x*s in the $y = f(x)$ equation. We may know and understand what the *x*s are (for example, temperature, pressure, humidity), but not necessarily their specific setting or level. Once the settings are determined or specified, they can be set to achieve a consistent y.

- CTSs are stated customer needs regarding Critical-to-Safety (CTS), the safety of the product or process. Once more, consider the previous example of product weight. Product weight is the CTS variable. Though identical to the CTQ and the CTC, it is identified by the customer because any product heavier than 20 pounds has been shown to introduce a high rate of back injuries. Product weights less than 15 pounds have been shown to introduce a high rate of back injuries as well since the lower weight induces the operator to pick up and transport multiple products. So far, we have seen that CT*x*s may be identical across categories, but for different reasons.

- CTDs represent those customers with Critical-to-Delivery (CTD) stated needs regarding delivery. Typically, one thinks of late deliveries. However, delivering too early may be a problem for some customers, as it represents excess inventory requiring payment before the need for such inventory arises. CTDs are translated into critical customer requirements through the quantification of these impact areas. For example, the customer may specify that products be received no earlier than two days prior to the delivery date and no later than one day after the delivery date.

Projects aligned with CT*x*s, and subsequently critical customer requirements, have the biggest impact on the customer and often on the business directly. A useful tool for ensuring this alignment or translation from critical customer requirement to CT*x* is the tree diagram—an example of which is shown in Figure 5.7.

Kano Model

One method used to analyze customer satisfaction data is the Kano model (Figure 5.8). The model was developed by Noriaki Kano to show the relationship between three types of product characteristics, or qualities. These include qualities that must be present, those that are one-dimensional, and those that are delighters. The presence or absence of must-be characteristics is shown by the curved line in the lower-right quadrant. When a must-be characteristic is not present or is not present in sufficient quantity, dissatisfaction exists. As the characteristic becomes

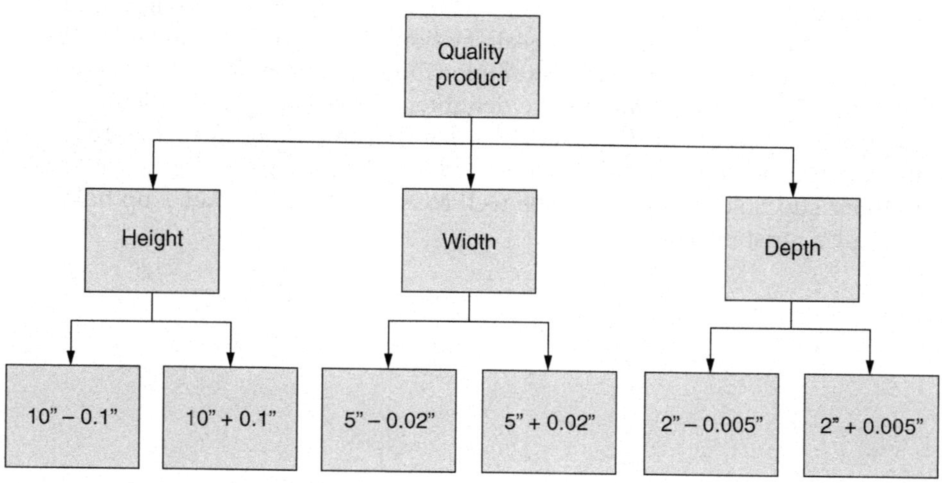

Figure 5.7 Example of a CTQ tree diagram.
Source: Donald W. Benbow and T.M. Kubiak, *The Certified Six Sigma Black Belt Handbook*, 2nd ed. (Milwaukee: ASQ Quality Press, 2007).

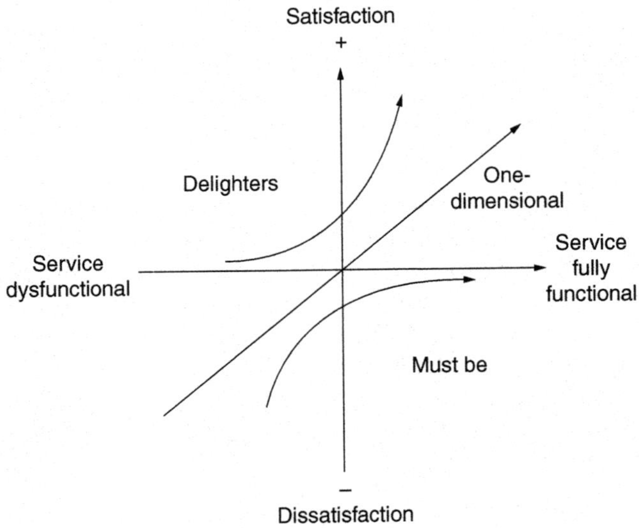

Figure 5.8 The Kano model.
Source: R. T. Westcott, *The Certified Manager of Quality/Organizational Excellence Handbook*, 3rd ed. (Milwaukee: ASQ Quality Press, 2006): 477.

more available or of a higher quality, customer satisfaction increases, but only to a neutral state, represented by the horizontal line. In other words, the factor can only satisfy, not delight, the customer.

One-dimensional factors drive satisfaction in direct relationship to their presence, and thus they are represented by a straight line.

For example, as the price of fuel for automobiles goes down, the consumer's satisfaction goes up with the additional availability of fuel. If fuel is readily available at a low price, the Service Fully Functional and Satisfaction will go higher. On the other hand, as fuel prices go up, this becomes a Dissatisfier for many consumers. The curved line in the upper left to center area represents delighters. If absent, these factors have no impact on satisfaction. When present, however, these features delight the customer. An example would be automobile cup holders. When first introduced, cup holders were delighters. This example also demonstrates an important point about delighters. As they become more available, they tend to move from the upper-left side of the model to the lower left. Today, an automobile without cup holders would not be well received in the market. Cup holders have become a must-be feature.

Chapter 6

Project Management Basics

PROJECT METHODOLOGY

> Define and apply agile and top-down project management methods. (Apply)
>
> Body of Knowledge II.C.1

OVERVIEW OF MANAGEMENT THEORIES AND STYLES

The personal style one uses in working with others depends on many factors. One is the background and personal perspective of the individual manager, especially the assumptions made about what makes people function. The characteristics exhibited by employees, individually and as a group, may also impact how a manager will approach an issue. A final factor is the particular situation. For example, the manager may act quite differently when carrying out the following two tasks: (1) a one-on-one discussion with an employee about a performance problem, and (2) communicating to the entire organization the opportunities for learning that will be embedded in a new company initiative.

Several theories of management have been proposed over the past century that can help to explain different management styles. Following are some of the more widely known theories.

Scientific Management. Frederick Taylor was one of the first contributors to the scientific theory of management, which had the aim of finding the one best way to perform a task so as to increase productivity. Taylor therefore emphasized the efficiency, not the satisfaction, of workers. He did this by breaking a job down into small-task components that could be studied to find a more efficient way of doing it.

Classical Organizational Theory. As organizations became larger and more complex, the study of management began to look at the management of not only individual tasks, but also the larger enterprise. Henri Fayol developed 14 principles of management, which included issues such as how power and authority needed to be subdivided into reporting relationships, and the subordination of individual interests to the common good.

Human Relations Theory. In the 1930s, researchers performed a series of studies at the Hawthorne Plant of Western Electric Company to determine the best level of lighting, length of workday, and length of rest periods to maximize worker productivity. As the lighting in the plant was increased, production levels rose. Surprisingly, production levels also rose as light levels decreased. Please see Richard Gillespie, *A History of the Hawthorne Experiments*, Cambridge University Press, New York, 1991, for a description of the Hawthorne effect.

Note: Walter Shewhart was aware of the studies as they were occurring, and there was a young quality engineer at the plant named Joseph M. Juran. There was also a young intern at the plant one summer named W. Edwards Deming. So, a lot of Juran's and Deming's management writing started with observing firsthand the Hawthorne Studies.

Behavioral Theories. Behavioral theory is the background on which many motivational theories are based. Two landmark theories related to motivation are Abraham Maslow's hierarchy of needs and Frederick Herzberg's two-factor theory (satisfiers and dissatisfiers).

Douglas McGregor formulated the theory *X* versus theory *Y* model. Theory *X* is a negative view of human nature, assuming that most employees do not want to work and will require continual prodding. Theory *Y* is a positive view, believing that employees want to work and will seek responsibilities that can offer solutions to organizational problems and their own personal growth.

William Ouchi's theory *Z* refers to a Japanese style of management characterized by long-term employment, slow promotions, considerable job rotation, consensus-style decision making, and concern for the employee as a whole.

Learning Theories. There are several theories of how people learn differently. David Kolb formulated the experiential learning model whereby a person: (1) has an experience, (2) reflects on the experience, (3) develops abstract theories based on the experience, and (4) applies what he or she has learned.

Behavior Management. Behavior management is an adaptation of B. F. Skinner's behavior modification/operant conditioning studies. Performance management is a technique involving an analysis of the antecedents of a behavior and the consequences of a behavior for the purpose of reinforcing desirable behavior to obtain or maintain positive consequences.

Situational Leadership/Management. Paul Hersey and Ken Blanchard's Situational Leadership model and instrument is useful for the leader/manager to appropriately apply the right amount of direction and supporting behavior to followers as they progress through the four maturity levels.

Systems Thinking. In *The Fifth Discipline*, Peter M. Senge discusses the disciplines of the learning organization: systems thinking, personal mastery, mental models, building shared vision, and team learning.[1] He points out that the five disciplines should develop as an ensemble. Systems thinking is the fifth discipline; it blends the disciplines into a coherent body of theory and practice. Systems thinking shifts our mental focus from perceiving ourselves as separate from the rest of the world to being connected to the world.

Complexity Theory. The growing body of knowledge on complexity theory and complex adaptive systems signals the emergence of a meaningful management concept. As a generalization, complexity theory is concerned with the interaction among the parts of a system as well as the interaction between the system and its environment.

Managerial Grid. Blake and Mouton proposed the managerial grid, a two-axis perspective of managerial attention. The *x*-axis indicates the amount of focus on accomplishing a task, and the *y*-axis a measure of concern for people.

Seven Habits. Stephen R. Covey's model and instruments amplify his *Seven Habits of Highly Effective People*.[2] Progressing from dependence through independence to interdependence, the basic model addresses:

- Being proactive
- Beginning with the end in mind
- Putting first things first
- Thinking win–win
- Seeking first to understand, then to be understood
- Synergizing
- Sharpening the saw (exercising the four dimensions of renewal)
  - Physical (exercise, nutrition, stress management)
  - Social/emotional (service, empathy, synergy, intrinsic security)
  - Spiritual (value clarification and commitment, study and meditation)
  - Mental (reading, visualizing, planning, writing)

Top-Down Management. Can be any combination of the above strategies where the management team (or project manager) seeks to direct and control the outcome of the staff or process. Most associated with Theory X management, the manager controls all aspects of the decision-making process.

Agile Approach. *Agile* was a term coined in the lean approach/lean thinking and is a focus of reducing cycle time and waste using a number of different techniques and tools, for example, value stream mapping, and identifying and eliminating monuments and non-value-added steps. *Lean* and *agile* are often used interchangeably. Please see Chapter 2 to see some of the tools used to call a process *agile*.

PROJECT CHARTER

> Define and describe elements of a project charter and develop a problem statement that includes baseline data or current status to be improved and the project's goals. (Apply)
>
> **Body of Knowledge II.C.2**

A *charter* is a document stating the purposes of the project. It serves as an informal contract that helps the team stay on track with the goals of the organization. Each charter should contain the following points:

a. *Problem statement.* This is a statement of what needs to be improved.

b. *Purpose.* Establishes goals and objectives of the team.

c. *Benefits.* States how the enterprise will fare better when the project reaches its goals.

d. *Scope.* Provides project limitations in terms of budget, time, and other resources.

e. *Results.* Defines the criteria and metrics for project success—including the baseline measures and improvement expectations.

The problem statement is a summation of what requires improvement. Examples include:

"The computer modem circuit card assembly takes too long to produce."

"The response time to Internet inquiries from potential customers is too long, and the responses are not adequate based on customer survey feedback."

Another term sometimes used for the project charter is the Statement of Work (SOW) and usually describes the actual work to be accomplished. It is derived from the work breakdown structure (WBS), which is described later in this section, and when combined with the project specification, becomes the basis for the contractual agreement on the project. The Statement of Work is also known as the Scope of Work (SOW).

Project Planning

Project planning is a disciplined approach to monitoring how and when a project will be accomplished, with recognition of the system that the project is working in. If you want to go on a long trip, do you just jump in a car and go? Most of us plan the trip—some to more detail than others—so that we know what needs to be taken, how long we will be gone, and any number of other details that have to be handled.

As with many of the tools and processes listed in this book, there is a lot of information available in various references on project planning and project management. There is even a professional certification available just for project management. Here, only the basics involved are discussed to assist in daily work.

Effective project planning requires skills in the following areas:

- Information processing
- Communication
- Resource negotiations
- Securing commitments
- Incremental and modular planning
- Assuring measurable milestones
- Facilitating top management involvement[3]

PROJECT SCOPE

> Help define the scope of the project using process maps, Pareto charts, and other quality tools. (Apply)
>
> Body of Knowledge II.C.3

As noted above, the purpose of documenting the project scope, or boundaries, is to ensure a common understanding of what the project team and its associated resources will work on, and what is outside those defined boundaries. The scope is usually defined, at least in part, based on the problem statement, which gives the project its initial focus. Using the problem statement, experience, and tools like SIPOC, brainstorming, Pareto charts, and so on, the scope of the project can be defined and documented.

PROJECT METRICS

> Help develop primary metrics (reduce defect levels by x-amount) and consequential metrics (the negative effects that making the planned improvement might cause). (Apply)
>
> Body of Knowledge II.C.4

Project timelines and activity plans can become little more than paperwork if meaningful performance measurements are not included. These measurements or metrics should link directly to the project goals and through them to the benefits for the enterprise. For example, if a goal is to increase process throughput by 25 percent, a key metric might be cycle time. The project's intermediate objectives and documentation need to include incremental cycle time reduction measures. Project measures typically are linked to a specific project, usually for the life of just that specific project, and they are often manifested as:

- Percentage of work accomplished on time (schedule performance index)
- Percentage of work accomplished on budget (cost performance index)
- Other similar measures (for example, availability of resources, quality of key deliverables)

A key aspect of measuring a process is to start as early as possible in tracking the data to benchmark the current state. It is common that as soon as personnel in the process start realizing that you are measuring something, the numbers will start improving almost immediately as the process is now aware that someone is watching.

In selecting metrics for your project, it is advisable to review what the key performance indicators (KPI) (sometimes called *key process indicators*) are for the topical area that you are going to investigate. What does management feel is important as it relates to your area of study? What subset of information are you directly working on that will be an important aspect of the KPI? Are your targets in a SMART format?

PROJECT PLANNING TOOLS

> Use work breakdown structures (WBS), Gantt charts, critical path method (CPM), program evaluation and review technique (PERT) charts, and toll-gate reviews to plan projects and monitor their progress. (Apply)
>
> **Body of Knowledge II.C.5**

Project planning tools include project charters, project management plans, milestone charts, Gantt charts, project schedules, goal and objective statements or bullets, and so on. The number of tools used and depth of data contained in the tools varies based on the project size and scope. Usually, larger projects have more documentation as they take more time and expend more resources. Project documentation is controlled and maintained (that is, updated) during the life cycle of the improvement project. There are many software programs that can assist in this effort.

For a beginning project planning tool, starting with an Excel spreadsheet and listing the projected dates of the project may be sufficient to show the team the intended project process flow. Another good starting point is to brainstorm a flowchart of how you and the team envision the project might develop and then assign some projected dates to the flowchart.

Work Breakdown Structure (WBS)

As noted in *The ASQ Certified Quality Process Analyst Handbook, Third Edition,* once the objectives of the project have been defined, it is necessary to break the project down into its component tasks. To this end, a work breakdown structure (WBS) is useful. The WBS does two things: It guides a logical development of tasks to be performed, and it provides a structure that summarizes project costs. Work breakdown structures can be used to break projects down to many levels. Each project activity is broken down into smaller activities until the project planner can easily and convincingly estimate the time it will take to do the individual task and how much it will cost to perform it. The WBS then combines the individual task costs into a total project estimated cost (see Figure 6.1).

Toll Gate Review

Many engineering and project management groups have models for ensuring that the design and development process runs smoothly. One common model in the automotive industry is the APQP process. This is very similar to the path of a Six Sigma project using the DMAIC method. The use of Toll Gates has been developed to fix certain points during the process where checks (could be formal validation and verification) are conducted by engineers or managers to ensure that the project is on the right path to ensure the desired outcomes.

The term *Toll Gate, Tool Gate Review,* or *Toolgate* may be used for such review points and can incorporate your Champion to a full cross-function team to review the progress of the Six Sigma project. The ultimate goal is to get feedback and

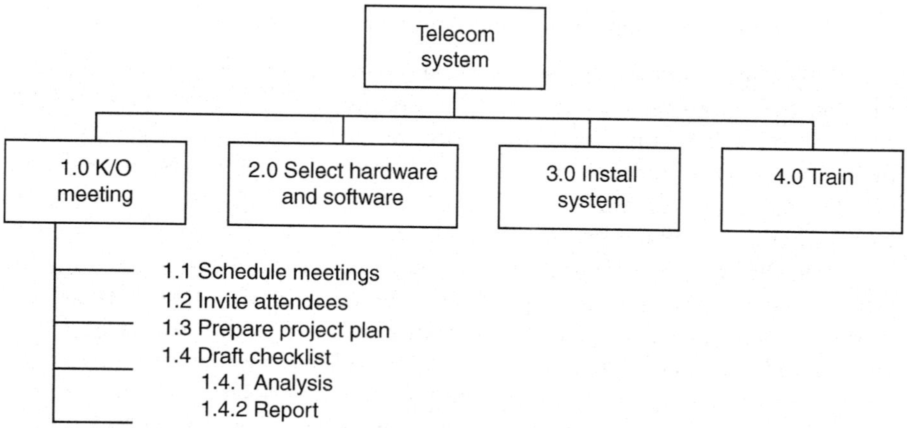

Figure 6.1 An example work breakdown structure.

other inputs to help ensure the process is working as expected. The Champion may even need to engage other senior managers to help in this process.

PROJECT DOCUMENTATION

> Describe the types of data and input needed to document a project. Identify and help develop appropriate presentation tools (storyboards, spreadsheet summary of results) for phase reviews and management updates. (Apply)
>
> **Body of Knowledge II.C.6**

Failure to provide project documentation leads to miscommunication and misunderstanding. Examples abound of projects that solved a problem other than the one they were intended to solve or answered a question no one was asking.

Project documentation usually includes:

- Goals and objectives
- Project sponsors and stakeholders
- Project plans and schedules—usually for larger projects
- Key project milestones
- Project budget
- Project boundaries
- Roles and responsibilities of project team members
- List of deliverables
- Metrics to be used to assess the project's performance

Smaller projects include these critical areas in the project charter; larger projects may require more detailed documentation and plans. A project charter can be in many formats but is usually limited to one or two pages to serve as a top-level, quick overview of the project. At various stages throughout your project, you may also be expected to conduct gate reviews with your Black Belt or Project Champion to ensure that the project is meeting expected outcomes at those points.

In addition to the project charter and project management plan, additional charts and diagrams can be produced for each of the activities listed. These graphical tools are also useful for tracking and evaluating the project at various phases and at final management reviews. Storyboards are sometimes used to convey project information involving changes that are easier to draw or photograph than to explain in words. A common application of storyboards is before-and-after pictures, often called *current state* and *future state*, for a proposed project. This approach is appropriate for facility or product redesign projects. Storyboards for Six Sigma projects are often formatted into five sections labeled *define, measure, analyze, improve,* and *control* (DMAIC) with charts, figures, and documents illustrating the activity in each area.

PROJECT RISK ANALYSIS AND MANAGEMENT

> Describe the elements of project risk analysis, including feasibility, potential impact, risk priority number (RPN), and risk management. Identify the potential effect risk can have on project goals and schedule, resources (materials and personnel), business continuity planning, costs and other financial measures, and stakeholders. (Understand)
>
> **Body of Knowledge II.C.7**

Project risk analysis is initially performed early in the project's life cycle—usually during the planning stage—to identify potential risks, the associated impacts, and potential mitigation plans. In addition to performing the initial risk analysis, it is recommended that the analysis be periodically reviewed and updated—especially if an identified risk is realized. Risk analysis is a formal process that can be performed using dedicated tools such as:

- Strengths–weaknesses–opportunities–threats (SWOT) analysis
- Risk priority number (RPN) or risk priority matrix
- Failure mode and effects analysis (FMEA)
- Formula for expected profit

These tools can be combined with other tools like brainstorming to ensure effective coverage for risk identification, analysis of potential impact if the risk is realized, and potential mitigation plans.

Successful risk analysis depends, in part, on ensuring that appropriate organizations are represented during risk analysis meetings and that all aspects of potential risk are considered. Aspects of risk to consider include, but are not limited to, the potential impact on:

- Meeting established goals or objectives
- The planned schedule
- Identified resources
- Safety
- Producibility

- Serviceability
- Reliability
- Meeting customer expectations and requirements

Risk assessment involves determining the impact severity if the risk occurs and the probability that the risk will occur. Determining the impact severity is usually done by performing analysis of like risks from other projects and historical data, and through the use of brainstorming, as well as other methods. Risk probability is determined based on the likelihood that the risk will occur during the execution of the project. A risk assessment assignment matrix, as shown in Figure 6.2, helps the project by having each risk individually ranked on a scale from 1 to 3 for severity and probability. The assignment number (a simpler version of the risk priority number [RPN] derived from FMEA data) is calculated as the severity rating multiplied by the probability rating. Risks with the highest assignment number require the most attention from the project team and the organization's leadership. After identification of the risks, risk mitigation and verification are performed throughout the project's life cycle. Risk mitigation begins with identifying activities the team or organization can perform to reduce the likelihood that an identified risk will occur and/or to reduce its impact (that is, reduce the delay in the schedule or the cost in dollars, product quality, or customer satisfaction) if the identified risk is realized. For example, if a risk is associated with a key supplier not delivering on time, mitigation activities might include developing another supplier, providing additional supplier oversight, colocation of personnel, shared design information, and so on, to help the supplier deliver on time.

Once risk mitigation planning occurs, the plans are implemented during the project and should be reviewed on a regular basis (usually weekly or monthly) to assess the status of existing risks (that is, are mitigation activities working?) and

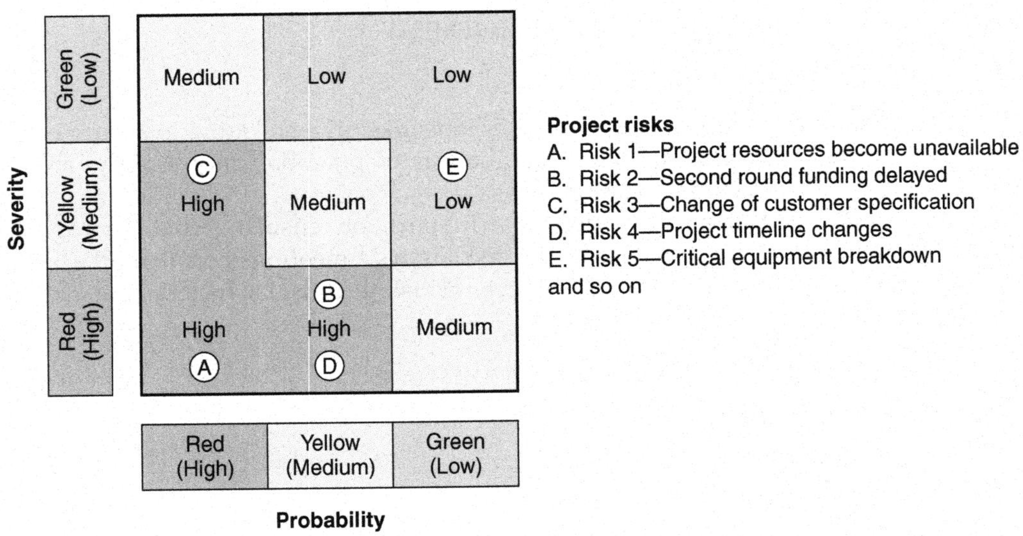

Figure 6.2 Risk assessment assignment matrix.

determine whether new risks have been identified or if identified risks were realized (that is, occurred). In some cases, mitigation is not always possible or only marginally successful and the risk is thus realized; the risk identification, mitigation, and review processes provide the team with knowledge of the worst-case scenario and what they should do. Risk verification is the process of ensuring that the risk mitigation activities reasonably prevent the risk from occurring. An example is the security risk for an automated teller machine. The risk is that someone else can improperly access your bank account. The mitigation is the additional requirement of a secured six- to 10-digit numeric password, along with a restriction of three false attempts before freezing the account. The verification is the attempt to access a secured account with an invalid or null password.

A risk management system will curtail the potential losses arising from quality problems and build quality into designs and processes more effectively than a reactive, trial-and-error approach of finding and fixing problems as they occur.

Types of Risks

Most everyone realizes that risk is present in our personal lives as well as in the organizations in which we work. Risks organizations face are:

- Events ranging from those causing minor disruption to those catastrophic events that affect the organization regardless of its location, financial stability, and so on.
- Events that can affect each organization differently; a minor disruption to one organization but a major disaster for another.

A risk that becomes a negative event can seriously disrupt your organization's processes and make it unable to meet customer requirements. Identifying and planning for potential risks is critical in the pursuit of customer satisfaction, organizational stability, and profitability, even under potentially adverse conditions.

The Risk Management Process

Risk management is defined in the ISO 31000:2018 (Risk management – guideline) clause 3.2 as the "coordinated activities to direct and control an organization with regard to risk." There can be any number of methods that you might use to understand the uncertainty of a situation. The FMEA process is one method used in the design of products, production, or machinery applications. In its simplest form, the PDCA cycle can be used to manage risks. Note: The ISO has created a series of documents around risk and mitigation processes.

Example:
The PDCA steps to take to establish an effective risk management process are:

1. *Plan.* Design the means by which your organization will:
    - Identify and define its potential exposures to loss.
    - Quantify the potential financial and nonfinancial risks.
    - Examine the feasibility of alternate risk management techniques.

- Select the best risk management technique(s).

2. *Do.* Implement a test of the chosen risk management technique(s).
3. *Check.* Assess the effectiveness of the technique and make necessary adjustments or select a new alternative.
4. *Act.* Implement the full process.
   - Roll out the full implementation of the tested technique(s).
   - Monitor the implemented techniques for adequacy of protection.
5. Improve the implemented techniques (return to the plan step).

Business Continuity Planning

With the worldwide pandemic of COVID-19, many organizations have become much more focused on ensuring that their operations can continue in adverse conditions. The automotive industry refers to this as contingency plans (IATF 16949:2016 clause 6.1.2.3). ISO released an auditable management system standard in 2019 called ISO 22301:2018 (Security and resilience—Business continuity management systems—Requirements) dealing with contingency, which is defined in clause 3.3 as the "capability of an organization to continue the delivery of products and services within acceptable time frames at predefined capacity during a disruption."

In the last few years, many organizations have been struggling with supply chain and available personnel issues. These and many other causes of disruptions to the business can be planned for in a worst-case scenario using many of the tools and methodologies that you are learning in the Green Belt process.

Some examples of topical areas that an organization might make plans for include:

- Pandemic – long-term health issues
- Major fire
- Destructive weather
- Loss of key personnel or the general workforce (retirements, injury, etc.)
- Supply chain disruptions (quality of goods or services)
- Legal impacts due to changing government regulations
- Sitewide safety accidents
- Environmental issues (spills, pollution, hazardous materials, etc.)
- Transportation disruptions (both incoming and shipping)

PROJECT CLOSURE

> Review with team members and sponsors the project objectives achieved in relation to the charter and ensure that documentation is completed and stored appropriately. Identify lessons learned and inform other parts of the organization about opportunities for improvement. (Apply)
>
> **Body of Knowledge II.C.8**

It is important to note that project closure is not a negotiation; it is a final step performed based on proving that the project met established goals and objectives, ensuring that required documentation is completed and appropriately stored, and conducting a closure meeting with the project sponsors to ensure agreement that the project is completed.

The project charter is an excellent tool to use as a measure of project completion as it established the scope, goals and objectives, and time frame for the project. Usually, a review of the charter against documented project results is sufficient for closing a project. At times, the project sponsor may want an independent review or assessment prior to formal project closure. Typically, this is done using an audit approach. Another method of proving that the project achieved its intent is analysis of project measures.

Lessons learned is a concept of identifying those things that are important to future projects to help get a quicker start in resolving those future issues. How can you capture things that the team has been learning on this project that can help future projects to find their conclusions quicker? If your company does not have a formal Lessons Learned database, you could consider adding information to the Corrective Actions database, which is common among companies who are ISO 9001 registered.

Chapter 7

Management and Planning Tools

> Define, select, and apply these tools:
> 1) affinity diagrams, 2) interrelationship digraphs, 3) tree diagrams, 4) prioritization matrices, 5) matrix diagrams, 6) process decision program charts (PDPC), 7) activity network diagrams, and 8) SWOT analysis. (Apply)
>
> **Body of Knowledge II.D**

In this section, the typical tools used during the *define* phase are highlighted. Many of these tools have templates and additional information available online.* Other highly recommended resources include:

- www.asq.org. See the "Quality Tools" page in the Knowledge Center for tools and explanations of their use.
- *The Quality Toolbox*, Second Edition, by Nancy R. Tague, ASQ Quality Press, 2005.
- *Customer Driven Company*, by Richard C. Whiteley, The Forum Corporation, 1991.
- www.iSixSigma.com. An improvement-based website that has numerous templates and tool guides, as well as case studies.
- www.freequality.org. This website has several free templates available for download.

ACTIVITY NETWORK DIAGRAM

The *activity network diagram* (AND) is similar to the PERT chart discussed later in this section because it graphically shows interdependencies between tasks. An AND highlights key tasks, the time to accomplish the tasks, flow paths (serial or

*https://asqassets.widencollective.com/portals/8ntnzkee/(H1469)SupplementalFiles-TheCertifiedSixSigmaGreenBeltHandbook,SecondEdition.

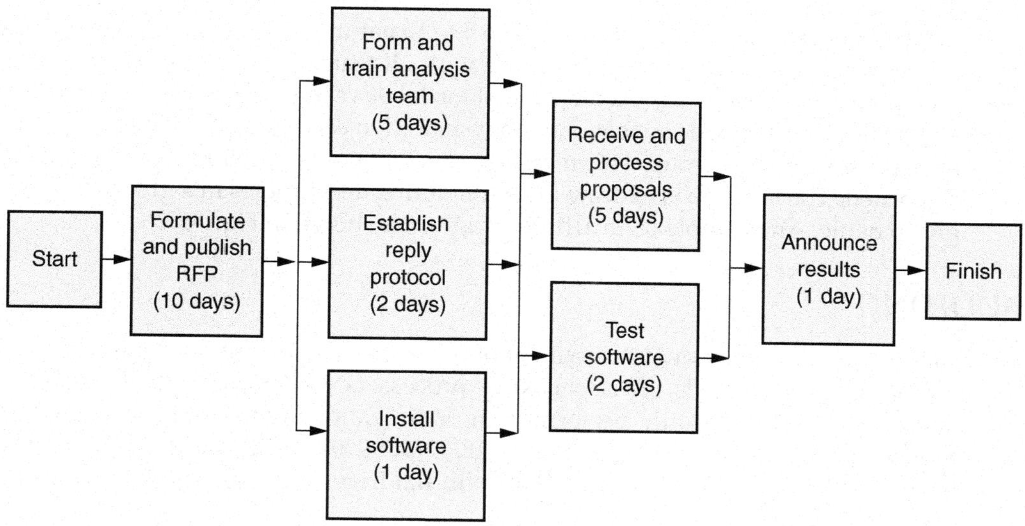

Figure 7.1 Example of an activity network diagram.

parallel), and so on. This tool, like PERT and critical path analysis, can provide a top-level overview or detailed data depending on the project need. An example of an AND is shown in Figure 7.1.

ADVANCED QUALITY PLANNING

Advanced quality planning (AQP) is founded on the idea that solid planning helps prevent surprises and saves valuable resources. Anything that is to be done can be first thought out, or written plans can actually be made to lay out a pattern or blueprint of what you are going to do.

AQP is a process where we first look at the parameters of what we are going to do. Do we have the right amount of material available? Do we have the right people to do the job or provide the service needed? Do we have the right tools to do the job well? Do we know the correct way of using everything we have to provide a safe working environment? All of these questions and many more should be answered before we start work! One of several tools that could be used here to help answer these questions is the cause-and-effect diagram. This forces us to ensure that we have thought of all the elements/inputs (causes) that will give us the desired output (effect). You can also think of AQP as the first step of the plan–do–study–act (PDSA) cycle, where the plan is done before work actually starts.

Note: Some industries now call this *advanced product quality planning* (APQP).

AFFINITY DIAGRAM

Affinity diagrams are used to produce many possible answers to an open-ended question such as "What are some of the ways to reduce cycle time for process A?" The first step is to brainstorm as many responses/ideas as possible to create a pool of ideas. Each response is clearly written on sticky note paper. The next step is to move the notes around until they fall into 5 to 10 natural groups. Some suggest that this organization step be done in silence with all team members participating, as they prefer. If a note gets moved back and forth between two groups several times, a duplicate may be made so that one can be placed in each group. The next step is to find a name for each group. This may take several iterations before team consensus is reached. The last step is to draw lines containing all the notes in a group with the group name. An example of an affinity diagram is shown in Figure 7.2.

AUDITING

Auditing is an independent assessment of processes and/or projects against established plans and goals. This checking process is sometimes called "quality" auditing and is commonly performed in organizations with established quality management systems (for example, AS9100, ISO 9001, IATF 16949) or those that have applied for a quality award (Malcolm Baldrige National Quality Award or state-level quality awards).

The first thing to know when planning to audit something is what the standards or criteria are that we will be auditing against. If it is a process audit, the auditor needs to know what applicable standard operating procedures or other process sheets are used to ensure that things are done in a standardized manner.

Problem: What are some of the ways to reduce cycle time for process A?

Machine

- Run machine faster
- Get a new machine
- Apply new controls
- Reduce setup time
- Simplify machine operations

Personnel

- Assign more people
- Provide additional training
- Let Joe run the process/machine
- Provide help during setup

Infrastructure

- Reduce paperwork
- Improve forklift uptime
- Better conveyor
- New overhead crane
- Prompt delivery of route sheets

Vendor

- Improve vendor communication
- Use cell phone to contact vendor
- Have additional sources
- Work with vendor to improve quality and delivery
- Reduce resupply time

Maintenance

- Better lubricant
- Reliability-centered maintenance
- More frequent lubrication
- More prompt response to maintenance calls

Figure 7.2 Example of an affinity diagram.

If conducting a safety audit, then having the written safety rules would be important. Some other items that could be audited include project closure, cleanliness, product quality, system knowledge, emergency procedures, and so on.

An individual audit, like most other activities, is a sequence of processes and typically includes audit planning (or preparation), audit performance, audit reporting, and audit closure. *Audit planning* is the auditor's preparation phase to ensure familiarity with the item being audited, to schedule the audit, and to notify the auditee. The *audit performance* phase is where the auditor gathers the evidence of performance through interviews, observations, and so on. This phase also includes reconciliation to ensure the validity of any potential findings (that is, noncompliances or nonconformances) or opportunities for improvement (that is, items that could lead to a finding). The *audit report* phase includes drafting the report, which summarizes the audit scope, activities, and results, and the generation of associated corrective actions (findings) or preventive actions (opportunities for improvement). The *audit closure* phase includes any necessary follow-up activities to ensure that action plans are implemented and effective in eliminating their associated causes.

It should be noted that there are several types of audits (see ISO 19011 for more details), including first-party, second-party, third-party, internal (most likely used for Six Sigma), and external. Audits typically focus on products, systems, suppliers, or regulatory compliance. There are numerous books available through ASQ Quality Press and other publishers related to auditing.

When asked to be part of an audit team, there are some basic things to be aware of and know in conducting an audit:

- Be pleasant to the person being audited; you are not a cop looking to catch them doing something wrong.
- Be prepared. Understand the process to be audited and associated documentation, and verify document availability, revision, and references prior to audit conduct.
- Be factual in what you observe. Hiding things does not help the company to improve, and being too critical may harm personal objectivity.
- Be thorough in looking at the process as it relates to the standard. If you miss something, customer satisfaction may suffer, or worse yet, someone might get hurt.
- Ask questions for clarity. We want to be as positive as possible given the situation.
- Record your observations for the record and the next person who will audit the area.
- Follow what the internal lead auditor directs you to do.

Being an internal auditor for your organization can be both challenging and informative as we usually get a better view of what our companies are doing if we have a chance to look at other operations. For some of us it will also break the routine of what we do every day and give us a chance to see how others might be doing things and to benchmark (see the following section) that against what we do.

BENCHMARKING

Benchmarking is the process of looking at one system(s) and applying the concepts observed to another system. This usually involves some give and take between the organization doing the benchmarking and the organization that is the benchmark. The idea is to make it a win/win situation—how can the organizations learn from each other? Another critical point about benchmarking is that it can be done internally and externally. When performed internally, different departments assess each other's processes and take the best from each process to improve their own process. When external benchmarking is performed, it can be done within the same industry or with an organization from another industry. Usually, benchmarking against an organization in another industry is easier as it removes the competitive aspects of comparing processes. All information gained during the benchmarking process should be considered protected—for internal use only.

The basic process steps of benchmarking are:

- Flowchart the current process
- Identify the areas to be improved
- Brainstorm ideas
- Investigate how others (internal and external) perform similar processes
- Develop plans for application of ideas
- Pilot test ideas
- Initiate the new process
- Evaluate the new process

Before starting a benchmarking project, it is advisable to ensure that you know exactly how your current process works. That may sound unnecessary, but it has been shown that when people actually flowchart a process, there is usually a lot of disagreement as to the exact steps and/or the order of those steps. Thus, any time that there is more than one person who works on or around a machine, the process should be flowcharted.

Then it is time to do some research into how others (internal or external to the organization) do similar things. Once you have seen other ways of doing things, it is time to try to figure out how you can do things differently in your own operations. Establish a plan for the changes and acquire the needed resources from management. The plan should list what materials are needed, when and where new operations will be installed, identification and planning of any training that may be needed, and other details that will allow for a changeover to the new idea. Then prepare and run a pilot test to ensure that the plan will work. It is usually unwise to just jump right into the new idea without testing it first. Even the best-laid plans may have unforeseen bugs that cause problems, so run a small test first.

BRAINSTORMING

See the brainstorming discussion in Chapter 9 for additional information.

CAUSE-AND-EFFECT DIAGRAM (FISHBONE, ISHIKAWA DIAGRAM)

Originally developed in the 1940s by Kaoru Ishikawa in Japan, the *cause-and-effect diagram* is a graphical analysis tool that allows the user to display the factors involved in a given situation. Cause-and-effect diagrams are drawn to clearly illustrate the various causes (x) affecting the item being investigated. "A good cause-and-effect diagram is one that fits the purpose, and there is no one definite form."[1] These causes can be any item or occurrence that is related to the effect (y) that is being studied. Thus, the effect of a situation is the result of the function of the causes [$y = f(x)$]. Other names for this tool that are sometimes used include *Ishikawa diagram, fishbone diagram,* or even *feather diagram* given the shape of the graphic.

Key Point: Cause-and-effect diagrams can be used to analyze positive effects as well as undesirable ones.

Asking the five Ws and one H (what, why, when, where, who, and how) can be effective in developing the elements of the cause-and-effect diagram. Besides using the five Ws and one H in creating the cause-and-effect diagram, consider starting with the six Ms:

- Man (people/operator)
- Machine (equipment)
- Methods (operating procedures)
- Materials
- Measurement
- Mother Nature (environment)
- Money (optional—but an important consideration)
- Management (optional)

This tool is relatively simple to use and yet very powerful. Once it is completed, it is able to show graphically the factors of the system or process to management and other teams.

Figure 7.3 is a cause-and-effect diagram example in which a manufacturing team tries to understand the source of periodic iron contamination in product. The team used the six generic headings to prompt ideas. Layers of branches show thorough thinking about the causes of the problem.

For example, under the heading "machines," the idea "materials of construction" shows four kinds of equipment and then several specific machine numbers. Note that some ideas appear in two different places. "Calibration" shows up under

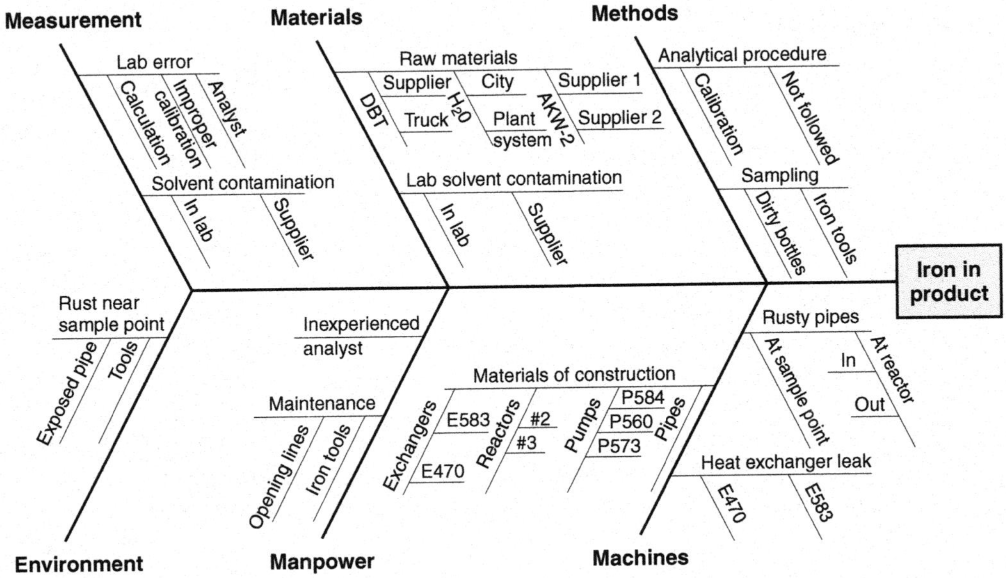

Figure 7.3 Example of a cause-and-effect diagram.
Source: N. R. Tague, *The Quality Toolbox*, 2nd ed. (Milwaukee: ASQ Quality Press, 2005): 87.

"methods" as a factor in the analytical procedure, and also under "measurement" as a cause of lab error. "Iron tools" can be considered a "methods" problem when taking samples or a "manpower" problem with maintenance personnel.

CHECK SHEETS

Check sheets are used to observe or review a process, usually during execution of the process. Check sheets pre-categorize potential outcomes for data collection using sets of words, tally lists, or graphics. Figure 7.4 is an example of a completed check sheet, in tabular format, used to collect data related to a paint mixing

Paint color defect causes

Reason	Day of the week							Weekly total
	Mon	Tue	Wed	Thu	Fri	Sat	Sun	
Operator misread instructions	I	I	0	II	卌	卌 II	III	19
Wrong pigment used	III	I	卌	III	IIII	IIII	卌 卌	30
Wrong color code from customer	II	I	II	0	0	卌	0	10
Outdated base paint	0	0	I	0	III	0	IIII	8
Daily total	6	3	8	5	12	16	17	67

Figure 7.4 Example of a check sheet.

process. This simple tool provides a method of easy collection of the data. By collecting data on a check sheet, common patterns or trends can be identified.

The basic steps in making a check sheet are:

1. Identify and agree to the causes or conditions that are to be collected.
2. Decide who will collect the data, over what time period(s), and how the data will be collected.
3. Create a check sheet that will work within the operation where it will be used.
4. Collect the data as designed to ensure consistency and accuracy of the information.

Check sheets can be the basis for other analytical tools and are incorporated into attribute process behavior charts. Creating and using a check sheet can help focus on continual improvement and may foster changes just because the check sheet is being used.

CUSTOMER FEEDBACK

Feedback is a method or process of finding out what the customer actually thinks of your products or services. Finding out what the customer thinks, wants, and needs is a very time-consuming effort, and many customers are getting tired of filling out paper surveys. Sometimes, when customers (internal or external) do tell us something, we either can't do anything about it, do not want to hear what they are saying, or they expect something other than what we offer.

There are many traditional techniques for talking with customers and, with the advent of e-mail and the Internet, even more so today. The problem with many surveys is that we really do not know how many or which customers actually bothered to send them back (i.e., what is the population sample that you received). Also, paper surveys tend to have a very low return rate, so we really do not get a very good overall picture. The point is that we must keep trying to talk with customers as often as possible to ensure that we know what their wants and needs are. The most effective method still seems to be actually getting out into the world with the customers to experience what they do and to interact with them as they use your products and services.

FAILURE MODE AND EFFECTS ANALYSIS (FMEA)

This tool is discussed in Chapter 3.

FLOWCHART

According to Nancy Tague in *The Quality Toolbox,* a flowchart is a picture of the separate steps of a process in sequential order, including materials or services entering or leaving the process (inputs and outputs), decisions that must be made, people who become involved, time involved at each step, and/or process measurements.

Tague describes the basic procedure for development of a flowchart as:

1. Define the process to be diagrammed. Write its title at the top of the work surface.
2. Discuss and decide on the boundaries of your process: Where or when does the process start? Where or when does it end? Discuss and decide on the level of detail to be included in the diagram.
3. Brainstorm the activities that take place. Write each on a card or sticky note. Sequence is not important at this point, although thinking in sequence may help people remember all the steps.
4. Arrange the activities in proper sequence.
5. When all activities are included and everyone agrees that the sequence is correct, draw arrows to show the flow of the process.
6. Review the flowchart with others involved in the process (workers, supervisors, suppliers, customers) to see if they agree that the process is drawn accurately.[2]

Figures 7.5 and 7.6 depict typical flowchart examples.

FOCUS GROUPS

Focus groups are an attempt to improve the depth and accuracy of customer responses. Focus groups generally provide more accurate answers with more depth than other techniques. However, the sample is often nonrandom and too small. As statisticians say, "The plural of *anecdote* is not *data.*" Understanding the limitations of focus groups, they can still be effective in gaining knowledge related to product quality and perceptions of the organization's ability to satisfy its customers (internal and external). Prior to conducting focus group sessions, planning for the sessions must occur. Focus group planning includes:

- Identifying the goals of the session
- Developing actual or prototype products
- Determining how the sessions will be conducted: brainstorming, survey, cause-and-effect (or question–response)
- Identify how you will encourage customers to attend the sessions

Once the planning is completed, the focus groups can be identified, the focus group sessions conducted, the data collected, and the data analyzed to determine

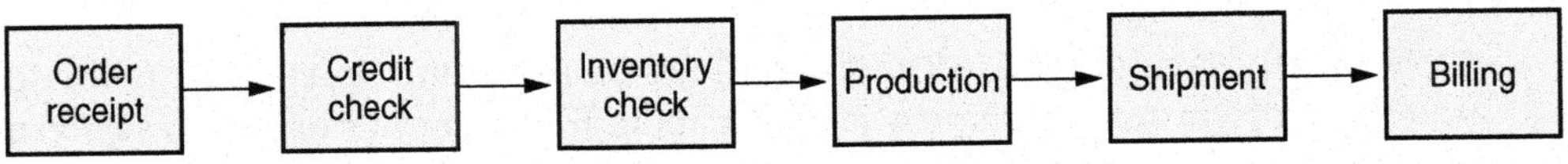

Figure 7.5 High-level flowchart for an order-filling process.
Source: N. R. Tague, *The Quality Toolbox*, 2nd ed. (Milwaukee: ASQ Quality Press, 2005): 257.

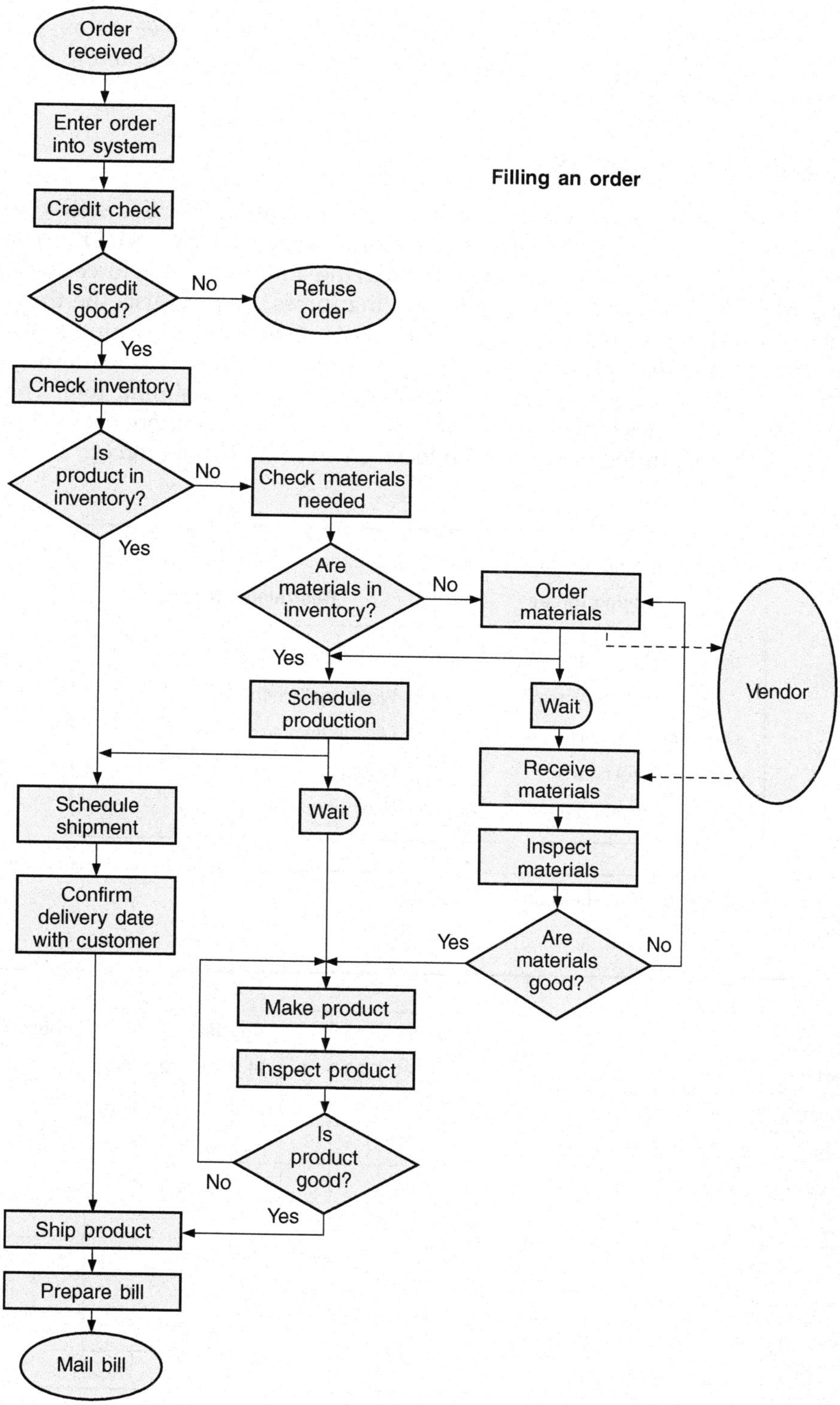

Figure 7.6 Detailed flowchart for the order-filling process.
Source: N. R. Tague, *The Quality Toolbox*, 2nd ed. (Milwaukee: ASQ Quality Press, 2005): 261.

how the process or product should be adjusted to better meet the expectations of the targeted groups. Focus group techniques are usually limited to consumer product–related activities.

FORCE-FIELD ANALYSIS

Force-field analysis is illustrated in Figure 7.7. A goal or objective is first listed as the future or desired state. Then, two columns are produced from brainstorming or a similar technique to determine the driving and restraining forces on achieving the future or desired state. The "driving forces" column lists the things that help make the future state occur, while the items in the "restraining forces" column are those that prevent the future state from occurring. The team then ranks the two lists using nominal group technique (NGT) or a similar tool. The team consensus provides guidance on how to proceed. It is often more useful to reduce or remove restraining forces than to focus entirely on driving forces.

Future state: number of errors is less than three per hundred documents.	
Driving forces	**Restraining forces**
Pressure from customer A	Ambient noise level
Group incentive system	Weak spell-checker
Operator enthusiasm	Lack of training
Use of control charts	Inertia
Supportive management	Poor input data

Figure 7.7 Example of force-field analysis.

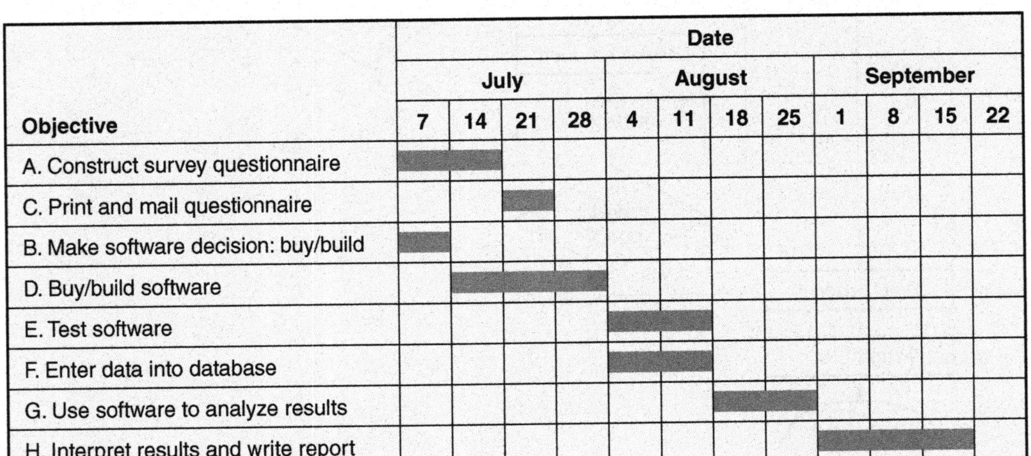

Figure 7.8 Gantt chart example.

GANTT CHART

A Gantt chart, as shown in Figure 7.8, is used to graphically display a project's key activities and the duration associated with those activities. These are then overlaid on a calendar to show when the activities occur. In addition to showing the key activities, the Gantt chart also shows the task milestones and the task relationships between predecessor and successor tasks. This provides a visual representation and allows for quick identification of the impact on the schedule (that is, delay or pull-in to tasks or the entire project) based on the delay or pull-in of a single task. The Gantt chart provides a quick look at the project planned activities and schedule, allows for assessment of key resources against the plan, and also allows for the assessment of the project's performance against the plan.

GRAPHICAL CHARTS, CONTROL CHARTS, AND OTHER STATISTICAL TOOLS

This information on control charts is provided as a quick overview; details are available in Chapter 7.

A control chart is a graph used to study how a process changes over time. Comparing current data to historical control limits leads to conclusions about whether the process variation is consistent (in control) or is unpredictable (out of control—affected by special causes of variation).

Different types of control charts can be used depending on the type of data. The two broadest groupings are for variables data and attributes data.

- *Variables data* are measured on a continuous scale. For example, time, weight, distance, or temperature can be measured in fractions or decimals. The possibility of measuring to greater precision defines variables data.
- *Attributes data* are counted and cannot have fractions or decimals. Attributes data are used when you are determining only the presence or absence of something: success or failure, accept or reject, correct or not correct. For example, a report can have four errors or five errors, but it cannot have four and a half errors.

Variables Charts

- $\bar{X}$ and R chart (also called averages and range chart)
- $\bar{X}$ and s chart
- Chart of individuals (also called X chart, $\bar{X}$ and R chart, IX-MR chart, XmR chart, moving range chart)
- Moving average–moving range chart (also called MA–MR chart)
- Target charts (also called difference charts, deviation charts, and nominal charts)
- CUSUM (also called cumulative sum chart)

- EWMA (also called exponentially weighted moving average chart)
- Multivariate chart (also called hotelling T2)

Attributes Charts

- *p*-chart (also called proportion chart)
- *np*-chart
- *c*-chart (also called count chart)
- *u*-chart
- *D*-chart (total weighted deficiencies)
- *U*-chart (average weighted deficiencies per unit)

Charts for Other Kinds of Data

- Short-run charts (also called stabilized charts or Z-charts)
- Group charts (also called multiple characteristic charts)
- Paynter charts[3]

More details are available in Chapter 22.

Graphical tools are an effective way to understand and convey customer feedback, process performance, defect data, and so on. The examples shown in Figure 7.9 are less complex graphical data representations.

The use of statistical tools to analyze data is explained in detail in various sections in this book. One cautionary note: Analysis of customer data almost always falls in the general category of "observational studies," which means that high correlation between variables doesn't imply cause-and-effect relationships. For example, suppose a survey shows a correlation coefficient of $r = 0.83$ between the number of television advertisements watched and the amount of money spent on product W. The analyst is not statistically justified in concluding that watching the ad causes an increase in the amount spent on the product.

It is useful to compare internal predictive metrics with customer feedback data. This helps shape internal prediction methods and procedures.

INTERRELATIONSHIP DIAGRAM (DIGRAPH)

Interrelationship digraphs are used to identify cause-and-effect relationships. A typical application would begin with the listing of a half dozen to a dozen concerns, one to a note sheet, arranged in no particular order around the perimeter of a flipchart or a whiteboard. For each pair of concerns draw an arrow from the one that is most influential on the other. Draw no arrow if there is no influential relationship. This is most easily accomplished by starting at the twelve o'clock position and comparing the item to the next item in a clockwise direction. Draw the appropriate arrow (or no arrow). Then compare the twelve o'clock note with the next

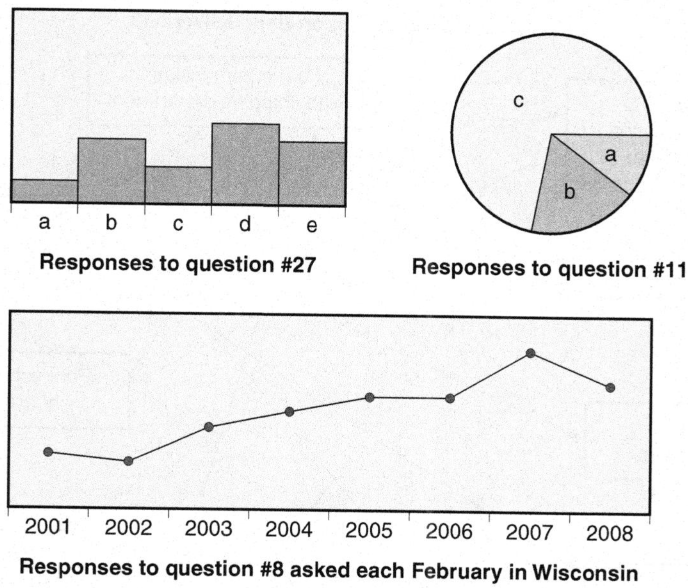

Figure 7.9 Graphical representations of question responses.

note, again moving clockwise. After the twelve o'clock note has been compared to all other notes, begin with the note clockwise from the twelve o'clock note and compare it to all other notes that it has not been compared to. Repeat this process until all pairs of notes have been compared. In each case ask, "Does A influence B more than B influences A?" Revise this diagram as necessary using additional information or data if needed. An example of an interrelationship digraph at this stage is shown in Figure 7.10. The next step is to find the note that has the most outgoing arrows. This note is called the driver. In the example shown in Figure 7.10 "poor scheduling practices" is the driver. The driver, or drivers if there is a tie or near tie, is often a key cause of the problem. The note with the greatest number of incoming arrows, "poor scheduling of trucker" in the example, is called the *outcome* and can often be used as a source of a metric for determining project success.

INTERVIEWS

Interviews by phone or in person permit a higher response rate than written surveys. Although interviews take more effort, the effort is usually returned with more accurate data gathered and additional information provided by the interviewee that may not be captured in written (hard copy or electronic) surveys. A skillful interviewer can record customer feelings that written surveys would not detect by noting interviewee emotions, body language, other nonverbal clues, and so on.

MATRIX DIAGRAM

A *matrix diagram* is typically used to discover and illustrate relationships between two groups of items. In Figure 7.11 the two groups are the units of a training

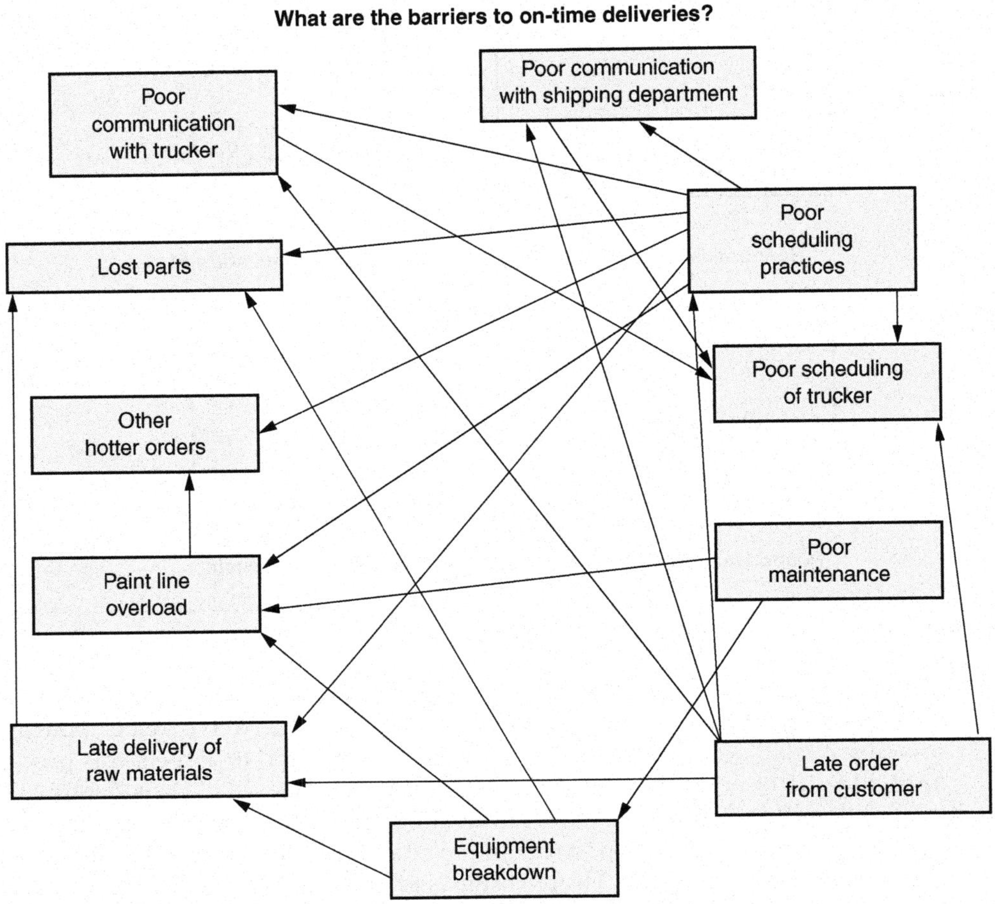

Figure 7.10 Example interrelationship digraph.

course and the objectives of the course. The items in one group are listed across the top of the chart, and the items in the other group are listed down one side. The team examines each square in the matrix and enters one of three symbols or leaves it blank depending on the relationship between the items in the row and column represented by the square. The most conventional symbols are shown in the example although letters and numbers are sometimes used. The team then examines the completed matrix and discusses possible conclusions.

MULTIVOTING

As discussed in Chapter 9, multivoting complements the *nominal group technique* (NGT). Even though this tool is typically used in combination with NGT, it can be used independently (for example, to prioritize brainstorming results).

		Unit								
		1	2	3	4	5	6	7	8	9
Objectives	Review basics	⊙		⊙	O			Δ		
	Math skills	O	Δ	⊙			⊙	O		
	Communication skills						O		Δ	
	Attitude/motivation								Δ	
	Sketching			⊙			O			
	Ohm's law				⊙	O	Δ			
	Kirkoff's law					⊙	O			
	Thevinev's law						Δ		O	
	Heisenberg's uncertainty principle									

⊙ = strong relationship O = moderate relationship Δ = weak relationship

Conclusions: Thevinev's law and communication skills covered only weakly
Attitude/motivation barely covered
Heisenberg's uncertainty principle not covered
Unit 2 contributes very little toward objectives
Unit 9 contributes nothing toward objectives

Figure 7.11 Example matrix diagram.

NOMINAL GROUP TECHNIQUE

This is also a type of brainstorming but with limited team vocal interaction. The tool is hence named "nominal" group technique (NGT). This technique has its application when some group members are much more vocal than others, to encourage equal participation from all members, or with controversial or sensitive topics, and so on. This technique helps to alleviate peer pressure and reduces the impact of such pressure on the generation of ideas.

Similarly to brainstorming, the facilitator explains the rules. The team leader presents the topic to the assembled members. The team is given a good 10 to 15 minutes so that they can silently sit, think, and generate ideas.

No verbal interactions are allowed during the session. The members' ideas are collected and posted in a space where all can read them. The members may also read the ideas aloud one by one in a round-robin format. At this stage no judgment or criticism is passed. The ideas are simply written down. The members are allowed to expand on existing ideas, provide clarity, and eliminate redundancy during the consolidation. For a controversial or sensitive subject, the team leader may opt to collect the ideas and write them down on the board, maintaining anonymity of the contributors.

PDCA (PDSA AND SDCA)

PDCA is an improvement cycle methodology that has evolved since 1939, starting with Shewhart and updated by Deming, with iterations of the model by Ishikawa and the Japanese Quality Circle movement. There are variations of the improvement model.

The model is a simple circle divided into four quadrants named plan–do–check–act. The improvement team brainstorms the activities that the four quadrants should encompass in the context of the problem at hand and populates them into the quadrant. Quality gurus like Deming and Ishikawa have already provided a foundation to this model to build upon, and both suggested that more time is needed in the plan quadrant to ensure that the others perform more smoothly.

The initial PDCA cycle (Figure 7.12) was created by solving shop floor problems and preventing recurrence of the problem. The seven tools of quality were used along with the PDCA model. The model is robust and applies to many applications including product design and development, service offering, educational curriculum, healthcare, and so on. No matter what the application is, we can map the activities to P-D-C-A. This forces a disciplined approach to program management and provides the team members with a big picture of what needs to be accomplished and where they are in the quadrant.

Plan: Define a problem and hypothesize possible causes and solutions.

Do: Implement a solution.

Check: Evaluate the results.

Act: Return to the plan step if the results are unsatisfactory, or standardize the solution if the results are satisfactory.

Near the end of Deming's career, someone suggested to him at a conference that what he was describing in the PDCA model was not totally correct! The question concerned the check quadrant/phase, which is a reactive state to issues that have happened. If in fact the process team is projecting what will happen, then a better descriptor for this phase would be to *study* (S) the issues as they are occurring

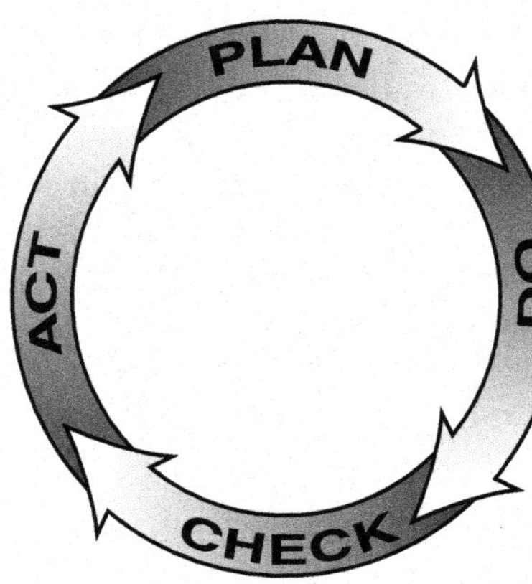

Figure 7.12 PDCA cycle.

instead of waiting until later—thus Deming modified the approach and the PDSA model was started.

The existence of both a reactive and proactive part of projects was noted by Munro and the standardize–do–check–act (SDCA) cycle was introduced to partner with the PDSA (see the e-book *Six Sigma for the Shop Floor*). Taken together, these allow for both sides of the improvement cycle: to both standardize the improvement for stabilization as well as allow for planning for future improvements.

PRIORITIZATION MATRIX

A *prioritization matrix* is used to aid in deciding between options. In the example shown in Figure 7.13 the options are four different software packages, A, B, C, and D. The team determines by consensus the criteria against which the options will be measured (for example, requirements document) and the relative importance of each of the criteria items. In the example, the criteria and their relative importance are compatibility (.25), cost (.30), ease of use (.40), and training time (.05).

Each option is ranked against the criteria, with the desirable numbers being larger. In the example, since there are four options, the highest-ranking option would be assigned a value of 4, the second-place item would be assigned a 3, and so on. Assigning each option the average values for the two places designates a tie. For example, if two options are tied for third place, the two places are third and fourth, which have values of 2 and 1, respectively, so each of the options would receive a value of 1.5. In the example in Figure 7.13, options A and C are the most desirable (lowest) cost, so each is assigned a value of 3.5 (A and C are tied). The package with the next lowest cost is option D and is assigned a value of 3. Option B has the highest cost and is assigned a value of 1.

Once the values are assigned, the next step is to multiply each of the option values by the criteria weights at the top of the column and calculate the row totals. The option with the highest total is the one most favored by the prioritization matrix.

Determine the most suitable software package

		Criteria				
		Compatibility 0.25	Cost 0.3	Ease of use 0.40	Training time 0.05	Total
Options	Package A	1.00	0.45	1.20	0.15	2.80
	Package B	0.25	1.20	0.80	0.05	2.30
	Package C	0.75	0.45	1.60	0.20	3.00
	Package D	0.50	0.90	0.40	0.10	1.90

Figure 7.13 Prioritization matrix example.

PROBLEM SOLVING

Some of the most successful attempts to solve problems have been accomplished through the use of a model or tools that outline the steps that should be followed in investigating and containing issues, and fixing the problems so that they will not return. Unfortunately, many of us have seen situations where someone or even a group or team will fix a problem, only for the same issue to crop up again in a week, month, or year. The question is, how do we permanently solve problems?

One common approach to problem solving is called the *eight discipline approach* (8D). The steps usually associated with this process are:

1. *Use a team approach.* Organize a small group (note that we did not say *team—it takes time to develop a group of people into a functional team*) of people with the process/product knowledge, allocated line authority, and skill in the required technical disciplines to solve the problem and implement corrective actions. The group must have a designated champion.

2. *Describe the problem.* Specify the internal/external customer problem by identifying the quantifiable terms for who, what, when, where, why, how, and how many (5W2H) for the problem. Use such methods as SIPOC, brainstorming, flowcharts, and any other methods that the group feels are appropriate.

3. *Start and check interim (containment) actions.* Define and implement containment actions to isolate the effect or issue from the current problem. Verify the effectiveness of this containment action to ensure that the internal or external customer does not see further problems.

4. *Define and check root causes.* Identify all potential causes that could explain why the problem occurred (a cause-and-effect chart is useful here). Isolate and verify the root cause by testing each potential cause (sampling is used here) against the problem description and test data (individual test or a design of experiments if needed). Identify alternative corrective actions to eliminate root causes using a process behavior chart to ensure that the process remains stable.

5. *Check corrective action.* Through a sampling plan, quantitatively confirm that the selected corrective actions will resolve the problem for the customer, and will not cause undesirable issues (FMEA and control plans).

6. *Start permanent corrective action.* Once it is verified that corrective action is working, update all procedures and processes to incorporate the new process. This should include training where appropriate.

7. *Stop future problems.* Modify the management systems, operating systems, preventive maintenance, practices and procedures, and documentation to prevent recurrence of this and all similar problems. Note: If similar processes are found in the organization, look closely at them also to ensure that they do not develop the same issue.

8. *Congratulate your team.* Improvements happen only because many people work together. Everyone deserves credit.[4]

Besides the method described above, your external customers may have a prescribed problem-solving methodology that they want the organization to use. Other methods that can be used in problem solving are the plan–do–study–act (PDSA) cycle or, although not recommended, the scientific method that you probably learned in high school.

PROCESS DECISION PROGRAM CHART

The *process decision program chart* (PDPC) is a tree diagram that is used to illustrate anticipated problems and list possible solutions. It may be used as a dynamic document to be updated as the project proceeds. An example PDPC is shown in Figure 7.14.

PROJECT EVALUATION AND REVIEW TECHNIQUE (PERT)

The *project evaluation and review technique* (PERT) and *critical path method* (CPM) have become essentially merged in current software packages. The *critical path* is the path from start to finish that requires the most time. In Figure 7.15 there are just two paths: ACEGH and BDFGH. Path ACEGH requires 10 + 5 + 10 + 10 + 15 = 50 days, and path BDFGH requires 5 + 15 + 10 + 10 + 15 = 55 days. Therefore, BDFGH is the critical path. Software packages are available to identify and calculate the critical path for projects with multiple paths. If activities on the critical path are delayed, the entire project will be delayed. The critical path time is the time required to complete the project. The only way to complete the project in less

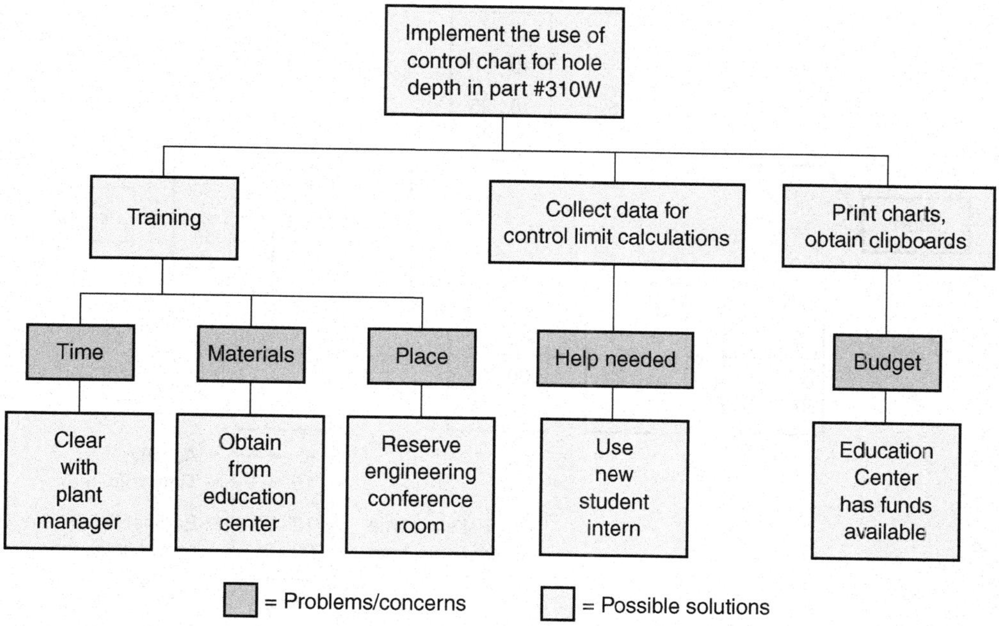

Figure 7.14 Example process decision program chart (PDPC).

time is to decrease the time for at least one of the activities. This is usually accomplished by putting more resources into one or more activities on the critical path. This is sometimes referred to as "crashing" the project.

QUALITY FUNCTION DEPLOYMENT

See Chapter 5 for details.

RISK PRIORITY NUMBER

The *risk priority number* (RPN) is calculated from FMEA data, specifically the severity rating (S), occurrence rating (O), and detection rating (D). RPN = S × D × O. The RPN is used to help determine the potentially highest-risk items to aid a project team in prioritizing the items to work on most aggressively.

RPN is covered in more detail in Chapter 3.

Key Point: The RPN is a factor of the process, not some external goal or prescribed numeral listing. The intent of the RPN is to use the full range (typically: 10 × 10 × 10 = 1000) to distinguish the most important risks for the team to work on!

SAMPLING PLAN

Dr. Joseph Juran said that 100 percent inspection is only 80 percent effective: "Collectively, these inspector errors result in a performance of about 80 percent

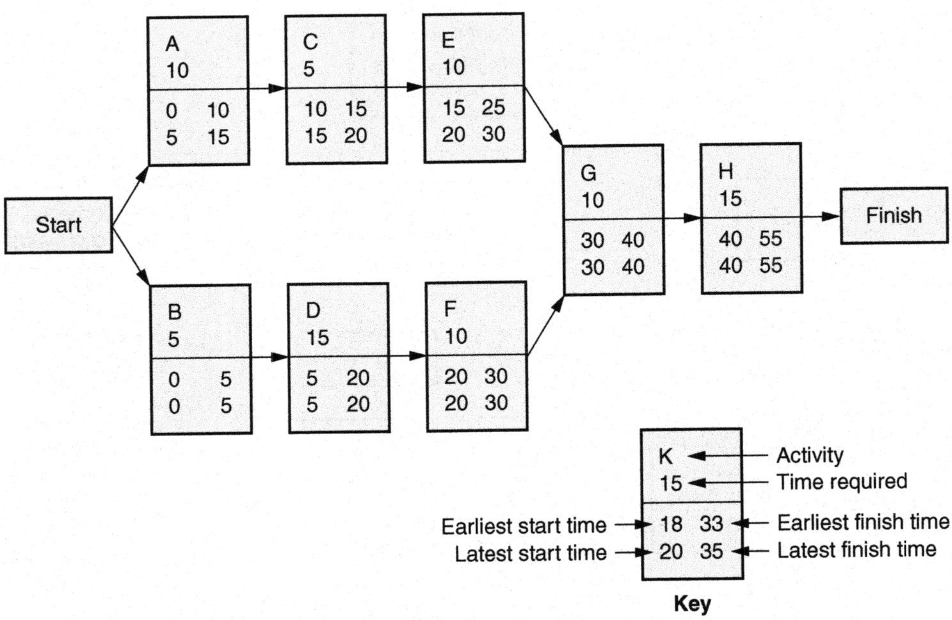

Figure 7.15 PERT/critical path chart example.

accuracy."[5] Whenever we try to inspect large quantities, any number of things can happen to distract us from the task at hand or cause problems in correctly identifying the items to be addressed. One solution to this is to take samples (preferably randomly) from the population in question to get an idea of what the population contains.

Usually, the quality organization will assist in developing a sampling plan for inspecting products. These plans take into account the various factors, for example, line speed, technology available, number of personnel available, customer expectations, and so on, to establish a sampling plan to ensure that the process, product, and customer requirements are met.

Randomness in sampling is when every part that could be measured has an equal chance or probability of actually being selected by the operator. Instead of saying "we only check the first parts after setup," we might say that we will check the first piece and then every *x* (specified number) part after that. Or we could say that once each hour we will check a certain number of parts. Then the operator must pick a random time each hour that fits into the production cycle to check the parts (versus doing it only at 15 minutes past each hour).

A check sheet to demonstrate that the checks were actually done and to simplify the collecting of the data will usually accompany sampling plans. An operator using a check sheet can quickly see if they are following the sample plan and when the next part(s) should be evaluated.

Sampling is a useful tool, and the sampling plan gives us a guide as to how and when we are to do the sampling. The sampling plan should be designed ahead of actual use and be available for inspection itself at any time.

SUPPLIERS-INPUTS-PROCESS-OUTPUTS-CUSTOMERS (SIPOC) DIAGRAM

To develop a suppliers–inputs–process–outputs–customers (SIPOC) diagram, start by defining the process and its boundaries (center of the diagram shown in Figure 7.16). Next, identify the outputs of the process, including data, services, products, information, records, and so on. For each identified output, identify all of the associated inputs. Then, move on to the internal and external customers—those that receive the identified outputs. Finally, move back to the supplier column to identify the internal and external suppliers for each identified input. Although it may seem odd to bounce back and forth from side to side on the chart, this is done to help stimulate thinking. For example, new outputs are often identified when discussing inputs or customers.

External suppliers to a process are those outside the enterprise that provide process inputs, including materials, purchased parts, contracted services, electrical power, and so on. *Internal suppliers* to a process are departments or processes inside the enterprise that provide process inputs. Similarly, a process's *external customers* are those outside the enterprise who receive process outputs, while *internal customers* are those inside the enterprise who receive process outputs.

Suppliers of either type are responsible for meeting the requirements of their customers. Customers of either type are responsible for communicating their requirements to their suppliers.

Suppliers	Inputs	Process		Outputs	Customers	
Providers of the required inputs/resources to ensure that the process executes as planned.	Resources required by the process to obtain the intended output.	Top-level description of activity.		Deliverables from the process. Note: deliverables can be hardware, software, systems, services, data, information, and so on.	Any organization that receives an output or deliverable from the process. Note: can also capture systems/databases that receive outputs, information, data, and so on.	
		Requirements			**Requirements**	
Development team	S/W size estimating guide	S/W size estimation methods/formulas	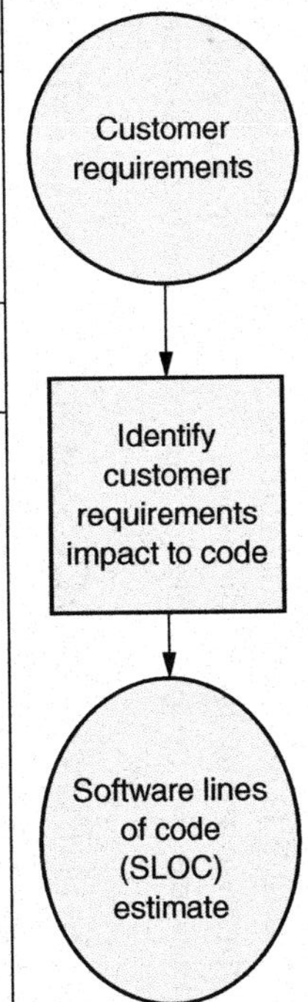			
External customer/ program manager	System specifications • Prime item development specification • System requirements doc • And so on	Total count of requirements allocated to S/W Preferred soft copy with requirements identified ("shall")		• New SLOC • Modified SLOC • Reused SLOC • Auto-generated SLOC	SLOC formatted for entry into price estimating software and organizational metrics collection system	Project/pursuit software lead
S/W development leads of past and current projects	Legacy systems knowledge	Legacy SLOC data from project assessment library and organizational metrics		Basis of estimate (BOE) for quote	Rational for SLOC estimates Information for fact finding	Proposal manager
Organization subject matter experts	Identification of most applicable/similar legacy S/W	Determine scope of similarities (number of requirements new, modified, reused, or deleted)		Legacy code product information	Reused S/W development information • Documentation • Version • Qualification test/results • Standards (498, DO178B, and so on)	Proposal manager

Figure 7.16 SIPOC diagram with requirements capture.

TREE DIAGRAM

Tree diagrams help to break a general topic into a number of activities that contribute to it. This is accomplished through a series of steps, each one digging deeper into detail than the previous one. A note listing the general topic is posted at the top of a flip-chart or whiteboard. Have the team suggest two to five slightly more specific topics that contribute to the general topic and write these on individual notes and post them in a horizontal row beneath the original general topic. For each of these new topics, have the team suggest two to five even more specific topics and post these on the next level down. Continue each branch of the tree as far as seems practical. Draw appropriate connecting lines. Review the tree by making sure that each item actually contributes to the item above it. The resulting diagram should provide specific activities that, when they occur, contribute to the general topic. An example is shown in Figure 7.17.

WRITTEN SURVEY

Written surveys can be sent to a randomly selected group of customers or potential customers, but getting responses from all those selected almost never occurs. In addition, the accuracy of the responses is questionable. A carefully worded and analyzed survey can, however, shed significant light on customer reactions.

TOOL REVIEW

The *define* phase focuses on determining the scope of the improvement project and the resources and schedule needed to execute the project. There are many tools to aid in the definition and management of the project, including the charter

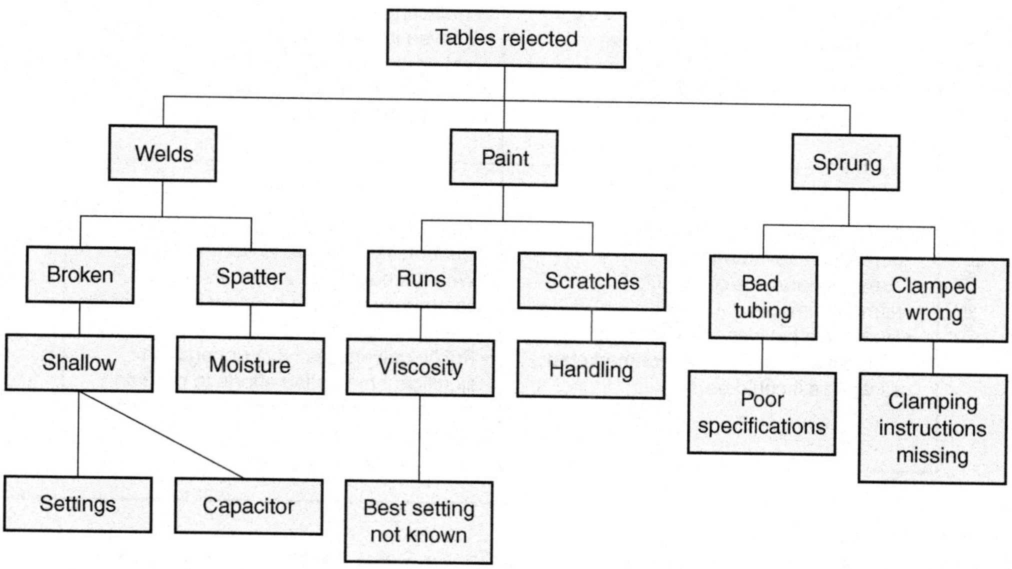

Figure 7.17 This example of a tree diagram is a fault tree (used to study defects and failures).

and Gantt chart. The tools and level of detail should be selected based on the size of the project and organization requirements. Additionally, the tools used to scope the problem and define the current state vary in detail, focus, and complexity. During the initial project planning, an initial determination of the tools to be used for the problem/current state definition should be identified, allowing for adjustment as the project matures. The primary goal of the define phase is to ensure that the problem and project are defined to provide focus for the remaining phases—measure, analyze, improve, and control.

SWOT

A SWOT (strengths, weaknesses, opportunities, and threats) is an effective strategic planning tool applicable to a business or project objective. Strengths and weaknesses are identified with respect to the internal capabilities of an organization. They provide an introspective view. Strengths and weaknesses should not be considered opposites. This can be readily seen in Figure 7.18. On the other hand, opportunities and threats look outside the organization. Quite simply, we are trying to identify opportunities for the organization and threats to the organization. When analyzing a SWOT, the question to ask is, "How can the organization leverage its strengths or improve its weaknesses to take advantage of the opportunities while mitigating or eliminating the threats?"

Internal	**Strengths** • High academic entrance requirements • Strong general education program • High percentage of full-time faculty • New endowments for faculty chairs • Fundraising goals are being met each year • Well-designed internship and study abroad opportunities	**Weaknesses** • Print communications for prospective students are bland and generic • Recruitment staff is too passive in contacting prospective students • Uneven implementation of assessment and program review methods • Student-centered customer focus is not recognized throughout the institution
External	**Opportunities** • Metropolitan area location has population advantages • Area corporations offer tuition reimbursements • Reputable business program with a global studies focus is not marketed as well as it could be	**Threats** • The number of tradional-age prospective students is expected to decline • With the cost of education escalating, community colleges have recently become viable competitors • Public universities have undertaken significant marketing efforts to position themselves as comparable in experience to private colleges

Figure 7.18 SWOT analysis.
Source: Donald W. Benbow and T.M. Kubiak, *The Certified Six Sigma Black Belt Handbook*, 2nd ed. (Milwaukee: ASQ Quality Press, 2007).

Chapter 8

Business Results for Projects

Business results can be shown in many forms and at many levels of detail. The measures for project results should be identified during the initial planning stages and be refined as the project progresses. Results, also known as *performance measures*, are usually related to the business, project, or process. Business performance measures are usually expressed through tools known as:

- Balanced scorecard
- Performance to established goals

To provide an approach to measuring multiple aspects of a business, not just the financial aspect, Dr. Robert Kaplan and Dr. David Norton developed the *balanced scorecard* in the early 1990s. Their research indicated that the financial aspect of measurement is restricted to past events and is only one part of the business. The balanced scorecard requires the organization to look at its performance in five primary areas: financial, customer, internal processes, and learning and growth. This approach helps the organization align its vision and goals with the objective of ensuring that no single area far outweighs the others—thus providing a balanced method of results measurement. Kaplan has written several books on activity-based costs and activity-based management, both central to the profitability emphasis of Six Sigma. These topics also provide other approaches for establishing business results for projects.

Performance to established goals is as simple as it sounds. The organization establishes goals and periodically measures its performance against them. The goals usually reflect desired results such as:

- Increase revenue (for example, sales) 10 percent over last year
- Improve net profit by eight percent over last quarter
- Ensure that all employees receive 40 hours of job-related training

Project performance measures usually include:

- *Cost performance index (CPI).* Measures the project's performance in dollar terms (for example, the ratio of value earned [budgeted cost of work performed] versus the actual cost of work performed). A ratio of 1 or higher is the desirable condition.
- *Schedule performance index (SPI).* Measure of the project's efficiency to schedule as expressed in the ratio of earned value to planned value.

- Other measures based on project or organizational requirements such as:
  - Defects per single line of code (SLOC) for software projects
  - Customer complaints or corrective action requests
  - Inquiry response time
  - Defect containment

PROCESS PERFORMANCE

> Calculate process performance metrics such as defects per unit (DPU), rolled throughput yield (RTY), cost of poor quality (COPQ), defects per million opportunities (DPMO), sigma levels, and process capability indices. Track process performance measures to drive project decisions. (Analyze)
>
> **Body of Knowledge II.E.1**

Note: Some industries have a legal obligation to make references to products or services that do not totally meet the specifications as nonconforming, nonconformities, or deficiencies! The word "defect" is not allowed to be used related to products or services. Check with your management team to see if this applies to your organization.

Process performance is usually a measure of how the process is executing against some established goals or statistical measure.

Process performance measures usually include:

- *Defects (deficiencies) per unit (DPU).* Calculated as the total number of defects divided by the total number of products produced in some time period (for example, per day).
- *Defects (deficiencies) per million opportunities (DPMO).*[1] To calculate the number of opportunities, it is necessary to find the number of ways each defect can occur on each item. In a hypothetical product, blurred printing occurs in only one way (the pencil slips in the fixture), so in the batch there are 40,000 opportunities for this defect to occur. There are three independent places where dimensions are checked, so in the batch there are 3 × 40,000 = 120,000 opportunities for dimensional defects. Rolled ends can occur at the top and/or the bottom of the pencil, so there are 40,000 × 2 = 80,000 opportunities for this defect to occur. The total number of opportunities for defects is 40,000 + 120,000 + 80,000 = 240,000. Let us consider that this product has an average 165 defects/unit:

$$DPO = DPU \div (\text{Total number of opportunities})$$

$$DPMO = DPO \times 10^6 = (165 \div 240{,}000) \times 10^6 = 687.5$$

- *Rolled throughput yield (RTY).*[2] RTY applies to the yield from a series of processes and is found by multiplying the individual process yields (Throughput yield = e^{-DPU}). If a product goes through four processes whose yields are .994, .987, .951, and .990, then RTY = .994 × .987 × .951 × .990 ≈ .924

- *Sigma levels.*[3] Suppose the tolerance limits on a dimension are 5.000 ± 0.012, that is, 4.988 to 5.012. Data collected from the process during second shift indicates that the process mean is 5.000 and its standard deviation $\sigma = 0.004$. Note that $\pm 3\sigma$ fits inside the tolerance because $\pm 3\sigma = \pm 3 \times 0.004 = \pm 0.012$. A capability calculation would show $C_p = C_{pk} = 1$. The traditional way to calculate yield in this situation is to use a standard normal table to determine the area under the normal curve between $\pm 3\sigma$. This gives a yield of about 0.9973. Experience indicates, however, that the process mean doesn't remain constant. There is general agreement on the somewhat arbitrary rule that the process mean may shift 1.5σ to the right or 1.5σ to the left. If we assume a 1.5σ shift to the right, the yield is the area under the normal curve to the right of -1.5σ or about 0.9332.

  Suppose, now, that process variation is reduced so that $\sigma = 0.002$. There is now $\pm 6\sigma$ between the tolerance limits, and the process can be called a 6σ process. To calculate the yield for a six sigma process, we allow the mean to shift $\pm 1.5\sigma$. Suppose the mean shifts 1.5σ to the right so the yield is the area under a normal curve to the right of -4.5σ. This turns out to be 0.9999966. The defect level is 1 – 0.9999966, which is 0.0000034 or 3.4 ppm, the oft-quoted defect level for six sigma processes. At best this is a rather theoretical number, because the mean may not shift exactly 1.5σ on each side, and no process is truly normal to the sixth decimal place.

- *Process capability indices.* There are various process capability indices, the most common being C_p, C_{pk}, and C_r.

  - C_p is the ratio of tolerance to six sigma, or the upper specification limit (USL) minus the lower specification limit (LSL) divided by six sigma.

  - C_{pk} is the lesser of the USL minus the mean divided by three sigma (or the mean) minus the LSL divided by three sigma. The greater the C_{pk} value, the better.

  - Capability ratio (C_r) is the ratio of 1 divided by C_p. The lower the value of C_r, the better, with 1 being the historical maximum.

Cost of Poor Quality (COPQ)

COPQ was originally called Cost of Quality (COQ) (sometimes today called Cost of Current Quality [COCQ]) and dates back to the original writing of Juran and

Feigenbaum. As a process performance measure, the concept usually follows four basic measures that your accounting department should assist with. These areas are:

- Prevention—costs associated with doing things right the first time (e.g., training courses, APQP efforts, etc.)
- Appraisal—costs associated with checking the current process (e.g., inspections, verifications, etc.)
- Internal failures—the cost of work incorrectly performed and caught before shipment to customers (e.g., rework, internal scrap, etc.)
- External failures—the costs of work incorrectly performed and caught by the customer (e.g., returned goods, external scrap, etc.)

One challenge for new Green Belts is the lack of an exact science around COPQ. There is no one exact way of calculating the costs, and every company seems to have its own method or categories that fit under the four basic themes. There are many references (search ASQ Quality Progress) and books available for more detail that can be reviewed, such as *Principles of Quality Costs, Fourth Edition* by Wood or *The Quality Toolbox, Second Edition* by Tague.

The key point to remember about quality costs is that, by focusing on the prevention of error (prevention costs), the overall total cost of quality will decrease! Most companies that start measuring their quality costs will find high values in the two failure areas. By focusing on prevention, the failures should be decreased.

COMMUNICATION

> Define and describe communication techniques used in organizations: top-down, bottom-up, and horizontal. (Apply)
>
> **Body of Knowledge II.E.2**

A message is sent to convey *information*; information is meant to change *behavior*. Hence, the effectiveness of a communication is vital if we expect desired action to happen. Communication is a key skill that is required for any individual, be they an operator on the assembly line or the CEO of an organization. The skill-level expectation increases as the individual moves up to higher positions and communicates information that can literally change the lives of employees, the future of the organization, and the consequences to society. For situations that can make an impact of this magnitude, it is important to "plan to communicate." In a project setting with a matrix organization structure, the importance of communication cannot be stressed enough. Go/no-go decisions for the project and support from stakeholders require effective communication.

There are several types of communication flow that are used in different situations:

- *Top-down flow or downward flow.* Used when top management or the executive sponsor of the project is providing instructions, communicating policies, or providing feedback on project performance. If this communication is passed from executive sponsor to champion to process owner to Black Belt to team members, there is a possibility of losing some information, as the message may get filtered or distorted. It is better for the top manager to send a written communication to the team and/or convene a meeting to communicate to the entire chain of command. Management may leave sensitive information to be communicated through the chain of command. This will help make the middle-level managers feel they are not losing their importance.
- *Bottom-up flow or upward flow.* When a line operator or front desk personnel want to provide feedback to the management, or a team member wants to communicate to the executive sponsor of the project, a bottom-up flow of communication occurs. Similarly to top-down, the bottom-up method may also see information distorted as it reaches the higher level. Top management should encourage an open-door policy, survey employees, and set up suggestion boxes and stand-up meetings. Luncheon meetings are good for providing bottom-up communication opportunities. All these ideas have to be planned and executed correctly or they will become a mockery of the system. Examples of poor implementation include suggestion boxes turning into complaint boxes, and the open-door policy being misused to disrupt work or point fingers.
- *Horizontal communication.* This is more effectively used in "flatter" organizations. This is very effective and quicker to get results. However, where there is a process to be followed for authorization and approval of management, horizontal communication is not suitable as it may tend to shortcut the process. In a vertical organization, the middle and higher management personnel may feel that they have been bypassed due to horizontal communication. It is better for management to provide guidelines as to where horizontal communication is encouraged.

Communication delivery method and meeting room setting are helpful to the effectiveness of the message as well. Reprimanding must be done in private, whereas appreciation has to be given in public. Sensitive issues have to be communicated in person, as body language plays a key role between the sender and receiver of the message. E-mails are not good at communicating emotions, and hence disagreements are best handled face to face.

To be a good communicator, you must also learn to be a very good listener. The BoK does not include listening as a topic for the exam and thus it is not presented in this handbook; however, there is a lot of useful information available on listening skills that could be useful in learning to be a better communicator.

Chapter 9

Team Dynamics and Performance

TEAM BASICS

Remember the famous quote, "There is no 'I' in 'team'"? The essence of it is to imply that a team is a collective effort of individuals. To harness the best of each individual, the team members need to understand each other's strengths, roles, and responsibilities, and the scope of the task. There are several books that go into detail about how to form a team, organize meetings, manage projects, and accomplish the desired goals. In the context of Six Sigma, we will cover areas important to a Green Belt. Protocols such as setting the team agenda, recording the minutes of the meeting with actions, sticking to meeting time, and enforcing meeting attendance need to be followed for an effective team meeting. An initial meeting to kick off the team, with introductions and high-level discussion on the goal, objective, milestones, and so on, will provide an opportunity for the team to get to know each other and understand the expectations. A team agenda can be flexible, but you need to have one.

Some teams have their team goals, objective, and scope/boundaries visibly displayed in every meeting to keep the members on track. Management presence during kickoff and with regular frequency during the project helps enforce the importance of the team objective.

TEAM FORMATION

A team usually comprises five to nine members (seven is considered an ideal size) with complementary skills to achieve the goals and objectives of the team. Team composition should be driven by the size and scope of the project; it is possible to have a team of one or two for a smaller project and a large team with subteams for a big project. The team includes subject matter experts and stakeholders. Subject matter experts sometimes remain outside the team as resource or extended team members. Stakeholders are always part of the team. The team will not be able to implement their ideas and solutions without having stakeholders or their representation on the team. Teams smaller than five reduce the opportunity for interaction problems and are easier to manage, whereas teams greater than nine produce a lot of interaction that can be counterproductive to a team's progress. Teams with greater diversity tend to produce better interaction between team members. Some teams also bring in individuals who are neither subject matter experts nor stakeholders but are outsiders to the team. The outsider helps the team ask questions that were never explored by the team members closer to the process. This needs to be moderated, as the outsider might ask too many questions and frustrate the core members. Typically, Six Sigma teams are cross-functional to address the issues from every angle.

VIRTUAL TEAMS

This is an interesting innovation that has evolved in the last several decades due to the development of technology in communication tools and the Internet, which have led to the ability to meet and share data virtually. Virtual teams enable people from all over the globe to meet via teleconferences, videoconferences, and Internet tools such as shared computers. There are many benefits to virtual teaming, the most prevalent being reduced costs and real-time data sharing and updating. However, virtual teams also face challenges that include slowing of the progression of normal team-building, inability to get true commitment and buy-in, and the potential for miscommunication—especially with voice-only teleconferencing, as the important factor of nonverbal communication is lost. Virtual teaming has its place in every organization and can be very effective, especially if team members are familiar with each other.

TEAM STAGES AND DYNAMICS

> Define and describe the stages of team evolution, including forming, storming, norming, performing, adjourning, and recognition. Identify and help resolve negative dynamics such as overbearing, dominant, or reluctant participants, the unquestioned acceptance of opinions as facts, groupthink, feuding, floundering, the rush to accomplishment, attribution, discounts, digressions, and tangents. (Understand)
>
> **Body of Knowledge II.F.1**

It is important to understand team dynamics and performance. There are many projects that have failed miserably because of lack of teamwork and not understanding the roles and responsibilities of the team members. It is important to note that the team members were technically competent and had complementary skill sets to succeed in those projects.

According to B. W. Tuckman's "Developmental Sequence in Small Groups," teams typically go through the stages of *forming, storming, norming,* and *performing.* Let us explore each stage and identify the appropriate management approach required for that stage.[1] We will also discuss two additional stages.

Stage 1: Forming

1. Team members are getting to know each other.
2. Group is immature.
3. There is a sense of belonging to the group.
4. Members take pride in membership with the group.
5. Members are trying to please each other.
6. Members may tend to agree too much on initial discussion topics.
7. Not much work is accomplished.
8. Members' orientation on the team goals.
9. Members understand the roles and responsibilities.
10. Group is going through the "honeymoon" period.

Stage 2: Storming

1. Team members voice their ideas.
2. Understanding of the scope and members' roles and responsibilities will be put to the test.
3. Ideas and understanding start to conflict.
4. Disagreements start to slow down the team.
5. Not much work is accomplished.
6. This is a necessary evil that every team member has to go through to position themselves on the team.
7. Take care to be aware of negative interactions between team members as too much disagreement can completely stall the team progress.

Stage 3: Norming

1. Team members resolve their conflicts.
2. Team members agree on mutually acceptable ideas to move forward.
3. Some amount of work gets accomplished.
4. The group is starting to function as a team.
5. Team members start to trust each other and share their ideas and work products without hesitation.

Stage 4: Performing

1. Team is effective, skills complementary, synergy is created.
2. Team members realize interdependence.
3. The group develops ability to solve problem as a team.
4. Large amount of work gets accomplished.

Stage 5: Transitioning/Adjourning

1. Team is disbanded.
2. Team members go on with other activities of their work.
3. If the project is continued with additional scope, some team members may be changed.
4. Team dynamic changes and tends to go back to one of the earlier stages.
5. Major changes can result in going back to forming stage.

Stage 6: Recognition

Recognition is the often forgotten piece of team dynamics, or rather, often taken for granted. Even though team members are salaried or compensated monetarily for their time and skill, it does not mean that the team is already recognized. Teams can be recognized in many ways, from a simple pat on the back by senior management, to thank-you notes, bulletin boards, organization-wide e-mails, newsletters, all-employee meetings, certificates of accomplishment, bonuses, stock options, and many other ways.

This is the typical evolution of team stages. Depending on organizational cultural issues, some stages may shorten or lengthen, but the team still goes through them. It is healthy for the team to go through these stages as they set ground rules and expectations for themselves. These stages also depend on team maturity, complexity of the task (project), and team leadership.

Team Leadership

The team leadership may vary depending on the maturity of the team and the stage the team is at based on the leader's perception. Examples of leadership activities during these stages include:

Forming. Appropriate leadership style during this stage is *directing*:

- Leader provides close supervision, exhibits directive behavior.
- Leader instructs the team as to what to do when, where, and how.
- Leader also listens to team's feedback.
- Leader encourages and welcomes the team.

- Leader explains the roles, responsibilities, and goals of team members.
- Leader identifies opportunities for developing skills to meet team goals.

Storming. Appropriate leadership style is *coaching*:

- Leader still continues close supervision, exhibits directive behavior.
- Leader also starts some supportive behavior.
- Leader increases the listening level to solicit the team's feedback.

As discussed earlier, to keep the storming at an acceptable level (not detrimental to the task at hand) the leader may use conflict resolution approaches.

Norming. Appropriate leadership style is *supporting*:

- Leader reduces the level of directive behavior and increases the supportive behavior.
- Leader encourages the team on decision-taking responsibilities.
- Leader helps move to a performing stage before the team reverts to earlier stages.
- Leader emphasizes ground rules, scope, roles and responsibilities.

Performing. Appropriate leadership style is *delegating*:

- Since the team is mature, the leader reduces the level of being directive and supportive in day-to-day functions.
- Team leader still monitors the goals and performance of the team.
- Leader watches for any change in dynamics due to major changes.

Negative Team Dynamics

Several negative team dynamics are pretty much reflective of the organizational culture rather than personalities of individuals. If something is "acceptable" within the organization as a norm, that becomes the way of running the business. In other words, the organizational culture becomes the "enabler" of the many team problems that organizations face.

Negative dynamics in the team can:

- Have a negative impact on team member motivation
- Hurt a team member's ego and self-esteem
- Intimidate team members
- Reduce the self-confidence of others
- Increase stress and exhaust patience
- Increase feelings of insecurity
- Foster a lack of morale

As a result, unchecked or unaddressed negative team dynamics may cause:

- Goals and objectives of the project/task to not be met
- Targets to be frequently revised to team's advantage
- The project to be canceled
- The team to miss project milestones and deadlines
- Project resources to not be effectively utilized
- The project to overrun its cost targets

Table 9.1 outlines common negative team dynamics and possible countermeasures. There are more facilitation tactics discussed in *The Certified Six Sigma Black Belt Handbook* and the *Team Handbook*.

Table 9.1 Common negative team dynamics and potential countermeasures.

Negative dynamic	Symptoms	Probable causes	Potential countermeasures
Overbearing member(s)	Team interaction is limited to a few individuals. The rest of the team is always in listening mode rather than participating in the discussion.	Team is composed of a few influential members (senior management staff, founders, inventors), members with legitimate authority (investor, major shareholder, owner), subject matter experts, and so on. This may intimidate other team members, who hesitate to voice their opinions.	With the support of the influential team member, the team leader reinforces round-robin voicing of opinions, using methods like nominal group technique, conducting the meeting in a more informal setting, keeping experts and influential members as an extended team, and so on.
Dominant member(s)	Meeting discussion getting chaotic and difficult to listen to or understand. Only a few members dominating the entire discussion.	Dominant team members keep interrupting the conversation of other team members.	Structure the agenda to provide equal participation for all team members. Effective moderation by team leader that allows other team members to finish their thoughts. Team leader initiates round-robin to provide opportunity for every team member.

Continued

Table 9.1 Common negative team dynamics and potential countermeasures. *(Continued)*

Negative dynamic	Symptoms	Probable causes	Potential countermeasures
Floundering	Team is currently proceeding or performing in an unsteady, faltering manner.	Lack of team direction. Some teams have high-profile team leaders from the organization but they hardly ever attend meetings or team discussions. There are situations where the organizations are going through major changes and no one is clear about the future of the team. Team members are overwhelmed. This can be due to multiple reasons. Organization going through a major change: leadership, downsizing, new mergers and acquisitions, offshore transfers, and so on. Postponing of team decisions. This is related to lack of direction from the team leadership. If there is no clear direction, decision-making gets difficult.	During early stages of the team more direction is required. Team leadership should be visibly present during the team meetings and decisions. Team leadership should keep the team focused by not getting distracted by events happening within the organization. Team leaders should address the concerns of the team members but not allow the team agenda to be hijacked by other events. Reinforce management support and commitment when team starts to challenge the purpose of the team.
Reluctant participants	Lack of participation, noncommittal feedback. Basically showing disinterest.	Team member may not have any stake in the team's outcome. Intimidated by other team members or leaders. In the process of moving out of the current job function or organization. Fear of losing job or position by voicing opinions.	Team leaders support the team members' active participation and protect the team members voicing their opinions.

Continued

Table 9.1 Common negative team dynamics and potential countermeasures. *(Continued)*

Negative dynamic	Symptoms	Probable causes	Potential countermeasures
Unquestioned acceptance of opinions as facts	Members present information without backing up data or analysis. Members present unfounded assumptions, and so on.	Mainly organization cultural reasons. Lack of management by facts.	Team leader requests supporting data, analysis, and conclusions that are statistically valid. Question the assumptions behind the analysis.
Groupthink	No public disagreements. Doubts expressed in private discussions. There are several other classical symptoms identified by researchers.	Members fear group cohesiveness will be at stake if there are any disagreements. Putting group harmony as paramount.	Bring independent members from outside to participate. Rotate roles and responsibilities of members at milestones. Management by fact.
Feuding	Hostilities resulting in heated arguments, slowed progress, low morale of the team.	Conflict resolution not effectively handled by the team leadership. Lack of mutual respect between team members. Team operating ground rules not enforced.	Confront the adversaries offline and not in the team meeting. Confronting in public can worsen the situation. Enforce discipline and emphasize mutual respect among team members. Restate the objective of the team as main focus.
Rush to accomplishment	Incomplete data collection. Inconsistent analysis. Trying to get conclusion faster.	Team under unrealistic deadline. Untrained team members. Looking for short-term gains.	Team leadership asks for data collection, analysis, and statistical significance. Ask for alternate solutions. Revise the deadline to a more realistic one based on resources.

Continued

Table 9.1 Common negative team dynamics and potential countermeasures. *(Continued)*

Negative dynamic	Symptoms	Probable causes	Potential countermeasures
Attribution	Members make casual references. Members don't seek explanations, preferring psychological and emotional judgments.	Similar to "rush to accomplishment" causes.	Team leaders challenge the assumptions made by team members. Use devil's advocate approach. Ask for analysis behind the conclusions drawn.
Discounts	Members' opinions are ignored. Members do not seem to listen to each other. Sarcasm, low team morale.		Encourage mutual respect. Enforce discipline. Ask for clarification from the members providing opinions.
Digressions and tangents	Discussion straying out of the scope/ agenda of the meetings. Distractions. Meeting time not properly utilized. Not much achieved from the meetings.	Organization going through major change. Cultural issues. Lack of focus from leadership.	Enforce compliance to agenda items and time allotment. Restate meeting ground rules. Redirect the discussions.

TEAM ROLES AND RESPONSIBILITIES

> Use tools, such as RACI, to describe and define the roles and responsibilities of participants on six sigma and other teams, including black belt, master black belt, green belt, champion, executive, coach, facilitator, team member, sponsor, and process owner. (Apply)
>
> **Body of Knowledge II.F.2**

RACI

A RACI Chart (Responsible, Accountable, Consulted, Informed) is meant to be a depiction of who on the team has what responsibilities as the project progresses through the various DMAIC tasks/steps. The basic idea is to list all of the team members on the vertical axis and all of the tasks of the process on the horizontal axis. The team can decide how many or what steps should be listed on the chart and can include Toll Gates if desired. Typically, the starting point for the tasks is the Statement of Work (SOW) and can be expanded as desired to show more steps in the process. A common example is listed in Table 9.2.

A sister chart to the RACI is the RASIC (Responsible, Accountable, Supporting, Consulted, Informed) chart. It is constructed the same way as the RACI with one additional factor for those team members who will provide support for the project.

Six Sigma successes are not just about application of statistical tools. A strong Six Sigma organization is necessary for sustainable success. Without this organization, there will be no accountability to the investment made in employees in terms of training, resources spent, and consistent approach of methodologies. Smaller organizations may combine some roles; however, the responsibilities should be maintained. Six Sigma organizational structures may range from the typical large Six Sigma organization shown in Figure 9.1 to a typical small Six Sigma organization as depicted in Figure 9.2, or anything in between.

One factor that has helped Six Sigma be successful is the structure it demands of organizations. Table 9.3 shows typical Six Sigma roles, the organizational members that typically fill the roles, their expected training or background, and the primary responsibilities for each role.

Table 9.2 Example of a RACI matrix

Task / Player	Set policy	Identify projects	Select projects	Work project	Achieve result	Maintain gain	Coach and mentor
Executive	A		A		I		
Sponsor		A	R				
Process owner		C	I			A	
MBB	I	C		C	R		A
BB	I		I	R	R		A
GB			I		A		
Council	R	C	I		C		

Source: Donald W. Benbow and T.M. Kubiak, *The Certified Six Sigma Black Belt Handbook*, 2nd ed. (Milwaukee: ASQ Quality Press, 2007).

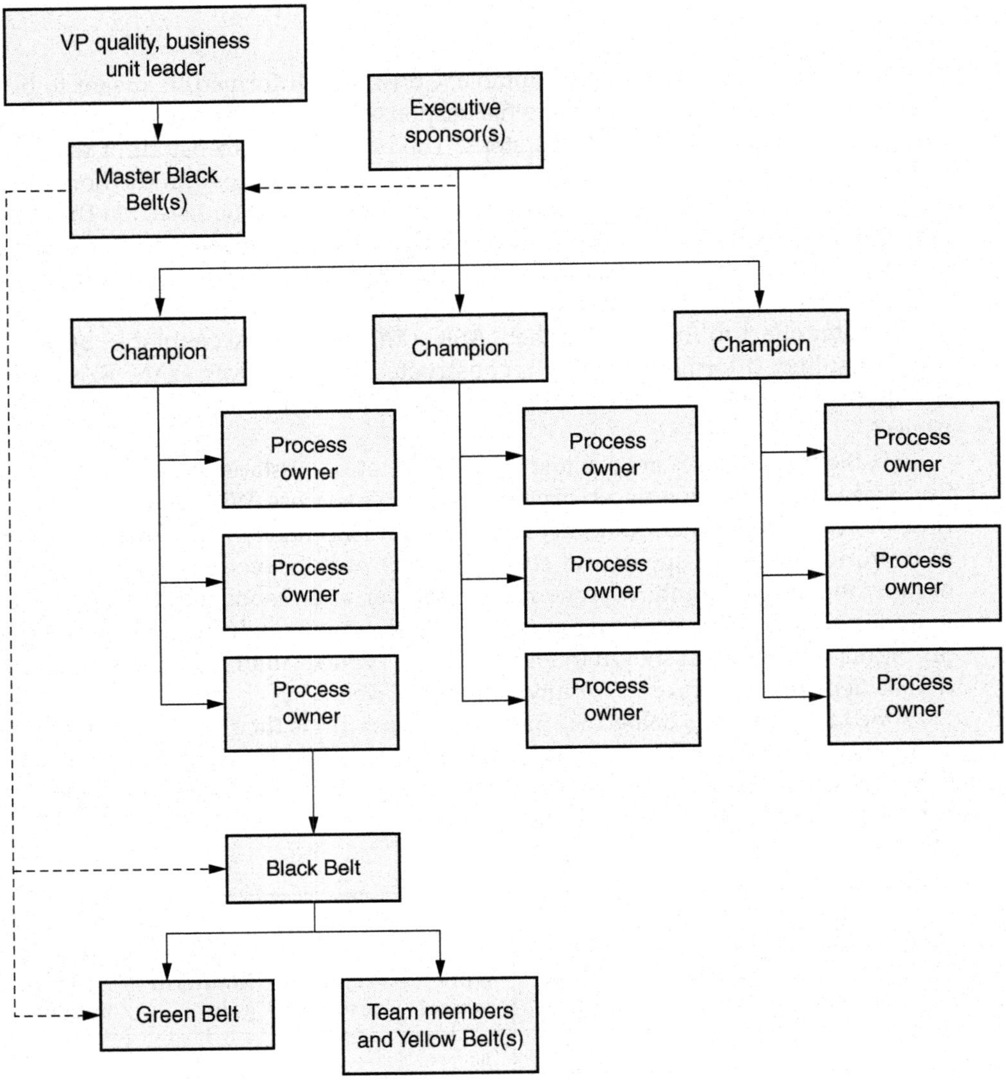

Figure 9.1 Typical large Six Sigma organization.

TEAM TOOLS AND DECISION-MAKING CONCEPTS

> Define and apply team tools such as brainstorming, and decision-making concepts such as nominal group technique, and multi-voting. (Apply)
>
> **Body of Knowledge II.F.3**

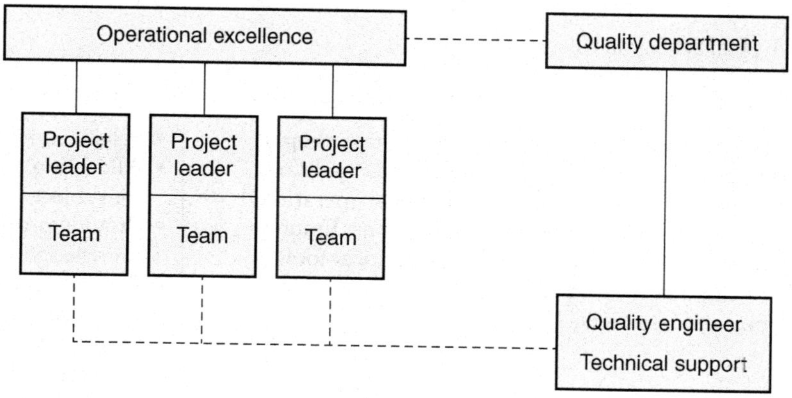

Figure 9.2 Typical small Six Sigma organization.

Table 9.3 Typical Six Sigma roles.

Role	Candidate	Training/background	Primary responsibilities
Executive sponsor	Business unit leader responsible for profit and loss (usually at director level or above).	Six Sigma concepts, strategies, overview, operational definitions.	• Set direction and priorities for the Six Sigma organization • Allocation of resources for projects • Set Six Sigma vision, overall objectives for the program • Monitor the progress of the overall program • Initiate incentive programs • Reward successful projects
Champion	Typically upper-level managers.	Six Sigma concepts, strategies, tools and methods, operational definitions. Emphasis on management tools.	• Liaison with senior management • Allocation of resources for projects • Determine project selection criteria • Remove barriers hindering the success of the project • Approve completed projects • Implement change

Continued

Table 9.3 Typical Six Sigma roles. *(Continued)*

Role	Candidate	Training/background	Primary responsibilities
Process owner	An individual responsible and accountable for the execution and results of the process. The sponsor or champion could also be a process owner.	Six Sigma concepts, strategies, tools and methods, operational definitions. Emphasis on statistical tools.	• Select team members • Allocation of resources for projects • Provide process knowledge • Review process changes • Approve changes/ support change management • Implement change • Ensure that improvements are sustained
Master Black Belt	Individuals trained in Six Sigma methodologies, statistical tools, basic financial tools, change management, risk assessment, project management, executive communication, and well experienced in teaching, coaching, and mentoring Black Belts and Green Belts. This is always a full-time position.	Six Sigma Body of Knowledge, lean enterprise synergy, finance for nonfinancial managers, risk assessment, project management, change agent skills, Master Black Belt train the trainer, presentation skills, communication skills, leadership skills, facilitation skills.	• Coach Six Sigma Black Belts and Green Belts • Utilize the resources provided by management effectively • Formulate overall business strategy linking to Six Sigma program • Monitor project progress closely • Typically between 15–20 projects overseen at a time • Provide coaching, mentoring for new Black Belts and Green Belts • Work with champions and process owners for selection of projects • Address issues of project stagnation • Remove barriers hindering the success of the project • Support as a subject matter expert for the organization • Review and approve completed projects • Share lessons learned with the extended team • Provide inputs to rewards committee

Continued

Table 9.3 Typical Six Sigma roles. *(Continued)*

Role	Candidate	Training/background	Primary responsibilities
Black Belt	Individuals trained in Six Sigma methodologies, statistical tools, basic financial tools, change management, risk assessment, project management, and well experienced in managing Black Belt projects. This is always a full-time position.	Six Sigma Black Belt Body of Knowledge, lean enterprise synergy, finance for nonfinancial managers, risk assessment, project management, change agent skills, presentation skills, communication skills, leadership and facilitation skills. Certified as Six Sigma Black Belt.	• Lead and manage Six Sigma projects • Utilize the resources provided by management effectively • Provide net present value, return on investment (ROI), payback calculations on projects • Work full-time on four to six projects per year • Monitor project progress closely • Follow DMAIC process, apply appropriate statistical methods • Work with champions, Master Black Belt, and process owners for selection of projects • Address issues of project stagnation/ consult Master Black Belt • Remove barriers hindering the success of the project • Update and present project progress to management • Review completed projects • Share lessons learned with the extended team

Continued

As we discussed earlier, teams go through several stages before performing, and may face obstacles due to human interactions. Hence, soliciting ideas from all team members by providing an equal opportunity and arriving at sound conclusions requires the use of a systematic and proven approach.

Team tools are useful for guiding team interaction in a systematic way. There are times when the topic of discussion is sensitive or controversial. There are times when a problem has not been explored enough, and the team leader is looking for as many ideas as possible. There also are times when team members have multiple

Table 9.3 Typical Six Sigma roles. *(Continued)*

Role	Candidate	Training/background	Primary responsibilities
Green Belt	Individuals trained in Six Sigma methodologies, basic statistical tools, and process improvement techniques. This is typically a full-time position. However, some organizations make this part of an existing job responsibility.	Six Sigma Green Belt Body of Knowledge, lean enterprise synergy, presentation skills, communication skills. Certified as Six Sigma Green Belt.	• Support Six Sigma projects with higher ROI • Lead smaller projects with moderate savings and ROI • Follow DMAIC process, apply appropriate statistical methods • Review the approach periodically with the experienced Black Belt and Master Black Belt • Provide inputs to Master Black Belt and Black Belt and process owners during selection of projects • Identify issues of project stagnation/ consult Black Belt, Master Black Belt • Identify and report barriers hindering the success of the project • Share lessons learned with the extended team
Project team member	Selected by process owner and trained in Six Sigma methodologies, quality, basic statistical tools, and process improvement techniques.	Six Sigma methodologies, quality tools, process improvement, teamwork.	• Support and contribute to Six Sigma projects • Participate in charter and scope definition • Provide inputs during project meeting, brainstorm ideas • Help collect data where responsible • Follow DMAIC process, apply appropriate tools • Review the approach periodically with the Green Belt and experienced Black Belt • Provide inputs to Green Belt and Black Belt and process owners during project

Continued

Table 9.3 Typical Six Sigma roles. *(Continued)*

Role	Candidate	Training/background	Primary responsibilities
Yellow Belt	Those new to the world of Six Sigma who have a small role, interest, or need to develop foundational knowledge.	Six Sigma methodologies, quality tools, process improvement.	• Follow DMAIC process, apply appropriate tools • Support management and teams as needed • Make suggestions for future projects • Use process and tools in the workplace • Be ready to talk knowledgeably with others

ideas and want to explore everything. These scenarios are not uncommon in a typical team setting. We will be discussing in this section some of the team tools and their application to solving these issues.

Brainstorming

Brainstorming is a process where an individual or team develops as many ideas concerning a topic as they can, using various creativity techniques or methods. There are two basic phases to brainstorming: the *creative* phase, which is used to generate a large number of ideas, and the *evaluation* phase, where the ideas generated are looked at for usefulness or applicability. There should be a time break between the two phases as different parts of the brain are used in each phase. At minimum, a 10-minute stretch break should be taken versus going directly into evaluation after being creative.

During the creative phase, there should be no criticism or other distractions allowed. During this phase, the team should maintain open minds to all the possibilities no matter how wild the idea—the goal is to get as many ideas as possible. If ideas are being put on a flip-chart with a large group, you should have two or more available to capture all of the ideas as they develop. Otherwise, you could have each person say what they are thinking and have them or someone else record the idea on a sticky note and put it on the wall. Facilitation can be used during the creative phase, but freewheeling also works well. Some basic guidelines that should be followed in the creativity phase of brainstorming include:

- No criticism, compliments, or questions
- Wild ideas are welcome
- Don't wait
- Quantity is important (versus quality)
- Hitchhike—build on previous ideas

During the evaluation phase, at some point after the creativity phase, it is best to have a facilitator work with the group to look over the ideas in a sequence. There are many ways to go about evaluating the ideas generated. One good starting point is to organize the list of things into like groups or categories (that is, build an *affinity diagram*) to help in the evaluation process. The caution here is to not get overly critical, as there may be something in one of those "crazy" ideas that might actually work for the given situation. This is often true because of new technology or different ways of doing things that are not common in our organizations.

To make brainstorming most effective, prior to starting the activity review, help the team understand the importance of avoiding these idea-stopping thoughts or behaviors:

- Don't be ridiculous
- Let's shelve it for right now
- It won't work here
- Our business is different
- Let's think about it some more
- We did all right without it
- It's too radical a change
- Management won't like it
- Where did you dig up that idea?
- It's not practical
- It's too expensive
- You can't be serious
- You can't do that
- The technology will not allow that
- Where will you get . . .
- We've never done it before
- I have something better
- It's too risky
- Let's be sensible
- We'll never get it approved
- The employees won't like it
- It's good, but . . .
- Let's check on it later
- It's too much work

- Let's get back to reality
- That's been tried before
- That's not my job
- You do not know how we do things around here
- That's too high-tech for us
- It will never work

As stated earlier in this section, brainstorming is a method of generating a large number of creative ideas in a short period of time. This tool is used when broad ranges of options and creative and original ideas are desired. This tool also encourages team participation.

In practical application, the team identifies the subject or problem at hand and writes it down on a whiteboard. It is important to clearly define the problem. This will keep the ideas on topic. Sometimes for a totally unfamiliar issue, it is acceptable to keep it open so that we get a very wide range of ideas. The team leader explains the problem or subject to the team members.

Following are example topics with the scope defined to facilitate the majority of ideas focusing in the defined area:

- Contamination of polished surfaces before optical subassembly
- Low attendance at ASQ section program meetings
- Food menu for Thanksgiving dinner

Following are examples with the scope wide open:

- Global warming
- Unemployment
- Organizational culture

The team is given a few minutes to think about the subject. In structured brainstorming the team leader opens up a round-robin discussion. This way everyone gets the opportunity to contribute. If someone doesn't have an idea at this time, they are allowed to pass and contribute during the next round. The team members are not allowed to criticize each other or evaluate the ideas at this stage. The recording individual can ask for clarity on an idea and phrases it the same way as the idea contributor. Rephrasing without the consent of the idea owner is not allowed. Everyone is allowed one idea at a time. Some members will have the urge to provide multiple ideas during their turn. The team leader should facilitate such situations. Members are allowed to develop an idea already cited by a fellow member. Quantity is more important than quality so that the ideas keep flowing. All ideas are recorded on the whiteboard or flip-chart.

Let us examine an example of defined-scope brainstorming: How can member attendance of ASQ section programs be improved? (Problem rephrased as a question.)

Every major city in North America has a local ASQ section run by volunteers. One of the main benefits of this section is the monthly program meeting.

Unfortunately, the section monthly program meetings draw a very low attendance (about seven to 10 percent) of members from the region, with at least 20 percent of the members attending once throughout the year.

The program chair (responsible for ASQ section monthly meetings) chairs the brainstorming session as a team leader. The section chair may act as a facilitator.

A team has been assembled with other section executives, past section chairs, and/or executives, section senior members, and members who were randomly selected from the membership database.

One of the members volunteers as a recorder, and the team is given three minutes to think about the subject in a focused manner, then the session is started in a round-robin style.

Ideas start flowing. Keep in mind, it is about quantity and not quality at this point! No judgment or evaluation is allowed.

How can member attendance of ASQ section programs be improved?

1. Bring in reputed speakers.
2. Present topics that are current.
3. Provide value for time and money.
4. Keep program interactive; debate, quiz.
5. Survey members for desired topics.
6. Rotate program locations based on member concentration.
7. Conduct some programs in the organizations with most members.
8. Not charge for meeting.
9. Offer pizza, snacks, sandwiches, and coffee.
10. Offer time for networking.
11. Section chair and executives mix with members and attendees during break (rather than talking among themselves as a small group).
12. Check weather forecast before planning meetings.
13. Update members on other section events.
14. Conduct less frequent but more effective meetings.
15. Not waste meeting time with logistics issues—be prepared.
16. Offer the meeting virtually—webcast, teleconference.
17. Draw name cards from fishbowl and offer a small gift.
18. Make the process easier for program attendance, recertification units claim.
19. Present two diverse topics so that members do not choose to attend only some meetings.
20. Active members provide carpool to meeting location for new or potential members.

21. Liaise with other professional organizations to offer combined program meeting.
22. Attract more students from universities.
23. Conduct some meetings on the local community college or university campus to attract students.
24. Provide "back to basics" programs with applications for students and small business owners.
25. Interview random sample of members who never attended a single meeting and find out why.
26. Interview random sample of members who always attend every meeting and find out why.
27. Introduce first-time attendee members/nonmembers in the group to make them feel wanted.
28. Program chair to survey every program for attendee satisfaction and review feedback.
29. Appoint marketing chair to reach wider member base and potential new members.
30. Keep the section website updated and easily accessible.
31. Upload archive presentations to the website.
32. Communicate at least twice about monthly program—three weeks before and one week before.
33. Announce and recognize newly certified professionals.
34. Record and archive the program events and make available to local libraries and online for free.

Wow, isn't this quite a collection of ideas? Now the team leader looks for any redundancy or any ideas that require further expansion for clarity.

Some teams will break after a few rounds and revisit the list with any additional thoughts. However, this should not be prolonged as the team may get bored, and ideas will start to be counterproductive or too critical.

There are other team tools used to take these ideas to the next step:

- *Multivoting*, to short-list the ideas as a group.
- *Cause-and-effect diagram*, to assign each idea under one category, namely, person–machine–material–method–measurement–environment, and further analyze why.

Decision-Making Concepts

There are any number of ways to make decisions on the most appropriate path or approach to be taken for a given situation. *Top-down management* is a method whereby the leader will decide what is to be done. With *consensus decision-making*, the team works through a process (two examples follow) to come to a conclusion regarding the topic at hand. Consensus means that the decision is acceptable enough that all team members will support it. To achieve consensus, all team members must be allowed to express their views and must also objectively listen to the views of others. The decision and alternatives should be discussed until consensus is reached.

In many of the ISO management system audits, you will also hear the term *fact-based decision-making,* which involves ensuring that data is being used to lead the decision-making process.

Nominal Group Technique

This is also a type of brainstorming but with limited team vocal interaction. The tool is thus named "nominal" group technique (NGT). This technique is applied when some group members are much more vocal than others to encourage equal participation from all members, or with a controversial or sensitive topic, and so on. This technique helps to alleviate peer pressure and reduces the impact of such pressure on the generation of ideas.

Similarly to brainstorming, the facilitator explains the rules, and the team leader presents the topic to the assembled members. The team is given a good 10 to 15 minutes so that they can silently sit, think, and generate ideas.

No verbal interactions are allowed during the session. The member ideas are collected and posted in a space where all can read them. The members may also read the ideas aloud one by one in a round-robin format. At this stage no judgment or criticism is passed. The ideas are simply written down. The members are allowed to expand on existing ideas, provide clarity, and eliminate redundancy during the consolidation. For a controversial or sensitive subject, the team leader may opt to collect the ideas and write them down on the board, maintaining anonymity of the contributors.

Multivoting

Multivoting complements nominal group technique. This can also be successfully used with brainstorming results. Even though this tool is typically used in combination with NGT, it can be a technique on its own. The consolidated ideas are numbered or identified by an alphabetical letter, and the team members are asked to prioritize the top five or 10 items that can be of significant influence on the problem.

The team members are given five to 10 minutes to prioritize, and the results are tabulated. Let us extend the previous example of "How can member attendance of ASQ section programs be improved?" The members were asked to submit and prioritize ideas. As we see, there were 34 ideas provided from the diversified member group. Even though many of these ideas are good, the section may not have resources to address them all at one time. The section chair wants to select the five most important ideas to address in the next three years and implement them in order of priority.

Every team member selects the five most important ideas by placing check marks by the idea. It is important for the facilitator to restate the objective and refocus the team to select ideas from the ASQ section point of view. If this facilitation is not done, you may end up with multiple ideas with an equal number of check marks. Once this is done and you have the five ideas that most team members have selected as significant to improving the attendance of a section program, the prioritization process is begun. This can be done through either a non-weighted (ranking) or weighted approach.

The members select the following five ideas as having the most significant impact on improving section attendance:

A. *Value.* Bring in reputed speakers and present topics that are current.

B. *Logistics.* Rotate program locations based on member concentration.

C. *Affordability.* Not charge for meeting and offer pizza, snacks, sandwiches, and coffee.

D. *Outreach.* Conduct some meetings on the local community college or university campus to attract students.

E. *Communication.* E-mails twice per month, updated section calendar event web page.

The multivoting ranked approach outcome is shown in Figure 9.3.

In the weighted multivoting approach, the team rates rather than ranks the choices. This is like the $100 or 100 points approach where the team member is asked to split $100 or 100 points between five choices. The multivoting weighted approach outcome is shown in Figure 9.4.

As can be seen by examining the data in Figure 9.3 and Figure 9.4, the two approaches produce similar ranking in this example. However, this is not always the case, which is why using both approaches can help a team to focus on the most critical items. If the values get too close to each other, another round of voting can be conducted between the close choices to select a clear winner.

Note: The problem chosen for this example and ideas generated are realities for most ASQ sections. However, this example is not targeted to a specific ASQ section. The top choices and ranking were created to demonstrate the example rather than to provide solutions to the existing problem.

Venue: Caribou meeting room
Date: 3-Feb-15
Subject: How can member attendance of ASQ section programs be improved?
Scale: 1 (least important) to 5 (most important)

	Member 1	Member 2	Member 3	Member 4	Member 5	Member 6	Member 7	Member 8	Member 9	Total
A	5	3	4	5	5	3	4	5	5	39
B	2	4	3	3	4	4	3	3	4	30
C	1	5	5	4	3	5	5	4	3	35
D	3	3	1	2	2	3	1	2	2	19
E	4	1	2	1	1	1	2	1	1	14

Figure 9.3 Multivoting ranked approach example.

Venue: Caribou meeting room
Date: 3-Feb-15
Subject: How can member attendance of ASQ section programs be improved?

	Member 1	Member 2	Member 3	Member 4	Member 5	Member 6	Member 7	Member 8	Member 9	Total
A	30	20	25	35	20	25	25	35	30	245
B	15	25	20	20	25	20	20	20	15	180
C	10	30	30	25	30	30	30	25	10	220
D	20	15	10	15	15	10	10	15	20	130
E	25	10	15	5	10	15	15	5	25	125

There is NO ranking scale applicable to this approach. The column total should add up to 100 for all individual columns and the relative importance of A to E to be understood by the points allotted by each member (from that member's point of view). Overall relative importance is understood from reviewing the "Total" column. Based on consolidated input from all members, in this example, A is most important, followed by C, B, D, and E.

Figure 9.4 Multivoting weighted approach example.

TEAM COMMUNICATION

> Identify and use appropriate communication methods (both within the team and from the team to various stakeholders) to report progress, conduct reviews, and support the overall success of the project. (Apply)
>
> **Body of Knowledge II.F.4**

Basics of communication were discussed in Chapter 8.

When defining team communication, a subset of communication is needed to identify how to talk with other potential project groups that may be working around you. Here again, the main challenge is to identify what you are doing that may impact what they are doing and vice versa. Simply talking to them as partners in the overall continual improvement process is key to effective business operations.

Using your project report (this could be the formal company system, A3 report, or some form of a scoreboard), you can allow non–team members to be aware of your progress and possibly your next steps in your project path.

Part III

Measure Phase

Part III is an overview of the *measure* phase, including summaries of those Six Sigma methods and practices designed and intended to determine and prioritize improvements to products, processes, systems, and organizations. It covers 20 of the 100 questions that will be asked on the ASQ CSSGB exam. While there are no major additions to the measure phase in the 2022 revised CSSGB BOK, there are minor changes to Chapter 13.

Chapter 10

Process Analysis and Documentation

> Develop process maps and review written procedures, work instructions, and flowcharts to identify any gaps or areas of the process that are misaligned. (Create)
>
> **Body of Knowledge III.A**

PROCESS MAPS AND FLOWCHARTS

ISO 9000 (*Quality management systems—Fundamentals and vocabulary*) defines a *process* as a set of interrelated or interacting activities that transforms inputs into outputs. A process is easily understood by visually presenting the process using common flowcharting shapes and symbols. Practitioners use process mapping and flowcharting interchangeably, however, there are differences. Namely, process mapping includes additional process details with the flowchart. Organizations often send their process information in the form of process map documentation to their suppliers and customers for contractual reasons. From my personal experience, I have seen Japanese organizations use process maps and flowcharts more extensively. They call it "QC process flow." It is typically an end-to-end process flow starting from contract review and approval through to delivery of goods. The flowchart is presented at the left of the page, continuously running for multiple pages in one column, and the space on the right is utilized to describe the process, responsibility, control points, metrics, and so on. Using consistent mapping icons helps different individuals to interpret the maps in the same way. International standard ISO 5807:1985 *Information processing—Documentation symbols and conventions for data, program and system flowcharts, program network charts and system resources charts* helps accomplish just that. Process mapping is often the first step in improving a process. Risk analysis tools such as process failure mode and effects analysis (PFMEA) start with process mapping. Value stream mapping, used in lean enterprise projects, is also a type of process mapping but uses different mapping icons.

Flowcharts show each step in a process, including inputs, decision points, and outputs. Process maps usually contain additional information about the steps, including costs, setup time, cycle time, inventory, types of defects that can occur, probability of defects, and other relevant information that helps in understanding the process better.

Process maps and flowcharts enable a broader perspective of potential problems and opportunities for process improvement. Teams using these tools get a better understanding of process steps and sequence of operations. Figure 10.1 shows some of the most frequently used process mapping symbols from the international standard ISO 5807:1985. Figure 10.2 gives a basic flowchart example. There are a number of mapping icons available within most widely used office productivity software applications.

Process mapping involving multiple departments or functions is more easily understood using "swim lane" mapping. Imagine different departments, functions, or stakeholders involved in a process as being in different swim lanes either horizontally or vertically. A swim lane process map is similar to a typical process map except that the process blocks are arranged in alignment with the lane of the department or function that performs a given process step.

Let us re-map the previous example (Figure 10.2) using the swim lane flowchart approach. For simplicity we have taken out the decision loops from the previous chart. In a real business scenario, the swim lane flowchart (see Figure 10.3) contains all the components presented in a basic flowchart. It uses the same flowchart symbols and guidelines (Figure 10.1) for creating the chart.

Creating a Flowchart (Process Map or Process Flow Diagram). When creating a flowchart, we are creating a picture of the actual steps in a process or system as it actually operates or is supposed to operate. Given the old adage that a picture is worth a thousand words, this tool allows us to communicate using standard

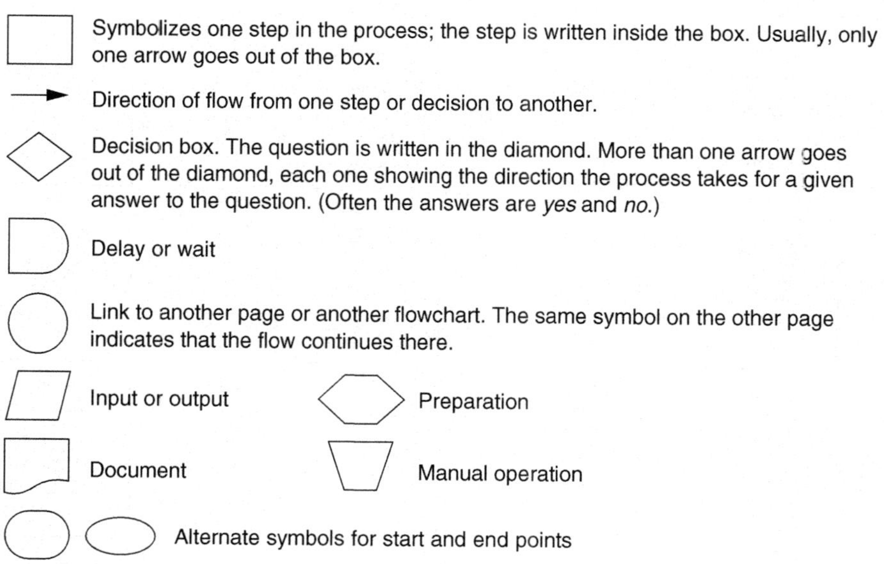

Figure 10.1 Symbols commonly used in flowcharts and process maps.
Source: N. R. Tague, *The Quality Toolbox*, 2nd ed.

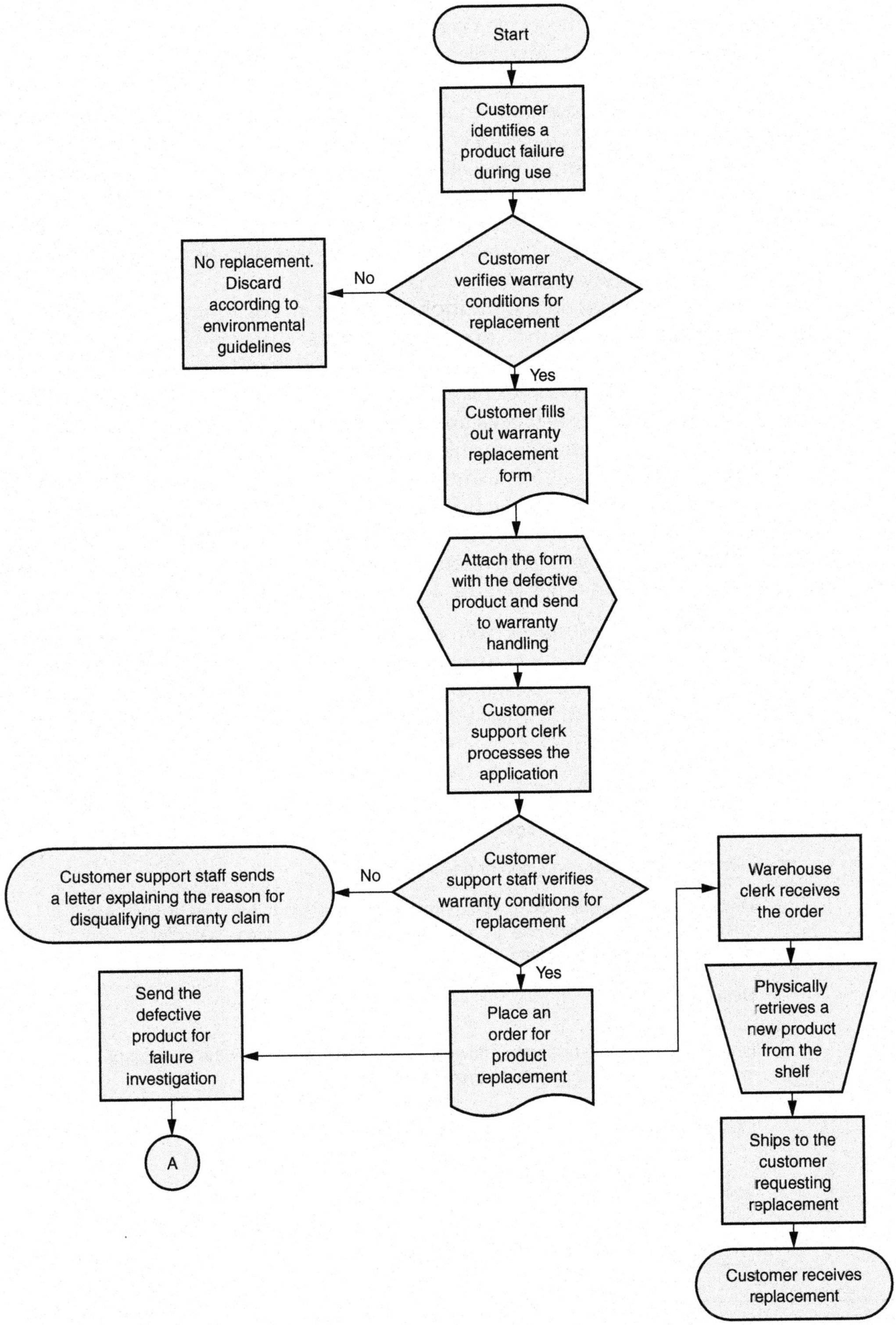

Figure 10.2 Basic flowchart for warranty product replacement.

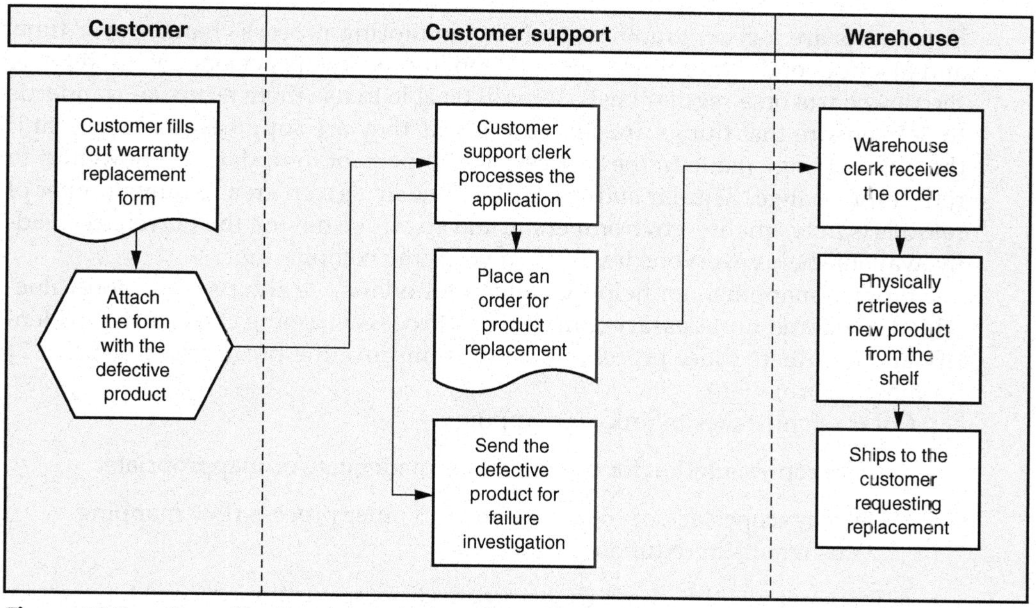

Figure 10.3 Cross-functional or swim lane flowchart.

symbols. The flowchart is very useful when looking at a process that we want to improve.

A flowchart should be developed for a new process before implementing the process. However, there may be situations where you are documenting an already existing process using a flowchart. We can follow some basic steps to create the flowchart:

1. Create the boundaries of the process that we intend to flowchart. These might be the inputs and outputs of the process or the suppliers and customers of the process.

2. Determine the various steps in the process through team brainstorming or walking the process (for documenting an already existing process). At this point, we are not worried about sequence, only collecting all of the steps.

3. Build the sequence of the process, putting everything into the appropriate order. We have to also make sure we understand that some process steps happen in parallel, and the chart should reflect this accordingly. There are also alternative paths identified in some charts.

4. Draw the flowchart using the appropriate mapping symbols.

5. Verify that the flowchart is complete and appropriate for the given operation (for a new chart). Verify that the flowchart fully matches with the process (for an already established process). This can be very important if more than one group is working on a large process. Overlaps or deletions may occur between processes. Hence, this activity is best performed as a team.

Flowcharts are a good graphical tool for monitoring process changes over time, and also for conducting training of new operators or supervisors. By referencing the flowcharts on a regular basis, we will be able to use them as visual standards to help ensure that things are still running as they are supposed to. Note that if there is a change made to the process, it is important to update the flowchart to reflect the change. Regular audits may be done in a given area for any number of reasons (safety, quality, environmental, and so on), so having the flowcharts readily available helps everyone involved in verifying compliance.

Process mapping can help visualize redundancy in the process, non-value-added steps, and unnecessary complexities. Process mapping can be used to identify and eliminate those process issues and improve the process (see Chapter 21 for more information).

Common mistakes in process mapping:

- Team representation for the process is inadequate or inappropriate.
- Unclear scope, lack of coordination with other process flow mapping teams results in redundant flows.
- Team not walking through the process (for an existing process) to capture what exactly is followed.
- Team spending too much time trying to create a "perfect" process flow diagram rather than focusing on the process.

Key Point: The biggest mistake of all in process mapping is not trying to do one!

Written Procedures and Work Instructions

Due to worldwide recognition and demand for ISO 9001 quality management systems compliance, the necessity of written procedures and work instructions has become very important. Irrespective of whether or not an organization pursues ISO 9001 registration, having written procedures and work instructions for business and manufacturing processes helps drive consistency. A consistent approach to process management helps with yield improvement, root cause analysis, traceability, and so on.

Procedures are written to describe:

- What is done during the process
- Why it is done (business reason, purpose)
- Where it is done (location/process step)
- When it is done (trigger)

Work instructions explain two other important aspects:

- Who does what (personnel with specific skill set)
- How it is done (step by step)

Where the organization has no specific internal procedure for a particular activity, and is not required to by a standard, it is acceptable for this activity to be conducted as a "method." A method is an unwritten process but must be followed consistently. In determining which processes should be documented, the organization should consider factors such as:

- Effect on quality
- Risk of customer dissatisfaction
- Statutory and/or regulatory requirements
- Economic risk
- Effectiveness and efficiency
- Competence of personnel
- Complexity of processes

Work instructions can be documented as:

- Written instructions
- Checklists
- Flowcharts
- Photographs
- Drawn pictures
- Videos
- Electronic screen shots
- Electronic software-driven process steps

PROCESS INPUTS AND OUTPUTS

Every process has input variables, output responses, and feedback loops. The feedback is required to improve the process.

Examples of inputs include:

- Needs
- Ideas
- Expectations
- Requirements
- Information
- Data
- Documents
- Resources

Examples of outputs include:

- Designs

- Decisions
- Results
- Measurements
- Products
- Services
- Proposals
- Solutions
- Authorizations
- Action

Identification of inputs and outputs is important before we start with analysis of relationships. Earlier, we dealt with process mapping, which provides a clear visual view of inputs to processes, interrelated process steps, and outputs of the process. Process maps provide a detailed view of the process. If we had to create a process map for a value stream or product line, or at the organizational level, we might end up creating a complex map with several hundreds of process blocks, decision boxes, inspection points, and storage locations. Hence, it is important to create a high-level process map that can encompass *suppliers, inputs, processes, outputs,* and *customers*. This high-level map, called *SIPOC* for short, provides a bird's-eye view at an enterprise level. It is recommended that in the process portion of a SIPOC chart you limit the number of process blocks to between four and seven high-level process steps to ease the complexity.

A SIPOC chart helps a team to quickly familiarize themselves with the process at an organizational level and visually understand the scope of the project.

The process input variables (x) are measured and variations are controlled so that resulting variations in output response (y) are correspondingly reduced. See Six Sigma philosophy in the earlier chapter on $y = f(x)$ (Chapter 1). The effects of variations of input on output are explored through quality tools like cause-and-effect diagrams, cause-and-effect diagrams with addition of cards (CEDAC), relationship matrices, cause-and-effect matrices, scatter diagrams, design of experiments, and so on.

A thorough understanding of process inputs and outputs and their relationships is a key step in process improvement. The *cause-and-effect diagram* (also called the *Ishikawa diagram* or *fishbone diagram*) traditionally divides causes into several generic categories. In use, a large empty diagram is often drawn on a whiteboard or flip-chart as shown in Figure 10.4.

This diagram is then used to populate the final list of causes from a brainstorming session. The participants in the session should include people with a working knowledge of the process as well as those with a theoretical background. For example, suppose a machining operation is producing surface defects. After a few steps of typical brainstorming, the cause-and-effect diagram would look like Figure 10.5.

Brainstorming is a powerful technique for soliciting ideas. Brainstorming intentionally encourages divergent thinking through which most possible

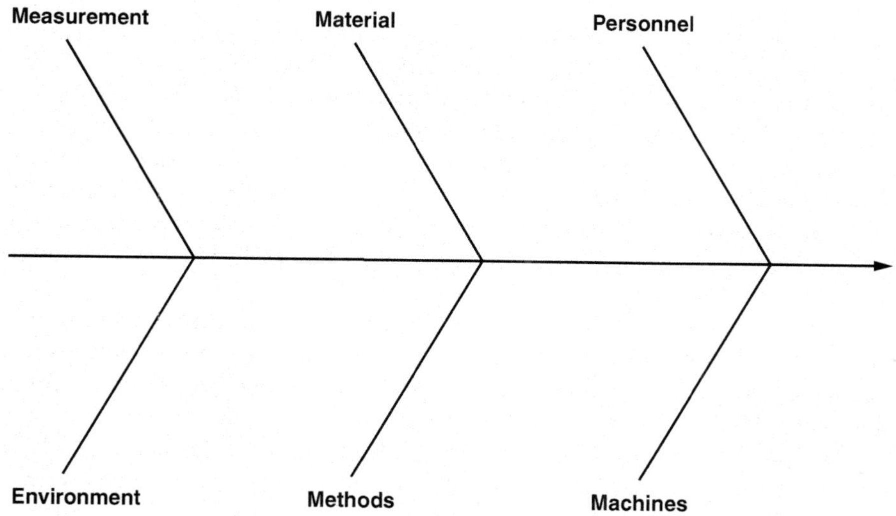

Figure 10.4 Empty cause-and-effect diagram.

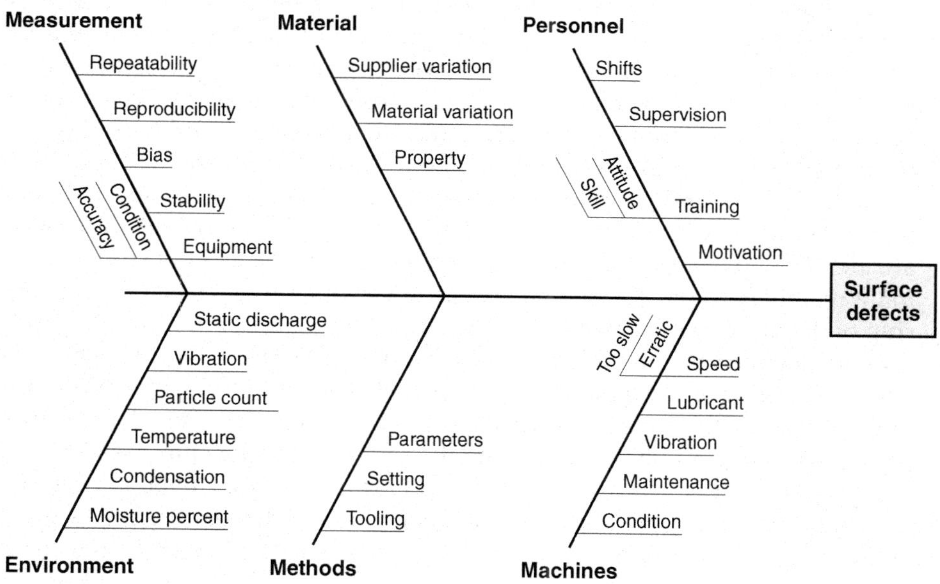

Figure 10.5 Cause-and-effect diagram after a few steps of a brainstorming session.

causes are identified. This is a team exercise and requires a trained facilitator to get the ideas flowing without hesitation. The facilitator's job is to enforce ground rules and encourage ideas during brainstorming. A common tendency within the brainstorming team is to criticize ideas instantly and discard them during the session. This will discourage team members from contributing for fear of being judged or their ideas being rejected. There are no bad ideas. At this stage, quantity of ideas is given priority. A typical brainstorming session can generate between 25 to 40 ideas. Once the ideas are collected, the team can review them for redundancy and feasibility and prioritize the ideas. The selected ideas are categorized under people (personnel)–machine–material–methods–measurement–environment. Sometimes the team includes *measurement* under *methods*. Cause-and-effect diagram categories are flexible depending on the operation (for example, software development uses people–processes–products–resources–miscellaneous). Try not to force-fit categories like machine or environment for a service type situation.

It is not uncommon for the team to continue the brainstorming in a second sitting to add more ideas to the existing list. There are other variations of brainstorming, like nominal group technique, idea mapping, and mind mapping that are used in different scenarios.

CEDAC (cause-and-effect diagram with addition of cards) is an alternative approach tried out by some organizations, where the fishbone diagram is displayed on a huge wall or board, and employees are encouraged to identify causes by writing on it or using sticky notes. The success of this approach depends on organizational culture and communication.

Once the ideas are collected by performing brainstorming, the next step is to condense the list into those causes most likely to impact the effect. *Multivoting* and *nominal group technique* (NGT) are two convergence tools that also rank the priorities. NGT is particularly effective when the issue being discussed is sensitive and emotional or when the team is composed of members from several layers of the organization (rank and file to senior management), and encourages quiet team members to contribute, think without pressure, and so on. Once the ideas are collected, redundant items are removed and the rest are displayed, and the team silently ranks the items. The ranks are tallied, and priority items to work on are identified. The prioritization obtained by this approach is not swayed by dominant members (typical in some teams). Hence, this provides better team ownership of the identified items. Multivoting is similar to this approach; 100 points are divided between the choices based on relative importance. Brainstorming, NGT, and multivoting are explained in Chapter 9. The *affinity diagram* (explained in more detail in Chapter 7) also complements brainstorming. The affinity diagram organizes a large number of ideas into their natural relationships. When the team is confronted with overwhelming facts and ideas, issues seem too large and complex to grasp, or when group consensus is necessary, the team arranges the ideas in the form of an affinity diagram. This tool is very helpful in analyzing customer qualitative data and feedback.

Convergence can also be accomplished by asking brainstorming participants to collect data on the various causes for a future reporting session. In some cases, the data might come directly from the process, for example, "I used two different coolants and found no difference in the surface defects." In other situations, the sources of the data might be exterior to the process, for example, "We found that the manufacturer recommends a relative humidity of 55 to 60 percent." As data are collected, the various causes are prioritized on a Pareto chart, as shown in Figure 10.6.

The Pareto chart has been so widely used in recent years that "Pareto" is sometimes used as a verb. It is not uncommon to hear from managers to "Pareto" data for presentation. Some people who are not familiar with Pareto charts

Defect code	Defect description	Occurrences
A	Scratches	15
B	Stains	17
C	Label smudge	12
D	Dent	14
E	Device nonfunctional	5
F	Broken LED	7
G	Missing screw	3

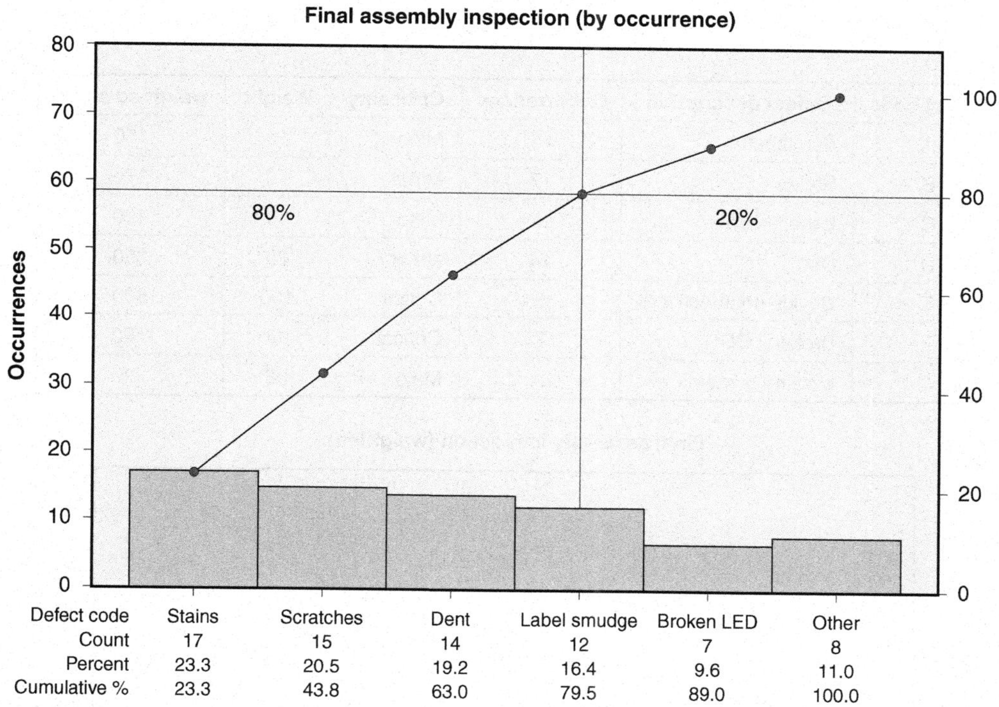

Figure 10.6 Pareto chart of final assembly inspection defect codes.

interchangeably use a bar graph to "Pareto" data. The true Pareto chart, however, has uniqueness to it. It shows the data arranged in descending order of frequency of occurrence (or other chosen measures like cost), the "trivial many" data are often pooled together as "miscellaneous" or "other," and the chart contains a secondary axis with percentages, and a cumulative percentage line plotted.

These characteristics make the Pareto chart more informative and useful compared to an ordinary bar graph that only displays the frequency of categories no matter how it is presented. The Pareto chart helps us to visualize the items charted as "vital few" and "trivial many" using the famous 20th-century Italian economist Vilfredo Pareto's principle of 80:20. Credit has been given to Dr. Joseph Juran for first applying this principle in quality improvement.

In the final assembly inspection example shown in Figure 10.6, the data are presented as a Pareto chart based on frequency of occurrence. While these data are important, one might want to put their resources into issues critical to customers, or issues that have more financial impact. So, the data are assigned weights based on criticality and multiplied by occurrence, and a Pareto diagram is created based on the weighted score. The table in Figure 10.7 shows the reprioritized defects based on criticality. Cost of repair or rework can also be used in place of weight, and the Pareto chart can be expressed in cost.

One important point to remember before performing a Pareto analysis is to make sure that the data are not too specific, with few occurrences for each specific

Defect code	Defect description	Occurrences	Criticality	Weight	Weighted score
A	Scratches	15	Minor	10	150
B	Stains	17	Minor	10	170
C	Label smudge	12	Minor	10	120
D	Dent	14	Major	25	350
E	Device nonfunctional	5	Critical	100	500
F	Broken LED	7	Critical	100	700
G	Missing screw	3	Major	25	75

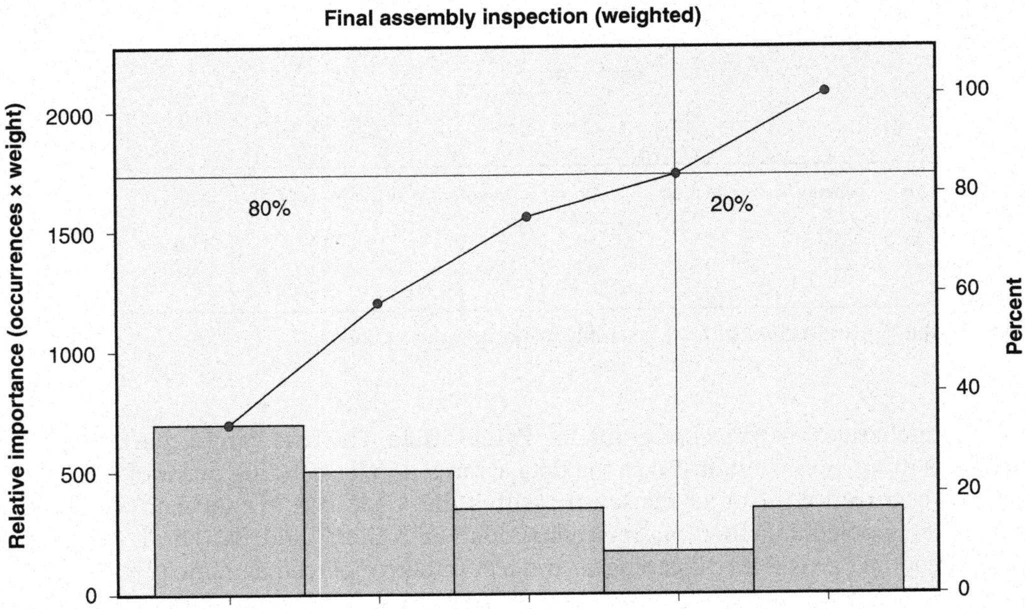

Defect code	Broken LED	Device nonfunctional	Dent	Stains	Other
Weighted score	700	500	350	170	345
Percent	33.9	24.2	16.9	8.2	16.7
Cumulative %	33.9	58.1	75.1	83.3	100.0

Figure 10.7 Pareto chart of final assembly inspection defect codes (weighted).

item. This will result in a poor Pareto chart resulting in vital many and trivial few—the exact opposite of the intended purpose (see Figure 10.8).

While defect data with specific locations or appearance are important, they may not serve the purpose "as is" in a Pareto chart. You may have to understand

Defect description	Occurrences	Defect description	Occurrences
Broken red LED	4	Label with edges smudge	3
Broken green LED	3	Label with fonts smudge	5
Device with no display	1	Stains on the cover	7
Device with erroneous reading	1	Stains on the display monitor	10
Device with error display	2	Scratches on the body	3
Device with fading display	1	Scratches on the display screen	7
Missing screw on lid	1	Scratches on the name plate	5
Missing screw on the stand	1	Dent on the casing	5
Missing screw on the handle	1	Dent on the front panel	7
Label with logo smudge	4	Dent on the name plate	2

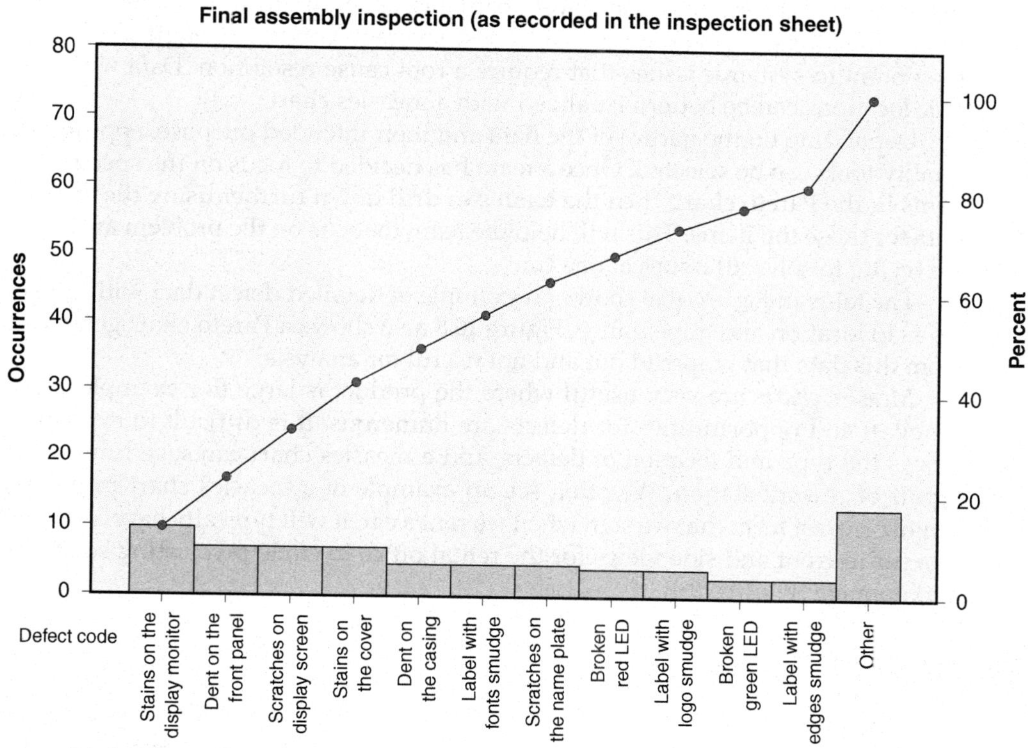

Count	10	7	7	7	5	5	5	4	4	3	3	13
Percent	13.7	9.6	9.6	9.6	6.8	6.8	6.8	5.5	5.5	4.1	4.1	17.8
Cumulative %	13.7	23.3	32.9	42.5	49.3	56.2	63.0	68.5	74.0	78.1	82.2	100.0

Figure 10.8 Example of a Pareto chart of a too-detailed defect summary.

Figure 10.9 Example of a measles chart (pictorial check sheet).

the defect or data and move one level up (that is, generalizing as scratches or dents or nonfunctional device) to be able to best leverage Pareto's strength. This way, they point to systemic issues that require a root cause resolution. Data with specific locations can be better visualized with a measles chart.

Depending on the nature of the data and their intended purpose, appropriate quality tools can be selected. Once a team has decided to focus on the specific top items in the Pareto chart, then the team can drill down further using the specific data for those top items. This will help the team to focus on the problem and obviate trying to solve all issues at one time.

The table in Figure 10.8 shows an example of detailed defect data with specifics as to location and appearance. Figure 10.8 also shows a Pareto chart generated from this data that is spread out and not useful for analysis.

Measles charts are very useful where the product is large (for example, automobiles) and opportunities for defects are numerous. It is difficult to explain in words the type and location of defects, and a measles chart can save time at the repair or rework station. We often see an example of a measles chart in the car rental contract form that we sign when we rent a car. It will typically have a picture of a car in front and side views for the rental office to circle preexisting scratches and damage (Figure 10.9).

Relationship Diagram

The relationship diagram is used to display the degree of the relationship between variables, causes and effects, and so on. The degree is often expressed as strong (⊙), medium (**O**), and weak (▽) or numerically, 9 for strong, 3 for medium, and 1 for weak. The totals of rows and columns are added and prioritized based on the total score. This analysis is further strengthened by adding a "weights" column and/or row to the matrix. In this case, the overall weighted score is used for

Relationships △ = weak O = moderate ⊙ = strong	Pizza not hot enough	Delivered late	Toppings not as per order	Wrong pizza delivered	Burnt crust
Traffic jam	O	⊙			
Oven heater error	O				⊙
Heat insulation quality	⊙				
Difficulty finding address	△	⊙			
Did not understand customer accent			△	△	
Clerical error during order receipt			⊙	⊙	
Order not clearly identified		△	⊙	⊙	
Mix-up on the delivery shelf		△	O	O	O

Figure 10.10 Relationship matrix.

prioritization. The relationships between causes and effects can be shown in a relationship matrix, as shown in Figure 10.10. In this example causes are listed on the left side of the matrix, and various customer issues are placed along the top. For example, cause "traffic jam" has a strong relationship to customer issue "delivered late." A team can use brainstorming and/or data collection techniques to determine the list on the left side and the relationship symbols shown in the matrix.

In the example in Figure 10.10, "traffic jam" is an unpredictable scenario that is beyond the caterer's control. However, the pizza caterer could invest in a better quality heat-insulated container, get the address correct, and travel in a path less prone to traffic incidents so that the pizza is still delivered hot.

By taking actions to identify pizzas by customer information on the delivery shelf and reduce the clerical errors, the caterer can reduce customer issues like delivering wrong toppings and wrong pizza.

The relationship matrix can also be used to describe the connection between process inputs and desirable process outputs. Refer to the discussion of QFD (Chapter 5) for a more elaborate example.

Chapter 11

Probability and Statistics

BASIC PROBABILITY CONCEPTS

> Describe and interpret basic probability concepts: independent events, mutually exclusive events, multiplication rules, permutations, and combinations. (Understand)
>
> **Body of Knowledge III.B.1**

Probability is probably the word most often used when people expect something to happen based on data or historical knowledge or experience:

"It is probably going to rain today." (Based on observation)

"The flight is probably going to be late." (Based on historical knowledge)

Hence, probability is closely attached to an event. Saying "It is not going to rain today" is probability 0, while "It will rain today" with complete certainty is probability of 1. In real life we can provide a complete certainty to events only very rarely. Most of the time the probability of an event happening is between 0 and 1. The sum of the probabilities of all possible outcomes of an event is 1.

The probability that a particular event will occur is a number between 0 and 1 inclusive. For example, if an urn containing 100 marbles has five red marbles, we would say the probability of randomly drawing a red marble is .05 or 5%. Symbolically this is written P (Red) = .05.

The word "random" implies that each marble has an equal chance of being drawn. If the urn had no red marbles, the probability would be 0 or zero percent. If the urn had all red marbles, the probability would be 1 or 100 percent.

Simple Events Probability

Probability of getting a head or tail in a fair coin = 1/2

Probability of getting 2 in a single toss of a die = 1/6

SUMMARY OF KEY PROBABILITY RULES

For events A and B:

Special addition rule: P(A or B) = P(A) + P(B) [Use only if A and B are mutually exclusive]

General addition rule: P(A or B) = P(A) + P(B) – P(A & B) [Always true]

Special multiplication rule: P(A & B) = P(A) × P(B) [Use only if A and B are independent]

General multiplication rule: P(A & B) = P(A) × P(B|A) [Always true]

Conditional probability: P(B|A) = P(A & B) ÷ P(A)

Mutually exclusive (or disjoint):

1. A and B are mutually exclusive if they can't occur simultaneously
2. A and B are mutually exclusive if P(A & B) = 0
3. A and B are mutually exclusive if P(A or B) = P(A) + P(B)

Independence:

1. A and B are independent events if the occurrence of one does not change the probability that the other occurs
2. A and B are independent events if P(B|A) = P(B)
3. A and B are independent events if P(A & B) = P(A) × P(B)

Compound Events Probability

Compound events are formed by two or more events.

Compound events can be better explained by the concept of the Venn diagram (see Figure 11.1).

Relations between Events

Complementation Rule. The probability that event A will not occur is 1 – (the probability that A does occur). Stated symbolically, P (not A) = 1 – P(A). Some texts use other symbols for "not A" including –A, ~A, A', and sometimes A with a bar over it.

Figure 11.1 Venn diagram.

EXAMPLE

If the probability of a car starting on a rainy day is 0.4, the complement of not starting would be 0.6. Pa = 0.4, (1 – Pa) = 0.6

Conditional Probability. Conditional probability is the probability of an event happening given that another event has happened. This concept is used extensively in reliability calculations. A formal definition for conditional probability is

$$P(B|A) = P(A \,\&\, B) / P(A)$$

EXAMPLE

Probability of getting 4 from throwing a die is 1/6.

Probability of getting 6 from throwing a die is 1/6.

Probability of getting either 4 or 6 is 1/6 + 1/6 = 1/3.

Probability of getting both 4 and 6 in a single throw is 0 (because we can get either of those but not both simultaneously). This means the two events are mutually exclusive.

If the two events are not mutually exclusive:

$$P(A \cup B) = P(A) + P(B) - P(A \cap B)$$

If an organization has two injection-molding machines, each having a probability of working 0.8, what is the probability the organization will meet the weekly production?

$$P(A \cup B) = 0.8 + 0.8 - (0.8 \times 0.8) = 1.6 - 0.64 = 0.96$$

(This is like building a redundancy for improving the reliability.)

Since both injection-molding machines can work simultaneously, this situation is not mutually exclusive.

If the two events are mutually exclusive, the additive law reduces to:

$$P(A \cup B) = P(A) + P(B)$$

Suppose an urn contains three white marbles and two black marbles. The probability that the first marble drawn is black is

$$2/(2 + 3) = 2/5$$

The probability that the second marble drawn is black (given that the first marble drawn is black) is

$$1/(1 + 3) = 1/4$$

The probability that both marbles drawn are black is

$$2/5 \times 1/4 = 2/20 = 1/10$$

(Without replacement.)

Mutually Exclusive Events

If occurrence of any one of the events excludes the occurrence of others, the events are mutually exclusive (see Figure 11.2). If two events A and B are mutually exclusive, then $P(A \cup B) = P(A) + P(B)$.

EXAMPLE

The assembly of product requires an electronic board. This electronic board is supplied by three different suppliers. Probability of the board from supplier A working on the product is 0.2, board from supplier B is 0.3, board from supplier C is 0.5. What is the probability that either board from B or C is working on the product?

$$P(B \cup C) = 0.3 + 0.5 = 0.8$$

The Multiplicative Law

If the events are dependent (no replacement):

EXAMPLE

A production lot of 50 units has 10 defective units. Three units were sampled at random from the production lot. What is the probability that all three are defective?

$$P(A \cap B) = 10/50 \times 9/49 \times 8/48 = 720/117{,}600 = 0.0061 \text{ or } 0.6\%$$

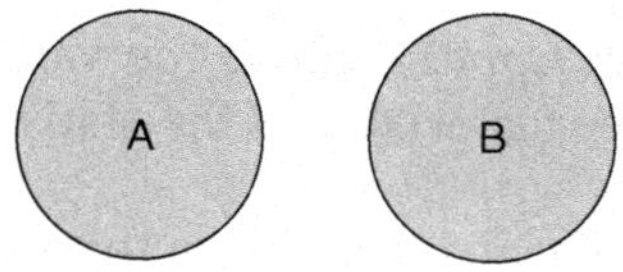

Figure 11.2 Mutually exclusive events.

If events are independent (replacement):

> **EXAMPLE**
>
> The assembly of an electronic board has two major components. Probability of component A working is 0.7, component B is 0.8. What is the probability the assembly will work?
>
> $$P(A \cap B) = 0.7 \times 0.8 = 0.56$$
>
> (The probability of component A and B working to make the assembly work.)

Permutations and Combinations

Permutations. *Permutation* is an ordered arrangement of n distinct objects. The number of ways of ordering the arrangement of n objects taken r at a time is designated by ${}_nP_r$. Permutations are an important concept that we use in our everyday life.

> **EXAMPLE**
>
> One might think AB and BA are same. It matters when arranging people for seating. In an airplane, AB seating is not the same as BA. One has a window and the other the aisle! Maybe A is a right-hander and B is a left-hander. Seating AB may cause inconvenience as their elbows interfere, whereas BA is a convenient seating arrangement.

The counting rule:

$${}_nP_r = n(n-1)(n-2)\ldots(n-r+1) = {}_nP_r = \frac{n!}{(n-r)!}$$

Important factorials to remember during calculations (proof beyond the scope of this book):

$$0! = 1, {}_nP_n = n!, {}_nP_0 = 1$$

$$n! = 1 \times 2 \times 3 \times \ldots n \qquad \text{(! is pronounced "factorial")}$$

Note: Calculators have an upper limit to the value that can use the x! key. If a problem requires a higher factorial, use the statistical function in a spreadsheet program.

How many words can be made by using the letters of the word "sigma" taken all at a time?

There are five different letters in the word *sigma*.

Number of permutations taking all the letters at a time = ${}_5P_5$

We know that ${}_nP_n = n! = 5! = 120$.

Combinations. The number of distinct combinations of n distinct objects taken r at a time. This is denoted by ${}_nC_r$. Combinations are used when order is not significant. Example: AB and BA are the same, and hence the result shows only AB. (Unlike permutation, where the result will have both AB and BA.)

The counting rule:

Number of combinations of r objects from a collection of n objects =

$$ {}_nC_r = \frac{n!}{r!(n-r)!} $$

Note: Another symbol for number of combinations is $\binom{n}{r}$

Important factorials to remember during calculations (proof beyond the scope of this book):

$$ 0! = 1,\ {}_nC_n = 1!,\ {}_nC_0 = 1 $$

Let us consider an example where order arrangement is not a concern: selection of r people from n available people.

EXAMPLE

A local ASQ section with 10 volunteers wants to form a task force of three volunteers to send for proctor training to potentially become exam proctors. How many different three-person combinations could be formed?

The combinations formula will be used to calculate the number of combinations of three objects from a collection of seven objects.

$$ {}_{10}C_3 = \frac{10!}{(10-3)!3!} = 120 $$

Now with the 120 different combinations, the section chair can find out which combination of people are available on a given date.

Spreadsheet formula: =COMBIN(10,3)

CENTRAL LIMIT THEOREM

> Define the central limit theorem and describe its significance in relation to confidence intervals, hypothesis testing, and control charts. (Understand)
>
> **Body of Knowledge III.B.2**

Key Point: The central limit theorem (CLT) is an important principle used in statistical process control.

Definition and Description

The central limit theorem is the foundation for several statistical procedures. In a nutshell, the distribution of averages tends to be normal, even when the distribution from which the average data are computed is from nonnormal distributions. Mathematically, if a random variable X has a mean μ and variance σ^2, as the sample size n increases, the sample mean $\bar{x}$ approaches a normal distribution with mean μ and variance $\sigma_{\bar{x}}^2$:

$$\sigma_{\bar{x}}^2 = \frac{\sigma_x^2}{n} \qquad \text{(See number 2 below)}$$

$$\sigma_{\bar{x}} = \frac{\sigma_x}{\sqrt{n}}$$

The central limit theorem consists of three statements:

1. The mean of the sampling distribution of means is equal to the mean of the population from which the samples were drawn.
2. The variance of the sampling distribution of means is equal to the variance of the population from which the samples were drawn divided by the size of the samples.
3. If the original population is distributed normally (that is, it is bell shaped), the sampling distribution of means will also be normal.

If the original population is not normally distributed, the sampling distribution of means will increasingly approximate a normal distribution as sample size increases (that is, when increasingly large samples are drawn). Weirder populations will require larger sample sizes for the sampling distribution of the mean to be nearly normal. Statisticians usually consider a sample size of 30 or more to be sufficiently large. See Figure 11.3.

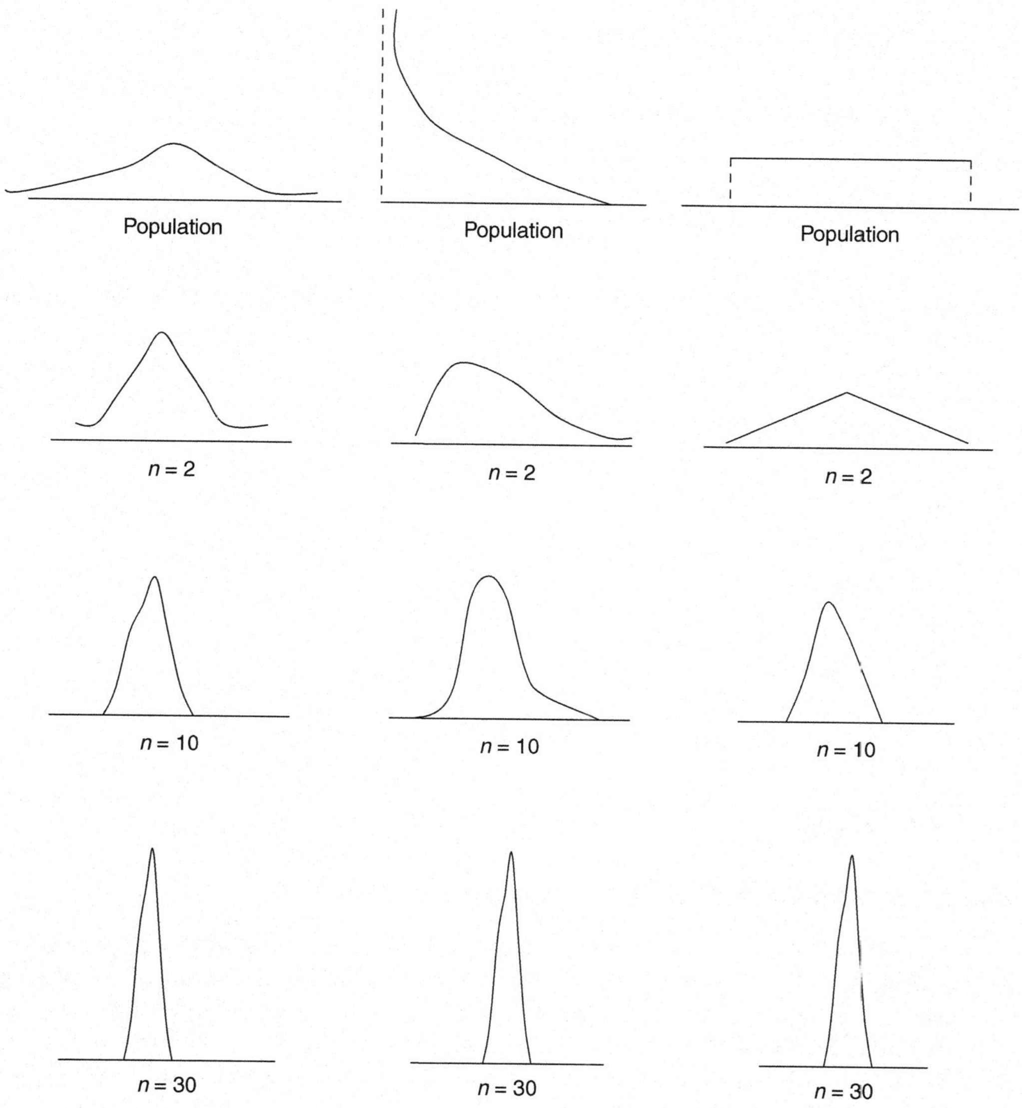

Figure 11.3 Various populations and sampling distributions of the mean for selected sample sizes.

Source: D. W. Benbow and T. M. Kubiak, *The Certified Six Sigma Black Belt Handbook* (Milwaukee: ASQ Quality Press, 2005): 58.

The *standard error of the mean* is expressed as:

$$\frac{\sigma_x}{\sqrt{n}}$$

It is used extensively to calculate the margin of error, which is used to calculate confidence intervals:

a. The *sampling distribution of the mean* roughly follows a *normal* distribution

b. 95% two-sided confidence interval on μ is:

$$pr\left[\left(\bar{x}-1.96\frac{\sigma}{\sqrt{n}}\right)\leq\mu\leq\left(\bar{x}+1.96\frac{\sigma}{\sqrt{n}}\right)\right]=0.95$$

Or, 95 percent of the time the true mean should lie within $\pm1.96(\sigma/\sqrt{n})$ of the interval:

$$pr\left[\left(\bar{x}-z_{\alpha}\frac{\sigma}{\sqrt{n}}\right)\leq\mu\leq\left(\bar{x}+z_{\alpha}\frac{\sigma}{\sqrt{n}}\right)\right]=1-\alpha$$

where

$$\frac{\sigma}{\sqrt{n}}$$

is standard error and

$$z_{\alpha}\frac{\sigma}{\sqrt{n}}$$

is margin of error.

Use of Central Limit Theorem in Control Charts

In the real world, not all processes are normally distributed. By applying the central limit theorem when taking measurement samples, the status of the process can be monitored by averaging the measurement values in subgroups, for example, in SPC control charts. Since control charts like $\bar{X}$ and R charts and $\bar{X}$ and s charts are plotted with averages of the individual readings, the charts are robust to departures from normality.

Use of Central Limit Theorem in Hypothesis Testing

Exercise Using the Central Limit Theorem. Historical standard deviation of a chemical filling process σ is 0.012 milligram. Estimate the sample standard deviation for a sample size of 16 fillings.

$$\sigma_{\bar{x}}=\frac{\sigma_{x}}{\sqrt{n}}=\frac{0.012}{\sqrt{16}}=0.003 \text{ milligrams}$$

Calculate the 95 percent confidence interval for the mean if the process average is 10 milligrams.

$$= 10 \pm 1.96\left(\sigma / \sqrt{n}\right) = 10 \pm 1.96(0.003) = 10 \pm 0.0059$$
$$= 9.9941 \text{ to } 10.0059 \text{ milligrams}$$

Drawing Valid Statistical Conclusions

Does this sound familiar?

- Is process A better than process B? (Manufacturing organization)
- Can we guarantee our customers "15 minutes or free?" (Restaurant)
- Is same-day delivery feasible? (Package forwarding company)
- Does medicine X reduce cholesterol? (Healthcare)
- How many missiles are required to destroy the enemy target? (Army)

In our everyday life, we come across many situations that demand decision-making. Whether in the office or at home or in society, decisions made can impact the organizational bottom line, personal finances, and the economics of society. Decisions of such criticality cannot be made just by "gut feeling." Comparing two scenarios merely by looking at the numbers may be better than gut feeling but still not good enough. The question is "Is the difference statistically significant?" Hence, statistics is used to draw valid conclusions.

In this area of study, what we do is to draw representative samples from a homogenous population. By analyzing the samples, we draw conclusions about the population. There are two types of studies used for drawing statistical conclusions, namely *descriptive* and *analytical (inferential)* (see Table 11.1).

In a statistical study, the word "population" refers to the collection of all items or data under consideration. A descriptive study typically uses all the data from a population. Significant values from a population are referred to as *population parameters*. A *parameter* is a numerical value that provides information about the entire population being studied. Examples of population parameters are

Table 11.1 Descriptive versus analytical statistics.

Descriptive (or enumerative) statistics	Analytical (or inferential) statistics
This consists of a set of collecting, organizing, summarizing, and presenting the data.	This consists of a set of making inferences, hypothesis testing, and making predictions.
A descriptive study shows various properties of a set of data such as mean, median, mode, dispersion, shape, and so on.	Uses data from a sample to make estimates or inferences about the population from which the sample was drawn.
Graphical tools include histograms, pie charts, box plots, and others.	Uses tools such as hypothesis testing and scatter diagrams to determine the relationships between variables and make predictions using regression equations.

Table 11.2 Sample versus population notations.

	Sample	**Population**
Size	n	N
Mean	$\bar{x}$	μ
Standard deviation	s	σ

population mean and *population standard deviation.* A *statistic* is a numerical value that provides information about a sample. A *sample* is a subset of the population. Samples are selected randomly so that they represent the population from which they are drawn.

It is traditional to denote sample statistics using Latin letters and population parameters using Greek letters. An exception is made for the size of the set under consideration. The symbols shown in Table 11.2 are the most commonly used in textbooks.

The Greek letter μ is pronounced "mew." The Greek letter σ is pronounced "sigma." This is a lowercase sigma. The capital sigma, Σ, is used to designate summation in formulas.

Chapter 12

Statistical Distributions

> Define and describe various distributions as they apply to statistical process control and probability: normal, binomial, Poisson, chi square, Student's t, and F. (Understand)
>
> **Body of Knowledge III.C**

Formulas for some of the probability distributions are shown in Table 12.1.

BINOMIAL

The "bi-" prefix indicates that a binomial distribution should be applied in situations where each part has just two states, typically:

- Good or bad

Table 12.1 Formula, mean, and variance of certain distributions.

Name	Formula	Mean	Variance
Normal	$P(x)=\dfrac{e^{-\frac{(x-\mu)^2}{2\sigma^2}}}{\sigma\sqrt{2\pi}}$	μ	σ^2
Exponential	$P(x)=\lambda e^{-\lambda x}$	$\dfrac{1}{\lambda}$	$\dfrac{1}{\lambda^2}$
Binomial	$P(x)=\dfrac{n!}{x!(n-x)!}p^x(1-p)^{n-x}$	np	$np(1-p)$
Poisson	$P(x)=\dfrac{e^{-\lambda}\lambda^x}{x!}$	λ	λ
Hypergeometric	$P(x)=\dfrac{{}_dC_x[{}_{(N-d)}C_{(n-x)}]}{{}_NC_x}$	$\dfrac{nd}{N}$	$\dfrac{nd(N-d)(N-n)}{N^3-N^2}$

- Accept or reject
- Conformance or nonconformance
- Success or failure

The binomial distribution (Figure 12.1) is used to model discrete data. Examples of binomial data that are frequently used in everyday life are:

- The number of defectives in a manufacturing lot
- The number of defective quotes sent by an insurance company
- The number of wrong patient prescriptions issued by a healthcare professional
- The number of goods shipped to a wrong address by a forwarding company

The binomial distribution has some conditions. It is applicable when the population denoted by N is greater than 50. In other words, for smaller lots, binomial modeling will not be accurate.

Another important condition is the ratio of the sample n to population N. The binomial model best applies when $n < 0.1N$ (that is, sample size is less than 10 percent of the population).

In one type of problem that is frequently encountered, the Six Sigma Green Belt needs to determine the probability of obtaining a certain number of defectives in a sample of known size from a population with known percentage defective. The symbols are: n = sample size, x = number of defectives, p = defective rate in the population.

The binomial formula is:

$$P(x) = \frac{n!}{x!(n-x)!} p^x (1-p)^{n-x}$$

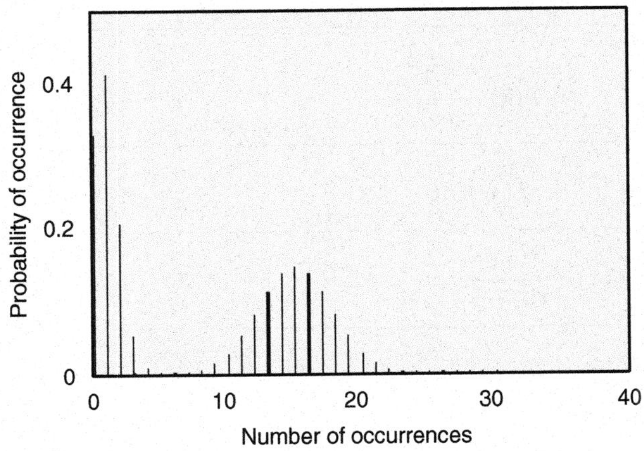

Figure 12.1 Binomial distribution.

As discussed in Chapter 11, $x!$ is pronounced "x factorial" and is defined as $x(x-1)(x-2)\ldots(1)$. Most scientific calculators have a factorial key.

EXAMPLE

A sample of size five is randomly selected from a batch with 10 percent defective. Find the probability that the sample has exactly one defective. Substitute $n = 5$, $x = 1$, $p = .10$ into the above formula:

$$P(1) = [5!/(1!(5-1)!)](0.10)^1$$

$$(0.9)^{5-1} = [120/(1\times 24)](0.10)(0.6561) \approx 0.328$$

Probability that the sample contains exactly one defective is 0.328.

The same can be calculated using a spreadsheet formula (see Figure 12.2):

=BINOMDIST(1,5,0.1,FALSE)

We can also use several online Java interactive calculators.

Teaser: Try the following example using the calculator above:

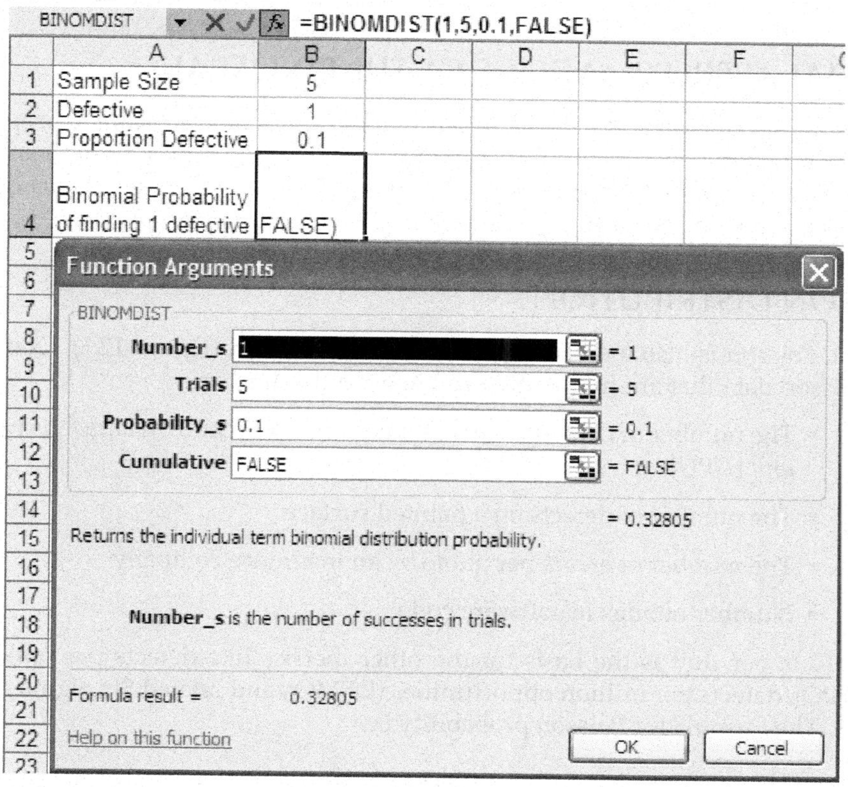

Figure 12.2 Binomial distribution using spreadsheet.

An ASQ Six Sigma Green Belt exam has 100 questions and four choices per question. Assuming the exam requires 80 right answers, what is the probability of a student passing the exam if he/she randomly chose from the four choices for all 100 questions (let us believe that this student doesn't have an iota of a clue about any question—no knowledge bias). Also find out up to how many questions on which the student may get lucky with maximum binomial probability.

As per Table 12.1, the mean and variance can also be calculated for the binomial distribution.

EXAMPLE

If we take an unbiased coin and toss 60 times, what is the average and standard deviation of the number of tails?

Unbiased coin: Having equal probability to be heads or tails.

$$p = (1/2) \text{ or } 0.5$$

$$n = 60$$

$$\mu = np, \mu = (60 \times 0.5) = 30 \text{ tails}$$

$$\sigma = \sqrt{np(1-p)}, \sigma = (30(1-0.5))^{1/2} = 3.872$$

NORMAL APPROXIMATIONS OF THE BINOMIAL

For large values of n, the distributions of the count X and the sample proportion p are approximately normal. This is understood from the central limit theorem. The normal approximation is not accurate for small values of n; a good rule of thumb is to use the normal approximation only if $np \geq 10$ and $np(1-p) \geq 10$.

POISSON DISTRIBUTION

The Poisson is also a discrete probability distribution (Figure 12.3). Examples of Poisson data that are frequently used in everyday life are:

- The number of defects in an assembly unit (also known as *defects per unit* [DPU])
- The number of defects on a painted surface
- The number of errors per quote by an insurance company
- Number of bugs in software code

Defects per unit is the basis for the other metrics like defects per opportunity (DPO), defects per million opportunities (DPMO), and related Six Sigma metrics.

The formula for Poisson probability is

$$P(x) = \frac{e^{-\lambda}\lambda^x}{x!}$$

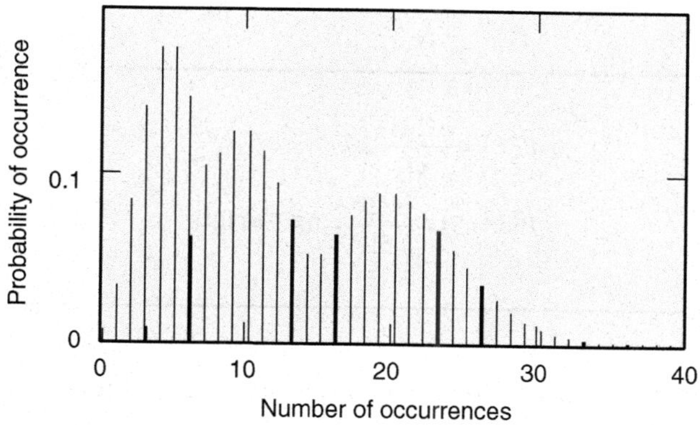

Figure 12.3 Poisson distribution.

Note: DPU monitoring may be performed through *c*-charts. See Chapter 22.

EXAMPLE

The number of defects on an assembly unit has a Poisson distribution with $\lambda = 5$. Find the probability that the second unit produces fewer than two defects.

The probability that the second unit has fewer than two defects is the sum of the probability of zero defects and the probability of one defect.

$$P(x<2) = P(x=0) + P(x=1)$$

$$P(x=0) = \frac{e^{-5}5^0}{0!} \approx 0.006$$

$$P(x=1) = \frac{e^{-5}5^1}{1!} \approx 0.034$$

$$P(x<2) = 0.04$$

The Poisson distribution also has a mean and standard deviation. An interesting fact is that the mean and variance of a Poisson distribution are the same! (λ)

EXAMPLE

A mechanical assembly process has a historical defective rate of 10 percent. What is the probability that a lot of 50 units will contain exactly five defectives?

$n = 50$

Proportion defective p = 10%, that is, 0.1

$\lambda = np = 50 \times 0.1 = 5$

$x = 5$ defectives as per the problem

Continued

Continued

$$P(x) = \frac{e^{-\lambda}\lambda^x}{x!}$$

$$P(x=5) = \frac{e^{-5}5^5}{5!} = 0.175 \approx 18\%$$

The same can be calculated using a simple spreadsheet formula (see Figure 12.4):

=POISSON(B2,B4,FALSE)

Additional exercises:

a. Try calculating the Poisson probability for defectives = 0, 1, 2, 3, 4

b. Try calculating the Poisson probability for historical defective rates of two percent, five percent, and seven percent

c. Now try creating the matrix of Poisson probability for the historical defective rate in (b) and defectives 0 to 5 in (a).

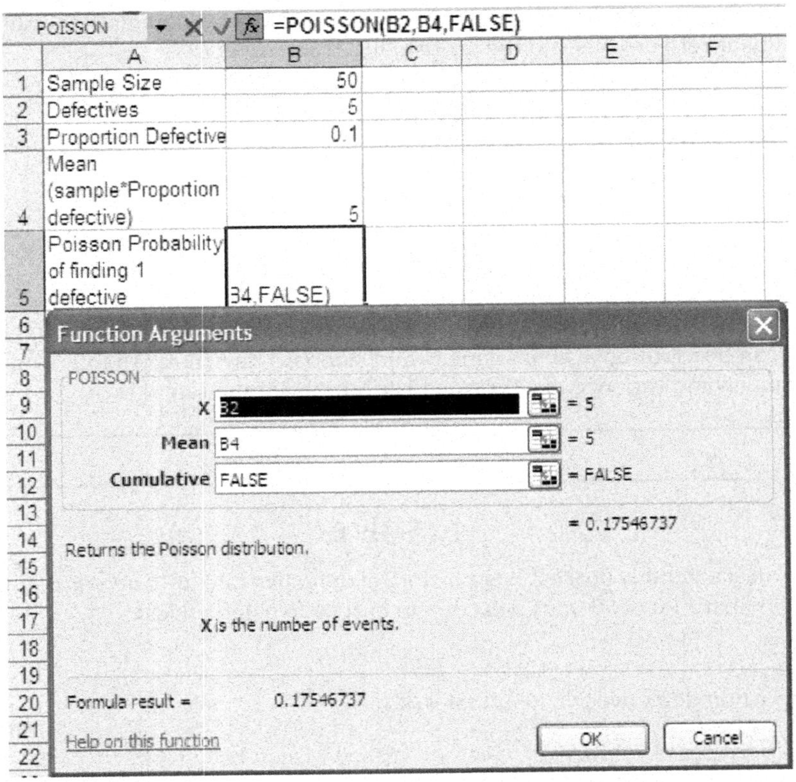

Figure 12.4 Poisson distribution using spreadsheet.

NORMAL DISTRIBUTIONS

The normal distribution is the one most frequently used by various professionals.

This is a continuous distribution used for variable data like measurement of length, mass, time, and so on. Several statistical analyses make an assumption that the data are following a normal distribution. According to the central limit theorem, the averages of measurements of individual data follow a normal distribution even if the individual data are from a different distribution. Since the distribution is in the shape of a bell, it is often referred to as a *bell curve* (Figure 12.5).

Mathematically, the formula for the normal distribution probability density function is

$$P(x) = \frac{e^{-\frac{(x-\mu)^2}{2\sigma^2}}}{\sigma\sqrt{2\pi}}$$

See Figure 12.6.

The area under the curve between any two points, expressed in standard deviation units (Z scores), can be determined from the statistical tables shown in Appendix F. The standard normal distribution has mean = 0 and standard deviation = 1. (See details on mean and standard deviation from Table 11.2.)

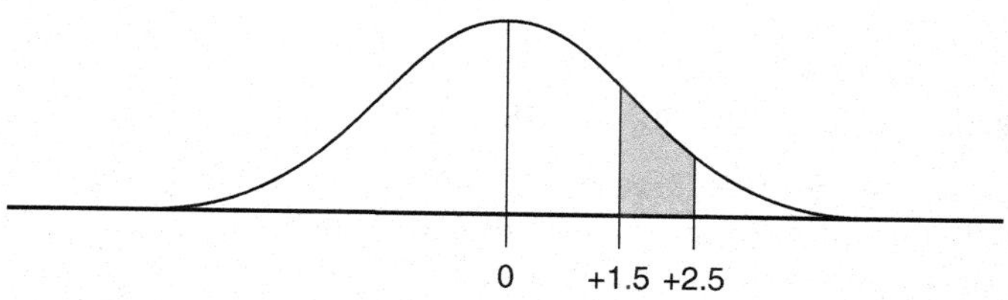

Figure 12.5 Normal distribution (bell) curve.

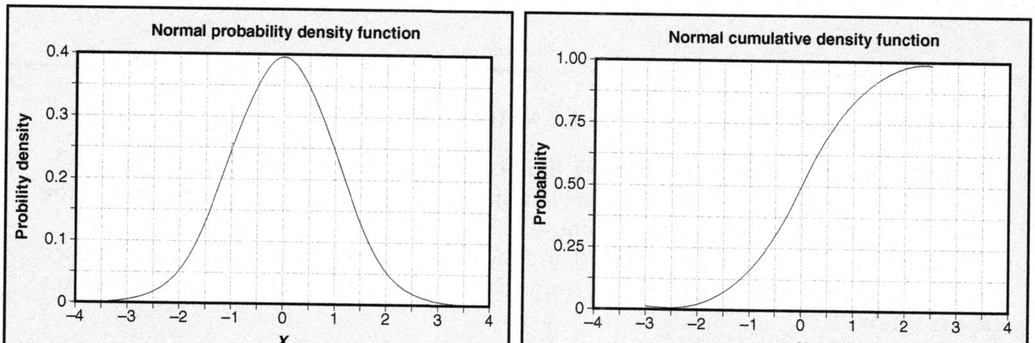

Figure 12.6 Normal probability density function and cumulative density function.
Source: NIST/SEMATECH e-Handbook of Statistical Methods.

EXAMPLE

Find the area under the standard normal curve between +1.50 standard deviations and +2.50 standard deviations.

Solution: Refer to Figure 12.5. Find the area to the right of 1.50 and subtract the area to the right of 2.50:

Using the standard normal tables, the area to the right of 1.50 = 0.0668 and

the area to the right of 2.50 = 0.0062

subtracting: 0.0606

Using Statistical Software and Spreadsheet Capabilities, the analysis results are:

Cumulative Distribution Function

Normal with mean = 0 and standard deviation = 1

$xP(X \leq x)$

2.5 0.993790

Spreadsheet function =NORMDIST(2.5,0,1,TRUE)

Cumulative Distribution Function

Normal with mean = 0 and standard deviation = 1

$xP(X \leq x)$

1.5 0.933193

Spreadsheet function =NORMDIST(1.5,0,1,TRUE)

Therefore, the area under the curve between the two values is 0.0606. The total area under the standard normal curve is 1 so the area under the curve between the two vertical lines is 6.06 percent of the area under the curve. Hence, we can mathematically calculate the area between any two *Z*-scores of interest. This is a very important concept as this calculation is used for process capability measurement. In this example, the *Z*-score is provided directly. Let us explore a real-life example where we have to compute the *Z*-score and find out the area under the curve.

EXAMPLE

A pizza restaurant's order processing time is normally distributed. A random sample has mean 30 minutes and standard deviation five minutes. Estimate the percent of the orders that are between 35 and 20 minutes.

Solution: Find the *Z*-score for 20 and 35. The *Z*-score is the number of standard deviations that the measurement is from the mean and is calculated by the formula $Z = (x - \mu)/\sigma$.

Continued

Continued

$Z(20) = (20 - 30)/5 = -2.00$

$Z(35) = (35 - 30)/5 = 1$

Area to the right of −2.00 = 0.97724

Area to the right of +1.00 = 0.15865

Subtracting: 0.8186

Approximately 82 percent of the orders are processed between 35 minutes and 20 minutes. Put another way, the probability that a randomly selected order will have a processing time between 35 minutes and 20 minutes is approximately 0.82.

Extended exercise:

> If the pizza restaurant promises their customers "35-minute delivery or free" and average order cost is $30, estimate the total cost of free food the restaurant has to give away with the current process variation.

Distributions like chi-square (χ^2), *t*, and F are used for decision-making using hypothesis testing.

CHI-SQUARE DISTRIBUTION

If *w*, *x*, *y*, and *z* are random variables with standard normal distributions, then the random variable defined as $f = w^2 + x^2 + y^2 + z^2$ has a chi-square distribution.

The chi-square (χ^2) distribution is obtained from the values of the ratio of the sample variance and population variance multiplied by the degrees of freedom. This occurs when the population is normally distributed with population variance σ^2.

The most common application of the chi-square distribution is testing proportions. As the degrees of freedom increase, the chi-square distribution approaches a normal distribution. See Figure 12.7.

Properties of the chi-square distribution:

- Chi-square is nonnegative. (It is the ratio of two nonnegative values, therefore must be nonnegative itself.)
- Chi-square is nonsymmetric.
- There are many different chi-square distributions, one for each degree of freedom.
- The degrees of freedom when working with a single population variance is $n - 1$.

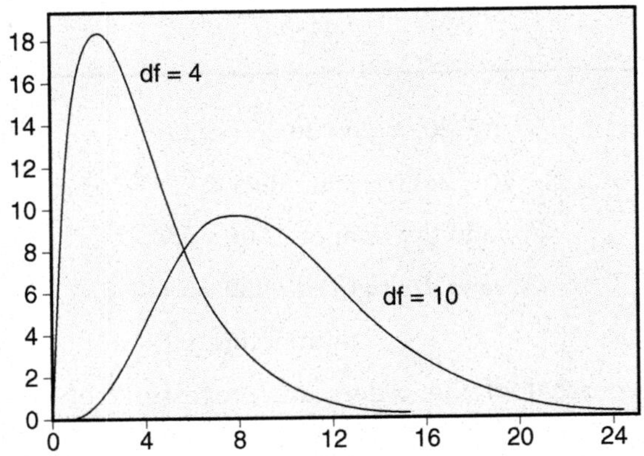

Figure 12.7 Chi-square distribution example.

DEGREES OF FREEDOM

The amount of information your data provide that you can "apply" to estimate the values of unknown population parameters, and calculate the variability of these estimates.

> **EXERCISE**
>
> Find the critical value for one percent of the area under the chi-square probability density for a random variable that has five degrees of freedom.
>
> From the chi-square table (Appendix I), df = 5 and $\chi^2_{0.01}$ is 15.09.

Let us also look at another example of chi-square distribution using tabulated data.

A Green Belt is interested in exploring the relationship between age group and social network usage hours. After interviewing 100 people about their social networking online hours and dividing them into groups of either 3 hours/day or 5 hours/day for ages 13–18 (column 1) and ages 21–35 (column 2) the worksheet data are tabulated as follows. Did the Green Belt see any difference between the two groups (that is, number of users from 3 hours/day and 5 hours/day)? Assume an alpha risk of 5%.

	Age 13–18	Age 21–35
3 hours/day	21	31
5 hours/day	29	19

Continued

Continued

	Age 13–18	Age 21–35	All
1	21 26 0.9615	31 26 0.9615	52
2	29 24 1.0417	19 24 1.0417	48
1: 3 hours/day, 2: 5 hours/day			
All	50	50	100

Cell Contents: Count
Expected count
Contribution to chi-square

Pearson Chi-Square = 4.006, DF = 1, P-Value = 0.045
Likelihood Ratio Chi-Square = 4.034, DF = 1, P-Value = 0.045

Since the p-value is < the assumed alpha value of 0.05, we can conclude statistically there is a difference between the two groups.

Now for an additional exercise, did the Green Belt see any difference between the two age groups?

t-DISTRIBUTION

If x is a random variable with a standard normal distribution and y is a random variable with a χ^2 distribution, then the random variable defined as

$$t = \frac{x}{\sqrt{\frac{y}{k}}}$$

is the t-distribution with k degrees of freedom, where

k = the degrees of freedom for the χ^2 variable

Notice that as $k \to \infty$, t approaches the normal distribution. This distribution is used in hypothesis tests as illustrated in Chapter 17.

Following are the important properties of Student's t-distribution:

1. Student's t-distribution is different for different sample sizes.
2. Student's t-distribution is generally bell-shaped, but with smaller sample sizes shows increased variability (flatter). In other words, the distribution is less peaked than a normal distribution and with thicker tails. As the sample size increases, the distribution approaches a normal distribution. For $n > 30$, the differences are negligible.
3. The mean is zero (much like the standard normal distribution).
4. The distribution is symmetrical about the mean.

5. The variance is greater than one, but approaches one from above as the sample size increases ($\sigma^2 = 1$ for the standard normal distribution).
6. The population standard deviation is unknown.
7. The population is essentially normal (unimodal and basically symmetric).

EXAMPLE

Twelve randomly selected chemical packs were measured before mixing with the raw material. The weights in grams of chemicals supplied by a vendor to an organization are as follows:

7.3, 7.9, 7.1, 7.3, 7.4, 7.3, 7.0, 7.3, 7.7, 7.3, 7.1, 7.8

The weight on the pack says 7.5 grams.

What is the probability that the weight of the rest of the packs in storage is greater than 7.5?

Solution:

The mean of the 12 packs is 7.375

The sample standard deviation of the 12 packs is 0.2832

To find the area under the curve:

$$t = \frac{\bar{x} - \mu}{s / \sqrt{n}}$$

Minitab analysis:

Using statistical software capabilities, the analysis results are:

Test of μ = 7.5 versus > 7.5

Variable	N	Mean	StDev	SE Mean	95% Lower Bound	T	P
Weight	12	7.37500	0.28324	0.08177	7.22816	–1.53	0.923

Approximately 7.8 percent of the packs could be greater than 7.5.

F-DISTRIBUTION

The *F*-distribution is the ratio of two chi-square distributions with degrees of freedom v_1 and v_2, respectively, where each chi-square has first been divided by its degrees of freedom. The *F*-distribution is commonly used for *analysis of variance* (ANOVA), to test whether the variances of two or more populations are equal. This distribution is used in hypothesis tests.

(See more details on ANOVA in Chapter 17.)

$$f(x) = \frac{\Gamma(\frac{\nu_1+\nu_2}{2})(\frac{\nu_1}{\nu_2})^{\frac{\nu_1}{2}} x^{\frac{\nu_1}{2}-1}}{\Gamma(\frac{\nu_1}{2})\Gamma(\frac{\nu_2}{2})(1+\frac{\nu_1 x}{\nu_2})^{\frac{\nu_1+\nu_2}{2}}}$$

where ν_1 and ν_2 are the shape parameters and Γ is the gamma function. The formula for the gamma function is

$$\Gamma(\alpha) = \int_0^\infty t^{\alpha-1} e^{-t} dt$$

The F probability density function for four different values of the shape parameters is shown in Figure 12.8.

EXERCISE

Find the F ratio given that $F_{0.05}$ with degrees of freedom (ν_1) 4 and (ν_2) 6 is 4.53. Find $F_{0.95}$ with degrees of freedom (ν_1) 6 and (ν_2)4.

$$F_{0.95,6,4} = 1 / F_{0.05,4,6} = 1 / 4.53 = 0.22$$

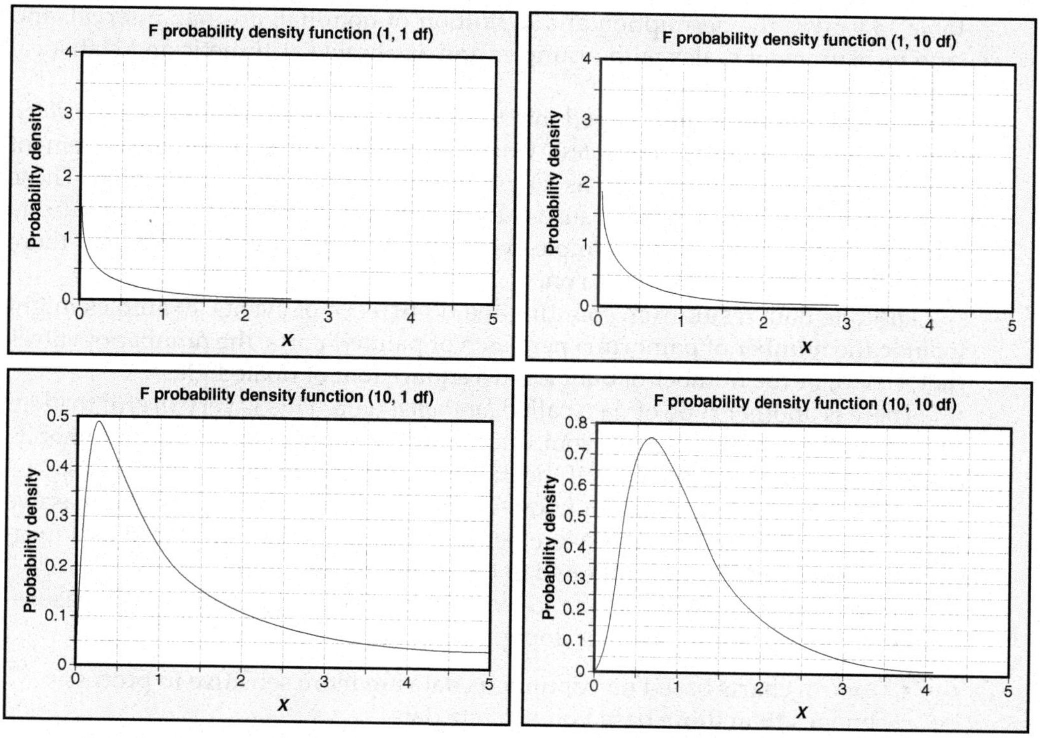

Figure 12.8 F-distribution with varying degrees of freedom.

Chapter 13

Collecting and Summarizing Data

TYPES OF DATA AND MEASUREMENT SCALES

> Identify and classify continuous (variables) and discrete (attributes) data. Describe and define nominal, ordinal, interval, and ratio measurement scales. (Analyze)
>
> **Body of Knowledge III.D.1**

Table 13.1 gives the description and definition of nominal, ordinal, interval, and ratio measurement scales with examples and applicable arithmetic and statistical operations.

Quantitative data are grouped into two types, *continuous* (also called *variables*) and *discrete* (also called *attributes*). Continuous data result from measurement on some continuous scale such as length, weight, temperature, and so on. These scales are called continuous because between any two values there are an infinite number of other values. For example, between 1.537 inches and 1.538 inches there are 1.5372, 1.5373, 1.53724, and so on.

Discrete data result from counting the occurrence of events. Examples might include the number of paint runs per batch of painted parts, the number of valves that leaked, or the number of bubbles in a square foot of floated glass.

There is another type of data called *locational* data. This is very useful to identify where the data are coming from. An example is paint defects in an automobile assembly line. It is not adequate if the data are collected as continuous or discrete. The rework technician needs to know where the defect is found on the massive surface area of the automobile. More details on location data displayed as a measles chart (Figure 10.9) are explained in Chapter 10.

Effort should always be made to move from discrete to continuous measurements. There are two reasons for doing this:

- Control charts based on continuous data are more sensitive to process changes than those based on discrete data.

Table 13.1 Measurement scales.

The measurement scales

Nominal	Ordinal	Interval	Ratio
The values of the scale have no "numeric" meaning in the way that usually applies with numbers, and no ordering scheme. Example: Color-coded wires by quantity in an electrical cable.	The intervals between adjacent scale values are indeterminate. Example: Categorization of defects by criticality. Critical: functional failures. Major: performance degradation. Minor: cosmetic defects, and so on.	Intervals between adjacent scale values are equal with respect to the attribute being measured. Example: The difference between 20 °C and 40 °C is the same as the difference between –10 °C and –30 °C.	There is a rational zero point for the scale. Example: Ratios are equivalent; for example, the ratio of 10 to 5 is the same as the ratio of 64 to 32.

Applicable arithmetic and statistical operations for the measurement scales

Nominal	Ordinal	Interval	Ratio
Counting Mode Chi square	"Greater than" or "less than" operations Median Interquartile range Sign test	Addition and subtraction of scale values Arithmetic mean Standard deviation *t*-test, *F*-test	Multiplication and division of scale values Geometric mean Coefficient of variation

For more details please refer to any standard statistical basics book.

- When designed experiments are used for process improvement, changes in continuous data may be observed even though the discrete measurement hasn't changed.

Discrete data require larger statistically valid sample sizes for decision-making than continuous data for the same consumer risk protection (type II error). See Chapter 17 for consumer risk.

SAMPLING AND DATA COLLECTION PLANS AND METHODS

> Define and apply various sampling methods (random and stratified) and data collection methods (check sheets and data coding). Prepare data collection plans that include gathering data and performing quality checks (e.g., minimum/maximum values, erroneous data, null values). (Apply)
>
> **Body of Knowledge III.D.2**

Data collection is performed in an organization for various reasons:

- Legal, regulatory, or statutory requirements
- Analysis, improvement, and knowledge management
- Contractual requirements of customers

Irrespective of reasons, data collection can be very expensive if the data collection is not planned and effectively implemented. Many organizations tend to collect more data than required. Answering some basic questions before actually starting to collect the data, such as what, why, where, when, who, and how (5W1H), can help make planning the data collection more effective. Where possible, real-time data acquisition from equipment is more effective and reduces human errors and data transfer errors. Also, with real-time data collection, action can be taken swiftly without significantly impacting goods produced or service rendered. Data that are not collected in real time will have a lag between data collection and decisions made. This can sometimes result in expensive corrections. One has to weigh the cost benefit. Real-time data acquisition can require additional costs for infrastructure upgrading.

Where manual data entry is involved, it is more efficient to use data coding to avoid repetitive recording of numbers and errors due to fatigue. Decoding (see Table 13.2) may be applied depending on the analysis to be performed.

There are several methods for collecting data:

- Surveys
- Face-to-face interviews
- Focus groups
- Mystery shopping
- Customer feedback
- Automatic data capture
- Manual data capture

Data collection methods that are one-on-one like focus groups and face-to-face interviews have higher integrity of data and provide opportunities to clarify with the respondents, while data collection methods like surveys have low response rates (approximately 10 to 15 percent), and improperly constructed surveys can result in misleading responses.

Customer feedback after a product failure or service issue is reactive. Hence, organizations should strive to gather as much up-front information as possible before designing the product or service.

TECHNIQUES FOR ASSURING DATA ACCURACY AND INTEGRITY

Even sophisticated data collection and analysis techniques can be defeated if the data are entered with errors. Common causes of errors include:

- Units of measure not defined (for example, feet or meters?)
- Closeness of handwritten characters/legibility (for example, 2 or Z?)
- Inadequate measurement system resolution/discrimination
- Rounding off measurements and losing precision
- Emotional bias resulting in distortion of data (for example, flinching)
- Inadequate use of validation techniques—using guesswork or personal bias
- Multiple points of data entry—opportunity for inconsistency and errors
- Poor instructions or training causing erroneous data entry
- Ambiguous terminology
- Clerical or typographical errors

To minimize error:

- Have a carefully constructed data collection plan following 5W1H—what, where, who, when, why, and how
- Maintain a calibration schedule for data collection equipment
- Conduct repeatability and reproducibility (R & R) studies on measurement system
- Record appropriate auxiliary information regarding units, time of collection, conditions, measurement equipment used, name of data recorder, and so on
- Use appropriate statistical tests to remove outliers
- If data are transmitted or stored digitally, use an appropriate redundant error correction system
- Provide clear and complete instruction and training for collection, transformation, analysis, and interpretation

Types of Sampling

Random Sampling. Every sample randomly picked from the lot has equal probability of getting picked. If effectively administered, sampling can save money for the organization. The lot being sampled has to be homogeneous for random sampling.

Sequential Sampling. Sequential sampling is used in destructive testing and reliability testing applications where higher cost is involved in testing the unit. The samples are tested one by one sequentially until the desired results are reached.

Stratified Sampling. When there is a mixture of parts from different machines, different streams, different raw material lots, or different process settings, there is no homogeneity of the lot. Hence, random sampling will not yield the right results. It will be more effective to stratify the lot based on the criteria (by machine, stream, lot, or settings) and pick random samples from the stratified group. Many quality practitioners do not take into consideration the lack of homogeneity of the lots coming out of their end of the manufacturing line or from the supplier, and attempt random sampling. This may result in proportion nonconformities greater than anticipated by applying a specific acceptable quality limit (AQL) sampling plan.

Random and stratified sampling are applicable to many industry sectors. In a service application like a call center, calls can be sampled using one of the above methods, in a healthcare setting, drugs in inventory, and in education, students and faculty can be sampled.

Where manual data entry is involved, it is more efficient to use data coding to avoid repetitive recording of numbers, and errors due to fatigue. Decoding may be applied depending on the analysis to be performed.

EXAMPLE

An inspector is measuring the diameter of a machine part. If the data are expected to fall between the upper and lower specification limits, typing the measurement data repetitively may result in clerical or administrative errors. In this case, the measurement values can be coded in a single-digit number representing the full measurement value. An important aspect of this approach is that we should still be able to arrive at the measurement value after decoding (see Table 13.2).

Table 13.2 Coding–decoding.

Coding		Decoding	
Actual measurements	**Coded value**	**Coded value**	**Actual measurements**
10.120	1	1	10.120
10.121	2	2	10.121
10.122	3	3	10.122
10.123	4	4	10.123

Decoding is turning the code back to actual measurements. Coding is also used to communicate the distribution and general trend of the data and protect confidentiality.

CHECK SHEETS

Check sheets are used to observe or review a process, usually during execution of the process. Check sheets pre-categorize potential outcomes for data collection using sets of words, tally lists, or graphics. Figure 13.1 is an example of a completed

Paint color defect causes

Reason	Day of the week							Weekly total
	Mon	Tue	Wed	Thu	Fri	Sat	Sun	
Operator misread instructions	I	I	0	II	卌	卌 II	III	19
Wrong pigment used	III	I	卌	III	IIII	IIII	卌 卌	30
Wrong color code from customer	II	I	II	0	0	卌	0	10
Outdated base paint	0	0	I	0	III	0	IIII	8
Daily total	6	3	8	5	12	16	17	67

Figure 13.1 Example of a check sheet.

check sheet, in tabular format, used to collect data related to a paint mixing process. This simple tool provides a method of easy collection of the data. By collecting data on a check sheet, common patterns or trends can be identified.

The basic steps in making a check sheet are:

1. Identify and agree to the causes or conditions that are to be collected.
2. Decide who will collect the data, over what time period(s), and how the data will be collected.
3. Create a check sheet that will work within the operation where it will be used.
4. Collect the data as designed to ensure consistency and accuracy of the information.

Check sheets can be the basis for other analytical tools and are incorporated into attribute statistical process control charts. Creating and using a check sheet can help focus on continual improvement and may foster changes just because the check sheet is being used.

Data collection is performed in an organization for various reasons:

- Legal, regulatory, or statutory requirements
- Analysis, improvement, and knowledge management
- Contractual requirements of customers

Irrespective of the reasons, data collection can be very expensive if it is not effectively planned and implemented. Many organizations tend to collect more data than required. Answering some basic questions—such as what, why, where, when, who, and how (5W1H)—before starting to collect the data can help make the data collection more effective. Where possible, real-time data acquisition from equipment is more effective and reduces human errors and data transfer errors.

A typical data collection plan (Table 13.3) includes:

What to measure

- What do you want to measure?
- Type of measure

Table 13.3 Data collection plan.

Define what to measure			Define how to measure			Who will do it?	Sample plan			
Measure	**Type of measure**	**Operational definition**	**Measurement or test method**	**Data tags needed to stratify the data**	**Data collection method**	**Person(s) assigned**	**What?**	**Where?**	**When?**	**How many?**
Name of parameter or condition. to be measured	X or Y attribute or discrete data, product, or process data	Measurement defined in such a way as to achieve repeatable results from multiple observers	Visual inspection or automated test? Test instruments are defined Procedures for data collection are defined	Data tags (e.g., time, date, location, tester, line, customer, buyer, and operator) are defined for the measure	Manual? Spreadsheet? Computer based?	State who has the responsibility	What measure is being collected	Location for data collection	How often the data are collected	The number of data points collected per sample

Measurement systems analysis
How will you know the measures are accurate, linear, stable, repeatable, and reproducible?

Responsibility for measurement system qualification and release? Responsibility for measurement training? Who will measure?

Control plan, FMEA, customer specification, manufacturing work instructions CTQ vs critical to process (CTP) matrix are typical sources of input for "what to measure"

- Operational definition (e.g., cycle time, lead time, process time; these terms can have different meanings in different organizations. Also within the organization, stakeholders could interpret the metric KPI differently).

How to measure

- Measurement or test method
- Data tags need to stratify the data
- Data collection method
- Data validation for quality and integrity (identify methods to identify and screen common errors like clerical, rounding off, truncation, bias, etc.)

When? (Frequency)

- Timing (beginning and end of day, beginning of every new shift, when the new raw material is loaded into equipment, etc.) and frequency (once a minute, once a day, weekly, monthly, annually)

Where? (Data collection point)

- Process step or physical location

Who will measure?

- Responsibility to measure (or collect) the data

Sampling plan

- How many? (Number of data points to be statistically valid)

DESCRIPTIVE STATISTICS

> Define, calculate, and interpret measures of dispersion and central tendency. Develop and interpret frequency distributions and cumulative frequency distributions. (Evaluate)
>
> **Body of Knowledge III.D.3**

Two principal types of statistical studies are *descriptive* and *inferential*. Inferential studies analyze data from a sample to infer properties of the population from which the sample was drawn. The purpose of descriptive statistics is to present data in a way that will facilitate understanding.

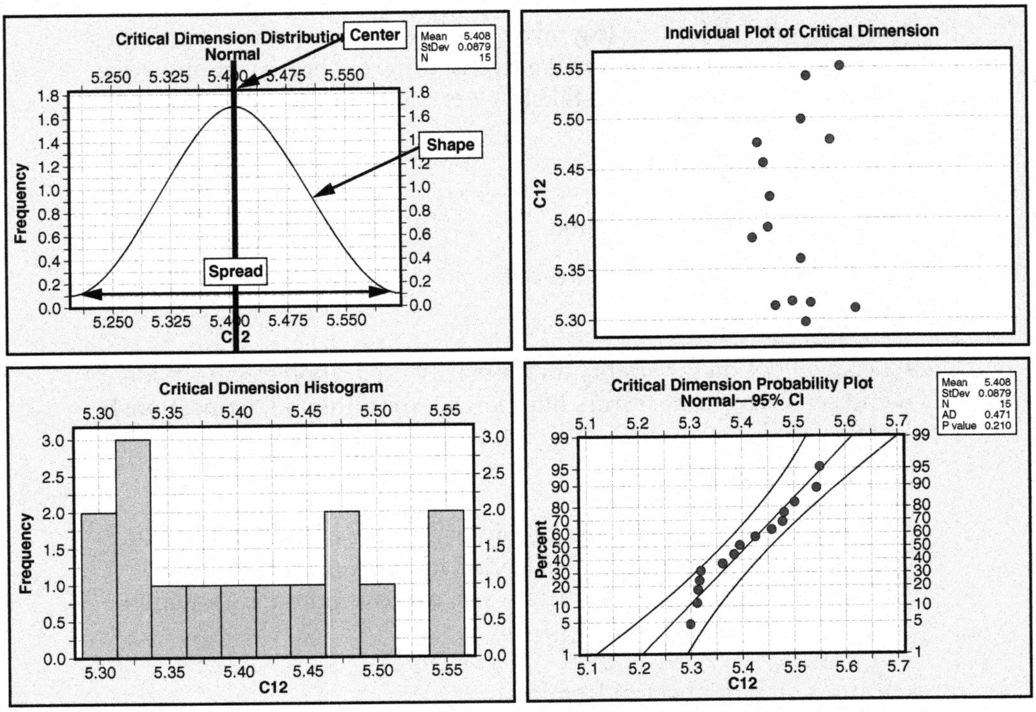

Figure 13.2 Example of a data set as illustrated by a frequency distribution, individual plot, histogram, and probability plot.

The following data represent a sample of critical dimensions of a chemical deposition operation. What conclusions can be reached by looking at the data set?

5.551, 5.361, 5.392, 5.479, 5.456, 5.542, 5.423, 5.476,
5.298, 5.499, 5.312, 5.319, 5.317, 5.314, 5.382

The charts in Figure 13.2 reveal information about the sample data that was not obvious from the data list, such as:

- The spread of the sample
- An indication of the shape of the sample
- Center of the sample
- Fitting of normal distribution of the sample (explained later in this chapter)

The first three attributes—spread, shape, and center—are key to understanding the data and the process that generated them.

The *spread* of the sample is also referred to as *dispersion* or *variation* and is usually quantified with either the sample range (defined as the highest value minus the lowest value) or the sample standard deviation. The sample standard deviation is the more sophisticated metric and is defined as

$$s = \sqrt{\frac{\Sigma(x-\bar{x})^2}{n-1}}$$

where

$\bar{x}$ = The sample mean or average

n = Sample size

This formula produces an estimate of the standard deviation of the population from which the sample was drawn. If data for the entire population are used (rare in practical applications), the population standard deviation is defined as

$$\sigma = \sqrt{\frac{\Sigma(x-\mu)^2}{N}}$$

where

μ = The population mean or average

N = Population size

Due to the complexity of these formulas, one should use a calculator with standard deviation capabilities.

The *shape* of the sample refers to a smooth curve that serves as a sort of umbrella approximately covering the tops of the bars in the histogram. In this case, it appears that the sample came from a normally distributed population. Other descriptors of shape include *kurtosis*, *symmetry*, and *skewness*. The *center* of the sample may be quantified in three ways:

- The *mean*, statistical jargon for the more common word "average"
- The *median*, which is defined as the value that is in the middle of a sorted list of data
- The *mode*, which is the value that appears most frequently in the sample

See more on the median in "Understanding Medians," *Quality Progress*, July 2014.

In the example in Figure 13.2:

- The mean = (Sum of the values) ÷ (Number of values) = $\Sigma x/n$ = 81.121/15 = 5.408
- The median of the 15 values would be the eighth value when the sample is sorted in ascending order, in this case 5.392. If there are an even number of values, the median is obtained by averaging the two middle values.

Of these three measures, the mean is the most useful in quality engineering applications. The sample mean is often denoted as an x with a bar above it and pronounced "x-bar."

SUMMARY OF DESCRIPTIVE MEASURES

Name	Symbol	Formula/Description
Measures of central tendency		
Mean	$\bar{x}$	$\frac{\Sigma x}{n}$
Median	$\tilde{x}$	Middle number in sorted list
Mode		Most frequent number
Measures of dispersion		
Range	R	High value–low value
Sample standard deviation	s	$\sqrt{\frac{\Sigma(x-\bar{x})^2}{n-1}}$

The population standard deviation σ uses the same formula as sample standard deviation with a denominator n.

Cumulative Frequency Distribution

If a column showing totals of the frequencies to that point is added to the frequency distribution, the result is called a cumulative frequency distribution. An example is shown in Figure 13.3.

x	Frequency	Cumulative frequency
.120	1	1
.121	3	4
.122	4	8
.123	7	15
.124	4	19
.125	4	23
.126	3	26
.127	1	27

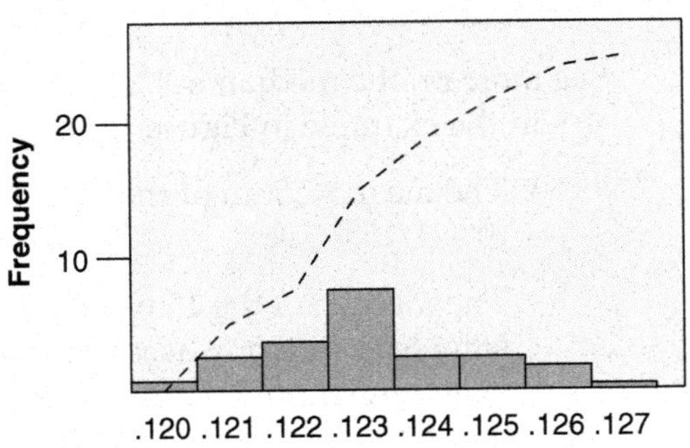

Figure 13.3 Cumulative frequency distribution in table and graph form.

GRAPHICAL METHODS

> Construct and interpret diagrams and charts that are designed to communicate numerical analysis efficiently, including scatter diagrams, normal probability plots, histograms, stem-and-leaf plots, box-and-whisker plots. (Create)
>
> **Body of Knowledge III.D.4**

Table 13.4 provides an overview of the graphical methods discussed in this section. The following paragraphs provide more information about those not already discussed.

Table 13.4 Comparison of various graphical methods.

Name	Purpose	Application	Interpretation	Ease of use
Tally	Provides a quick tally of total quantity and by class interval. Provides visual idea of the distribution shape.	Used to count defect quantity by type, class, and/or category	Tally mark concentration and spread roughly indicate distribution shape. Tally marks of five are crossed out as a group for easy counting. Isolated groups of tally marks indicate uneven distribution.	Very easy to create and interpret
Frequency distribution	Provides a pictorial view of numerical data about location and spread.	Especially useful if tally column cells have a large number of marks	Concentration of data is seen as a peak, and spread of the data is demonstrated by the width of the curve. Thinner distribution indicates lesser variation. Distribution can be unimodal (with one peak), bimodal (two peaks), or multimodal (multiple peaks) indicating a mixture of populations. Distribution with no peak and flat curve indicates rectangular distribution.	Not so easy to create but easier to interpret

Continued

Table 13.4 Comparison of various graphical methods. *(Continued)*

Name	Purpose	Application	Interpretation	Ease of use
Stem-and-leaf plot	Provides numerical data information about the contents of the cells in a frequency distribution.	Useful to quickly identify any repetitive data within the class interval	If data values within cells are not fairly evenly distributed, measurement errors or other anomalous conditions may be present	Easy to create but difficult to interpret
Box-and-whisker plot	Provides a pictorial view of minimum, maximum, median, and interquartile range in one graph.	Provides more information than distribution plot but easier to interpret. Outliers are easily identified on the graph.	If the location of the center line of the box is right in the middle, the data may be normally distributed. If moved to one of the sides, the data may be skewed. The data points outside the whiskers indicate outliers. Unequal whiskers indicate skewness of the distribution.	Easy to create and interpret
Scatter diagram	Detects possible correlation or association between two variables, or cause and effect.	Used for root cause analysis, estimation of correlation coefficient, making prediction using a regression line fitted to the data	To estimate correlation, the relationship has to be linear. Nonlinear relationship may also exist between variables. If the plotted data flow upward left to right, the relationship is positively correlated. If the plotted data flow downward from left to right, the relationship is negatively correlated. If data are spread about the center with no inclination to right or left, there may not be any correlation.	Easy to create and interpret

Continued

Table 13.4 Comparison of various graphical methods. *(Continued)*

Name	Purpose	Application	Interpretation	Ease of use
Run chart	Provides a visual indicator of any nonrandom patterns.	Used when real-time feedback is required for variables data	Patterns like cluster, mixture, trend, and oscillation are spotted based on the number of runs above and below the mean or median. *p*-value identifies the statistical significance of a nonrandom pattern. *p*-value less than 0.05 identifies a stronger significance.*	Easy to create and interpret

**p*-value explained in Chapter 17.

Stem-and-Leaf Plot

A *stem-and-leaf plot* is constructed much like a tally column except that the last digit of the data value is recorded instead of the tally mark. This plot is often used when the data are grouped. Consider the following example:

These are the weight values in grams collected by weighing a batch of mixed chemical products:

10.3, 11.4, 10.9, 9.7, 10.4, 10.6, 10.0, 10.8, 11.1, 11.9,
10.9, 10.8, 11.7, 12.3, 10.6, 12.2, 11.6, 11.2, 10.7, 11.4

The normal histogram would look like the first chart in Figure 13.4.

The stem-and-leaf plot on the right conveys more information than a tally column or the associated histogram would. Note that the ordered stem-and-leaf sorts the data and permits easy determination of the median.

The display has three columns:

1. *The leaves* (right). Each value in the leaf column represents a digit from one observation. The "leaf unit" (declared above the plot) specifies which digit is used. In the example, the leaf unit is 0.1. Thus, the leaf value for an observation of 7 is 7, while the leaf value for an observation of 10 is 0.

2. *The stem* (middle). The stem value represents the digit immediately to the left of the leaf digit. In the example, the stem value of 10 indicates that the leaves in that row are from observations with values greater than or equal to 10 (for example, 10.0, 10.3, 10.4).

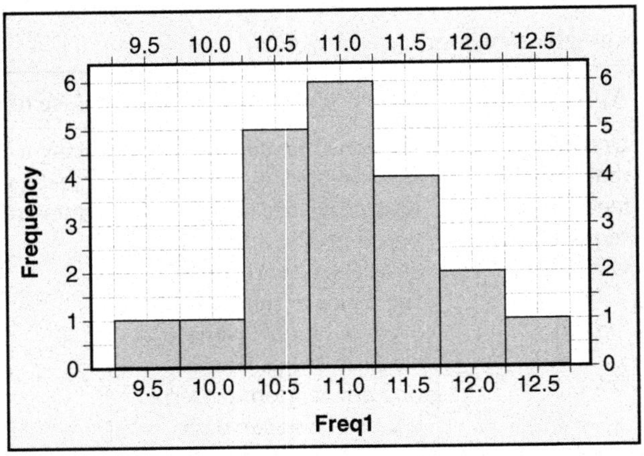


```
Stem-and-leaf plot
N = 20
Leaf unit = 0.10

  1    9    7
  4   10    034
 (7)  10    6678899
  9   11    1244
  5   11    679
  2   12    23
```

Figure 13.4 Histogram and stem-and-leaf plot comparison.

3. *Counts* (left). If the median value for the sample is included in a row, the count for that row is enclosed in parentheses. The values for rows above and below the median are cumulative. The count for a row above the median represents the total count for that row and the rows above it. The value for a row below the median represents the total count for that row and the rows below it.

In the example, the median for the sample is 10.9, so the count for the third row is enclosed in parentheses. The count for the second row represents the total number of observations in the first two rows. Similarly, the fourth row provides the count of the fourth, fifth, and sixth rows.[1]

Box Plots

The *box plot* (also called a *box-and-whisker plot*), developed by Professor John Tukey of Princeton University, uses the high and low values of the data as well as the quartiles.

The quartiles of a set of data divide the sorted data values into four approximately equal subsets. The quartiles are denoted Q_1, Q_2, and Q_3. Second quartile Q_2 is the median. Q_1 is the median of the set of values at or below Q_2. Q_3 is the median of the set of values at or above Q_2. This is illustrated in Figure 13.5.

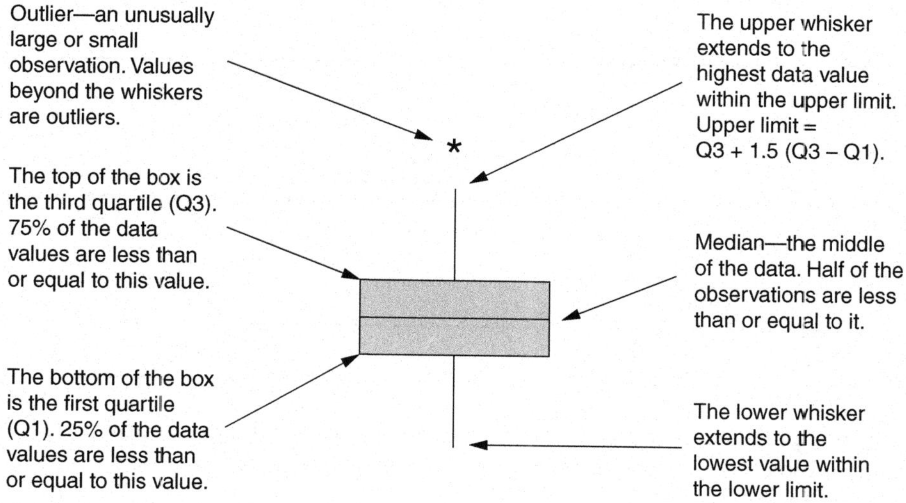

Figure 13.5 Box plot.

EXAMPLE

Following are the pull test data from testing bond strength of a consumer product:

8.250, 8.085, 8.795, 9.565, 11.880, 9.180, 9.950, 9.630, 8.150

The data after sorting:

8.085, <u>8.150, 8. 250</u>, 8.795, **9.180**, 9.565, <u>9.630, 9.950</u>, 11.880

Low value is 8.085, high value is 11.88, $Q_1 = 8.20$, $Q_2 = 9.18$, and $Q_3 = 9.79$. (Note that quartiles need not be values in the data set itself.) The resulting plot is shown in Figure 13.6. Figure 13.7 shows how the shape of the dot plot is reflected in the box plot.

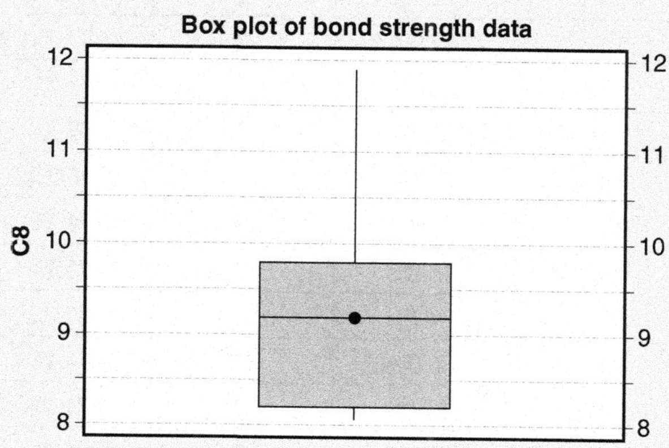

Figure 13.6 Example box plot.

Continued

Continued

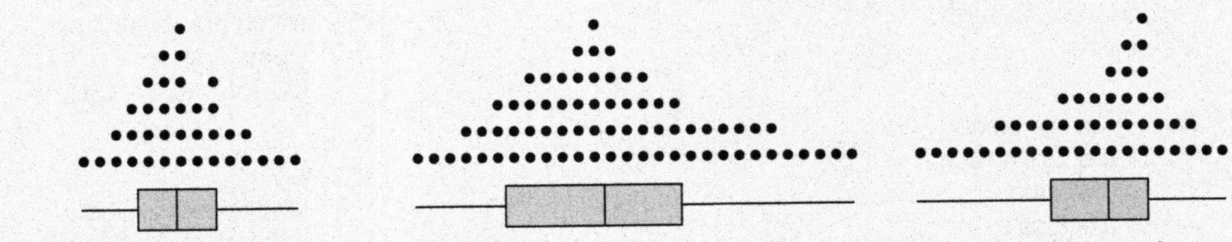

Figure 13.7 Box plots versus dot plots.

Box plots can be used to mine information from a database. Box plots can also be used to compare two or more populations visually and see the shift in median and variation.

EXAMPLE

An experiment was conducted in an assembly process with three different process settings. The Six Sigma Green Belt wants to know visually if there is any variation between processes. (Statistical methods more appropriate for this application, such as ANOVA, are discussed in Chapter 17.)

Partial data from the experiment are shown in Table 13.5.

Table 13.5 Example of yield data from three periods.

Yield	Period	Yield	Period	Yield	Period
87.0%	1	57.1%	2	77.1%	3
84.4%	1	62.5%	2	74.2%	3
76.9%	1	60.0%	2	62.3%	3
90.0%	1	72.3%	2	66.7%	3
84.4%	1	42.9%	2	80.0%	3
86.2%	1	69.1%	2	71.4%	3
89.6%	1	59.5%	2	55.1%	3
94.2%	1	71.3%	2	36.2%	3
95.6%	1	79.5%	2	94.3%	3
85.2%	1	43.8%	2	71.4%	3
89.7%	1	94.7%	2	67.1%	3
100.0%	1	66.0%	2	91.4%	3
89.1%	1	40.0%	2	71.4%	3
96.9%	1	67.8%	2	61.4%	3
93.6%	1	65.0%	2	75.4%	3

Continued

Continued

The resultant data and box plots are shown in Figure 13.8.

The period can be further divided into major process settings like gas flow or temperature between the three periods and further analyzed to get even greater insight. The box plots in Figure 13.8 show that the largest variation is in period 2. The Green Belt would probe further to see what may have caused the largest variation in the second setting using more appropriate root cause analysis tools.

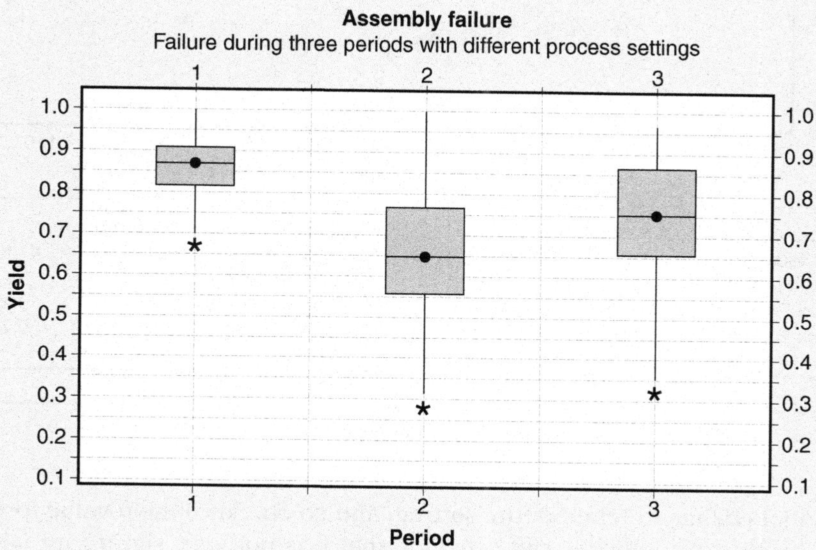

Figure 13.8 Box plot by period.

The Run Chart

The *run chart* is used to identify patterns in process data. There are also related statistical tests that can be performed to detect any nonrandom behavior. All of the individual observations are plotted in time sequence, and a horizontal reference line is drawn at the median. Typically, a run chart is used when the subgroup size is one. When the subgroup size is greater than one, the subgroup means or medians are calculated and connected with a line, similarly to control charts. However, run charts are different from control charts (for example, $\bar{X}$ and R charts); run charts do not have statistical control limits to monitor variation.

Run chart tests can detect trends, oscillation, mixtures, and clustering. These are nonrandom patterns and suggest that the variation observed can be attributed to special causes. Common cause variation is variation that is inherent in the process. A process is in control when only common causes are present. (See more details on common and special causes in Chapter 22).

Lack of steadiness in a process can cause oscillation. In the example shown in Figure 13.9, the p-value for oscillation is 0.78, indicating that it is not significant. A trend can be either upward or downward due to tool wear, loosening of a fixture,

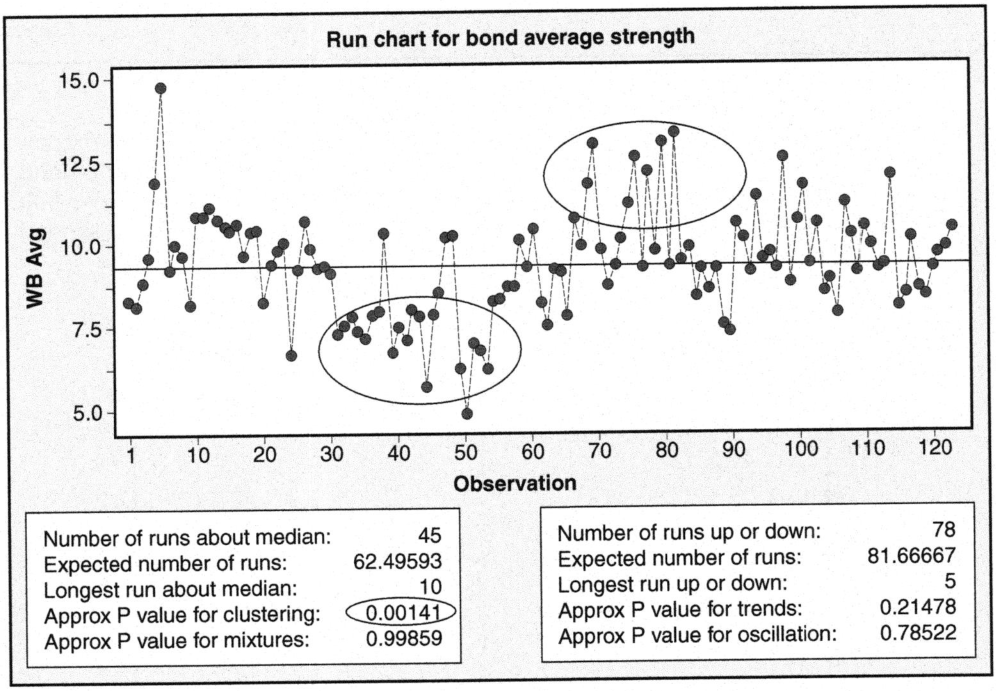

Figure 13.9 Run chart analysis (using statistical software).

gradual change in temperature setting, and so on. Since the *p*-value for trends is 0.21 in this example, we can conclude that it is not very significant (although a visual trending can be seen in two regions). When there is a mix-up between parts from two different machines, two different operators, or two lots of materials, the process points tend to appear on either side of the chart with nothing closer to the centerline. We don't see that in this example as the *p*-value is 0.998 (almost so high that we can rule out this possibility). Now we have a problem with clustering, which is highly significant and may be due to measurement problems or lot-to-lot or setup variability. Carefully reviewing the measurement system analysis reports and procedure for setting up the machine, and verifying whether the operator is trained for machine-setting, and so on, can reveal some insight into clustering. (See Chapter 17 for definition of *p*-value.)

Key Point: The common run chart is an extremely powerful tool for showing how stable a process is behaving. This assumes, of course, that you want to see process behavior. Otherwise, use a bar graph.

Scatter Diagrams

The *scatter diagram* is a powerful visual tool used to display relationships or associations between two variables, cause and effect, and so on. While plotting the scatter diagram, the independent variable corresponds to the *x* axis, or horizontal axis, with the dependent variable on the *y* axis, or vertical axis. The plot pattern identifies whether there is any positive or negative correlation, or no correlation. There is also the possibility for a nonlinear relationship between the variables

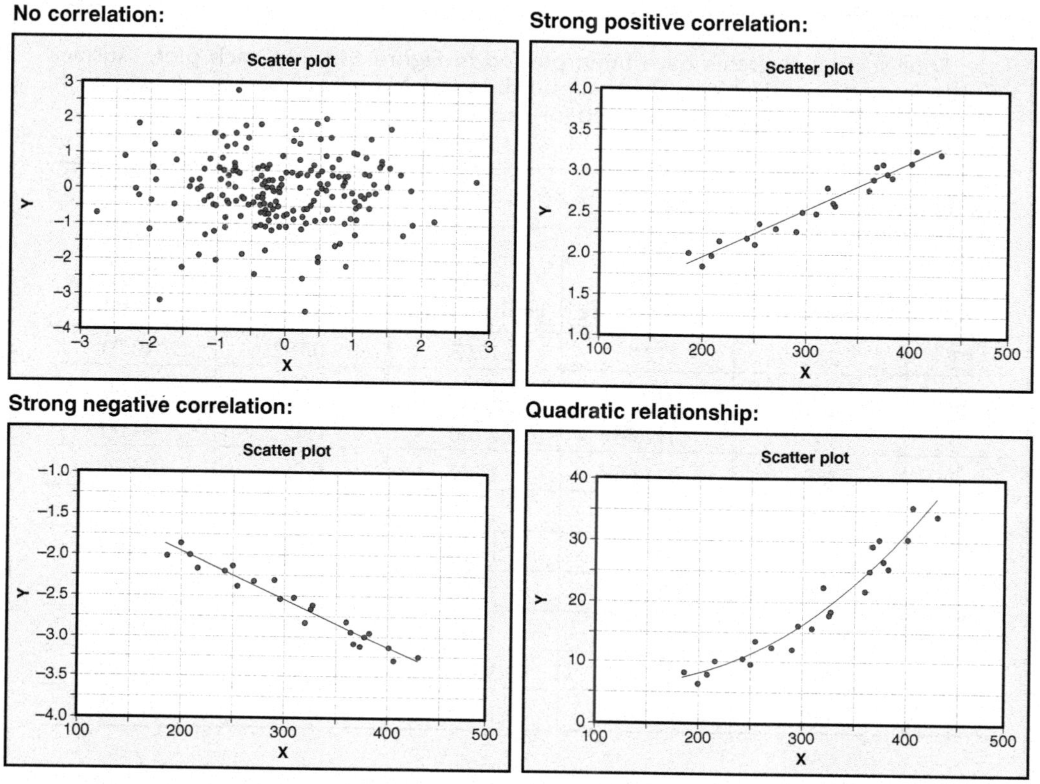

Figure 13.10 Scatter diagrams.

(these are explored through more advanced statistical techniques). Also note that positive correlation between variables does not mean there is a cause-and-effect relationship. See correlation versus causation explanation in Chapter 16.

Figure 13.10 shows different types of relationships between two variables or between causes and effects.

EXAMPLE

Let us take an example from an injection-molding process. The following potential causes have been suggested using engineering judgment by a cross-functional team during a brainstorming session:

- Mold compression pressure
- Coolant temperature
- Mold cooling time
- Mold hold time

The team is trying to identify the relationship of these variables to the quality characteristic "part surface finish." The data are shown in Table 13.6.

Continued

Four scatter diagrams have been plotted in Figure 13.11. In each plot, "surface finish" is on the vertical axis. The first plot shows mold pressure versus surface finish. On each diagram, one point is plotted for each batch.

Table 13.6 Mold process data.

Batch no.	Mold compression pressure	Coolant temperature	Mold cooling time	Mold hold time	Part surface finish
1	242	112.75	15.95	0.792	40.70
2	220	110.88	17.60	1.001	33.00
3	451	112.86	16.50	0.99	44.00
4	385	111.65	17.82	0.748	35.20
5	539	110.88	18.48	0.935	29.70
6	396	111.54	16.28	0.836	38.50
7	407	112.75	15.73	1.034	47.30
8	363	109.78	18.15	0.781	25.30
9	308	110.88	16.50	0.715	35.20
10	440	111.32	18.26	1.056	33.00

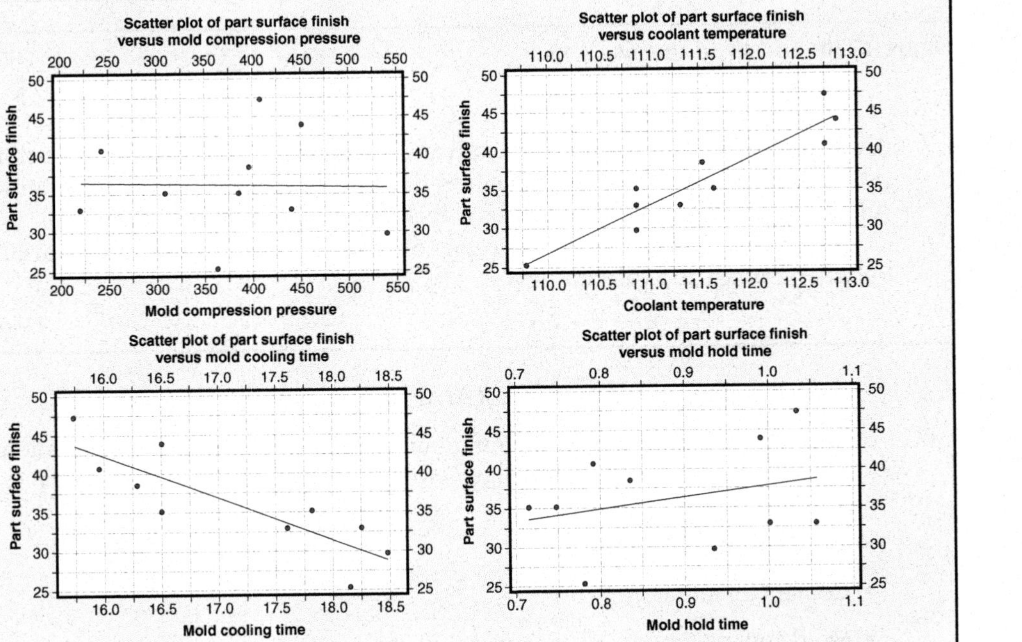

Figure 13.11 Examples of scatter diagrams (variables versus effects).

Continued

Continued

A best-fit line is drawn to cover the plotted points across the axes. In the manual approach, practitioners use an "eyeball" estimation to draw a line approximately in the middle of the plotted points covering end to end. Statistical software does a more thorough job in fitting a line. If the points are closer to each other, the fitted line identifies a lesser variation in the relationship estimation. The relationship between the two variables or causes and effects (Figure 13.12) can be mathematically expressed and represented by a letter "*r*," called the *correlation coefficient* or *Pearson correlation*. The value of *r* is always between –1 and +1 inclusive. This may be stated symbolically as $-1 \leq r \leq +1$.

$$r = \frac{\Sigma xy - \dfrac{\Sigma x \Sigma y}{N}}{\sqrt{\left(\Sigma x^2 - \dfrac{(\Sigma x)^2}{N}\right)\left(\Sigma y^2 - \dfrac{(\Sigma y)^2}{N}\right)}}$$

	Mold compression	Coolant temperature	Mold cooling time	Mold hold time
Coolant temperature	0.028 0.940			
Mold cooling time	0.336 0.342	–0.712 0.021		
Mold hold time	0.328 0.355	0.292 0.412	0.082 0.821	
Part surface finish	–0.033 0.927	0.935 0.000	–0.855 0.002	0.281 0.432

Pearson correlation

P value

Figure 13.12 Correlation between variables and effects (using statistical software).

A *correlation coefficient* measures the extent to which two variables tend to correlate. To be able to calculate the correlation coefficient, a linear relationship is required. This can be visually verified without any sophisticated software. For example, suppose a relationship is suspected between exercise time in minutes/day and weight loss in pounds. To check that relationship, four readings are taken from a weight loss center's data, although in an actual application much more data would be desirable to reduce the error in estimation.

Exercise time in minutes/day	30	45	60	75
Weight loss in pounds	1	2	4	4.5

The first step is to plot the data as shown in Figure 13.13 to see if it seems reasonable to approximate it with a straight line.

Although a straight line can't be drawn through these four points, the trend looks linear. The next step would be to calculate the coefficient of linear correlation. This can be done using the statistical functions in a spreadsheet or the following formula:

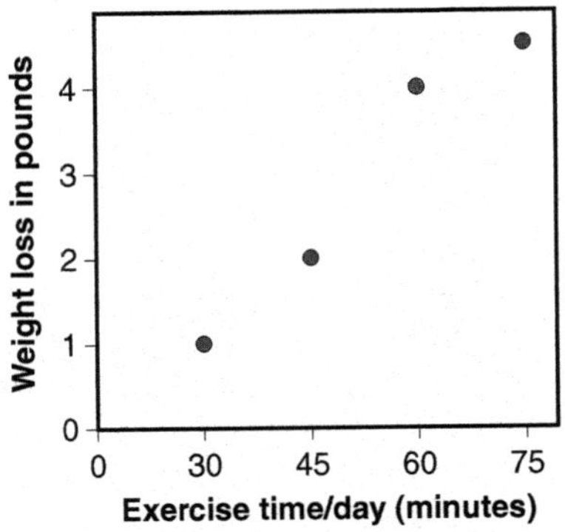

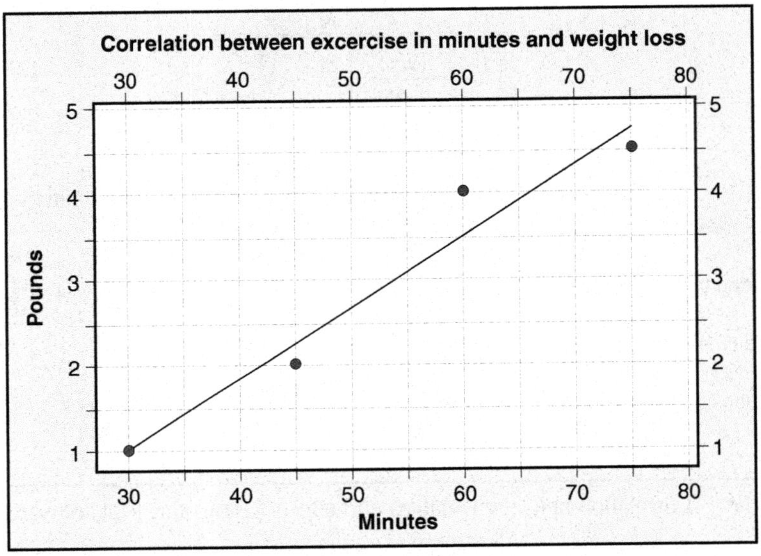

Pearson correlation of pounds and minutes = 0.977
P value = 0.023

Figure 13.13 Example scatter plots—exercise versus weight loss.

$$r = \frac{S_{xy}}{\sqrt{S_{xx}S_{yy}}}$$

where

n = number of points

$S_{xx} = \Sigma x^2 - (\Sigma x)^2 / n$

$S_{xy} = \Sigma xy - \Sigma x \Sigma y / n$

$S_{yy} = \Sigma y^2 - (\Sigma y)^2 / n$

In the above example, using x for minutes and y for pounds:

	x	y	x^2	xy	y^2
	30	1	900	30	1
	45	2	2,025	90	4
	60	4	3,600	240	16
	75	4.5	5,625	337.5	20.25
Σ	210	11.5	12,150	697.5	41.25

$S_{xx} = 12{,}150 - 210^2/4 = 1125$

$S_{xy} = 697.5 - 210 \times 11.5/4 = 93.75$

$S_{yy} = 41.25 - 11.5^2/4 = 8.1875$

So

$$r = \frac{93.75}{\sqrt{(1125)(8.1875)}} \approx 0.9768$$

The value of r will always satisfy the inequality $-1 \leq r \leq 1$.

When r is positive, the scatter diagram displays a positive slope, and when r is negative it displays a negative slope as per the figures displayed earlier. The closer r is to 1 or –1, the stronger the association between the two variables and the higher the likelihood that the variables are related. A key issue here is the distinction between association and causality. When engineering judgment is used to select the variables, relationships between variables can reveal opportunities for improvement. Similarly, scatter diagrams also help as a root cause analysis tool.

The reason engineering judgment is required is because mathematical relationships can be identified even between irrelevant variables. For example, a relationship could exist between two unrelated variables like the price of gold and the infant mortality rate from 1930 to 2000. We should not conclude, however, that as gold prices increase over the years, infant mortality decreases. This can misguide rather than help with root cause identification. This is an example of correlation that does not imply causation. Hence, engineering judgment should be solicited before exploring relationships.

The fact that there is a strong association or correlation between exercise time and weight loss in the above example might lead one to believe that weight loss could be controlled by increasing or decreasing exercise time. This is not necessarily always true. Many variable pairs have a strong association with no causal relationship.

Another related value is the *coefficient of determination*, denoted by r^2 or R. It is defined as the square of the correlation coefficient, as the notation implies. The coefficient of determination is a measure to indicate how well the regression line fits the data. In other words, r^2 explains how much of the variability in the y's can be explained by the fact that they are related to x.

Normal Probability Plots

Normal probability plots are constructed to test whether random data come from a normal probability distribution. Several statistical analyses have a base assumption of normality. Hence, it is important to test for normality before proceeding with further analysis. Normal probability plots can be constructed either manually or by statistical software. Software packages are a more efficient and accurate way of generating probability plots. Normal probability graph paper is designed so that a random sample from a normally distributed population will form an approximately straight line (using the manual construction).

EXAMPLE

Following are the data for bond strength tested on an assembly. Does it appear that the following randomly selected measurements came from a normal population?

8.250, 8.085, 8.795, 9.565, 11.880, 9.180, 9.950, 9.630, 8.150, 10.800, 10.800, 11.080, 10.730, 10.520, 10.380, 10.535, 9.600, 10.340, 10.410

The analysis shown in Figure 13.14 was performed using statistical software with capabilities to conduct tests with the data using the Anderson-Darling formula. Notice the term "AD" in the graphic output. When the Anderson-Darling values are smaller, the distribution fits the data better. This is also reflected in the higher *p*-value. As the *p*-value is greater than 0.05 (alpha risk), we can conclude that these data come from a normal distribution.

Assuming that the specification (or tolerance) for the bond strength is 9 to 11 as indicated by the vertical dashed lines, the horizontal dashed lines show that about 20 percent of the parts will be below the lower specification limit and about 17 percent will be above the upper specification limit (Figure 13.14).

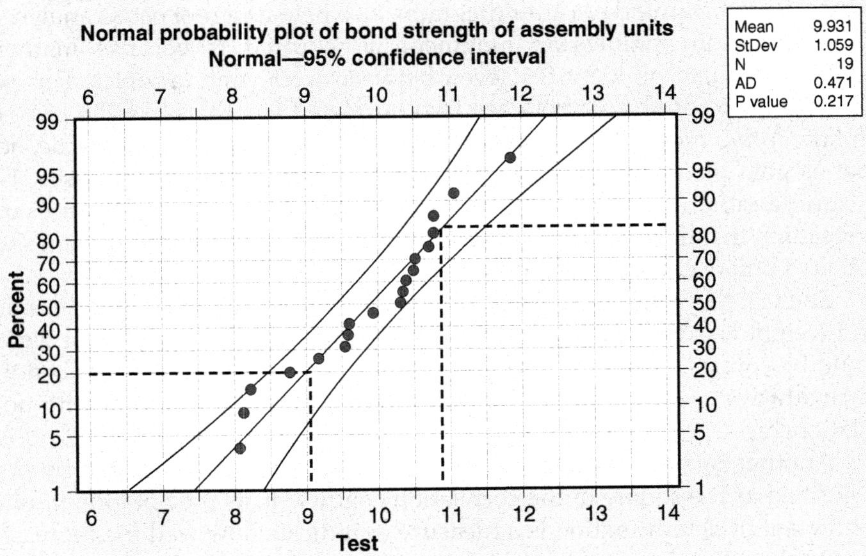

Figure 13.14 Normal probability plot.

There are several tests to check normality of random data. To name a few:

- The Anderson-Darling test for normality, an ECDF (empirical cumulative distribution function)–based test
- The Ryan-Joiner test, a correlation–based test
- The Shapiro-Wilk test, similar to the Ryan-Joiner test
- The Kolmogorov-Smirnov test for normality, also an ECDF-based test

Of these tests, Anderson-Darling is most widely used by statistical software.

Weibull Plots

The Weibull distribution has the general form

$$P(x) = \alpha\beta(x - \gamma)^{\beta-1}e^{-\alpha(x-\gamma)^{\beta}}$$

where

α = Scale parameter

β = Shape parameter

γ = Location parameter

The Weibull function is mainly used for reliability data when the underlying distribution is unknown. Weibull probability paper can be used to estimate the shape parameter β and *mean time between failures* (MTBF), or *failure rate*. Weibull plots can be generated manually as well as by using computerized software packages. Software packages are a more efficient and accurate way of generating Weibull plots. There is relevance between all these parameters and the life cycle of a product.

Weibull distributions with beta < 1 have a failure rate decreasing with time. This is also known as *infant mortality* or *early life failures*. Beta of close to 1 or 1 is the *useful life* or *random failure* period of Weibull distributions, with beta > 1 having failure rates increasing with time. This is also known as *wear-out failure*.

EXAMPLE

Fifteen units were tested for environmental stress, and the number of hours the units managed to remain in operation under testing was collected from a precision timer. Estimate the value of the shape parameter β, MTBF, and the reliability at 3.9 hours. (See Figure 13.15.)

Fail duration 4.011, 3.646, 5.226, 4.740, 4.739, 5.833,
4.861, 4.618, 4.012, 3.646, 4.497, 3.646, 4.49, 3.281, 3.889

In this example, we need to find β (shape parameters) to know if these units are failing at a particular period of the life cycle such as infant mortality, constant failure, or wear-out failure. Secondly, MTBF is required from the problem. For a normal distribution,

Continued

Continued

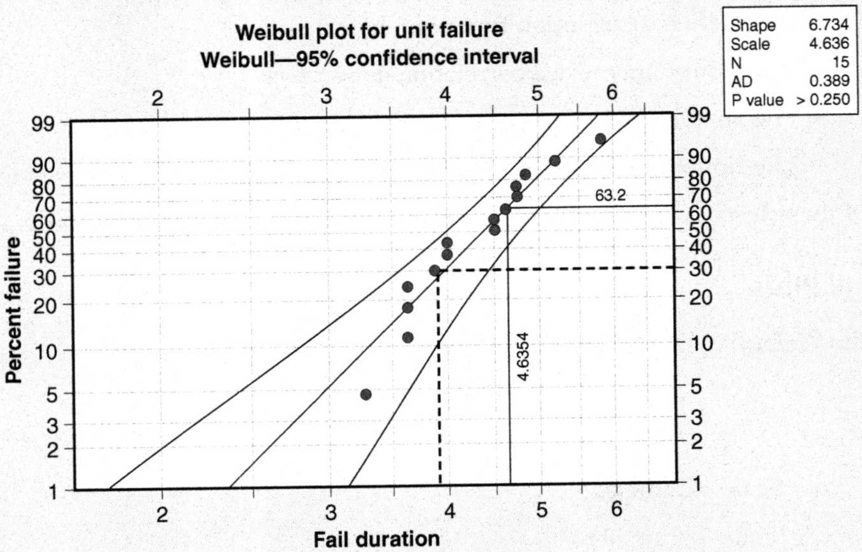

Figure 13.15 Example of Weibull plot.

the mean is the 50th percentile, whereas for a Weibull distribution, the mean is at the 63.2 percentile. The problem also requires the reliability at 3.9 hours.

The data are input into statistical software, and the Weibull distribution is chosen to perform this analysis. Several commercially available software packages can perform this function.

The software generates the Weibull plot on probability paper. The *p*-value is > 0.25 and hence there is higher probability that the data come from a Weibull distribution.

The software also creates a 95 percent confidence interval for the data analyzed. This is useful for creating an interval estimate at a desired point.

The vertical axis on most Weibull paper is labeled "percent failure." Since MTBF is located at 36.8 percent on a reliability scale, it is, in other words, at 63.2 percent on a failure scale. The horizontal coordinate of the point where the 63.2 percent line crosses the best-fit line is the estimate for MTBF. Interpolation on this curve provides estimates for other values. From the above plot, by finding the point of intersection for 63.2 percent, we can find the MTBF as 4.635 hours.

The β (shape) value is 6.734. We can conclude that the unit is failing at wear-out period.

A vertical line drawn through the 3.9-hour point on the horizontal axis crosses the fitted line at about 30 percent on the percent failure scale, so the estimate for reliability at 3.9 hours is $R(3.9) \approx 0.70$.

Chapter 14

Measurement System Analysis

> Calculate, analyze, and interpret measurement system capability using gauge repeatability and reproducibility (GR&R) studies, measurement correlation, bias, linearity, percent agreement, and precision/tolerance (P/T). (Evaluate)
>
> **Body of Knowledge III.E**

Measurement system analysis (MSA) is an area of statistical study that explores the variation in measurement data due to:

- *Calibration.* Drift in average measurements of an absolute value. (See Glossary for ASQ definition of calibration.)
- *Stability.* Drift in absolute value over time.
- *Repeatability.* Variation in measurement when measured by one appraiser on the same equipment in the same measurement setting at the same time.
- *Reproducibility.* Variation in measurement when measured by two or more appraisers multiple times.
- *Linearity.* Accuracy of measurement at various measurement points of measuring range in the equipment.
- *Bias.* Bias (difference between absolute value and true value) with respect to a standard master at various measurement points of the measuring range.
- *Accuracy.* "Closeness" to the true value, or to an accepted reference value.
- *Precision.* "Closeness" of repeated readings to each other. A random error component of the measurement system.

Until the early 1990s, MSA was used extensively in measurement laboratories, and less known to the industrial world. Since the inception of the QS-9000 (now IATF 16949) standard in the automobile industry, the importance of MSA has been well understood by other sectors as well.

An important issue for the practitioner here is that in the quest to reduce variation, the measurement system should be one of the first things analyzed because all data from the process are, in effect, filtered through the measurement system.

Even statistical process control experts have started to rewrite their SPC flow with conducting an MSA study as a starting step. MSA is actually that important. It is not uncommon for measurement systems to have an error of 40 to 50 percent of the process specification.

Repeatability is the equipment measurement variation expressed as standard deviation. Measurements are taken from the same equipment by one appraiser over a short period of time. See Figure 14.1. (See Govind Ramu, "Evaluating Repeatability," Expert Answers in *Quality Progress*, November 2014).

Reproducibility is the appraiser measurement variation expressed as standard deviation. Measurements are taken from the same equipment by more than one appraiser. See Figure 14.2.

The repeatability portion of the measurement variation is attributed to the inherent variation of the measurement equipment. Factors that influence this portion of variation include the design of the measurement system itself. In the case of reproducibility, the influential factors are the setting of the work piece (any special loading and unloading), operator training, skill, and knowledge, consistency in measurement, and so on.

Following are the steps in conducting a gage repeatability and reproducibility (GR&R) study:

1. Plan the study in detail by communicating to the line supervisor, making sure the equipment and appraisers are available for the study, equipment is calibrated and in good working condition, the samples are in good condition, and so on. Some studies can take a very long time to complete all the trials due to the

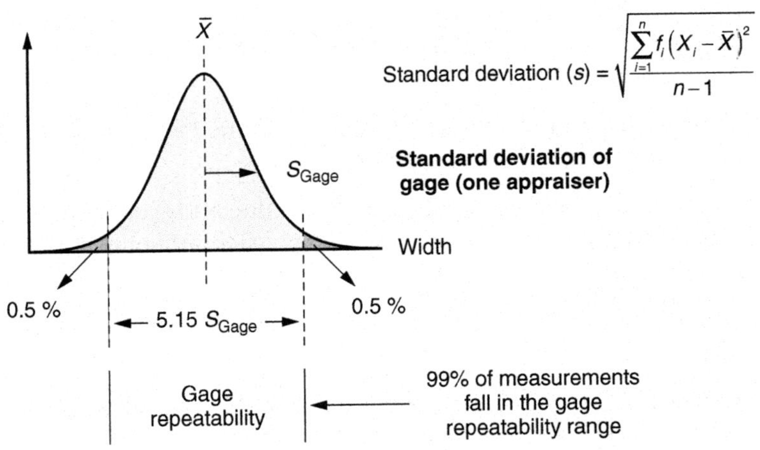

Figure 14.1 Gage repeatability.

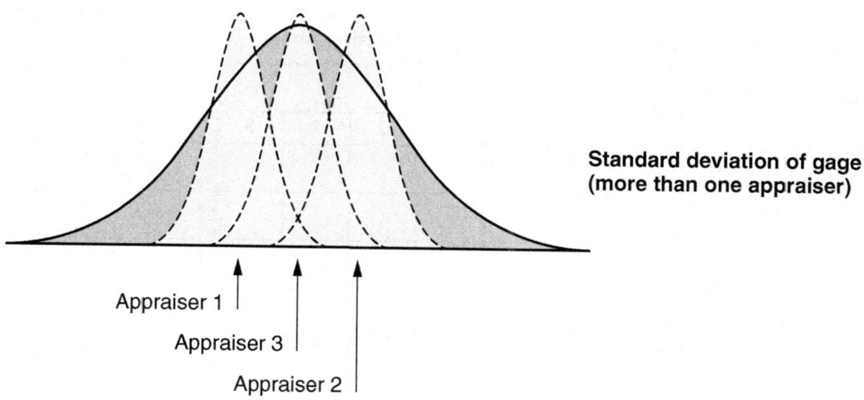

Figure 14.2 Gage reproducibility.

measurement duration of a specific characteristic. Make sure that the appraisers are aware of the measurement criteria and inspection method and are trained to perform this measurement. These should be the appraisers who perform these measurements in the process on a regular basis. Time is money. Think of everything that can go wrong and plan for contingencies during the study.

2. The first step is to select and identify samples for the GR&R study. It is important to handpick the samples covering the spread rather than picking random samples from the production bin. It is recommended that the experimenter identify the samples in a location that is not obviously visible to the appraisers. It is also recommended that the experimenter be present during the R&R study.

3. The next step is to create a table for experimentation purposes with randomized samples between trials and appraisers. Table 14.1 is an example of running a GR&R experiment in a randomized manner (10 samples × 3 operators × 3 trials).

4. Appraisers are called one after the other as per the randomized table and requested to perform measurements. This includes multiple trials by every appraiser. It is important that each appraiser complete the study by measuring all samples for every trial. The calculations assume a complete study. An incomplete study can cause imbalances in data, and most statistical software will indicate an error message.

5. The complete study conducted based on the randomized experiment (see Table 14.1) is entered into the calculation tabular sheet (Figure 14.3). This sheet is arranged in the sample number sequence. The experimenter may also directly input the data on the tabular calculation sheet. However, care should be taken not to mistakenly fill in the wrong worksheet cells.

6. Calculations: Calculate the values for row 16 by averaging the values in rows 4, 9, and 14. Calculate the values in the far right-hand column of rows 4, 5, 9, 10, 14, and 15 by averaging the 10 values in their respective rows. Calculate the two entries in the right-hand column of row 16 by finding the average and ranges

	Gage Repeatability and Reproducibility Data Collection Sheet											
	Appraiser/ trial #	Part										
		1	2	3	4	5	6	7	8	9	10	Average
1	A 1											
2	2											
3	3											
4	Average											
5	Range											
6	B 1											
7	2											
8	3											
9	Average											
10	Range											
11	C 1											
12	2											
13	3											
14	Average											
15	Range											
16	Part average											
17	$([\bar{R}_a =\ \ ] + [\bar{R}_b =\ \ ]) + [\bar{R}_c =\ \ ]) / [\#\text{ of appraisers} =\ \ ] =$											$\bar{\bar{R}} =$
18	$\bar{X}_{DIFF} = [Max\ \bar{X} =\ \ ] - [Min\ \bar{X} =\ \ ] =$											$\bar{X}_{DIFF} =$
19	$^*UCL_R = [\bar{\bar{R}} =\ \ ] \times [D_4 =\ \ ] =$											
	$^*D_4 = 3.27$ for two trials and 2.58 for three trials. UCL_R represents the limit of individual *R*s. Circle those that are beyond this limit. Identify the cause and correct. Repeat these readings using the same appraiser and unit as originally used, or discard values and re-average and recompute $\bar{\bar{R}}$ and the limiting value from the remaining observations.											

Figure 14.3 Gage R&R data collection sheet.

Source: Used with permission of the Automotive Industry Action Group (AIAG).

of the 10 values in that row. Substitute from the right-hand columns of rows 5, 10, and 15 into the formula in row 17 to calculate $\bar{\bar{R}}$. For clarity, the formula is repeated here:

$$\bar{\bar{R}} = \frac{\bar{R}_a + \bar{R}_b + \bar{R}_c}{k}$$

where k = number of appraisers. Record this value in the right-hand column of row 17.

Let max $\bar{X}$ = the largest of $\bar{X}_a$, $\bar{X}_b$, and $\bar{X}_c$

Let min $\bar{X}$ = the smallest of $\bar{X}_a$, $\bar{X}_b$, and $\bar{X}_c$

Calculate $\bar{X}_{DIFF}$ = max $\bar{X}$ – min $\bar{X}$ and place the value in the right-hand column of line 18.

Calculate the upper control limit for the R values using the formula shown in line 19. The $\bar{\bar{R}}$ value is from the right-hand column of row 17, and the D_4 value is 2.58 if each part was measured three times as outlined above. If, instead, each part was measured only twice, the D_4 value is 3.27.

Note the instructions at the bottom of the form. They indicate that each of the 10 R values in row 5 should be compared to the UCL_R value calculated in row 19. Any R value that exceeds UCL_R should be circled. Repeat this for the R values in rows 10 and 15. The circled R values are significantly different from the others, and the cause of this difference should be identified and corrected. Once this has been done, the appropriate parts can be remeasured using the same appraiser, equipment, and so on, as the original measurements. All impacted values must be recomputed.

Recall that *repeatability* is the variation in measurements that occurs when the same measuring system, including equipment, material, appraiser, and so on, are used. Repeatability, then, is reflected in the R values as recorded in rows 5, 9, and 15 and summarized in row 17. Repeatability is often referred to as *equipment variation*, but the individual R averages may indicate differences between *appraisers*. In the example in Figure 14.4, $\bar{R}_a$ is somewhat smaller than $\bar{R}_b$ or $\bar{R}_c$. This indicates that appraiser A may have done better at repeated measurements of the same part than the other two appraisers. Further analysis may be required to investigate why a certain appraiser has wider variation than others.

Reproducibility is the variation that occurs between the overall average measurements for the three appraisers. It is reflected by the $\bar{X}$ values in rows 4, 9, and 14 and summarized in the value of $\bar{X}_{DIFF}$ in row 18. If, for instance, $\bar{X}_a$ and $\bar{X}_b$ had been quite close and $\bar{X}_c$ were significantly different, it would appear that appraiser C's measurements have some sort of bias. Again, further investigation can be very productive.

The next step in the study is to complete the Gage Repeatability and Reproducibility Report as shown in Figure 14.5. A completed report based on the data from Figure 14.4 is shown in Figure 14.6. The quantity labeled EV for *equipment variation* is an estimate of the standard deviation of the variation due to repeatability. It is sometimes denoted σ_E or σ_{rpt} (repeatability error). The quantity labeled AV for *appraiser variation* is an estimate of the standard deviation of the variation due to reproducibility and is sometimes denoted σ_A or σ_{rpd} (reproducibility error). The quantity labeled GRR is an estimate of the standard deviation of the variation due to the measurement system and is sometimes denoted σ_M. The quantity labeled PV is an estimate of the standard deviation of the part-to-part variation and is sometimes denoted σ_P. PV is calculated by multiplying with appropriate constant

	Gage Repeatability and Reproducibility Data Collection Sheet											
	Appraiser/ trial #	Part										
		1	2	3	4	5	6	7	8	9	10	Average
1	A 1	0.29	−0.56	1.34	0.47	−0.80	0.02	0.59	−0.31	2.26	−1.36	0.194
2	2	0.41	−0.68	1.17	0.50	−0.92	−0.11	0.75	−0.20	1.99	−1.25	0.166
3	3	0.64	−0.58	1.27	0.64	−0.84	−0.21	0.66	−0.17	2.01	−1.31	0.211
4	Average	0.447	−0.607	1.260	0.537	−0.853	−0.100	0.667	−0.227	2.087	−1.307	$\bar{X}_a = 0.1903$
5	Range	0.35	0.12	0.17	0.17	0.12	0.23	0.16	0.14	0.27	0.11	$\bar{R}_a = 0.184$
6	B 1	0.08	−0.47	1.19	0.01	−0.56	−0.20	0.47	−0.63	1.80	−1.68	0.001
7	2	0.25	−1.22	0.94	1.03	−1.20	0.22	0.55	0.08	2.12	−1.62	0.115
8	3	0.07	−0.68	1.34	0.20	−1.28	0.06	0.83	−0.34	2.19	−1.50	0.089
9	Average	0.133	−0.790	1.157	0.413	−1.013	0.027	0.617	−0.297	2.037	−1.600	$\bar{X}_b = 0.068$
10	Range	0.18	0.75	0.40	1.02	0.72	0.42	0.36	0.71	0.39	0.18	$\bar{R}_b = 0.513$
11	C 1	0.04	−1.38	0.88	0.14	−1.46	−0.29	0.02	−0.46	1.77	−1.49	−0.223
12	2	−0.11	−1.13	1.09	0.20	−1.07	−0.67	0.01	−0.56	1.45	−1.77	−0.256
13	3	−0.15	−0.96	0.67	0.11	−1.45	−0.49	0.21	−0.49	1.87	−2.16	−0.284
14	Average	−0.073	−1.157	0.880	0.150	−1.327	−0.483	0.080	−0.503	1.697	−1.807	$\bar{X}_c = -0.2543$
15	Range	0.19	0.42	0.42	0.09	0.39	0.38	0.20	0.10	0.42	0.67	$\bar{R}_c = 0.328$
16	Part average	0.169	−0.851	1.099	0.367	−1.064	−0.186	0.454	−0.342	1.940	−1.571	$\bar{\bar{X}} = 0.0014$ $\bar{R}_p = 3.511$
17	$([\bar{R}_a = 0.184] + [\bar{R}_b = 0.513]) + [\bar{R}_c = 0.328]) / [\text{\# of appraisers} = 3] =$											$\bar{\bar{R}}$ **= 0.3417**
18	$\bar{X}_{DIFF} = [Max\ \bar{X} = 0.1903] - [Min\ \bar{X} = -0.2543] =$											$\bar{X}_{DIFF}$ **= 0.4446**
19	$^*UCL_R = [\bar{\bar{R}} = 0.3417] \times [D_4 = 2.58] = 0.8816$											
	$^*D_4 = 3.27$ for two trials and 2.58 for three trials. UCL_R represents the limit of individual R's. Circle those that are beyond this limit. Identify the cause and correct. Repeat these readings using the same appraiser and unit as originally used, or discard values and re-average and recompute $\bar{\bar{R}}$ and the limiting value from the remaining observations.											

Figure 14.4 Gage R&R data collection sheet with data entered and calculations completed.
Source: Used with permission of the Automotive Industry Action Group (AIAG).

K3 depending on number of parts selected for experiment. The quantity labeled TV is an estimate of the standard deviation of the total in the study and is sometimes denoted σ_T. The right-hand column in Figure 14.6 shows for each type of variation the percentage of total variation it consumes. Sometimes the right-hand column is based on the tolerance for the dimension.

Gage Repeatability and Reproducibility Report

Part no. & name: | Gage name: | Date:
Characteristics: | Gage no.: | Performed by:
Specifications: | Gage type:

From data sheet: $\bar{\bar{R}}$ = | $\bar{X}_{DIFF}$ = | R_p =

Measurement Unit Analysis	% Total Variation

Repeatability—Equipment Variation (EV)

$EV = \bar{\bar{R}} \times K_1$

= ___ × ___

= ___

Trials	K_1
2	0.8862
3	0.5908

% EV = 100[EV / TV]

= 100[___ / ___]

= ___%

Reproducibility—Appraiser Variation (AV)

$AV = \sqrt{(\bar{X}_{DIFF} \times K_2)^2 - (EV^2 / (nr))}$

$\sqrt{(\_\_\_ \times \_\_\_)^2 - (\_\_\_^2 / (\_\_\_ \times \_\_\_))}$

= ___

n = parts r = trials

Appraisers	2	3
K_2	0.7071	0.5231

% AV = 100[AV / TV]

= 100[___ / ___]

= ___%

Repeatability & Reproducibility (GRR)

GRR $= \sqrt{EV^2 + AV^2}$

$= \sqrt{\_\_\_^2 + \_\_\_^2}$

= ___

% GRR = 100[GRR / TV]

= 100[___ / ___]

= ___%

Parts	K_3
2	0.7071
3	0.5231
4	0.4467
5	0.4030
6	0.3742
7	0.3534
8	0.3375
9	0.3249
10	0.3146

Part Variation (PV)

$PV = R_p \times K_3$

= ___ × ___

= ___

% PV = 100[PV/ TV]

= 100[___ / ___]

= ___%

Total Variation (TV)

$TV = \sqrt{GRR^2 + PV^2}$

$= \sqrt{\_\_\_^2 + \_\_\_^2}$

= ___

ndc = 1.41 (PV / GRR)

= 1.41 (___ / ___)

= ___

Figure 14.5 Gage repeatability and reproducibility report.

Source: Used with permission of the Automotive Industry Action Group (AIAG).

Gage Repeatability and Reproducibility Report

Part no. & name:	Gage name:	Date:
Characteristics:	Gage no.:	Performed by:
Specifications:	Gage type:	
From data sheet: $\bar{\bar{R}} = 0.3417$	$\bar{X}_{DIFF} = 0.4446$	$R_p = 3.511$

Measurement Unit Analysis	% Total Variation

Repeatability—Equipment Variation (EV)

$EV = \bar{\bar{R}} \times K_1$

$= 0.3417 \times 0.5908$

$= 0.20188$

Trials	K_1
2	0.8862
3	0.5908

% EV $= 100[EV / TV]$

$= 100[0.20188 / 1.14610]$

$= 17.62\%$

Reproducibility—Appraiser Variation (AV)

$AV = \sqrt{(\bar{X}_{DIFF} \times K_2)^2 - (EV^2 / (nr))}$

$\sqrt{(0.4446 \times 0.5231)^2 - (0.20188^2 / (10 \times 3))}$

$= 0.22963$

n = parts r = trials

Appraisers	2	3
K_2	0.7071	0.5231

% AV $= 100[AV / TV]$

$= 100[0.22963 / 1.14610]$

$= 20.04\%$

Repeatability & Reproducibility (GRR)

$GRR = \sqrt{EV^2 + AV^2}$

$= \sqrt{0.20188^2 + 0.22963^2}$

$= 0.30575$

% GRR $= 100[GRR / TV]$

$= 100[0.30575 / 1.14610]$

$= 26.68\%$

Parts	K_3
2	0.7071
3	0.5231
4	0.4467
5	0.4030
6	0.3742
7	0.3534
8	0.3375
9	0.3249
10	0.3146

Part Variation (PV)

$PV = R_p \times K_3$

$= 3.511 \times 0.3146$

$= 1.10456$

% PV $= 100[PV / TV]$

$= 100[1.10456 / 1.14610]$

$= 96.38\%$

Total Variation (TV)

$TV = \sqrt{GRR^2 + PV^2}$

$= \sqrt{0.30575^2 + 1.10456^2}$

$= 1.14610$

ndc $= 1.41\ (PV / GRR)$

$= 1.41\ (1.10456 / 0.30575)$

$= 5.094 = 5$

Figure 14.6 Gage repeatability and reproducibility report with calculations.

Source: Used with permission of the Automotive Industry Action Group (AIAG).

Precision to tolerance ratio (P/T): In this case the value of the divisor TV is replaced by one-sixth of the tolerance, that is, (Tolerance) ÷ 6. Most authorities agree that in this situation the %GRR is defined as:

$$(100 \times \text{GRR error}) \div (\text{Tolerance}/6)$$

Six is used to cover 99.73 percent of variation. Some practitioners also use 5.15 to cover 99 percent of the variation. For information on the theory and constants used in this form see *MSA Reference Manual.*

There are also many inexpensive spreadsheet-based macro applications available as off-the-shelf software. Figure 14.7 shows the same data analyzed using statistical software.

Notice that there are several points outside the control limits in the sample mean chart. This is the way it is supposed to be, as we intentionally picked the samples to cover the spread of the process specification. Since parts used in the study represent the process variation, approximately one-half or more of the averages plotted on the $\bar{X}$ chart should fall outside the control limits. If less than half of the plotted average points fall outside the control limits, then either the measurement system lacks adequate effective resolution or the sample does not represent the expected process variation. On the other hand, the points in the sample range charts should be within the control limits. If all the calculated ranges are within control, all appraisers are consistent. If one appraiser is out-of-control, the method used differs from the others. If all appraisers have some out-of-control ranges, the measurement system is sensitive to appraiser technique and needs improvement.

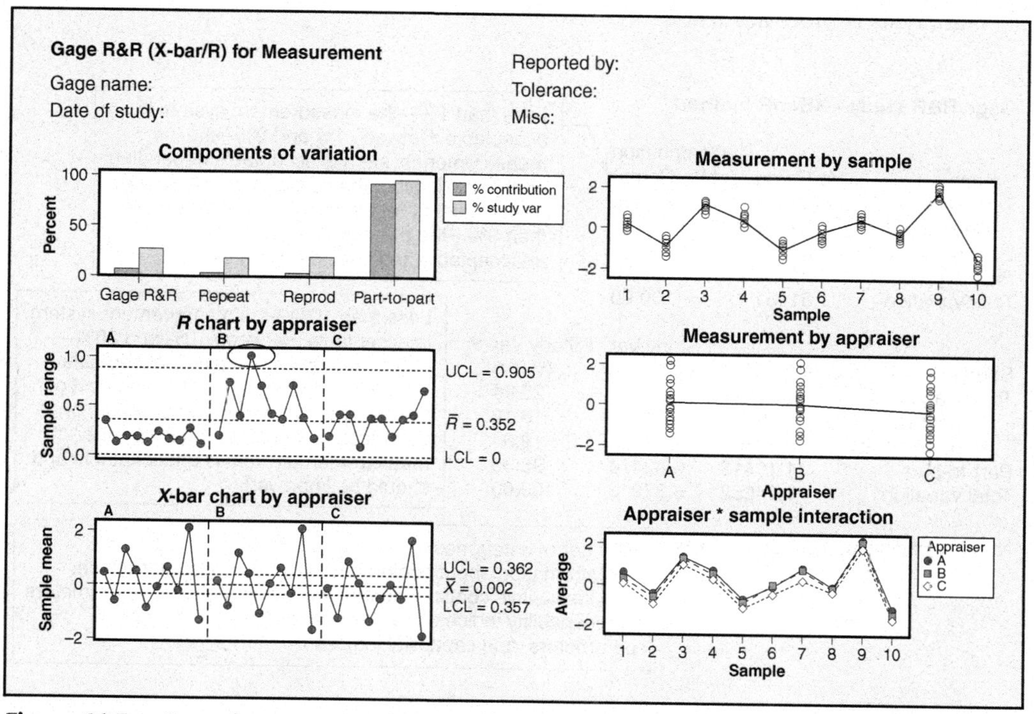

Figure 14.7 Example gage repeatability and reproducibility analysis.

We notice that the fourth measurement point of appraiser B is outside the control limits. Investigate to verify whether this is a recording error or due to any special causes.

The other charts visually indicate the spread of measurement points and any noticeable outliers, analyzed by sample and by appraiser. If a sample had issues with retaining certain measurements during the study period, this will probably show up in the chart (by sample). Also, if an appraiser is consistently measuring higher or lower than the others, it will be noticeable in the chart (by appraiser). In our example, appraiser C is consistently measuring lower. This is also visible in the appraiser * sample interaction graph (Figure 14.7—bottom graph to the right).

Sources of measurement variation are shown in Figure 14.8 (for the previous example) and graphically in Figure 14.9.

The AIAG MSA Manual explains, "Discrimination is the measurement resolution, scale limit, or smallest detectable unit of the measurement device and standard. It is an inherent property of gage design and reported as a unit of measurement or classification. The number of data categories is often referred to as the *discrimination ratio* since it describes how many classifications can be reliably distinguished given the observed process variation."

Number of distinct categories: Measurement system discrimination is the ability to detect changes in the measured characteristic. If a measurement system's discrimination is inadequate, it may not be possible to measure process variation or quantify characteristic values (such as the mean) of individual parts.

Number of distinct categories greater than or equal to 5 is considered acceptable for process monitoring applications. There are minor differences between the manual method and statistical software due to rounding errors. The statistical analysis is more accurate.

Gage R&R study—XBar/R method

Source	VarComp	%Contribution (of VarComp)
Total Gage R&R	0.09542	7.26
Repeatability	0.04315	3.28
Reproducibility	0.05228	3.98
Part-to-part	1.21909	92.74
Total variation	1.31451	100.00

Less than 1%—the measurement system is acceptable. Between 1% and 9%—the measurement system is acceptable depending on the application, the cost of the measuring device, cost of repair, or other factors. Greater than 9%—the measurement system is unacceptable and should be improved.

Source	StdDev (SD)	Study Var (6 * SD)	#Study Var (%SV)
Total Gage R&R	0.30891	1.85343	26.94
Repeatability	0.20772	1.24631	18.12
Reproducibility	0.22864	1.37183	19.94
Part-to-part	1.10412	6.62474	96.30
Total variation	1.14652	6.87913	100.00

Less than 10%—the measurement system is acceptable. Between 10% and 30%—the measurement system is acceptable depending on the application, the cost of the measuring device, cost of repair, or other factors. Greater than 30%—the measurement system is unacceptable and should be improved.

Number of distinct categories = 5

Number of categories
1—Information about conformance versus nonconformance,
2–4 Insensitive controls, coarse estimates of process parameters and capability indices, five or more control charts, process parameters, and capability indices.

Figure 14.8 GR&R report using statistical software.

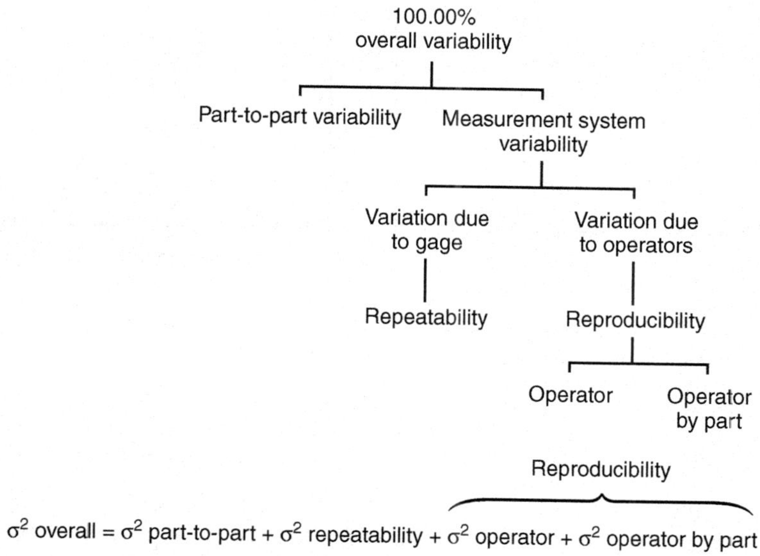

Figure 14.9 Sources of measurement variation.

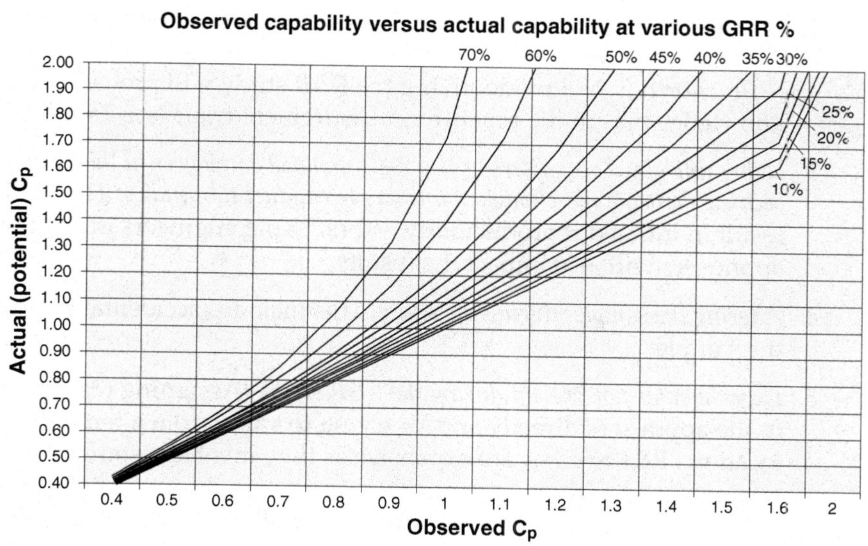

Figure 14.10 Observed versus actual capability.

Effect of R&R on capability: As mentioned earlier, the measurement system plays a major role in process capability (C_p) assessment. The higher the gage R&R, the higher the error in C_p assessment. This increases even more as the capability increases. Example: With an observed C_p of 1 and a GR&R of 50 percent, the actual C_p is 1.23. By bringing the GR&R to 10 percent, the actual C_p is more reflective of the actual process, that is, 1.01. See Figure 14.10. More details on this table, formula, and graphs are available in *Concepts of R&R Studies* by Larry B. Barrentine.

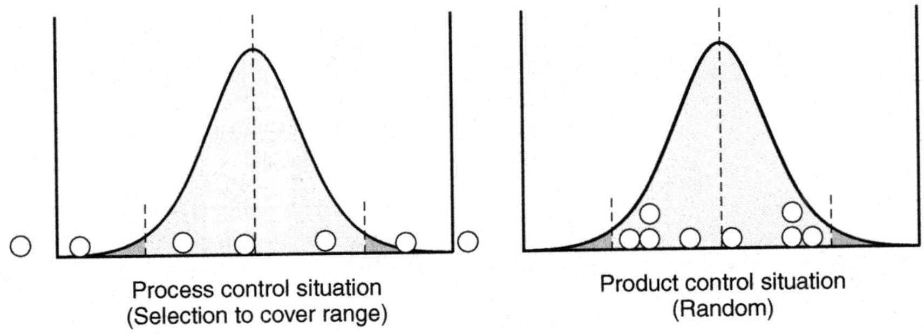

Figure 14.11 GR&R sample selection.

Following are some common errors made while performing GR&R:

1. *In process control situations, not selecting samples covering the tolerance spread* (Figure 14.11). In fact, it is recommended to even pick samples outside the specification limits. It is a common tendency for experimenters to pick some random samples from the process to study GR&R. See AIAG Reference section at the end of this chapter for rationale.

2. *Not randomizing the samples during measurement.* Randomizing the experiment takes some effort and care from the experimenter, but it is really worth it. Not randomizing the R&R study will probably introduce knowledge bias in the repetitive measurement trials (see Table 14.1).

3. *Using untrained appraisers or process-unrelated employees in the experiment because there are not enough appraisers to conduct the studies.* This will result in inflated reproducibility errors. Using engineers instead of the appraisers will also impact the results.

4. *Altering the samples during the study.* This includes accidental dropping of the samples.

5. *Experimenter not present during the R&R study.* (Assigning responsibility to the appraisers directly and/or trying to study from a remote location.) R&R studies are expensive as they involve resources like equipment and labor. Any mistakes performed during the study can invalidate the results and require starting all over. There are cases where the measurement is automated and duration long enough that the experimenter need not stay. However, it is important to be present during the human interaction portion of the measurement, that is, loading, setting, aligning, unloading, and so on.

6. *Publishing the results with appraisers' names.* It is important for the experimenters to know who has introduced more variation and analyze the root causes, assign additional training, and so on. It is not required to release the analysis with actual appraiser names for general viewing.

Table 14.1 GR&R random spreadsheet example.

Run order	Parts	Operators	Run order	Parts	Operators	Run order	Parts	Operators
1	2	B	31	1	B	61	5	A
2	6	B	32	4	C	62	9	C
3	5	C	33	8	B	63	10	B
4	3	B	34	9	B	64	10	C
5	2	A	35	3	B	65	1	A
6	6	C	36	5	A	66	8	B
7	4	C	37	8	B	67	4	A
8	3	B	38	4	C	68	1	C
9	8	A	39	9	A	69	8	C
10	1	A	40	4	B	70	7	B
11	10	B	41	3	A	71	8	C
12	6	A	42	5	B	72	7	B
13	7	B	43	10	B	73	2	B
14	6	A	44	7	A	74	10	A
15	3	C	45	5	B	75	5	C
16	9	C	46	2	A	76	10	C
17	6	A	47	8	C	77	10	A
18	9	B	48	5	A	78	3	A
19	1	B	49	7	C	79	9	C
20	2	C	50	9	A	80	8	A
21	9	A	51	2	C	81	3	A
22	2	A	52	6	C	82	4	B
23	5	B	53	4	A	83	6	C
24	5	C	54	6	B	84	1	B
25	10	C	55	6	B	85	7	A
26	10	A	56	1	A	86	3	C
27	4	A	57	3	C	87	2	C
28	4	B	58	8	A	88	1	C
29	2	B	59	9	B	89	7	A
30	1	C	60	7	C	90	7	C

This can create unhealthy comparisons between appraisers or make some appraisers uneasy. This may result in not cooperating for future studies. It is recommended to present the results as appraiser A, B, C, and so on.

7. *Assuming that the GR&R results are valid forever.* GR&R results have to be periodically validated, just as we do gage calibration at a regular frequency. Over a period of time, the measurement methods change, the appraisers change, settings change. Equipment that uses software/firmware (embedded software) may also change.
8. *Assuming that GR&R performed on a specific piece of equipment is the same as for all other equipment of that kind* (sometimes referred to as a "family of gages"). This may not be true. Equipment #1 may be used by a set of appraisers from three shifts and equipment #2 used by a different set of appraisers. One piece of equipment may be used under controlled environmental conditions and others used under rugged conditions on the shop floor. They may be the same type of equipment but from different manufacturers or used in very different ways or settings.

Linearity and Bias

Having discussed equipment variation and appraiser variation, we still have some unanswered questions. What if the equipment is accurate at one point of measurement and not at other points of measurement across the measurement range? We need to perform a linearity study to answer this question. Also, we would like to know how biased the measuring equipment is compared to a "master." Let us review using a statistical software analysis: Appraiser A measurements from the previous example were taken and compared with process variation to estimate percent linearity and compared with a master value of measurement at the point where the measurement was made in the measuring range. See Table 14.2.

Figure 14.12 shows the statistical software analysis of these data. Our first important observation is that the R-Sq value is 0.0%. This shows a nonlinearity issue with the measuring equipment.

Possible causes of nonlinearity include instrument not calibrated properly at both the high and low end of the operating range, error in one or more of the master part measurements, worn instrument, and characteristics of the instrument design.[1]

Percent linearity is smaller; however, due to the nonlinearity issue, the percent linearity is not valid.

In the same graph, we see that the average percent bias gives a *p*-value of zero, indicating a higher significance of bias issues with the instrument. The report also provides a breakdown of percent bias at the measurement points covered during the study.

The experimenter is advised to look into nonlinearity issues and also to try conducting the study on a broader range of measurements of the equipment to reassess the linearity.

Table 14.2 Measurement test data for linearity and bias.

Sample	Appraiser	Trial	Measurement	Master
1	A	1	0.29	0.4
1	A	2	0.41	0.4
1	A	3	0.64	0.4
2	A	1	–0.56	0.6
2	A	2	–0.68	0.6
2	A	3	–0.58	0.6
3	A	1	1.34	1.2
3	A	2	1.17	1.2
3	A	3	1.27	1.2
4	A	1	0.47	0.5
4	A	2	0.5	0.5
4	A	3	0.64	0.5
5	A	1	–0.8	–0.85
5	A	2	–0.92	–0.85
5	A	3	–0.84	–0.85
6	A	1	0.02	–0.1
6	A	2	–0.11	–0.1
6	A	3	–0.21	–0.1
7	A	1	0.59	0.7
7	A	2	0.75	0.7
7	A	3	0.66	0.7
8	A	1	–0.31	–0.2
8	A	2	–0.2	–0.2
8	A	3	–0.17	–0.2
9	A	1	2.26	2
9	A	2	1.99	2
9	A	3	2.01	2
10	A	1	–1.36	–1.2
10	A	2	–1.25	–1.2
10	A	3	–1.31	–1.2

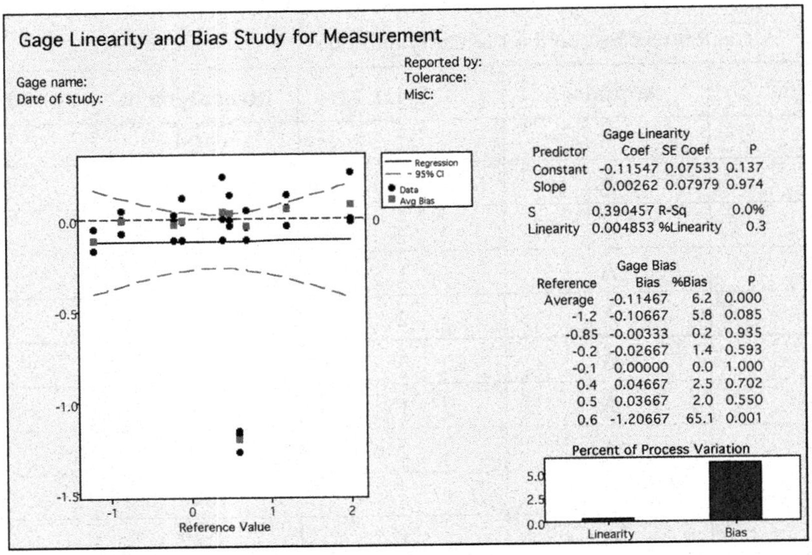

Figure 14.12 Linearity and bias analysis using statistical software.

Measurement correlation: Measurement correlation is used when measurements are taken simultaneously with multiple measurement devices of the same type for parts coming from multiple streams of manufacturing. Creating scatter diagrams and estimating the correlation coefficient between measurement systems can provide insight into whether the multiple measuring devices are contributing to a special cause. Statistical software tools provide "orthogonal regression" capability (Figure 14.13) to compare and analyze two pieces of measurement equipment at a time. A more sophisticated approach would be to conduct an experiment with multiple appraisers, multiple measuring devices, and samples and trials fully randomized, and analyze for "components of variance." If the variance between measuring equipment shows a significant *p*-value, then this is an area for the experimenter to investigate.

Equipment A	Equipment B
10.1	10.08
10.3	10.33
10.5	10.57
10.7	10.65
10.9	10.85
11.1	11.19
11.3	11.21
11.5	11.60
11.7	11.65
11.9	11.95

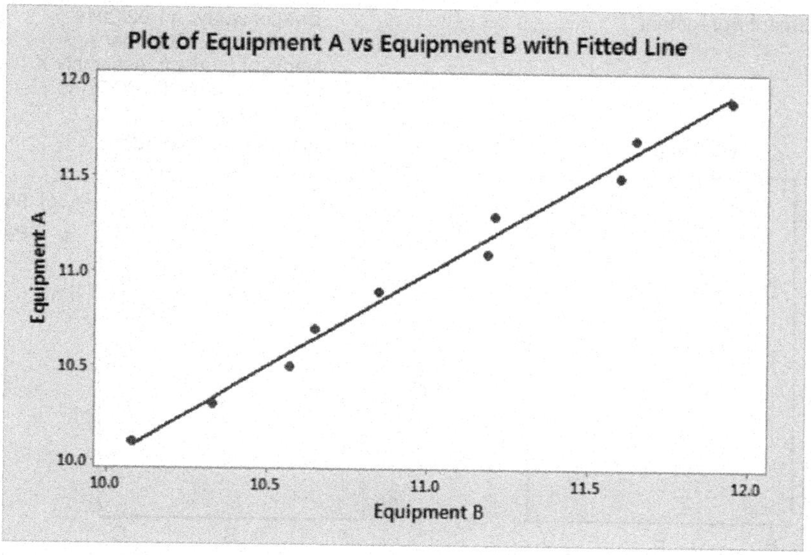

Figure 14.13 Orthogonal regression example.

```
Orthogonal Regression Analysis
Equipment A versus Equipment B

Error Variance Ratio (Equipment A/Equipment B): 0.9

Regression Equation
Equipment A = 0.136 + 0.987 Equipment B

Coefficients

Predictor        Coef    SE Coef        Z      P       Approx 95% CI
Constant      0.13608   0.428735   0.3174  0.751   (-0.704222, 0.97639)
Equipment B   0.98691   0.038894  25.3745  0.000    (0.910681, 1.06314)
```

In orthogonal regression, if the intercept is close to 0 and the slope is close to 1, the two methods most likely provide equivalent measurements. In this example 0.136 is intercept and 0.987 is slope. 0 is contained with the CI (-0.704222, 0.97639), 1 is contained within CI (0.910681, 1.06314)

Percent agreement: An R&R study can also be extended to attribute characteristics like go/no-go results. In the transactional process (service) industries, data may not always be continuous. Results may be expressed as yes/no, OK/not OK, or accept/reject, with rating scales from 1 to 5, and so on. In such cases, an attribute agreement study is used to assess the variation. In many cases, this is purely human variation in judgment and/or evaluation. In some cases, it is purely machine variation, for example, automatic measurement gauging where parts are automatically screened as good or bad by the machine.

A study was conducted with three appraisers inspecting 32 samples in a visual inspection criteria assessment of a polished glass surface. The appraisers

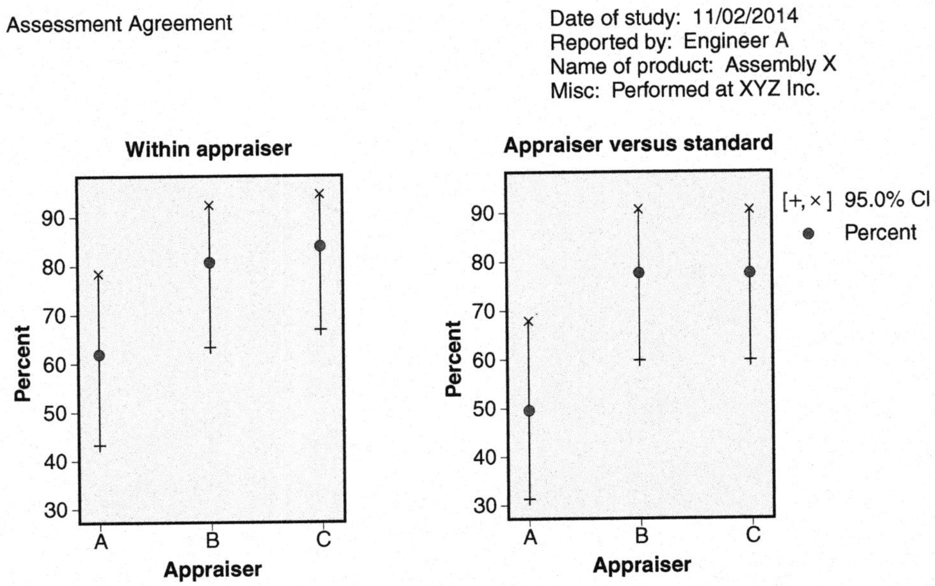

Figure 14.14 Attribute agreement analysis using statistical software.

inspected the surface and using the criteria made judgment of pass or fail. Following are the results and analysis (see Figure 14.14).

Assessment agreement:

Appraiser	# inspected	# matched	Percent (%)	95.0% CI
A	32	20	62.5	(43.7, 78.9)
B	32	26	81.3	(63.6, 92.8)
C	32	27	84.4	(67.2, 94.7)

Number matched: appraiser agrees with him/herself across trials.

"Within appraiser" shows the inconsistencies within an appraiser. Sometimes, the appraiser may judge the same sample as "pass" and another time as "fail." These inconsistencies are not uncommon when human judgment is used. Inconsistencies may be caused due to not understanding the criteria properly, human mood swings, ergonomics of the inspection area, fatigue, and many other reasons.

The same results are matched with the "master results" from a senior appraiser or the process expert. Let us call that a *standard.*

Assessment agreement:

Appraiser	# inspected	# matched	Percent (%)	95.0% CI
A	32	16	50.0	(31.9, 68.1)
B	32	25	78.1	(60.0, 90.7)
C	32	25	78.1	(60.0, 90.7)

Number matched: appraiser's assessment across trials agrees with standard.

Appraisers B and C have higher percent matching to standard be it accept or reject.

Assessment disagreement:

Appraiser	# R/A	Percent (%)	# A/R	Percent (%)	# Mixed	Percent (%)
A	3	11.5	1	16.7	12	37.5
B	0	0.0	1	16.7	6	18.8
C	0	0.0	2	33.3	5	15.6

Number R/A: assessments across trials = R/standard = A.
Number A/R: assessments across trials = A/standard = R.
Number mixed: assessments across trials are not identical.

Here, appraiser A is rejecting 11.5 percent of good parts whereas appraisers B and C are not rejecting any good parts. Appraisers A and B are accepting 16.7 percent of bad parts whereas appraiser C is accepting 33.3 percent of bad parts. This is not good either, as the customer may get bad parts. Since appraiser C is more consistent in judgment, it is easier to train appraiser C to reduce the risk of accepting bad parts.

More advanced statistics are available for this type of attribute study involving ranking scores. Examples of this application where objectivity is added to subjective measures include:

- Tasting tea, coffee, and wine and assigning a score for taste attributes
- Examiners correcting a paper and assigning a score
- Fabrics or polished surfaces where the score is assigned by feeling the surface

Practical challenges in GR&R studies:

1. Management commitment to conduct the study on a periodic basis and monitoring the GR&R percent for critical parameters. Since there is a lot of resource commitment in this type of study, management buy-in is required to sustain this practice on an ongoing basis.
2. GR&R can be challenging as all applications are not as straightforward as textbook examples. Some of the challenges that the author has experienced:
   a. One-sided specification
   b. Skewed distribution
   c. Fully automated equipment with no or minimal appraiser interaction
   d. Destructive testing
   e. New product introduction where only a few units are available
   f. Multiple station comparison

g. Equipment that requires resetting or calibration after every measurement

h. Incomplete GR&R data (units shipped during the study due to urgency)

This is an area of study that is developing new ideas. There are several technical papers that discuss GR&R for destructive testing using nested design. Green Belts and Black Belts are encouraged to expand their knowledge by understanding the new concepts for various challenging applications of GR&R.

AIAG Reference. For product control situations where the measurement result and decision criteria determine "conformance or nonconformance to the feature specification" (that is, 100 percent inspection or sampling), samples (or standards) must be selected but need not cover the entire process range. The assessment of the measurement system is based on the feature tolerance (that is, percent GR&R to tolerance).

For process control situations where the measurement result and decision criteria determine "process stability, direction, and compliance with the natural process variation" (that is, SPC, process monitoring, capability, and process improvement), the availability of samples over the entire operating range becomes very important. An independent estimate of process variation (process capability study) is recommended when assessing the adequacy of the measurement system for process control (that is, percent GR&R to process variation).[2]

When an independent estimate of process variation is not available, *or* to determine process direction and continued suitability of the measurement system for process control, *the sample parts must be selected from the process and represent the entire production operating range.* The variation in sample parts (PV) selected for MSA study is used to calculate the total variation (TV) of the study. The TV index (that is, percent GR&R to TV) is an indicator of process direction and continued suitability of the measurement system for process control. If the sample parts *do not* represent the production process, TV must be ignored in the assessment. Ignoring TV does not affect assessments using tolerance (product control) or an independent estimate of process variation (process control).

Key Point: The underlying reason to conduct ongoing measurement system analysis of your measurement equipment is to understand the uncertainty of the measurement system. That is, what exactly are you really measuring?

Chapter 15

Process and Performance Capability

PROCESS PERFORMANCE VS. PROCESS SPECIFICATIONS

> Define and distinguish between natural process limits and specification limits, and calculate process performance metrics. (Evaluate)
>
> **Body of Knowledge III.F.1**

Natural process limits are calculated from process variation. This is done after all special causes have been removed and the process has achieved statistical stability. Specifications, on the other hand, are expectations from the engineering or customer point of view. When the process variation is significantly lower than the width of the specification (upper limit–lower limit), then we call the process a *capable* process.

The matrix in Figure 15.1 explains in brief the type of action a Green Belt should take in a process based on one of the four given scenarios.

EXAMPLE

First quadrant. If the process is in a state of statistical control (within natural process control limits and follows other applicable rules) and meets specification, the situation is "good."

Fourth quadrant. If the process is not in a state of statistical control (within natural process control limits and follows other applicable rules) and does not meet specification, the Green Belt should stop the process and immediately investigate.

Engineers often get confused between natural process limits and specification limits. Many who are in the third quadrant scenario may even argue that they need not worry as they are producing products that meet specification. They may be hesitant to invest any time in investigating out-of-control conditions. It is

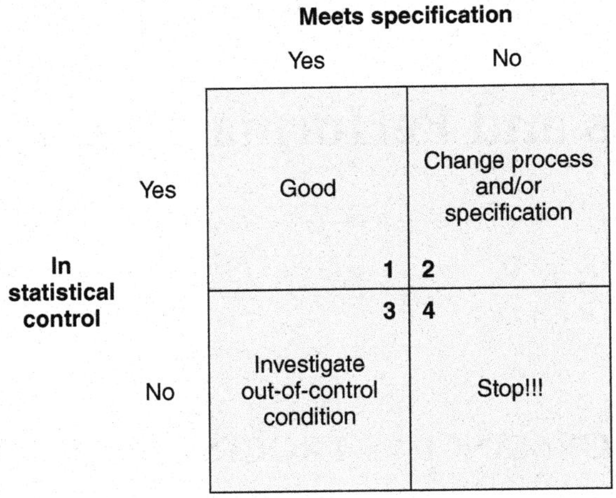

Figure 15.1 Control limit versus specification limit grid.

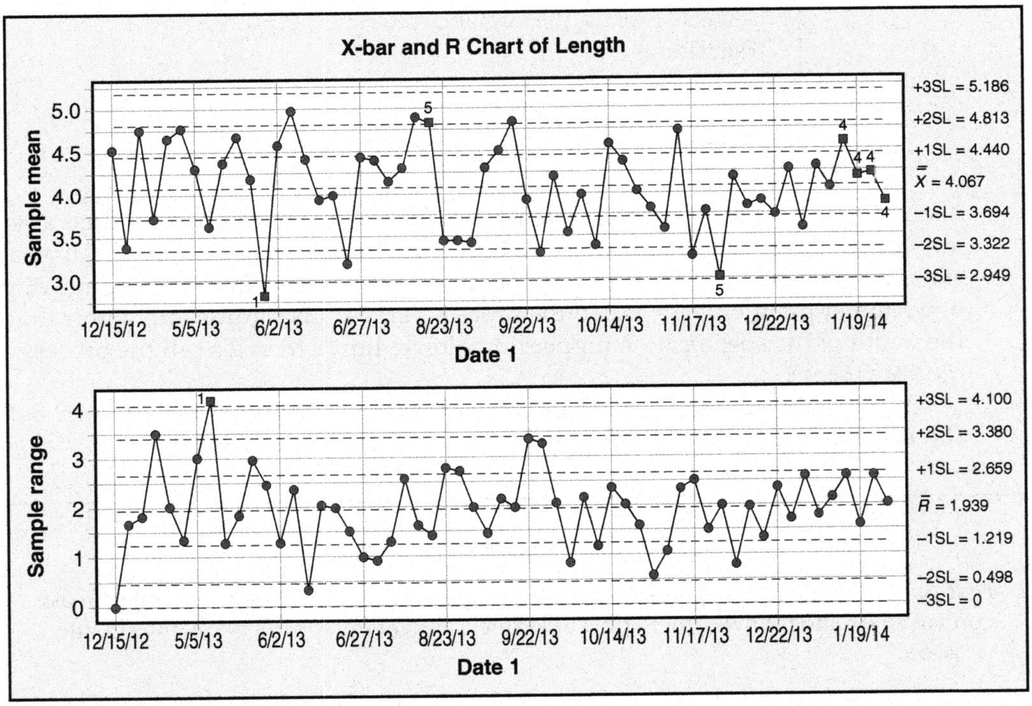

Figure 15.2 Example of $\bar{X}$ and R control chart.

important that Green Belts explain to those engineers that out-of-control situations cause lack of predictability in the process. Once a process is unpredictable, the process may go out of spec at any time.

The chart in Figure 15.2 identifies the natural process limits. As you can see, the square points are those subgroups that have assignable causes. The numbers adjacent to the square points are SPC test rule numbers being violated. Modern statistical software has made this analysis very easy. However, the interpretation of the chart and taking appropriate action to remove the assignable causes are still human endeavors. It is important to ensure that the range chart is within control before reviewing the average chart. This goes with the understanding that the variation should be in control before making adjustment to averages. Six Sigma Green Belts are expected to review those out-of-control violations, assign special causes, and recalculate the control limits before firming up the limits for implementation. Chapter 22 will cover this topic more extensively.

Test Results for $\bar{X}$ Chart of Length

Test 1. One point more than 3.00 standard deviations from centerline (see Figure 15.2). Test failed at point: 12

Test 4. 14 points in a row alternating up and down. Test failed at points: 54, 55, 56, 57

Test 5. Two out of three points more than two standard deviations from centerline (on one side of centerline). Test failed at points: 24, 45

Test Results for *R* Chart of Length

Test 1. One point more than 3.00 standard deviations from centerline (see Figure 15.2). Test failed at point: 8

EXAMPLE

A product after packing is specified to weigh between 1.00 to 1.10 lbs. Data collected from shipping over a period of 30 days indicates that the distribution is normally distributed and the associated control chart is stable. The control chart used a subgroup sample size of four every two hours. The control chart calculations show that the grand process average is 1.065 lbs. and the average range is 0.05. A Green Belt is assigned to estimate the percentage of product that could have been shipped underweight and overweight.

Solution: The average $\bar{\bar{X}}$ is 1.065 lbs. The point estimate for process standard deviation is given by the formula

$$\hat{\sigma} = \frac{\bar{R}}{d_2} = \frac{0.05}{2.059} \approx 0.024 \text{ lbs.}$$

(The value of statistical constant d_2 for a subgroup of size four is 2.059.)

Continued

Continued

The distance from the upper specification limit to the process average is 1.10 – 1.065 = 0.035. Dividing this by the standard deviation gives 0.035/.024 = 1.46, which may be thought of as the number of standard deviations between the process average and the upper specification limit. It is customary to label this quantity Z_U. The formula for Z_U would then be

$$Z_U = \frac{\text{USL} - \bar{\bar{X}}}{\hat{\sigma}}$$

where

USL = Upper specification limit

$\bar{\bar{X}}$ = The process average

$\hat{\sigma}$ = Estimated standard deviation

Similarly, Z_L, the number of standard deviations between the process average and the lower specification limit, is given by the formula

$$Z_L = \frac{\bar{\bar{X}} - \text{LSL}}{\hat{\sigma}}$$

where

LSL = Lower specification limit

$\bar{\bar{X}}$ = The process average

$\hat{\sigma}$ = Estimated standard deviation

In this example, the value of Z_L is about 2.70. The area beyond each of these Z values can be found using the areas under standard normal curve table (Appendix F). These areas correspond to the proportion of production that falls outside specifications. From the table, the area beyond 1.46 standard deviations is 0.0721, and the area beyond 2.70 standard deviations is 0.0035. The capability analysis indicates that approximately 7.21 percent of shipped packages could have been shipped overweight, and approximately 0.35 percent of shipped packages could have been shipped underweight.

The traditional definition of natural process limits is $\pm 3\sigma$. In the previous example, the natural process limits are 1.065 ± 3(0.024), or approximately 0.993 to 1.137. The fact that the natural process limits are outside the specification limits of 1.00–1.10 indicates that the process is not capable of meeting the specification. It is also not uncommon for organizations to create internal specifications tighter than the customer specifications to provide additional protection to customers. Please note that control charts like $\bar{X}$ and R and $\bar{X}$ and s are robust to nonnormality in data (courtesy of the central limit theorem). Users are advised to understand distribution of data and calculate limits appropriate for the distribution. Transformation of data is another approach to accomplish just that.

PROCESS CAPABILITY STUDIES

> Define, describe, and conduct process capability studies, including identifying characteristics, specifications, and tolerances, and verifying stability and normality. (Evaluate)
>
> **Body of Knowledge III.F.2**

Process capability is the ability of the process to meet the expected specifications. Every process has variation due to common causes and special causes, both internal and external to the process.

Examples of common causes include interaction between process steps caused by the way the processes are sequenced, manufacturing equipment design, natural variation in incoming material supply, measuring equipment design, and so on. Special cause examples include significant changes in incoming material quality, operators with varying skill levels, changes to process settings, environmental variations, equipment drift, and so on.

The random process variations caused by common causes influence the ability of the process to meet the expected specifications (process capability). Assignable (special) causes are investigated and removed before estimating the process capability.

In practice, people experienced with the process are usually able to identify the few characteristics that merit a full capability study. These are the characteristics that past experience has shown to be difficult to hold to specification. In some industries, the customers themselves identify critical characteristics that need to be monitored by SPC.

But, best practice is to perform a comprehensive process FMEA, identify the parameters or characteristics that require statistical process control, and create a control plan with detailed SPC planning. The reason is that implementing SPC and measuring process capability costs the organization money. Hence, selection of appropriate process parameters and characteristics is important. More details on FMEA and control plans can be found in Chapter 9.

Specifications and tolerances are obtained from engineering drawings and customer contracts.

Sometimes they are also publicly announced as guarantees to customers. Have you seen advertisements guaranteeing the following?

- Expediters: next-day delivery
- Restaurants: (wait time) 15 minutes or free
- Even rental apartments: emergency maintenance resolution in 24 hours or one-day rental free

All the above are examples of specifications and tolerances. Unlike manufacturing, these are areas where customers do not explicitly state their expectations. Service providers study the market and customer expectations through surveys and informal interviews and identify specifications and tolerances themselves to be competitive in the market.

Steps for Process Capability Studies

Measurement system verification: The first step in conducting a process capability study is to perform measurement system analysis. Measurement system variation can mislead the process capability assessment and process stability monitoring. As discussed in the previous chapter in relation to Figure 14.10, Larry Barrentine, in his book *Concepts for R&R Studies*, presents a graph that shows the relationship between actual process capability and observed process capability for various measurement system error percentages. Once the MSA is performed and sources of variation are identified and removed, the percentage of measurement variation reduces to a small proportion of overall process variation and process tolerance. Now we can proceed to performing the process capability studies.

The next step is to identify appropriate rational subgrouping of samples for control chart plotting. Subgroup size can be anywhere between two and 10. However, subgroups greater than five are uncommon. A typical SPC chart has five consecutive samples taken at equal intervals from a process and average/range plotted to observe the stability of the process. Care has to be taken that within-subgroup variation is smaller than between-subgroup variation. For low-volume processes, individual charts are plotted to monitor the stability where the subgroup size is one. Average standard deviation charts are used when the subgroup size is greater than eight. This is a more powerful chart for detecting shifts in processes but can be costly for data collection. An individual chart with subgroup size one is, on the other hand, less sensitive to detecting shifts. An average range chart provides an economic balance between the cost of running SPC and information that can be usefully derived.

Stability is a fairly sophisticated statistical concept, equivalent to the absence of special causes of variation. After 20 subgroups of points have been plotted, if the chart shows that no special causes are present, the process is considered to be stable. Although authorities disagree on the number of points needed, 20 or 30 points commonly are used. More points are plotted for higher confidence in the stability conclusion.

Control chart monitoring is not impacted even if the distribution of the data is nonnormal. However, to measure the process capability, normality is required for continuous data.

To do this, construct a histogram using the original readings (not the averages) from the control chart. If the histogram looks normal, with most points grouped around a single peak and fairly symmetric tails on each side, it is assumed that the data constitute a sample drawn from an approximately normal population. Again, the more data used, the greater the confidence one can have in this conclusion. All commercially sold statistical software can construct a probability plot with confidence intervals and test for normality.

If the process data are not normally distributed, techniques like the Box-Cox transformation and Johnson transformations are used for nonnormal to normal data transformations. If the data are normally distributed, the next step is to use a normal distribution table to estimate process capability. The most common method is to use the data from a control chart to estimate μ and σ.

The overall objective of the process capability study is to monitor whether a process is in statistical control and the process is capable of meeting specifications. If the process is capable, we move on to review other characteristics. If not, we take action to improve the capability. Given that the process is stable, the first obvious step is to try to center the process and review the percent nonconformance outside the specification limits. If the process variation is smaller than the specifications, this can reduce the number of nonconformances. The next important action is to reduce the variation. This is the crux of Six Sigma methodology and the major payback on effort. Sometimes for economic reasons it is unfortunately required to off-center the process distribution in a direction that creates rework and salvage rather than scrapping of parts (see Figure 15.3). This is a containment action until engineering figures out a means for reducing the variation. Another possibility is to revisit the specification limits from the customer and engineering standpoint. Surprisingly, it is not uncommon to see specifications that are set unrealistically tight by designers without reviewing the capability of the process and limitations of technology.

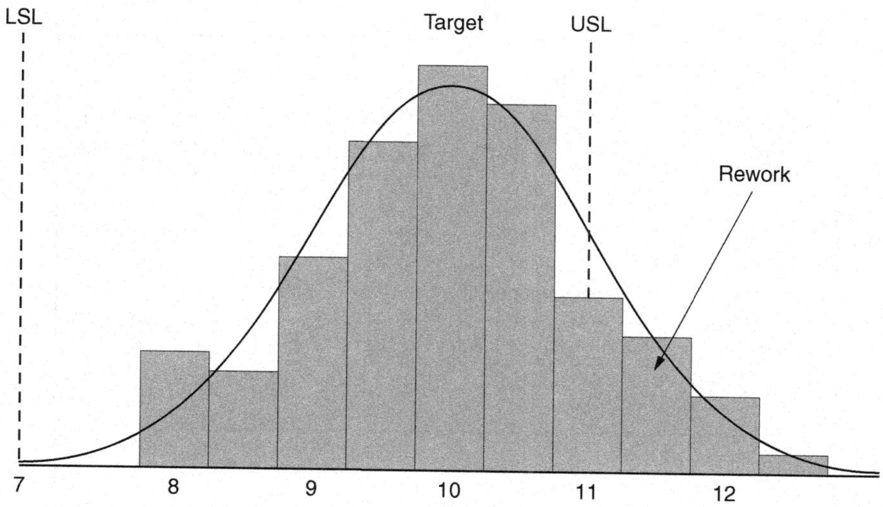

Process data	
LSL	7
Target	10
USL	11
Sample mean	9.98378
Sample N	100
StDev (Overall)	1.00121

	Performance observed	Performance expected overall
% < LSL	0.00	0.14
% > USL	15.00	15.51
% Total	15.00	15.65

Figure 15.3 Process capability report for shaft machining.

Sampling with Respect to Statistical Process Control

Random sampling is particularly useful if we have a batch process like oven baking, spray painting, heat treatment, group therapy, and so on. If in earlier experiments it is proven that the part selected at random from a batch is representative of the group, we can pick random samples, average the measurements of the samples, and plot them as one data point of a subgroup. Note that the measurements made within a batch are *not* a subgroup. Batch-to-batch variation can be represented as an average moving range. Within-batch variation can be represented with a range or standard deviation chart.

Systematic sampling can be used when performing individual moving range SPC monitoring by sampling every *n*th part. This is typically applied when parts are coming out of a conveyor line. In a transactional process situation such as a banking mortgage transaction, sampling every *n*th customer might be used to assess service quality.

The subgroup approach of sampling is the typical approach used for plotting $\bar{X}$ and R charts or $\bar{X}$ and s charts. An important factor to keep in mind is that the within-subgroup variation should contain only common causes. This is the reason that consecutive parts are sampled in the $\bar{X}$ chart approach. The subgroup intervals should be carefully planned to capture special causes, if any. See Figure 15.4 for a summary of these types of sampling.

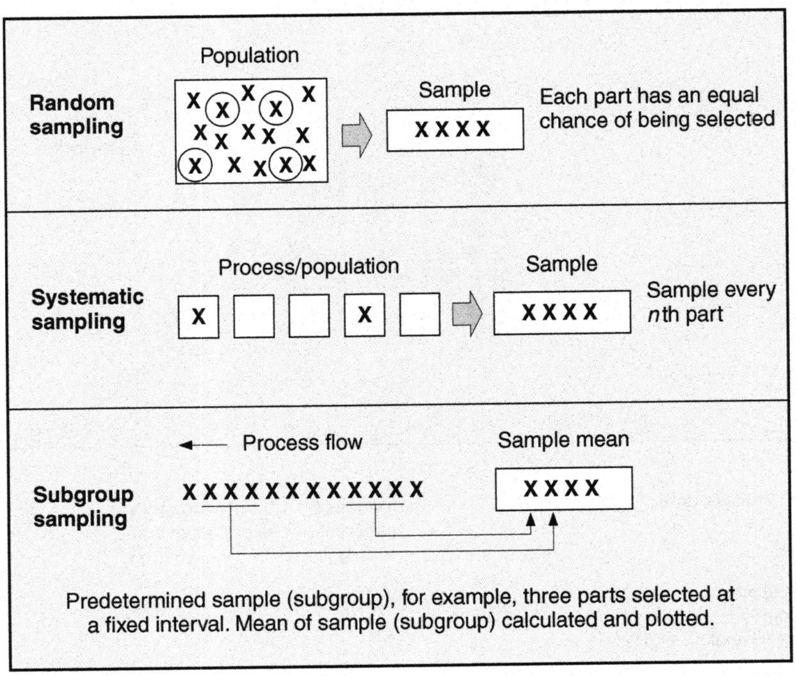

Figure 15.4 Types of sampling for SPC data collection.

PROCESS CAPABILITY (C_P, C_{PK}) AND PROCESS PERFORMANCE (P_P, P_{PK}) INDICES

> Describe the relationship between these types of indices. Define, select, and calculate process capability and process performance. Describe when C_{pm} measures can be used. Calculate the sigma level of a process. (Evaluate)
>
> **Body of Knowledge III.F.3**

Various capability indices have been developed in an attempt to quantify process capability in a single number. Three of the most common indices are C_{pk}, C_p, and C_r. These are defined and illustrated in the following paragraphs.

$$C_{pk} = \frac{\text{Min}(Z_U, Z_L)}{3}$$

where min (Z_U, Z_L) is defined as the value of the smallest Z value.

In the previous example, $C_{pk} = 1.46 \div 3 = 0.49$. The "min" in the formula for C_{pk} means that this index looks at the nearest specification limit.

Historically, a C_{pk} value of 1 or larger was considered "capable." This would be equivalent to stating that the natural process limits lie inside the specification limits. More recently, quality requirements have become more stringent, and many customers require C_{pk} values of 1.33, 1.66, or 2.00. Notice that this is the equivalent of requiring $\pm 4\sigma$, $\pm 5\sigma$, and $\pm 6\sigma$ to be inside the specification. It is the move toward $C_{pk} = 2$ or $\pm 6\sigma$ that inspired the Six Sigma terminology. Most authors currently define a 6σ process as one with $\sigma \leq 1/12$ (specification) and with the process average not drifting more than 1.5σ over time (advocated by Motorola). Therefore, the percentage violating each specification limit is based on values from the Z table corresponding to 4.5σ ($6\sigma - 1.5\sigma$).

$$C_p = \frac{\text{Tolerance zone}}{6\sigma}$$

In the previous example, $C_p = 0.1 \div 0.144 \approx 0.69$. The formula for C_p doesn't take into consideration whether the process is centered in the specification. In fact, it shows how good C_{pk} could be if the process were centered.

C_r is the inverse of C_p. C_r expressed as a percentage (by multiplying by 100) shows the percentage of specification used up by the process variation.

$$C_r = 1/C_p$$

In the previous example, $C_r \approx 1.45$. (145 percent of the specification is consumed by the process variation.) Lower C_r values are better.

Process Capability Calculations

There are typically two calculations done to identify how capable a process is. This is done so that we can determine if the possibility of improvement exists for the process in question. These two calculations are called C_p (capability index) and C_{pk} (process performance).

Assuming the processes are centered, Figure 15.5 is the illustration of four processes with different process capability (C_p).

Some examples of common values seen on the shop floor include:

1. $C_p = 2$ and $C_{pk} = 1.5$ are the values given when a process has achieved Six Sigma quality.
2. C_p, $C_{pk} \geq 1.33$ shows that the process is capable.
3. A C_p, C_{pk} value of 1.0 means that the process barely meets the specification. This will produce 0.27 percent defective units.
4. A C_p, C_{pk} value less than 1.0 means that the process is producing units outside engineering specifications.
5. Abnormally high C_p, C_{pk} (> 3) shows either that the specification is loose or identifies an opportunity to move to a less expensive process. (Often, people do nothing fearing that they may worsen the situation.)

Some common interpretations of C_p and C_{pk}:

1. In both C_p and C_{pk}, the higher the value, the better.

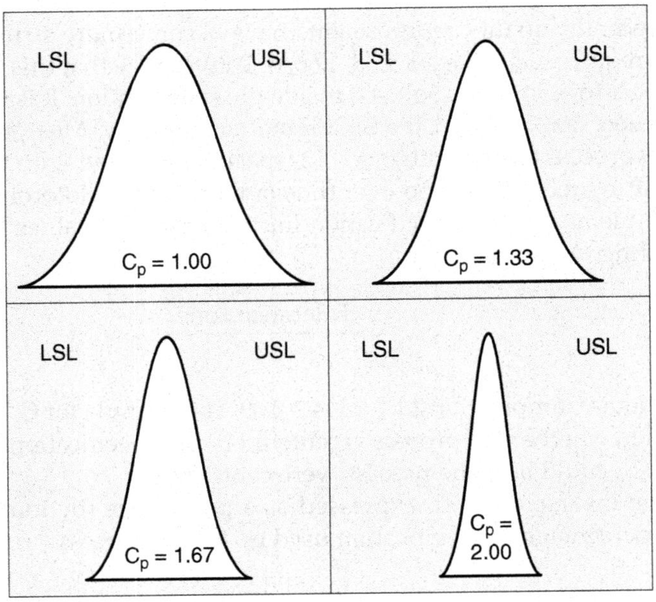

Figure 15.5 Process capability scenarios.

2. The C_p value does not change as the process is being centered to target unless something in the process changes.
3. The C_p and C_{pk} values will be equal if the process is perfectly centered.
4. C_{pk} is always equal to or smaller than C_p.
5. If the C_{pk} value becomes a negative number, then the process average is outside one of the engineering specification limits.
6. In a process with one-sided specification, either C_p upper limit or C_p lower limit is calculated (depending on whether the specification is one sided with maximum limit or minimum limit).
7. Higher C_{pk} observed from a small sample may not be of much use as the confidence interval for C_{pk} will be very wide.

SHORT-TERM VS. LONG-TERM CAPABILITY AND SIGMA SHIFT

> Describe the assumptions and conventions that are appropriate to use when only short-term data are used. Identify and calculate the sigma shift that occurs when long- and short-term data are compared. (Evaluate)
>
> **Body of Knowledge III.F.4**

Performance indices provide a picture of current process operation and can be used for comparison and prioritization of improvement efforts. Three such indices are P_{pk}, P_p, and C_{pm}, which require stability of the process as well. The formulas for P_{pk} and P_p are equivalent to the corresponding capability indices (C_{pk} , C_p) except that sample standard deviation is used instead of σ. The formulas are:

$$P_{pk} = \min\left(\frac{USL - \bar{\bar{X}}}{3s}, \frac{\bar{\bar{X}} - LSL}{3s}\right)$$

where s is the sample standard deviation

$$s = \sqrt{\frac{1}{N-1}\sum_{i=1}^{N}(x_i - \bar{x})^2}$$

$$P_p = \frac{\text{Tolerance zone}}{6s}$$

where s is the sample standard deviation

EXAMPLE

A manufacturing assembly line that ships built devices on a weekly basis performs random sampling of the devices' critical parameters. The quality engineer uses these data to measure performance indices for those critical parameters. Since the data also include long-term variations and shifts, the engineer should not use the metrics C_p and C_{pk}. Measures like C_p and C_{pk} are used when process stability is monitored through SPC and the standard deviation is derived from the mean range. The engineer measures a sample standard deviation of 0.004 and a mean of 10.014.

The upper and lower specification limits for this critical parameter are 9.99 to 10.03.

$$s=\sqrt{\frac{1}{N-1}\sum_{i=1}^{N}(x_i-\bar{x})^2}$$

$$s=0.004$$

$$P_p=\frac{\text{Tolerance}}{6s}=\frac{0.04}{6(0.004)}=\frac{0.04}{0.024}=1.667$$

$$P_{pk}=\text{Min}\left(\frac{USL-\bar{\bar{X}}}{3s},\frac{\bar{\bar{X}}-LSL}{3s}\right)=\text{Min}\left(\frac{10.03-10.014}{3(0.004)},\frac{10.014-9.99}{3(0.004)}\right)$$

$$P_{pk}=\text{Min}\left(\frac{0.016}{0.012},\frac{0.024}{0.012}\right)=\text{Min}(1.33,2.00)=1.33$$

The engineer now compares the contractual requirement from the customer on P_p and P_{pk} with the measured values of P_p of 1.667 and P_{pk} of 1.33.

The Automotive Industry Action Group (AIAG) and the American National Standards Institute (ANSI) recommend the use of P_p and P_{pk} when the process is not in control. This is a somewhat controversial position because an out-of-control process is by definition unpredictable. Montgomery[1] states that, "The process performance indices P_p and P_{pk} are actually more than a step backwards. They are a waste of engineering and management effort—they tell you nothing."

Figure 15.6 shows an example of P_p, P_{pk}, and C_{pm} analysis of a vendor-supplied product feature using statistical software.

$$C_{pm}=\frac{USL-LSL}{6s_{C_{pm}}}$$

where

USL= Upper specification limit

LSL = Lower specification limit

s = Sample standard deviation

$$s_{C_{pm}}=\sqrt{\sum_{i=1}^{n}\frac{(x_i-T)^2}{n-1}}$$

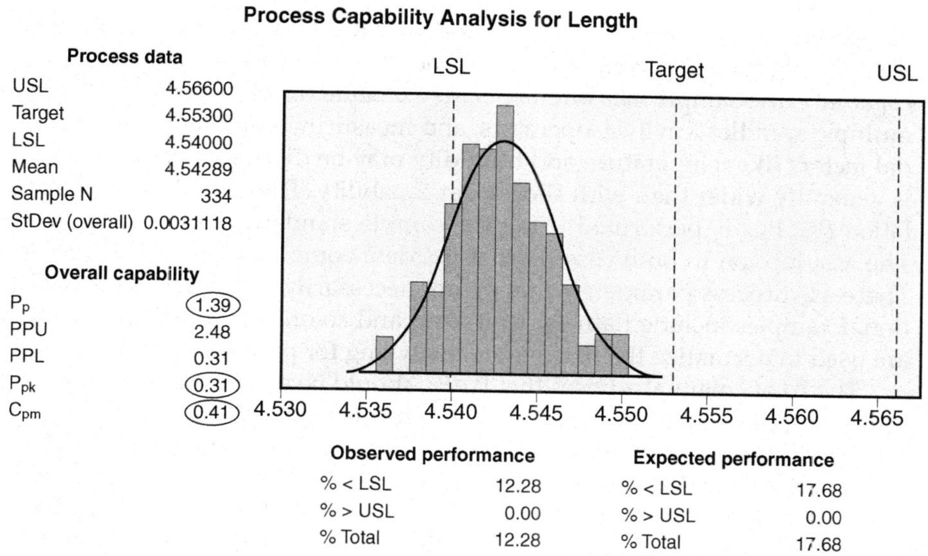

Figure 15.6 Process performance indices P_p, P_{pk}, and C_{pm}.

and

T = Specification target

x_i = Sample reading

n = Sample size

Undoubtedly, C_p, C_{pk} is a better measure as it reflects capability derived from common cause variation. P_p, P_{pk} can be applied to data collected from incoming inspection material to obtain an indication of process performance at the supplier end where the SPC data are not available. Given that the process data may contain special causes as well, as long as the data follow a normal distribution, P_p, P_{pk} can provide some idea about processes. It is unlikely that all suppliers have SPC implementation in place. It is also unlikely that SPC-implemented suppliers will be willing to share their process data in time sequence with their customers. Hence, P_p, P_{pk} can serve as an indicator for these situations.

C_{pm} is a useful measure where the process has a target value rather than a conventional nominal value, which is typically the midpoint of specifications. An example would be tolerances for shafts/holes where specific mechanical fit is involved, for example, clearance fit, interference fit, and transition fit.

Short-Term vs. Long-Term Capability

Short-term capability is calculated from the data collected from a stable process monitored for a short time of approximately 20 to 30 subgroups. For such a short interval, the variability is often relatively small due to a focus on specific equipment, a set of trained operators, homogeneous material, the same measuring equipment, and so on. The process capability (C_p, C_{pk}) calculated with these data uses the sigma value computed from the average process range. For long-term capability, the sample measurements may include different streams of machines, multiple spindles, cavities, operators, and measuring equipment, and even external factors like temperature and humidity may be different. Hence, the variation is generally wider than with short-term capability. This process capability calculation (P_p, P_{pk}) is performed using the sample standard deviation of the values. The assumption in both cases is that the data come from a normal distribution. There are process parameters that do not necessarily follow the normal distribution. Examples include flatness, wait time, and so on. Transformation techniques are used to normalize the data before analyzing for process capability.

The AIAG manual advises that P_p, P_{pk} should be used in combination with C_p, C_{pk} for comparison and prioritization of improvement efforts.

We discussed increase in variation of a process over the long term. It is also important to note that the mean also shifts over the long term. The concept of 1.5-sigma shift advocated by the late Bill Smith, a reliability engineer from Motorola, has been explained earlier in this book. In the context of short- and long-term variation, Figure 15.7 shows how the mean from short-term variation can shift over time.

The dynamic variation of a process with respect to time is explained further in Figure 15.8. The long-term capability of the process is due to changes to the mean and variation over time.

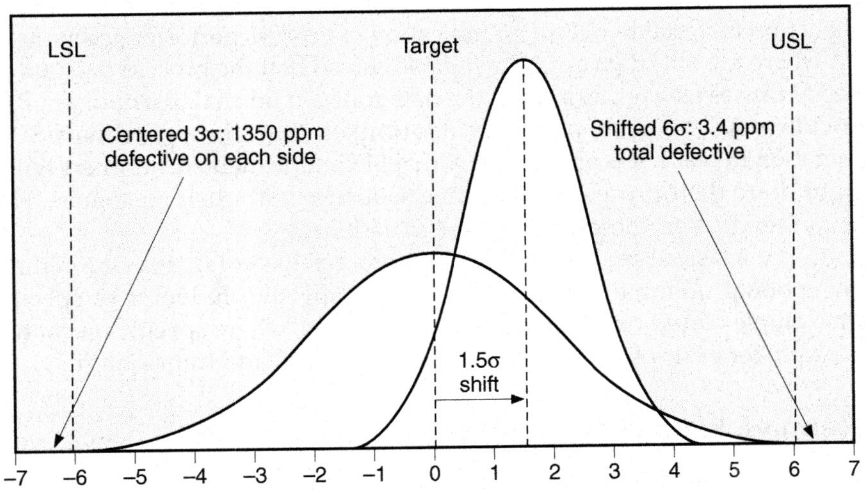

Figure 15.7 Process performance 1.5-sigma shift.

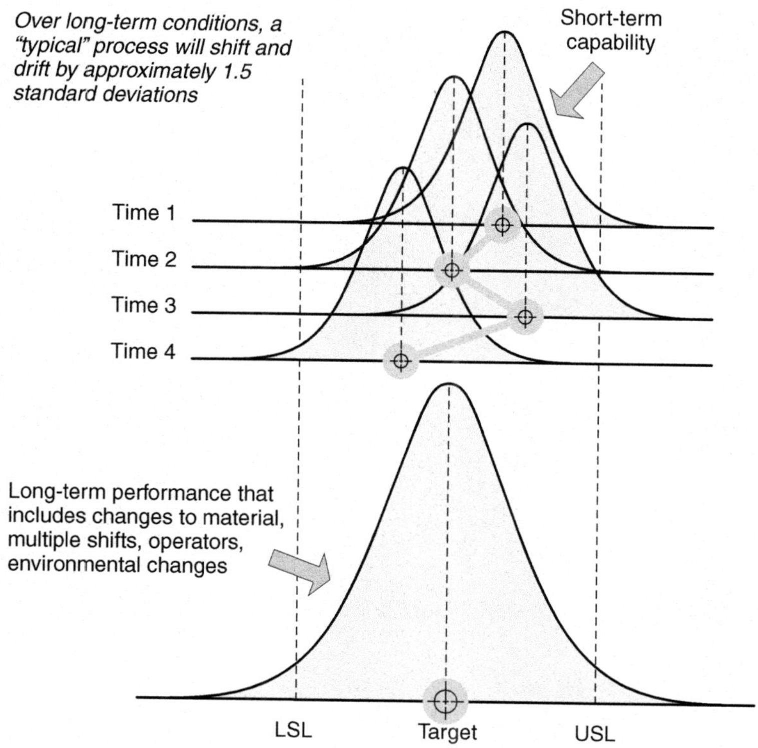

Figure 15.8 Short-term versus long-term performance with 1.5-sigma shift.

Part IV
Analyze Phase

Chapter 16	Exploratory Data Analysis
Chapter 17	Hypothesis Testing
Chapter 18	Additional Analysis Methods

Part IV is an overview of the *analyze* phase, including summaries of those Six Sigma methods and practices designed and intended to determine and prioritize improvements to products, processes, and organizations. It covers approximately 18 of the 100 questions that will be asked on the ASQ CSSGB exam. Sections have been improved with additional explanations, definitions, and examples.

OVERVIEW

In the *analyze* phase, statistical methods and tools are used to identify key pieces of information that are critical to explaining defective products. In this phase, practical business problems are analyzed using statistical tools. Data are collected to determine the relationships between the variable factors in the process and to determine the method for improvements. This phase determines how well or how poorly the process is currently performing and identifies possible root causes for variation in quality. The data analyzed can reveal the basic nature and behavior of the process and show how capable and stable the process is over an extended period of time. For example, is the problem sporadic or persistent? Or is the problem technology or process related?

The analyze phase covers two major sections. The first part covers exploratory data analysis, which includes multivariate studies to differentiate positional, cyclical, and temporal variation, and simple linear correlation and regression to determine the statistical significance (p-value) and difference between correlation and causation. The second part offers an introduction to hypothesis testing, tests for means, variances, and proportions, paired-comparison hypothesis tests, analysis of variance (ANOVA), and chi-square testing to determine statistical significance.

Chapter 16

Exploratory Data Analysis

MULTI-VARI STUDIES

> Select appropriate sampling plans to create multi-vari study charts and interpret the results for positional, cyclical, and temporal variation. (Create)
>
> **Body of Knowledge IV.A.1**

Variations in processes can occur in different ways, specifically for products and services that are complex in nature. Interest in exploring these variations depends on the impact these variations can cause on product and service performance. If we are interested in variation in the diameter of a shaft at one location on the shaft, or variation in a customer service call center in the first one hour of the business day, the answer is simple. However, a product where an assembly fits in one end of a shaft, but not in the middle or the other end of the shaft, could be a problem. Also, if the shaft produced in one manufacturing shift shows different variation than in the other two shifts of manufacturing, there will be a problem with the standardized assembly process. Similar issues may occur in the service sector, for example, variation between service centers, variation between different periods of the year. As a Green Belt we would want to analyze these variations, understand the sources, and plan to improve the process in the next phase. Multi-vari studies help us understand these variations both analytically and graphically.

The multi-vari chart is a useful tool for analyzing the three types of variation: *cyclical*, *temporal*, and *positional* variation (see Table 16.1). It also helps to minimize variation by identifying areas in which to look for excessive variation.

Multi-vari studies are the perfect tool for investigating the stability or consistency of a process. They help to determine where the variability is coming from within a process.

Often, *positional* variation is called *within-part* variation and refers to variation of a characteristic on the same product. *Cyclical* variation is also called *part-to-part* or *lot-to-lot* variation. *Temporal* variation, also called *shift-to-shift* variation, occurs as change over time.

Figure 16.1 illustrates a multi-vari line chart. The chart consists of a series of vertical lines, or other appropriate schematics, along a time scale. The length of each line or schematic shape represents the range of values found in each sample set. Figures 16.2 through 16.4 illustrate excessive variability, less variability, and a variability shift over time, respectively.

Procedure for Multi-Vari Sampling Plan

1. Select the process and the characteristics to be investigated.
2. Select a manageable sample size, for example, 3 to 5 samples with data collection time and identified frequency of collection. First, it is important to understand the variations within and between samples.
3. Record the time and values from each sample set in a table format.
4. Plot a graph with time along the horizontal scale and the measured values on the vertical scale.

Table 16.1 Types of variation.

	Product/process under consideration	
Types of variation	**Piece**	**Batch/lot**
Positional	Within-part	Within-batch/within-lot
Cyclical	Part-to-part	Batch-to-batch/lot-to-lot
Temporal	Over time (shift-to-shift)	Over time (shift-to-shift)

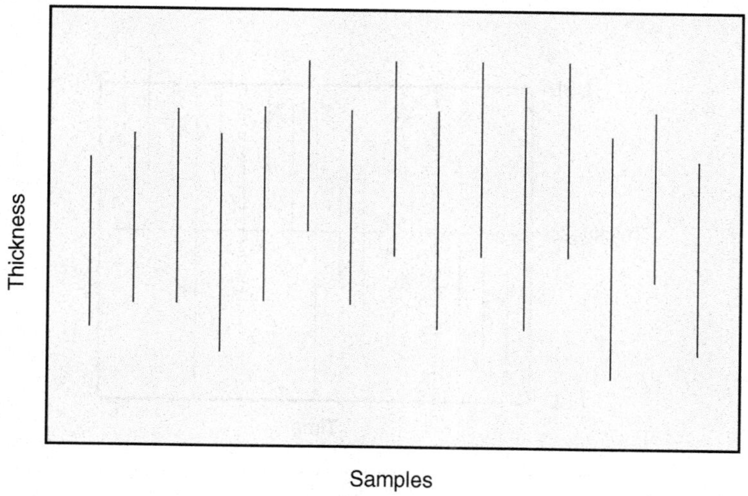

Figure 16.1 Variation within samples.

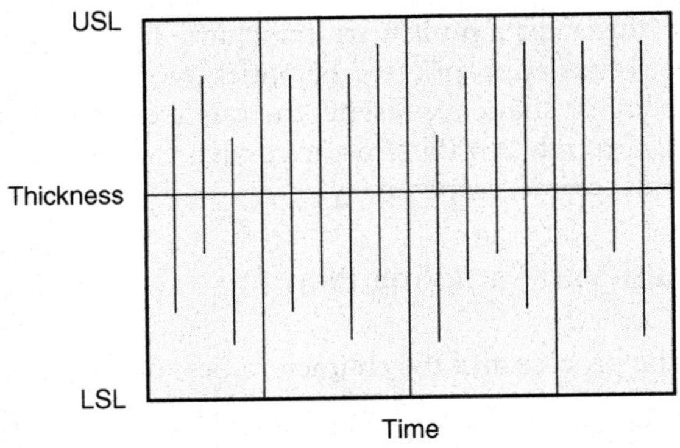

Figure 16.2 Excessive variability (part is tapered).

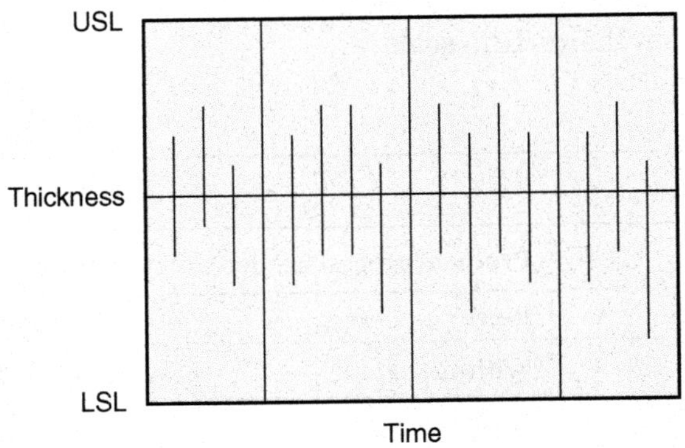

Figure 16.3 Less variability (center is thicker).

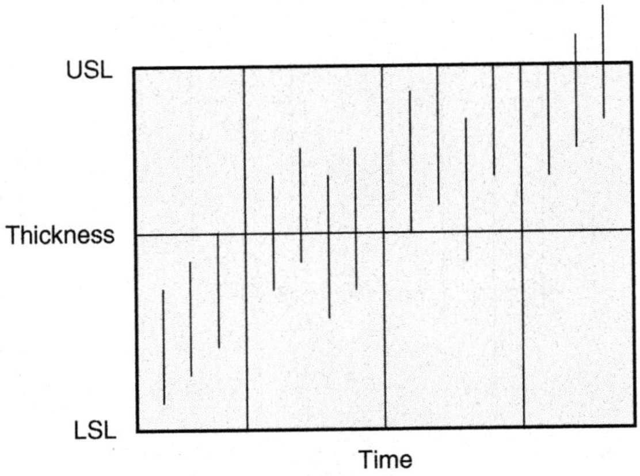

Figure 16.4 Variability shift over time (part is getting larger).

5. Connect the observed values with lines.
6. Observe and analyze chart for variation within the sample, sample-to-sample, and over time.
7. Conduct additional studies to concentrate on the area(s) of apparent maximum variation.
8. Make process improvements and repeat multi-vari study to confirm the results.

Note: You may further refine your experiment by calculating statistically valid sample sizes.

EXAMPLE

A stainless steel casting is used as an example for this study. This part has a tight tolerance on its machined inner diameter (ID) as a piston moves inside it that must make a good seal. The part is a closed-end cylinder as illustrated in Figure 16.5. The pistons leak as a result of a poor seal. For machinists, this is considered a challenging design for manufacturing since the walls are very thin, the hole depth is deep, and the dead end adds further to the problems.

This issue had been around for quite some time, and various team members from the manufacturing process had ideas they wanted to try. One of the engineers suggested that the lathe chucks (used to hold the part during machining) are clamping too tightly and squeezing the part, making it go out of round. Another would like to change the tooling so that machining is uniform, and so on. Many of the ideas have been tried in the past with little or no improvement. What was lacking was a structured approach to addressing the problem.

To solve this problem, a Six Sigma improvement team was put in place. A Green Belt was assigned to conduct a multi-vari study with a data collection scheme that will capture the various types of variation. The data are displayed on graphs that will aid in identifying the largest source of variation. To cover the within-part variation, the inner diameters (ID) are measured at the top, middle, and bottom as indicated by the section

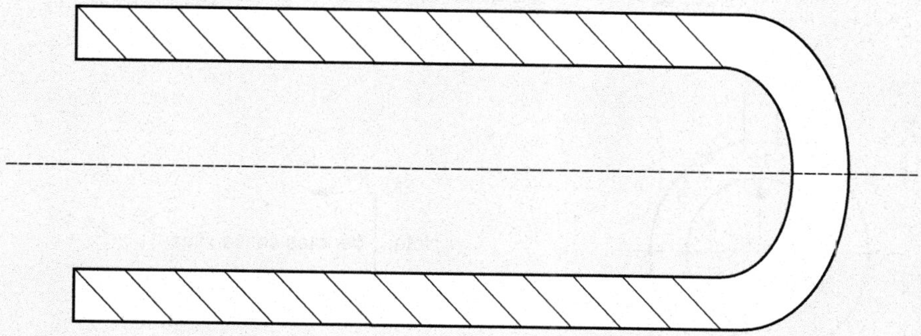

Figure 16.5 Stainless steel casting with critical inner diameter.

Continued

Continued

lines T, M, and B in Figure 16.6a. Additional within-part variation is measured by checking for out-of-round conditions. To detect this out-of-round variation, the ID is measured at three different angles, 12 o'clock, 2 o'clock, and 4 o'clock, with 12 o'clock on the top as shown in Figure 16.6b.

The 12 o'clock measurement is the diameter from 12 o'clock to 6 o'clock.

The 2 o'clock measurement is the diameter from 2 o'clock to 8 o'clock.

The 4 o'clock measurement is the diameter from 4 o'clock to 10 o'clock.

To capture the variation over time, five pieces are selected at approximately equal time intervals during a shift. All measurements are obtained with measuring equipment using a dial bore gage and recorded on the data sheet shown in Figure 16.7.
The measurement results from five parts from one shift are shown in Table 16.2. What can be learned by looking at these numbers? The answer is "very little." However, plotting the numbers on a multi-vari graph as shown in Figure 16.8 does reveal an interesting pattern.

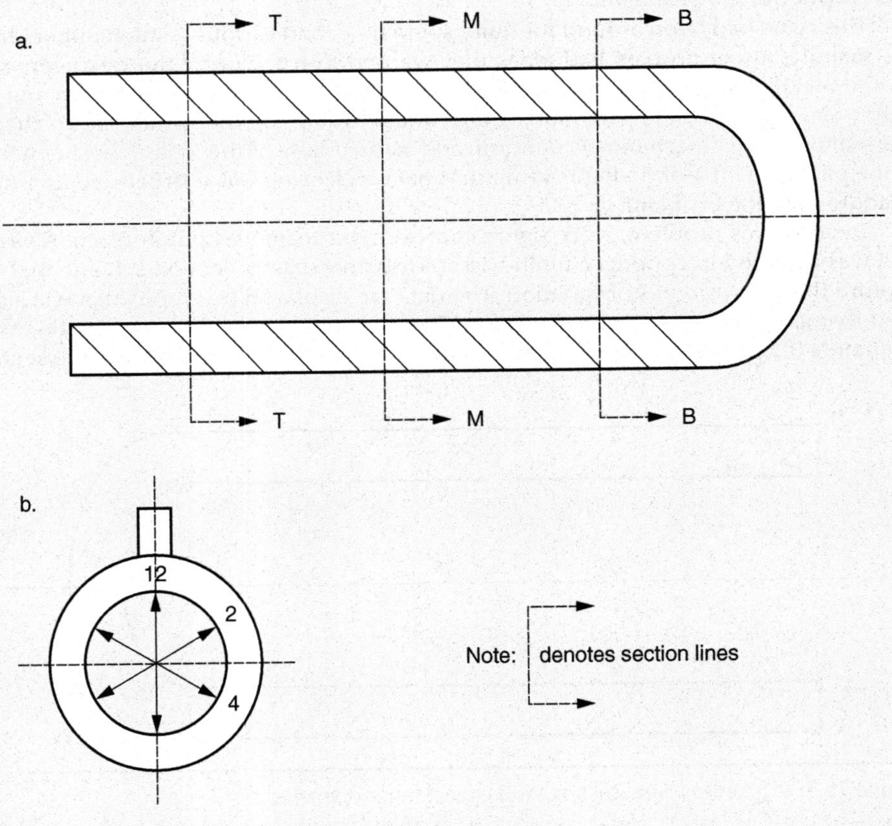

Figure 16.6 Sectional views of stainless steel casting.

Continued

Continued

Angle	Section T-T	Section M-M	Section B-B
12 o'clock			
2 o'clock			
4 o'clock			

Figure 16.7 Data collection sheet.

Table 16.2 Measurement data from five parts produced during one shift.

Part #1			Part #2			Part #3			Part #4			Part #5		
T	M	B	T	M	B	T	M	B	T	M	B	T	M	B
.998	.992	.996	.984	.982	.981	.998	.998	.997	.986	.987	.986	.975	.980	.976
.994	.996	.994	.982	.980	.982	.999	.998	.997	.985	.986	.986	.976	.976	.974
.996	.994	.995	.984	.983	.980	.996	.996	.996	.984	.985	.984	.978	.980	.974

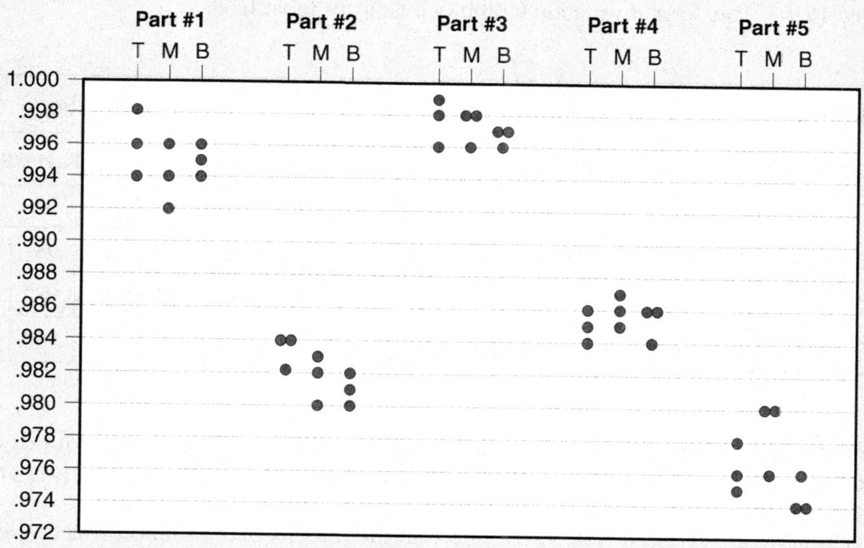

Figure 16.8 Graph of data from Table 16.2.

A visual study of Figure 16.8 may help determine what type of variation is most dominant—is it out-of-round, top-to-bottom, or part-to-part? This may be analyzed by drawing a different type of line for each type of variation, as shown in Figure 16.9.

This figure illustrates that the most significant variation is caused by part-to-part differences. The team brainstormed possible causes for part-to-part variation. Team

Continued

Continued

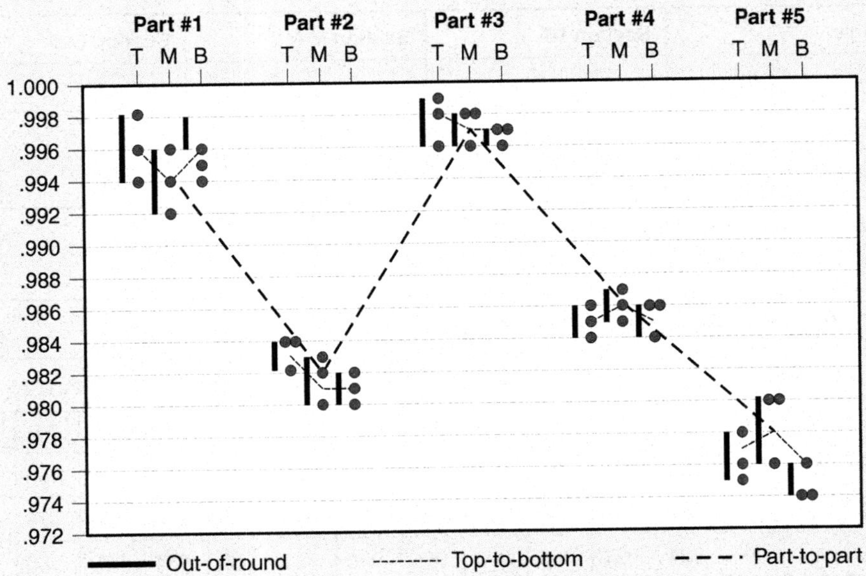

Figure 16.9 Part-to-part variation (graph of data from Table 16.2).

Table 16.3 Data from five parts produced during one shift using precision castings.

Part #1			Part # 2			Part # 3			Part # 4			Part # 5		
T	M	B	T	M	B	T	M	B	T	M	B	T	M	B
.985	.984	.980	.981	.985	.983	.982	.982	.982	.984	.984	.983	.981	.981	.984
.982	.981	.982	.985	.982	.981	.985	.984	.984	.981	.982	.982	.984	.982	.981
.985	.983	.983	.983	.981	.984	.982	.983	.982	.982	.981	.981	.983	.983	.982

members with ideas on how to reduce other variation, such as out-of-round, were asked to hold those ideas until later because they wouldn't reduce the large part-to-part variation.

Upon investigation it was concluded that the part-to-part variation was caused by the casting process. A new metal foundry that could do precision casting was sourced to reduce part-to-part variation.

The new batch of precision-cast parts arrived, and again five parts were selected from a shift. The machined parts were measured, and the results are shown in Table 16.3. Note that a more discriminating measurement system (see Chapter 14 discussion on number of distinct categories) was used for these castings, and the scale for the graph was also changed.

Table 16.3 data were graphed as shown in Figure 16.10. Does it appear that the part-to-part variation has been reduced? Assuming that the part-to-part variation remains

Continued

Continued

relatively small, which type of variation is now the most dominant? Figure 16.10 suggests that the out-of-round variation (within part) is now dominant.

The team now began to discuss possible causes for the out-of-round variation. At this point some team members returned to the theory that chuck clamping pressure was causing the part to be squeezed into an out-of-round contour, and a round hole was then machined, so that when the chuck pressure was released, the part snapped back to a round contour, which left an out-of-round hole. They suggested a better pressure regulator control for the air line feeding the pneumatic chuck (clamp). But one observer, focusing on the top of the hole where the out-of-round should be most prominent, noticed that sometimes the 12 o'clock dimension is longest as in part #4, and sometimes another dimension is longer (see the numbers in Table 16.3). Since the parts are always clamped with the same orientation, the chuck pressure could not be the cause. The team wondered if there might be a pattern relating the various orientations and the diameters, so they placed orientation numbers by each dot as shown in Figure 16.11 and tried to find a pattern. Unfortunately, no pattern was discovered, and the relationship seemed to be random.

So, the issue now is, why is there a random difference in diameters and what is causing it? At this point a second-shift operator mentioned the fact that after the ID is cut, a burnishing (smoothing the internal surface) tool is passed across the surface. He suggested that the machining chips generated from the previous lathe machining operation could cause this diameter difference as these chips may be burnished into the surface. He also suggested that the parts be removed from the chuck and pressure-washed before the burnish operation. The initial manufacturing sequence did not consider this, as it would result in additional steps and increase the part manufacturing cost. Some team members objected that this had been tried a year ago with no noticeable improvement. Our Green Belt pointed out, however, that at that time the large

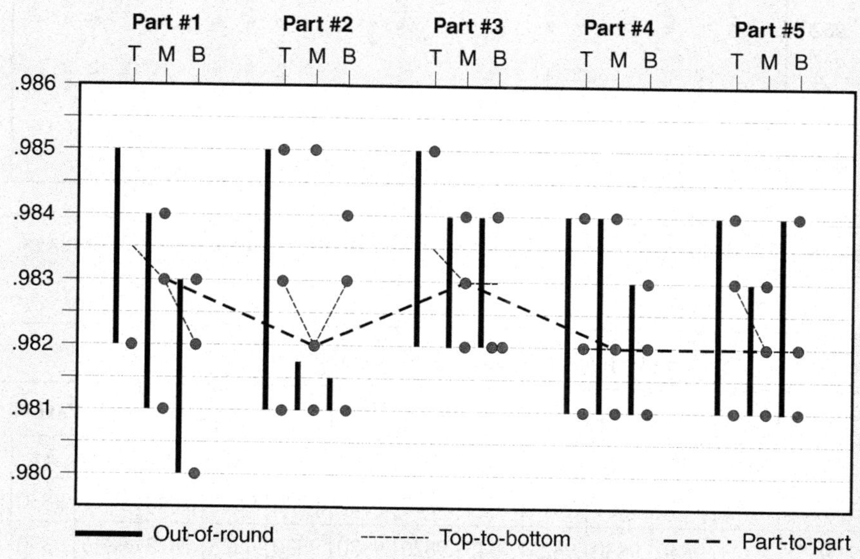

Figure 16.10 Graph of data from Table 16.3.

Continued

Continued

part-to-part variation would have masked any reduction in the minor variation now being seen. The team agreed that the part should be removed from the lathe and pressure-washed before performing the burnishing step. The measurements that resulted are shown in Table 16.4.

Plotting these data shows that the out-of-round variation has been noticeably reduced. One of the team members noticed a remaining pattern of shift-to-shift variation present and suggested that this might be due to tool wear. Upon confirmation, a tooling wear verification and change was initiated. This resulted in reducing variation even further.

A similar study can be conducted for a transaction process like bank teller transaction time per client or call center average handle time (AHT), or in healthcare, treatment time per patient, and so on. However, in a transactional process, the transaction differs by customer need. In order to understand the variation in a meaningful way, the data collection should first stratify the data by transaction type and complexity before applying a multi-vari study.

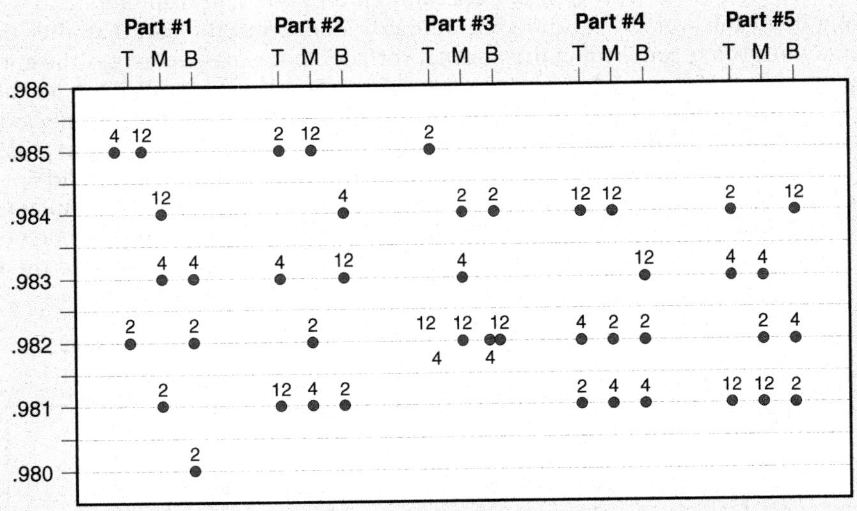

Figure 16.11 Graph of data from Table 16.3 with orientation measurement numbers.

Table 16.4 Data from five parts after pressure-washing.

Part #1			Part #2			Part #3			Part #4			Part #5		
T	M	B	T	M	B	T	M	B	T	M	B	T	M	B
.9825	.9825	.9815	.9830	.9830	.9815	.9830	.9820	.9820	.9820	.9825	.9820	.9835	.9830	.9820
.9835	.9851	.9825	.9840	.9830	.9825	.9845	.9820	.9830	.9830	.9825	.9825	.9845	.9830	.9825
.9820	.9820	.9820	.9832	.9820	.9820	.9825	.9812	.9820	.9820	.9820	.9820	.9835	.9830	.9820

CORRELATION AND LINEAR REGRESSION

> Describe the difference between correlation and causation. Calculate the correlation coefficient and linear regression and interpret the results in terms of statistical significance (p-value). Use regression models for estimation and prediction. (Evaluate)
>
> **Body of Knowledge IV.A.2**

Correlation

Correlation is finding a relationship between two or more sets of data. It measures the strength (strong, moderate, weak) and direction (positive, negative) of the relationship between variables. In order to find a correlation, one needs an independent variable (x) that causes an observed variation, which is considered the dependent variable (y). Table 16.5 lists a few examples of independent and dependent variable pairs.

Independent variables are not impacted by changes to other variables in a process. Often, an independent variable may be an effect of the dependent variable, which may be a cause.

Correlation versus Causation

A cause that produces an effect, or that which gives rise to any action or condition, is termed *causation*. For example, "if a change in X produces a change in Y, the X is said to be a cause of Y." One may also observe, however, that there is a W that caused X, a V that caused W, a U that caused V, and so on. Every cause is itself the result of some prior cause or causes. There is no such thing as an absolute cause for an event, the identification of which satisfies and completes all inquiry. The alphabetic example just given implies a "causal chain."

Two variables may be found to be causally associated depending on how the study was conducted. If two variables are found to be either associated or correlated, that doesn't mean that a cause-and-effect relationship exists between the two variables. This has to be proved by a well-designed experiment or several

Table 16.5 Examples of dependent and independent variables.

Independent variable (x)	Dependent variable (y)
Hours studied	Exam grade
Hours of exercise	Weight loss
Level of advertising	Volume of sales

different observational studies to show that an association or correlation crosses over into a cause-and-effect relationship.

A scatter plot provides a complete picture of the relationship between two variables. Figure 16.12 illustrates the four different types of correlation that exist in scatter plots. The convention is to place the independent x variable on the horizontal axis and the dependent y variable on the vertical axis.

Caution: Be careful when deciding which variable is independent and which is dependent. Examine the relationship from both directions to see which one makes the most sense. The use of a wrong choice may lead to misinterpretation by the users.

Correlation Coefficient

The correlation coefficient, r, provides both the strength and direction of the relationship between the independent and dependent variables. Values of r range between –1.0 and +1.0. When r is positive, the relationship between x and y is positive (Figure 16.12a), and when r is negative, the relationship is negative (Figure 16.12b). A correlation coefficient close to zero is evidence that there is no relationship between x and y (Figures 16.12c and 16.12d).

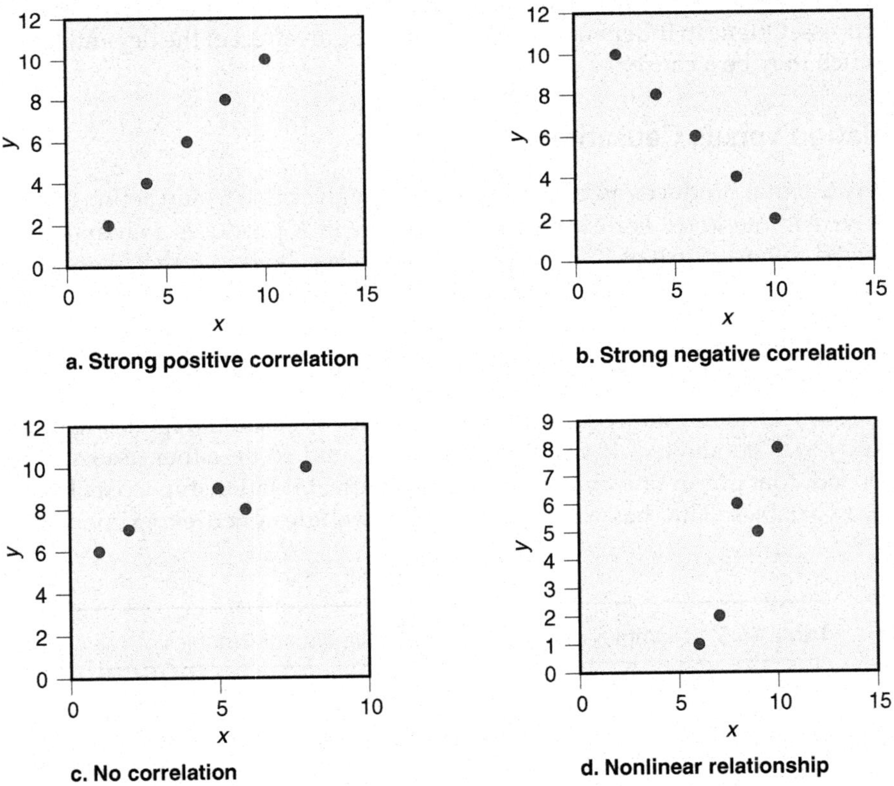

Figure 16.12 The four types of correlation in scatter plots.

The strength of the relationship between x and y is measured by how close the correlation coefficient is to +1.0 or –1.0 (Figures 16.12a and 16.12b).

We can calculate the correlation coefficient using the following formula:

$$r = \frac{1}{n-1}\sum\frac{(x-\bar{x})(y-\bar{y})}{S_x S_y}$$

where n = number of paired samples, $\bar{X}$ and $\bar{Y}$ are mean values of x and y, and S_x, S_y are standard deviations of samples x, y.

Procedure for Calculating the Correlation Coefficient

1. Calculate the mean for all x values ($\bar{x}$) and the mean for all y values ($\bar{y}$).
2. Calculate the standard deviation of all x values (S_x) and the standard deviation for y values (S_y).
3. Calculate $(x - \bar{x})$ and $(y - \bar{y})$ for each pair (x, y) and then multiply these differences together.
4. Get the sum by adding all these products of differences together.
5. Divide the sum by $S_x \times S_y$.
6. Divide the results of step 5 by $n - 1$, where n is the number of (x, y) pairs.

EXAMPLE

Let us say that the x values are 3, 3, and 6, and the y values are 2, 3, and 4, and the data sets are (3,2), (3,3), and (6,4). The correlation coefficient is given by

x values	y values	$(x-\bar{x})(y-\bar{y})$
3	2	1
3	3	0
6	4	2
$\bar{x} = 4$	$\bar{y} = 3$	$\Sigma(x-\bar{x})(y-\bar{y}) = 3$

$$S_x = 1.73,\ S_y = 1.00$$

$$r = \frac{1}{n-1}\Sigma\frac{(x-\bar{x})(y-\bar{y})}{S_x S_y} = \frac{1}{3-1}\times\frac{3}{(1.73)(1)} = 0.867$$

Statistical analysis for the above data:

Pearson correlation of x values and y values = 0.866

p-value = 0.333

What Is a p-*Value?*

A *p*-value is used in hypothesis tests to decide whether to reject a null hypothesis or fail to reject a null hypothesis.

Statistical software explains that "The *p*-value is the probability of obtaining a test statistic that is at least as extreme as the actual calculated value, if the null hypothesis is true."

A commonly used cut-off value for the *p*-value is 0.05. For example, if the calculated *p*-value of a test statistic is less than 0.05, you reject the null hypothesis.

In the above example, since the *p*-value is > 0.05, the positive correlation (0.866) calculated is not statistically significant. In fact, there is a 0.33 probability of mistakenly rejecting the hypothesis.

Statistical software applications provide "*p*-value" for most analyses that involve hypothesis testing. This helps users to make decisions on statistical strength for decision making. The above example is one such interesting scenario. Without a *p*-value, one would infer a high correlation. However, reviewing the analysis alongside the *p*-value provides better insight.

The Strength of the Relationship

- The graph in Figure 16.12a shows an example of positive linear correlation; as x increases, y also tends to increase in a linear (straight line) fashion.
- The graph in Figure 16.12b shows an example of negative linear correlation; as x increases, y tends to decrease linearly.
- The graph in Figure 16.12c shows an example of no correlation between x and y. This set of variables appears to have no impact on each other.
- The graph in Figure 16.12d shows an example of a nonlinear relationship between variables.

Figure 16.12a illustrates a perfect positive correlation (positive slope) between x and y with $r = +1.0$. Figure 16.12b shows a perfect negative correlation (negative slope) between x and y with $r = -1.0$. Figures 16.12c and 16.12d are examples of weaker relationships between the independent and dependent variables.

$$r = \frac{\sum_{i=1}^{n}(x_i - \bar{x})(y_i - \bar{y})}{(n-1)s_x s_y}$$

where:

$\bar{x}$ = Sample mean for the first variable

s_x = Standard deviation for the first variable

$\bar{y}$ = Sample mean for the second variable

s_y = Standard deviation for the second variable

n = Number of samples

EXAMPLE

Let's investigate the relationship between exercise time and weight loss for an individual. Table 16.6 shows sample data from six candidates who were randomly chosen.

Using these values, with $n = 6$, the number of ordered pairs, we have:

$$r = \frac{6(217) - (38)(24)}{\sqrt{\left[6(372) - (38)^2\right]\left[6(130) - (24)^2\right]}}$$

$$r = \frac{390}{\sqrt{(788)(204)}} = 0.972$$

Thus, one can see that there is a strong positive correlation between hours of exercise and weight lost.

Caution: Be careful when distinguishing between Σx^2 and $(\Sigma x)^2$. With Σx^2, we first square each value of x and then add each squared term. With $(\Sigma x)^2$, we first add each value of x, and then square this total. The two results are very different.

Table 16.6 Data table for correlation calculation.

Hours of exercise in a week (x)	Weight reduction in lbs (y)	xy	x^2	y^2
3	2	6	9	4
5	4	20	25	16
10	6	60	100	36
15	8	120	225	64
2	1	2	4	1
3	3	9	9	9
$\Sigma x = 38$	$\Sigma y = 24$	$\Sigma xy = 217$	$\Sigma x^2 = 372$	$\Sigma y^2 = 130$

Inferences in Correlation/Testing the Significance of the Correlation Coefficient

The letter r denotes the sample correlation coefficient. It is conventional to use the Greek letter ρ (small case rho) to denote the population correlation coefficient. A confidence interval for ρ can be obtained from the sample r statistic using

$$r \pm t_{\alpha/2}\sqrt{\frac{1-r^2}{n-2}}$$

where $1 - \alpha$ = confidence level and df = $n - 1$.

Procedure for Testing the Significance of the Correlation Coefficient

1. Conditions for using this test are: population regression equation $y = \beta_0 + \beta_1 x$; for a given value of x, the mean of the response variable y is $\beta_0 + \beta_1 x$; for a given value of x, the distribution of y-values is normal and independent; distributions of y-values have equal standard deviations. β_0 is the intercept and β_1 is the slope.
2. Decide the significance level α.
3. $H_0 : \rho = 0$; H_1 could be any of these: $\rho \neq 0$, $\rho < 0$, or $\rho > 0$ for a two-tail, left-tailed, or right-tailed test, respectively.
4. Critical values are obtained from the t-table (Appendix L) using $n - 1$ degrees of freedom: $\pm t_{\alpha/2}$ for the two-tail test, $-t_\alpha$ for the left-tail test, and t_α for the right-tail test.
5. The test statistic is given by the formula

$$t = \frac{r}{\sqrt{\dfrac{1-r^2}{n-2}}}$$

6. Now compare the test statistic with the critical value obtained in step 4. Reject the null hypothesis if the test statistics $-t_{\alpha/2}$ < critical value in a left-tailed test or $+ t_{\alpha/2}$ is > critical value in a right-tailed test. If not, do not reject the null hypothesis.
7. State the conclusion in terms of the problem context.

EXAMPLE

Using the above weight loss example, the calculated t-statistic becomes:

$$t = \frac{r}{\sqrt{\dfrac{1-r^2}{n-2}}} = \frac{0.972}{\sqrt{\dfrac{1-(0.972)^2}{6-2}}}$$

$$t = \frac{0.972}{\sqrt{\dfrac{.0552}{4}}} = 8.273$$

The null hypothesis is $t \leq t_c$; hours of exercise do not help weight loss or even have a negative effect. The alternate hypothesis is $t > t_c$.

The critical t-statistic is based on df = $n - 2$. We chose $\alpha = 0.05$, $t_c = 2.132$ from t-table (Appendix L) for a one-tail test. Because $t > t_c$, we reject H_0 and conclude that there is indeed a positive correlation between hours of exercise and weight loss.

Using Spreadsheets to Calculate Correlation Coefficients

We don't need sophisticated statistical software to calculate basic statistics like correlation. Spreadsheets can be used to calculate correlation coefficients. Use the CORREL function under formula, which has the following characteristics:

CORREL(array1,array2)

where

array1 = The range of data for the first variable

array2 = The range of data for the second variable

Figure 16.13 shows the CORREL function being used to calculate the correlation coefficient for the weight loss example. Cell C8 contains the spreadsheet formula =CORREL(A2:A7,B2:B7) with the result being 0.972717.

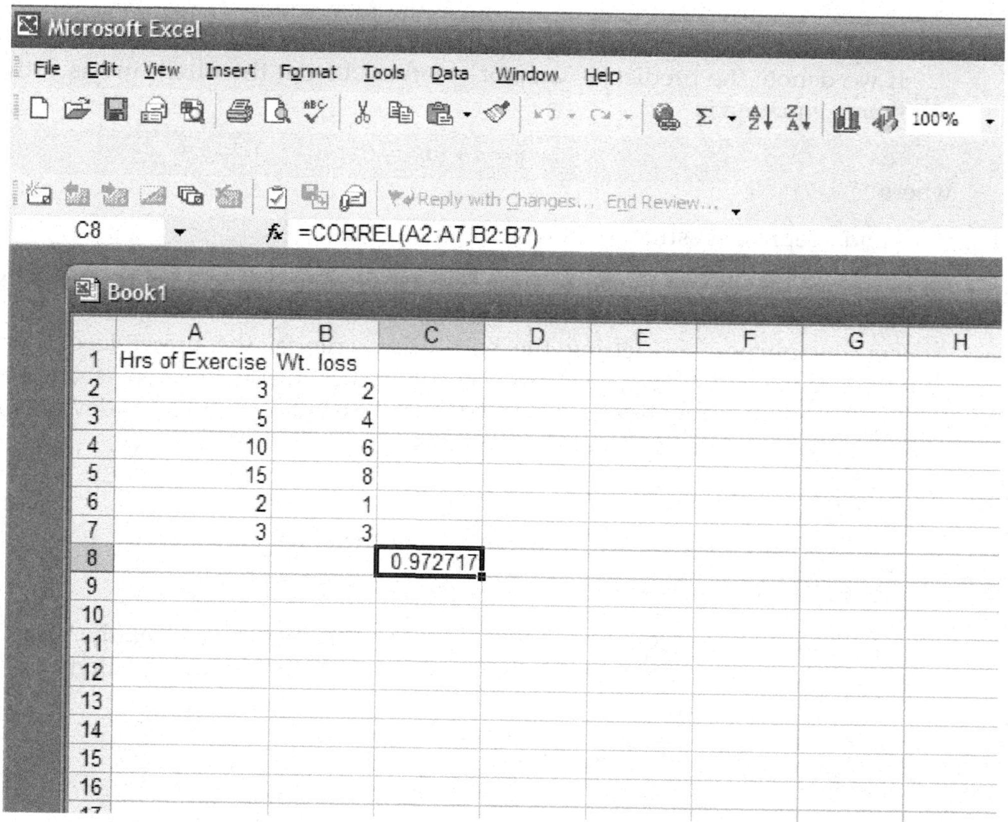

Figure 16.13 CORREL function in spreadsheet.

Simple Regression

Simple regression is used to describe a straight line that best fits a series of ordered pairs (x,y). An equation for a straight line, known as a linear equation, takes the form:

$$\hat{y} = a + bx$$

where

$\hat{y}$ = The predicted value of y, given a value of x

x = The independent variable

a = The y-intercept for the straight line

b = The slope of the straight line

The Least Squares Method

This is a mathematical procedure to identify the linear equation that best fits a set of ordered pairs by finding values for a, the y-intercept, and b, the slope. The goal of the least squares method is to minimize the total squared error between the values of y and $\hat{y}$.

If we denote the predicted value of y obtained from the fitted line as $\hat{y}$, the prediction equation is

$$\hat{y} = \hat{a} + \hat{b}x$$

where

$\hat{a}$ and $\hat{b}$ represent estimates of true a and b.

Since we need to choose the best-fitting line, we need to define what we mean by "best."

For the purpose of getting the best-fitting criteria, the principle of least squares is employed, that is, one has to choose the best-fitting line, that is, one has to choose the best-fitting line, that minimizes the sum of squares of the deviations of the observed values of y from those predicted.

Procedure for Least Squares Method

1. Calculate xy, x^2, and y^2 values and enter them in a table
2. Calculate the sums for x, y, xy, x^2, y^2, and $\bar{x}$ and $\bar{y}$
3. Find the linear equation that best fits the data by determining the value for a, the y-intercept, and b, the slope, using the following equations:

$$b = \frac{n\sum xy - \left(\sum x\right)\left(\sum y\right)}{n\sum x^2 - \left(\sum x\right)^2}$$

$$a = \bar{y} - b\bar{x}$$

where

$\bar{x}$ = The average value of x, the dependent variable

$\bar{y}$ = The average value of y, the independent variable

EXAMPLE

Obtain the least squares prediction for the table containing the month and number of complaints received in a manufacturing facility. The data are shown in Table 16.7.

$$b = \frac{n\Sigma x_i y_i - (\Sigma x_i)(\Sigma y_i)}{n\Sigma x_i^2 - (\Sigma x_i)^2} = \frac{8(353) - (36)(73)}{8(204) - (36)^2} = 0.5833$$

$$a = \bar{y} - b\bar{x} = 9.125 - 9.5833(4.5) = 6.50015$$

The regression line would be $\hat{y} = 6.50015 + 0.5833x$.

Because the slope of this equation is positive 0.5833, there is evidence that the number of complaints increases over time at an average rate of one per month. If someone wants to predict how many complaints there will be in another six months at this rate, the equation would be

$$\hat{y} = 6.50015 + 0.5833(14) = 14.666 \approx 14 \text{ complaints}$$

Table 16.7 Least squares example.

Month, x_i	Complaints, y_i	x_i^2	$x_i y_i$	y_i^2
1	8	1	8	64
2	6	4	12	36
3	10	9	30	100
4	6	16	24	36
5	10	25	50	100
6	13	36	78	169
7	9	49	63	81
8	11	64	88	121
$\Sigma x_i = 36$	$\Sigma y_i = 73$	$\Sigma x_i^2 = 204$	$\Sigma x_i y_i = 353$	$\Sigma y_i^2 = 707$
$\bar{x} = 4.5$	$\bar{y} = 9.125$			

Procedure for Testing Simple Regression

1. Set conditions for this test:
   - Population regression equation $y = mx + b$.
   - For a given specific value of x, the distribution of y-values is normal and independent and has equal standard deviations.
2. $H_0 : \beta_l = 0$ (that is, the equation is not useful as a predictor of values of y).

   $H_1 : \beta_l \neq 0$ (that is, the equation is useful as a predictor of values of y).
3. Decide on a value of α.

4. Find the critical values $\pm t_{\alpha/2}$ from a *t*-table using $n - 2$ degrees of freedom.
5. Calculate the value of the test statistic *t*:

$$t = \frac{b_i}{s / \sqrt{S_{xx}}}$$

6. If the test statistic is beyond one of the critical values (that is, greater than $t_{\alpha/2}$ or less than $-t_{\alpha/2}$), reject the null hypothesis; otherwise, do not reject.
7. State the result in terms of the problem.

EXAMPLE

Test the hypothesis for the temperature–viscosity variation as shown below.

Temperature °C	10	15	20	15
Viscosity, C_p	2	3	5	4

1. Assume that the conditions are met.
2. $H_0: \beta_l = 0$ (that is, the equation is not useful as a predictor of values of viscosity).

   $H_1: \beta_l \neq 0$ (that is, the equation is useful as a predictor of values of viscosity).
3. Let $\alpha = .05$.
4. The critical values from the *t*-table are 4.303 and –4.303.
5. $t = 0.3/\left(0.5/\sqrt{50}\right) = 0.3/0.071 = 4.24$
6. Compare test statistic (*t*) with critical values (t_c). Test statistic is not greater than critical value.
7. Conclusion—fail to reject null hypothesis.

Using a Spreadsheet for Simple Regression

1. Let us use the above example and sort the data into columns A and B in a blank spreadsheet.
2. Go to the Tools menu and select Data Analysis.
3. From the Data Analysis dialog box, select Regression as shown in Figure 16.14 and click OK.
4. Set up the regression dialog box as shown in Figure 16.15; enter input *x* and *y* range.
5. Click OK, which brings up the results shown in Figure 16.16.

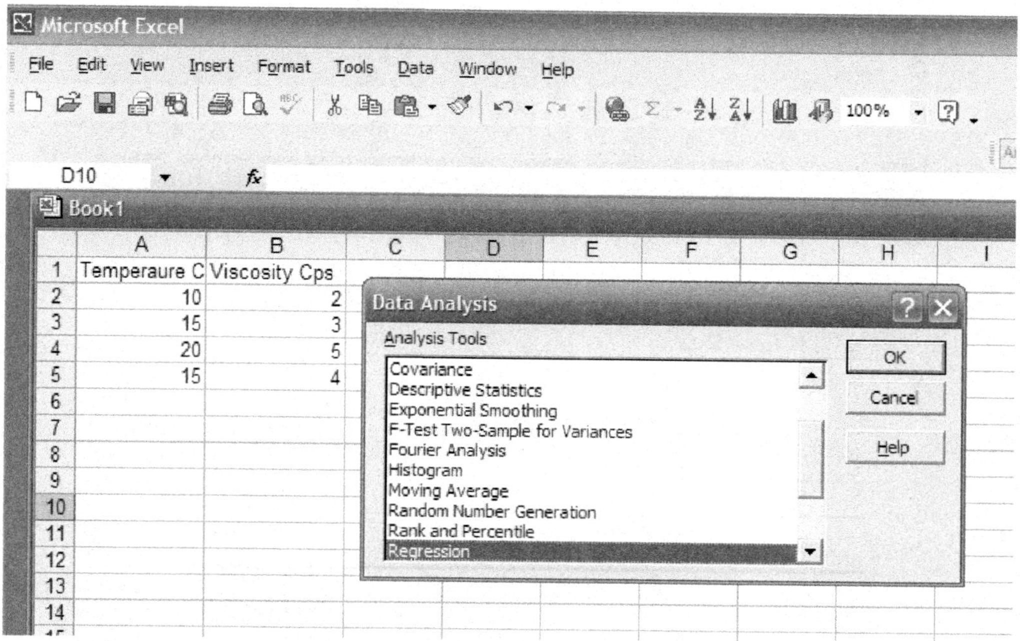

Figure 16.14 Regression dialog box.

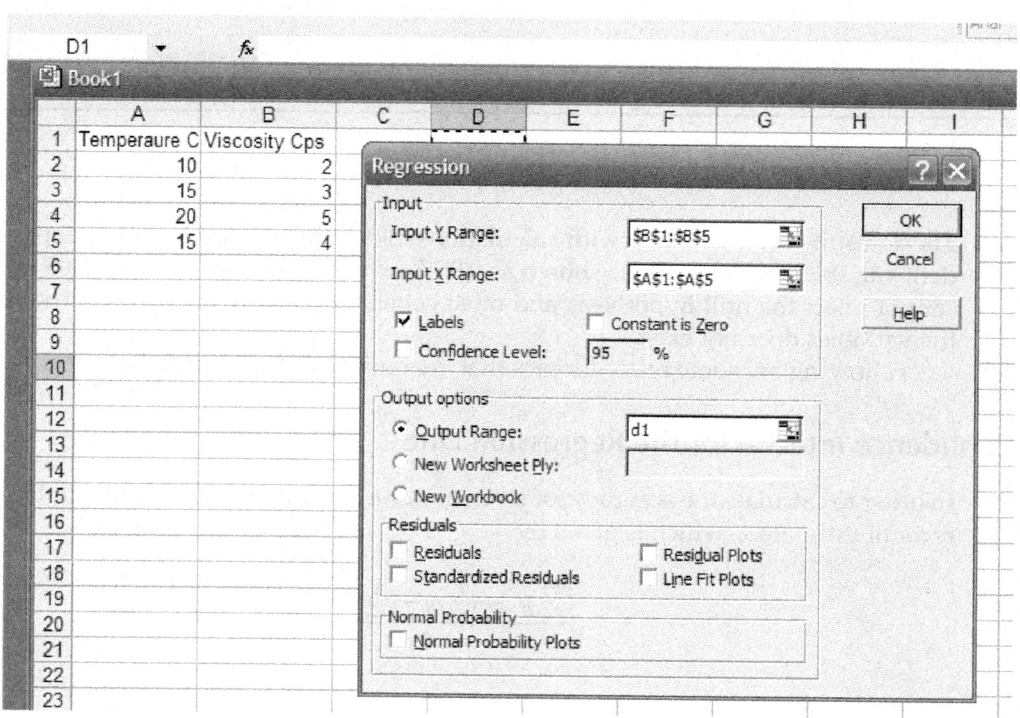

Figure 16.15 Regression data analysis.

SUMMARY OUTPUT

Regression Statistics	
Multiple R	0.9486833
R Square	0.9
Adjusted R Square	0.85
Standard Error	0.5
Observations	4

ANOVA

	df	SS	MS	F	Significance F
Regression	1	4.5	4.5	18	0.051316702
Residual	2	0.5	0.25		
Total	3	5			

	Coefficients	Standard Error	t Stat	P-value	Lower 95%	Upper 95%
Intercept	-1	1.089724736	-0.91766	0.455669	-5.688710374	3.68871
TempretaureC	0.3	0.070710678	4.242641	0.051317	-0.004243704	0.604244

Figure 16.16 Simple regression results.

These results are consistent with calculations. Since the p-value for the independent variable (temperature) is shown as 0.0513, which is greater than $\alpha = 0.05$, we cannot reject the null hypothesis and must conclude that a relationship between the variables does not exist.

Following are some related topics that are outside the current BoK.

Confidence Interval for the Regression Line

In order to calculate the accuracy for y versus x, we need to determine the standard error of estimate s_e, which is given by

$$s_e = \sqrt{\frac{\sum y^2 - a\sum y - b\sum xy}{n-2}}$$

The standard error of the estimate measures the amount of dispersion of the observed data around the regression line. The standard error of the estimate is relatively low if the data points are very close to the line, and vice versa.

The confidence interval around the mean of y given a specific value of x is given by

$$\mathrm{CI} = \hat{y} \pm t_c s_e \sqrt{\frac{1}{n} + \frac{(x-\bar{x})^2}{\left(\sum x^2\right) - \frac{\left(\sum x\right)^2}{n}}}$$

where

t_c = Critical t-statistic from Student's t-distribution

s_e = Standard error of the mean

n = Number of ordered pairs

Procedure for Confidence Interval

1. Test for the slope of the regression line.
2. Set conditions: if β is the slope of the true population, then the hypothesis would be

$$H_0 : \beta = 0, H_1 : \beta \neq 0$$

3. Decide on α significance level.
4. Calculate the standard error of slope, s_b, using

$$s_b = \frac{s_e}{\sqrt{\sum x^2 - n\bar{x}^2}}$$

where s_e is the standard error of the mean.

5. Test for the hypothesis using

$$t = \frac{b - \beta_{H_0}}{s_b}$$

where β_{H_0} is the value of the population slope according to the null hypothesis and b = slope of sample pairs.

6. Find the critical t-statistic value from Student's t-distribution with $n - 2$ degrees of freedom.
7. If $t > t_c$, reject the null hypothesis.

EXAMPLE

Use the data from Table 16.7 to calculate confidence interval:

$$s_e = \sqrt{\frac{\Sigma y^2 - a\Sigma y - b\Sigma xy}{n-2}}$$

$$s_e = \sqrt{\frac{(707) - 6.50015(73) - 0.5833(353)}{(6)}} = 2.105$$

For month 8, there are 11 complaints; the regression line would have been

$$\hat{y} = 6.50015 + 0.5833(8) = 11.1666$$

For a 95 percent confidence level and using the critical t-statistic from the table:

$$CI = \hat{y} \pm t_c s_e = \sqrt{\frac{1}{n} + \frac{(x-\bar{x})^2}{(\Sigma x^2) - \frac{(\Sigma x)^2}{n}}}$$

$$CI = 11.1666 \pm (2.447)(2.105)\sqrt{\frac{1}{8} + \frac{(8-4.5)^2}{(204) - \frac{(36)^2}{8}}}$$

$$CI = 14.49 \text{ and } 7.84$$

Our 95 percent confidence interval for the number of items in the table in month 8 is between 7.84 and 14.49 complaints.

Now let us calculate the slope of the regression line:

$$s_b = \frac{s_e}{\sqrt{\Sigma x^2 - n\bar{x}^2}}$$

$$s_b = \frac{2.105}{\sqrt{204 - 8(4.5)^2}} = 0.325$$

$$t = \frac{b - \beta_{H_0}}{s_b} = \frac{0.5833 - 0}{0.325} = 1.7948$$

From the equation, $\hat{y} = 6.50015 + 0.5833(8) = 11.1666$, $b = 0.5833$. The critical t-statistic is taken from Student's t-distribution with $n - 2 = 6$ degrees of freedom. With a two-tail test and $\alpha = 0.05$, $t_c = 2.447$ as per the t-table. Because $t < t_c$, we cannot reject the null hypothesis and must conclude that there is no relationship between the month and the number of complaints.

Multiple Linear Regression

Multiple linear regression is an extension of the methodology for linear regression to more than one independent variable. By including more than one independent variable, a higher proportion of the variation in y may be explained.

The general form for the equation is

$$y = b_0 + b_1x_1 + b_2x_2 + b_3x_3 + \ldots + b_kx_k$$

where

the b_i's are the coefficients and the x_i's are the variables. Statistical software is usually employed to find the values of the b_i's.

Inferences in Regression

Usually, the sample data are used to calculate the coefficients b_i's. One has to find out the closeness of the calculated *b*-values to the actual coefficient values for the population. For this purpose, we will refer to the values obtained from the sample data as b_i's and the coefficients for the population as β_i's. The b_i's are approximations for the β_i's. The accuracy of the approximation depends on sampling error. This discussion is restricted to simple linear regression that involves just one independent variable. It is assumed that the means of these distributions lie in a straight line whose equation is $y = \beta_0 + \beta_1 x$ and that the distributions are normal with equal standard deviations σ as indicated in Figure 16.17.

Under these assumptions, a confidence interval for β_1 can be calculated using the following formula (confidence intervals are discussed in detail in Chapter 17):

$$b_i \pm \frac{t_{\alpha/2} s_e}{\sqrt{s_{xx}}}$$

where

$$s_e = \sqrt{\frac{\sum y^2 - a\sum y - b\sum xy}{n-2}}$$

$s_{xx} = \Sigma x^2 - (\Sigma x)^2/n$

$1 - \alpha =$ Confidence level

$\text{df} = n - 2$

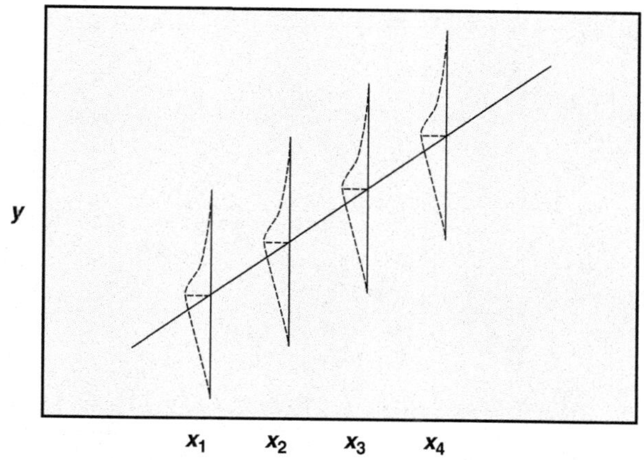

Figure 16.17 A schematic showing variation in *y* as a function of *x*.

To review, then, just like a sample mean is used to estimate a population mean, the values of b_0 and b_i are used to estimate the population values β_0 and β_1.

The formula for the best-fitting line (or regression line) is $y = mx + b$, where m is the slope of the line and b is the y-intercept. In other words, the best-fitting line does find the line that best fits the sample data, and if the sample is randomly chosen, that line should be close to the line that best fits the population data. A hypothesis test can be applied to determine whether the independent variable x is useful as a predictor for the dependent variable y.

Chapter 17

Hypothesis Testing

BASICS

> Distinguish between statistical and practical significance. Determine appropriate sample sizes and develop tests for significance level, power, and type I and type II errors. (Apply)
>
> Body of Knowledge IV.B.1

The Null and Alternative Hypotheses

A hypothesis is an assumption about a population parameter, for example:

The average adult drinks 1.7 cups of coffee per day.

No more than two percent of our products that we sell to customers are defective.

The above statements about a population may or may not be true. The purpose of hypothesis testing is to make a statistical conclusion about accepting or not accepting such statements.

All hypothesis tests have both a null hypothesis and an alternative hypothesis. A null hypothesis, denoted by H_0, represents the status quo and involves stating the belief that the mean of the population is $\geq$, $=$, or $\leq$ a specific value. The null hypothesis is believed to be true unless there is overwhelming evidence to the contrary. It is similar to a court trial. The hypothesis is that the defendant is not guilty until proven guilty. However, the term "innocent" does not apply to a null hypothesis. A null hypothesis can only be rejected or fail to be rejected; it cannot be accepted because of a lack of evidence to reject it. If the means of two populations are different, the null hypothesis of equality can be rejected if enough data are collected. When rejecting the null hypothesis, the alternative hypothesis must be considered. For example, the average weight of a component length is six grams. μ is the population mean. The null hypothesis would be stated as:

$$H_0 : \mu = 6.0,\ H_0 : \mu \leq 6.0,\ H_0 : \mu \geq 6.0$$

The alternative hypothesis, denoted by H_1, represents the opposite of the null hypothesis and holds true if the null hypothesis is found to be false. The alternative hypothesis always states that the mean of the population is <, ≠, or > a specific value. The alternative hypothesis would be stated as

$$H_1 : \mu \neq 6.0, H_1 : \mu < 6.0, H_1 : \mu > 6.0$$

In order to test a null hypothesis, a calculation must be made from sample information. This calculated value is called a *test statistic* and is compared to an appropriate critical value. A decision can then be made to reject or not reject the null hypothesis. The critical value is obtained from the *t*-distribution table in Appendix L against a chosen level of significance. The typical levels of significance are 1 percent, 5 percent, and 10 percent (both tails).

Types of Errors

There are two types of errors possible when formulating a conclusion regarding a population based on observations from a small sample.

Type I error. This type of error results when the null hypothesis is rejected when it is actually true. For example, incoming products are good but were labeled defective. This type of error is also called α (alpha) error and referred to as the *producer's risk* (for sampling).

Type II error. This type of error results when the null hypothesis is not rejected when it actually should have been rejected. For example, incoming products are defective, but labeled good. This type of error is also called β (beta) error and referred to as the *consumer's risk* (for sampling).

The types of errors are shown in Table 17.1.

One-Tail Test

Any type of test on a hypothesis comes with a risk associated with it, and it is generally associated with the α risk (type I error, which rejects the null hypothesis when it is true). The level of this α risk determines the level of confidence $(1 - \alpha)$ that we have in the conclusion. This risk factor is used to determine the critical value of the test statistic, which is compared to a calculated value.

If a null hypothesis is established to test whether a sample value is smaller or larger than a population value, then the entire α risk is placed on one end of a distribution curve. This constitutes a one-tail test (Figure 17.1).

$$H_0 : \text{Level} \geq 20\%, H_1 : \text{Level} < 20\%$$

Table 17.1 Error matrix.

	Null hypothesis is true	**Null hypothesis is false**
Fail to reject H_0	$p = 1 - \alpha$, correct outcome	$p = \beta$, type II error
Reject H_0	$p = \alpha$, type I error	$p = 1 - \beta$, correct outcome

Note: $p = 1 - \beta$ is also called *power*. Higher power is better in a hypothesis test.

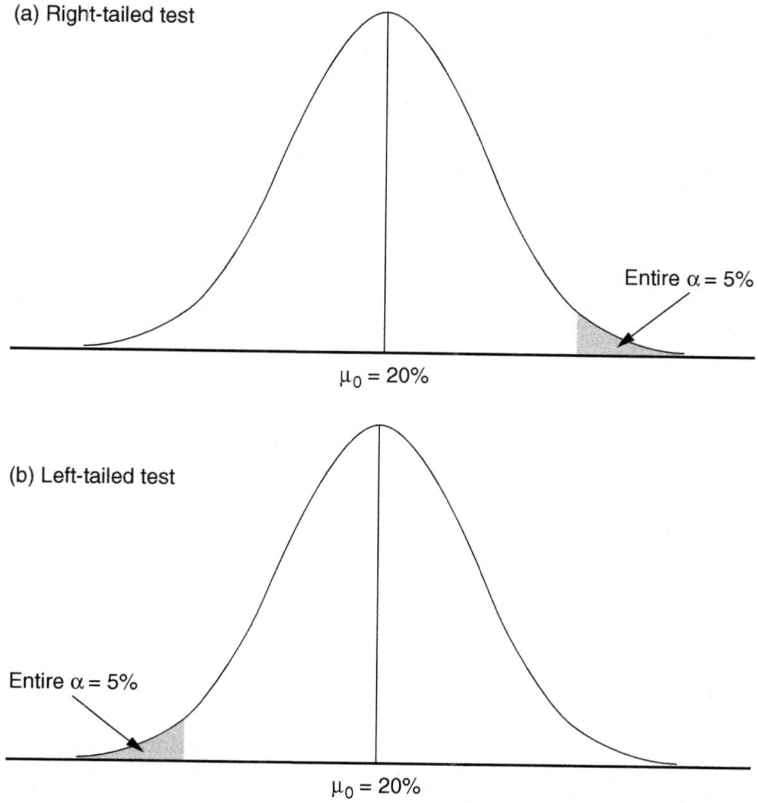

Figure 17.1 One-tail test: (a) right-tailed test and (b) left-tailed test.

EXAMPLE

If a company invents a golf ball that it claims will increase the driving distance off the tee by more than 20 yards, the hypothesis would be set up as follows:

$$H_0 : \mu \leq 20,\ H_1 : \mu > 20$$

In Figure 17.1, there is only one rejection region, which is the shaded region on the distribution. We follow the same procedure outlined below for the two-tail test and plot the sample mean, which represents the average increase in distance from the tee with the new golf ball. Two possible scenarios exist:

- If the mean sample falls within the white region, we do not reject H_0. That is, we do not have enough evidence to support H_1, the alternative hypothesis, which states that the new golf ball will increase distance off the tee by more than 20 yards.
- If the sample mean falls in the rejection region, we reject H_0. That is, we have enough evidence to support H_1, which confirms the claim that the new golf ball will increase distance off the tee by more than 20 yards.

Continued

Continued

Note: For a one-tail hypothesis test, the rejection region will always be consistent with the direction of the inequality for H_1. For $H_1 : \mu > 20$, the rejection region will be in the right tail of the sampling distribution. For $H_1 : \mu < 20$, the rejection region will be in the left tail.

Additional yards:

18, 20, 19, 18, 21, 19, 23, 18, 19, 22

```
One-sample t: C1

Test of μ = 20 vs. > 20

Variable  N    Mean StDev SE Mean 95% Lower Bound      T     P

C1        10 19.673 1.775   0.561          18.644 -0.58 0.713
```

The *p*-value > 0.05. The mean additional yards claimed by the company is not statistically validated.

Two-Tail Test

If a null hypothesis is established to test whether a population shift has occurred in either direction, then a two-tail test is required. In other words, a two-tail hypothesis test is used whenever the alternative hypothesis is expressed as ≠. The allowable α error is generally divided into two equal parts (see Figure 17.2).

For example:

- An economist must determine if unemployment levels have changed significantly over the past year.

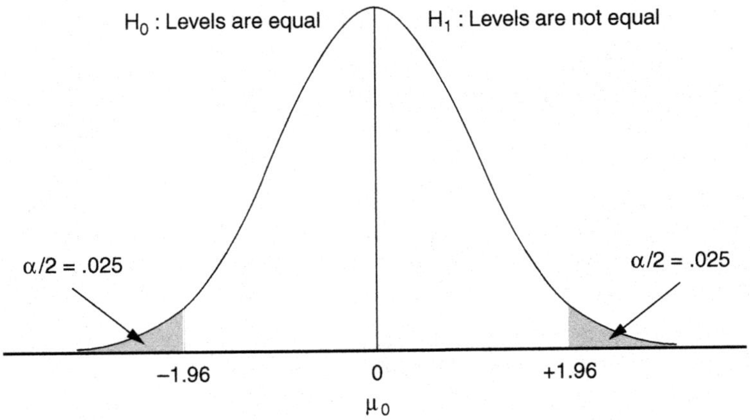

Figure 17.2 Two-tail test.

- A study is made to determine if the salary levels of company A differ significantly from those of company B.

Required Sample Size

So far, it has been assumed that the sample size n for hypothesis testing has been given and that the critical value of the test statistic will be determined based on the α error that can be tolerated. The ideal procedure, however, is to determine the α and β error desired and then to calculate the sample size necessary to obtain the desired decision confidence.

The sample size n needed for hypothesis testing depends on:

- The desired type I (α) and type II (β) risk
- The minimum value to be detected between the population means $(\mu - \mu_0)$
- The variation in the characteristic being measured (s or σ)

Variable data sample size, only using α, is illustrated by the following example.

EXAMPLE

Let us say we want to determine whether an operational adjustment in a shipyard will alter the process hourly mean by as much as 10 metric tons per hour. What is the minimum sample size that at the 95 percent confidence level ($Z = 1.96$) would confirm the significance of a mean shift greater than eight tons per hour? Historically, the standard deviation of the hourly output is 35 tons. The general sample size equation for variable data is

$$n = \frac{Z^2\sigma^2}{E^2} = \frac{(1.96)^2(35)^2}{10^2} = 470.59$$

where E = process mean.

Get 470 hourly yield values and determine the hourly average. If this mean deviates by more than eight tons from the previous hourly average, a significant change at the 95 percent confidence level has occurred. If the sample mean deviates by less than eight tons per hour, the observable mean shift can be explained by chance cause.

Statistical and Practical Significance

Statistical significance is used to evaluate whether the decision made in a hypothesis test is valid. We often encounter situations like comparing before and after improvements, test equipment, service quality level, defects from two processes, and so on. Statistical significance is expressed by a "p-value" (*see also* Chapter 16). Standard statistical references define p-value as "the probability of obtaining a test statistic result at least as extreme or as close to the one that was actually observed, assuming that the null hypothesis is true." If the calculated p-value is greater than the significance level considered (often 1% or 5%), then the p-value is not significant enough to conclude that the null hypothesis is true.

As to practical significance, a difference between two data sets may be statistically significant, but the question is, is it practically significant? As an example, the difference between two process defect levels is statistically significant. The difference is 0.5%. The customer is allowing a defect level of 3% for this process as the step is not critical. Similarly, a difference between two pieces of test equipment of up to 50 microns may be acceptable practically, while statistical significance may require a difference of 20 microns to determine that the null hypothesis is true.

It is important to ensure that the sample size calculated for the hypothesis testing is statistically adequate. Without a statistically significant sample size, the analysis may reveal a difference between data sets, but the probability of correctly detecting such a difference may be low due to low power in the experiment. This may lead to making incorrect decisions, taking risks, and costing the organization. By conducting statistical analysis, even if the sample size is inadequate to conclude the difference, we can estimate power in the decision and inform stakeholders of risks involved in the lower-than-desired power.

What Is a Desired Power?

A power of 0.8 and above is typically required for making a conclusion. A power of 0.9 or more may be required in some situations based on the risk to the organization.

A power of 0.8 means an experiment with the current sample size has an 80% likelihood to identify a significant difference (more than 1% defectives) when one truly exists, and a 20% likelihood it will incorrectly identify a significant difference when the difference does not exist.

When to Calculate Statistically Significant Sample Size

Statistically significant sample sizes using power and sample size are calculated before you design and run an experiment or improvement or after you perform an experiment or improvement.

How to Calculate Statistically Significant Sample Size

(Inputs): Calculating sample size depends on:

> σ—The variability in the population (or experimental variability). As σ decreases, power increases.
>
> δ—The size of the effect (difference to detect). As the size of the effect increases, power increases.
>
> α—The probability of a type I error (also called the *level of significance*).
>
> β—When H_0 is false and you fail to reject it, you make a type II error. The probability (p) of making a type II error is called *beta*.

(Output): Statistically significant sample size. Using the data from an example that is discussed for the *Z* test. Standard deviation of 0.03, difference to detect 0.04 (1.88–1.84), power of 0.9 (required by the experimenter). See Figure 17.3.

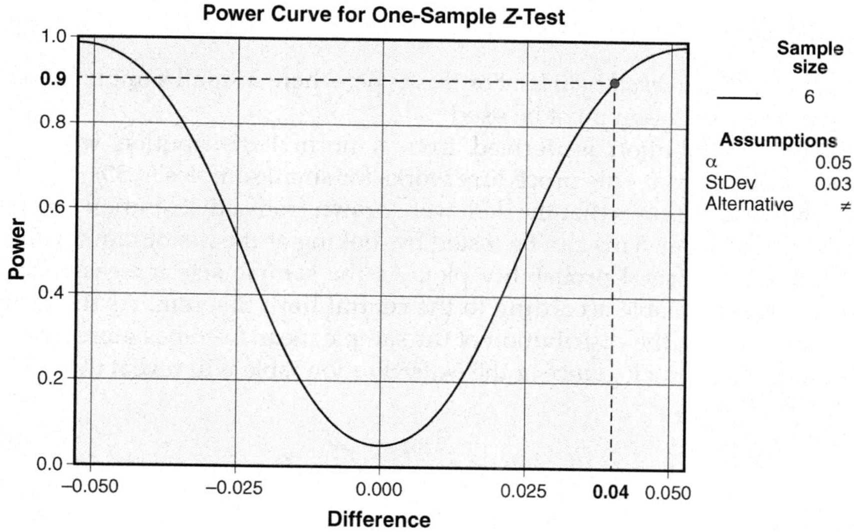

Figure 17.3 Power curve.

Power and Sample Size

```
One-sample Z-test

Testing mean = null (versus ≠ null)
Calculating power for mean = null + difference
α = 0.05 Assumed standard deviation = 0.03

Sample Target
Difference Size Power Actual Power
0.04            6    0.9      0.904228
```

The experimenter only requires 6 samples to detect a difference of 0.04 with a power of 0.9 (10% false acceptance risk).

TESTS FOR MEANS, VARIANCES, AND PROPORTIONS

> Conduct hypothesis tests to compare means, variances, and proportions (paired-comparison t-test, F-test, analysis of variance (ANOVA), chi square) and interpret the results. (Analyze)
>
> **Body of Knowledge IV.B.2**

Test for Means

Continuous data—small samples: For this type, where a small sample is used (< 30), then the *t*-distribution must be used.

The *t*-distribution is derived from a normal distribution with unknown standard deviation. This procedure works for small samples (< 30).

The assumption is that the data were drawn from a distribution that is normal or close to normal. This can be tested by looking at the histogram for the data or by plotting a normal probability plot. As the sample size increases, the results become more reliable according to the central limit theorem. As the sample size increases to ≥ 30, the distribution of the sample mean becomes more like a normal distribution. A quick glance at the *t*-distribution table will reveal this.

$$\bar{X} \pm t_{\alpha/2} \frac{s}{\sqrt{n}}$$

where:

$\bar{X}$ = Sample average

s = Sample standard deviation

n = Sample size

$t_{a/2}$ = *t*-distribution value (Appendix L) for a desired confidence level with $(n - 1)$ degrees of freedom

EXAMPLE

Let us say for the previous example the sample size is 25:

$$\mu = 78.25 \pm 2.064 \frac{(37.50)}{\sqrt{25}} = 78.25 \pm 15.48$$

$$93.73 \geq \mu \geq 62.77$$

Hypothesis Tests for Means

Z-Test. If the null hypothesis is denoted by H_0 and the alternative hypothesis is denoted by H_1, the test statistic is given by

$$Z = \frac{\bar{X} - \mu_0}{\sigma_{\bar{X}}} = \frac{X - \mu_0}{\frac{\sigma_x}{\sqrt{n}}}$$

where

$\bar{X}$ = Sample average

n = Number of samples

σ = Standard deviation of population (assumed as known)

If the population follows a normal distribution, and the population standard deviation $\sigma_{\bar{x}}$ is known, then the hypothesis tests for comparing a population mean μ with a fixed value μ_0 are given by the following:

$$H_0 : \mu = \mu_0,\ H_0 : \mu \leq \mu_0,\ H_0 : \mu \geq \mu_0$$

$$H_1 : \mu \neq \mu_0,\ H_1 : \mu > \mu_0,\ H_1 : \mu < \mu_0$$

If $n > 30$, the standard deviation s is often used as an estimate of the population standard deviation $\sigma_{\bar{x}}$. The test statistic Z is compared with a critical value Z_α or $Z_{\alpha/2}$, which is based on a significance level α for a one-tail test or $\alpha/2$ for a two-tail test. If the H_1 sign is $\neq$, it is a two-tail test. If the H_1 sign is <, it is a left one-tail test, and if the H_1 sign is >, it is a right one-tail test.

Procedure for Testing the Mean

1. Set conditions:

   a. Normal population or large sample ($n \geq 30$)

   b. σ known

2. $H_0 : \mu = \mu_0$, $H_1 : \mu \neq \mu_0$, or $\mu < \mu_0$ or $\mu > \mu$

   It has a two-tail test (Example Figure 17.2).

3. Determine the α value.

4. Determine the critical values:

   a. For a two-tail test, use a Z-table to find the value that has an area of $\alpha/2$ to its right. This value and its negative are the two critical values. The reject region is the area to the right of the positive value and the area to the left of the negative value.

   b. For a right-tailed test (see Figure 17.1a), use a Z-table to find the value that has an area of α to its right. This value is the critical value. The reject region is the area to the right of the positive value.

   c. For a left-tailed test (see Figure 17.1b), use a Z-table to find the value that has an area of α to its right. The negative of this value is the critical value. The reject region is the area to the left of the negative value.

5. Calculate the test statistic:

$$Z = (\bar{x} - \mu_0)\frac{\sqrt{n}}{\sigma}$$

6. If the test statistic is in the reject region, reject H_0. Otherwise, do not reject H_0.

7. State the conclusion in terms of the problem. That is, the population mean is equal to the specified value or not equal to the specified value with a significance of α.

EXAMPLE

A vendor claims that the average weight of a shipment of parts is 1.84. The customer randomly chooses 64 parts and finds that the sample has an average of 1.88. Suppose that the standard deviation of the population is known to be 0.03. Should the customer reject the lot? Assume that the customer wants to be 95 percent confident that the supplier's claim is incorrect before he or she rejects. Using the procedure just described:

1. Conditions (a) and (b) are met.
2. $H_0: \mu = 1.84$, $H_1: \mu \neq 1.84$; this is a two-tail test.
3. From the problem statement, $\alpha = .05$.
4. Critical values are the Z-value that has .025 to its right and the negative of this value. These values are 1.96 and –1.96. The reject region consists of the area to the right of 1.96 and the area to the left of –1.96.
5. $Z = (1.88 - 1.84)\sqrt{64}/(.03) = 10.7$
6. Since 10.7 is in the reject region, H_0 is rejected.
7. At the .05 significance level, the data suggest that the vendor's assertion that the average weight is 1.84 is false.

```
Statistical Software Analysis:
One-sample Z

Test of μ = 1.84 vs. ≠ 1.84
The assumed standard deviation = 0.03

N      Mean SE Mean            95% CI        Z      P
64 1.88000 0.00375 (1.87265, 1.88735) 10.67 0.000
```

Since the vendor's mean is outside of the confidence interval of the sample mean, we can conclude that the means are different.

There may be situations where the vendor may claim that the mean is less than 1.84 or greater than 1.84. Use the above data and conduct one-tail tests for both less-than and greater-than scenarios.

Student's *t*-Test

The *t*-test is usually used for making inferences about a population mean when the population variance σ^2 is unknown and the sample size n is small. Student's *t*-distribution applies to samples drawn from a normally distributed population. The use of the *t*-distribution is never wrong for any sample size. However, a sample size of 30 is normally the crossover point between the t- and Z-tests. The test statistic formula for this is

$$t = \frac{\bar{X} - \mu_0}{s/\sqrt{n}}$$

where

$\bar{X}$ = Sample mean

μ_0 = Target value or population mean

s = Sample standard deviation

n = Number of test samples

For the t-test the null and alternative hypotheses are the same as they were for the Z-test. The test statistic t is compared with a critical value t_α or $t_{\alpha/2}$, which is based on a significance level α for a one-tail test or $\alpha/2$ for a two-tail test, and the number of degrees of freedom, (df). Degrees of freedom is determined by the number of samples n and is given by

$$\text{df} = n - 1$$

Procedure for Calculating t-Test

1. Calculate the sample mean $\bar{X}$ and the sample standard deviation s
2. Calculate the $\bar{X} - \mu_0$
3. Calculate the standard error $s/\sqrt{n}$
4. Divide the results of step 2 by step 3
5. Calculate test statistic (t)
6. Compare with critical values (t_c)
7. Make conclusion using the null hypothesis statement.

EXAMPLE

A cutoff saw has been producing parts with a mean length of 4.125. A new blade is installed, and we want to know whether the mean has decreased. We select a random sample of 20, measure the length of each part, and find that the average length is 4.123 and the sample standard deviation is .008. Assume that the population is normally distributed. Use a significance level of .10 to determine whether the mean length has decreased.

Since the population standard deviation is unknown, the t-test will be used.

1. Condition is met (sample size threshold).
2. $H_0: \mu = 4.125$, $H_1: \mu < 4.125$, which is a left-tailed test.
3. From the problem statement, $\alpha = .10$.

Continued

Continued

4. The positive critical value is in the 19th row of the $t_{.10}$ column of the *t*-table. This value is 1.328. Since this is a left-tailed test, the critical value is –1.328. The reject region consists of the area to the left of –1.328.
5. $t = (4.123 - 4.125)\sqrt{20}/(.008) = -1.1$
6. Since –1.1 is not in the reject region, H_0 is not rejected.
7. At the .10 significance level, the data does not indicate that the average length has decreased.

Statistical Software Analysis:

One-Sample T

Test of μ = 4.125 vs < 4.125

N	Mean	StDev	SE Mean	90% Upper Bound	T	P
20	4.12300	0.00800	0.00179	4.12538	-1.12	0.139

Since the sample mean is not greater than the upper bound, *p*-value > 0.10, we fail to reject the null hypothesis and declare that that there is no change since the blade change.

Confidence Intervals for the Mean

Continuous data—large samples: For this type of data one uses the normal distribution to calculate the confidence interval for the mean:

$$\bar{X} \pm Z_{\alpha/2}\frac{\sigma}{\sqrt{n}}$$

where

$\bar{X}$ = Sample average

σ = Population standard deviation

n = Sample size

$Z_{\alpha/2}$ = Normal distribution value for a desired confidence level

Procedure for Calculating Confidence Intervals for the Mean

1. Find the confidence level from the table for normal distribution (see Appendix F) and determine the appropriate *Z*-value
2. Calculate the sample mean $\bar{X}$, sample standard deviation σ, and sample size n

3. Calculate margin of error by multiplying Z times σ and divide by the square root of n
4. Calculate $\bar{X}$ plus or minus the margin of error to obtain confidence intervals

EXAMPLE

We will use the home shopping channel population as an example. Let's say from a sample of 32 customers, the average order is $78.25, and the population standard deviation is $37.50. (This represents the variation among orders within the population.) We can calculate the 95 percent confidence interval for the population mean as follows:

$$\mu = 78.25 \pm 1.96 \frac{(37.50)}{\sqrt{32}} = 78.25 \pm 13.00$$

$$91.25 \geq \mu \geq 65.25$$

Test for Variance

This test calculates confidence intervals for the standard deviation and variance of a population, and performs a hypothesis test to determine whether the population variance equals a specified value per null hypothesis.

EXAMPLE

The variance for a sample of 35 parts is 46 units. Verify whether the variance of 36 for parts from a different processing method is statistically the same.

```
Statistical Software Analysis:
Test and CI for One Variance

Method:

Null hypothesis σ-squared = 36
Alternative hypothesis σ-squared ≠ 36

The chi-square method is only for the normal distribution.

Statistics:

N  StDev Variance
35  6.78     46.0

95% Confidence Intervals

                    CI for        CI for
Method              StDev       Variance
Chi-Square (5.49, 8.89) (30.1, 79.0)
```

Continued

Continued

The new process variance has to be either lower than 30 or higher than 79 to declare a difference in processing methods.

```
Tests

Method       Statistic DF P-Value
Chi-Square       43.44 34   0.257
```

Since the p-value is greater than 0.05 (alpha value assumed), we can conclude that there is no difference between the two processes.

Confidence Intervals for Variation

Confidence intervals for the mean are symmetrical about the average. This is not true for the variance, since it is based on the chi-square distribution, for which the formula is

$$\frac{(n-1)s^2}{x^2_{\alpha/2},\ n-1} \le \sigma^2 \le \frac{(n-1)s^2}{x^2_{1-\alpha/2}, n-1}$$

where

n = Sample size

s^2 = Point estimate of variation

$x^2_{\alpha/2}$ and $x^2_{1-\alpha/2}$ are the table values for $(n-1)$ degrees of freedom

Procedure for Calculating the Confidence Intervals for Variation

1. Find the critical chi-square value from the chi-square distribution table (Appendix I) for $n-1$ degrees of freedom.
2. Use the above formula and calculate the confidence interval for variance.
3. Report the results.

EXAMPLE

The sample variance for a set of 35 samples was found to be 46. Calculate the 90 percent confidence interval for the variance:

$$\frac{(34\times 46)}{48.60} \le \sigma^2 \le \frac{(34\times 46)}{21.66}$$

$$32.18 \le \sigma^2 \le 72.21$$

Test for Proportion

This is a hypothesis test to determine whether the proportion of trials that produce a certain event is equal to a target value.

EXAMPLE

This example examines the proportion of defectives from a product lot. The historical proportion defective of supplier A is 10%. A Green Belt was assigned to compare the performance of supplier B, who had 15 defectives from a sample size of 100.

Statistical Software Analysis:

Test and CI for One Proportion

```
Test of p = 0.1 vs p ≠ 0.1

Exact
Sample   X    N Sample p                95% CI P-Value
1       15  100 0.150000 (0.086454, 0.235308) 0.130
```

Based on the confidence interval, there is no statistical difference with a proportion defective between 8.6% and 23.5%. Since the p-value > 0.05, we can conclude there is no difference between suppliers.

Confidence Intervals for Proportion

For large sample sizes with $n(p)$ and $n(1-p) \geq 4$ or 5, the normal distribution can be used to calculate a confidence interval for proportion. The following formula is used:

$$p \pm Z_{\alpha/2}\sqrt{\frac{p(1-p)}{n}}$$

where

p = Population proportion estimate

$Z_{a/2}$ = Appropriate confidence level from a Z-table

n = Sample size

Procedure for Calculating the Confidence Intervals for Proportion

1. Determine the confidence level and find the appropriate Z value in the Z-table (Appendix F).
2. Find the sample proportion p by taking the number of people in the sample having the characteristic of interest divided by sample size n.

3. Multiply p by $(1 - p)$ and then divide that amount by n.
4. Take the square root of the result from step 3.
5. Multiply your answer by Z (margin of error).
6. Take p plus or minus the margin of error to obtain the confidence interval.

EXAMPLE

Let us say we want to estimate the proportion of female home shopping channel customers. Out of a sample of 175 random customers pulled, 110 were females. Calculate the 90 percent confidence interval for the proportion:

$$0.629 \pm 1.645\sqrt{\frac{(0.629 \times 0.371)}{175}} = 0.629 \pm 0.0600$$

$$0.689 \geq p \geq 0.569$$

Other confidence interval formulas exist for percent nonconforming, Poisson distribution data, and very small sample size data.

One Population Proportion (p-Test)

A p-test is used when testing a claim about a population proportion and we have a fixed number of independent trials having constant probabilities, with each trial having two outcome possibilities (a binomial distribution). When $np < 5$ or $n(1 - p) < 5$, the binomial distribution is used to test hypotheses relating to proportion.

Procedure for Calculating p*-Test*

1. Set conditions that $np \geq 5$ and $n(1 - p) \geq 5$ are to be met; then the binomial distribution of sample proportions can be approximated by a normal distribution.
2. The hypothesis tests for comparing a sample proportion p with a fixed value p_0 are given by the following:

$$H_0 : p = p_0,\ H_0 : p \leq p_0,\ H_0 : p \geq p_0$$

$$H_1 : p \neq p_0,\ H_1 : p > p_0,\ H_1 : p < p_0$$

The null hypothesis is denoted by H_0 and the alternative hypothesis is denoted by H_1.

3. Decide on α, the significance level.
4. Find the critical values in a standard normal table $-Z_\alpha$, Z_α, and $Z_{\alpha/2}$, (left-tail, right-tail, or two-tail test, respectively).
5. Calculate the test statistic using

$$Z = \frac{p' - p_0}{\sqrt{\frac{p_0(1-p_0)}{n}}}$$

where

p' = Sample proportion = x/n

n = Number of samples

x = Number of items in the sample with the defined attribute

p_0 = The hypothesized proportion

6. Reject H_0 if the test statistic is in the reject region. If not, do not reject H_0.
7. State the conclusion.

EXAMPLE

A vendor claims that at most two percent of a shipment of parts is defective. Receiving inspection chooses a random sample of 500 and finds 15 defectives. At the 0.05 significance level, do these data indicate that the vendor is wrong?

Here, $n = 500$, $x = 15$, $p' = 15/300 = 0.03$, and $p_0 = 0.02$.

1. Compute $np_0 = 500 \times 0.02 = 10$ and $n(1 - p_0) = 500 \times 0.98 = 490$.

   Both values are ≥ 5 so conditions are met.
2. $H_0 : p \leq 0.02$ and $H_1 : p > 0.02$ (right-tailed test).
3. $\alpha = 0.05$.
4. Critical value = 1.645, from a normal table.
5. $Z = \frac{0.03 - 0.02}{\sqrt{(0.02 \times 0.98) \div 500}} \approx 1.597$
6. Do not reject H_0.
7. At the 0.05 significance level, the data do not support a conclusion that the vendor is incorrect in asserting that at most two percent of the shipment is defective.

```
Statistical Software Analysis:
Test and CI for One Proportion

Test of p = 0.02 vs p > 0.02

Sample   X    N Sample p 95% Lower Bound Z-Value P-Value
1       15  500 0.030000        0.017452    1.60   0.055
```

Using the normal approximation.

PAIRED-COMPARISON TESTS

Two-Mean, Equal Variance *t*-Test

In a two-mean, equal variance *t*-test, the tests are between two sample means. ($\bar{X}$ versus $\bar{X}_2$), and σ_1, and σ_2 are unknown but considered equal.

$$H_0: \mu_1 = \mu_2, H_1: \mu_1 \neq \mu_2$$

$$t = \frac{\bar{X}_1 - \bar{X}_2}{s_p / \sqrt{\frac{1}{n_1} + \frac{1}{n_2}}}$$

where

s_p = pooled standard deviation

$$s_p = \sqrt{\frac{\left((n_1 - 1)s_1^2 + (n_2 - 1)s_2^2\right)}{n_1 + n_2 - 2}}$$

$$\text{df} = n_1 + n_2 - 2$$

EXAMPLE

Two operators are machining parts in two CNCs and they want to test the difference between two sample means. Samples are taken in pairs and their differences are calculated. For this, a paired *t*-test is used, where

$$H_0: \mu_1 = \mu_2, H_1: \mu_1 \neq \mu_2$$

	CNC 1	CNC 2
1	5.257	5.243
2	5.220	5.193
3	5.235	5.225
4	5.230	5.220
5	5.225	5.223
	$x_1 = 5.2334$	$x_2 = 5.2208$
	$s_1 = .0143$	$s_2 = .0179$

$$s_p = \sqrt{\frac{\left((n_1 - 1)s_1^2 + (n_2 - 1)s_2^2\right)}{n_1 + n_2 - 2}}, \text{df} = 5 + 5 - 2 = 8$$

$$s_p = \sqrt{\frac{4(.0143)^2 + 4(.0179)^2}{8}} = .0162$$

Continued

Continued

$$t = \frac{5.2334 - 5.2208}{.0162\sqrt{\frac{1}{5}+\frac{1}{5}}} = 1.230$$

The critical value for $t_{.025,\ 8} = 2.306$ (a two-sided test for $\alpha = 0.05$). Therefore, the null hypothesis H_0 cannot be rejected.

```
Statistical Software Analysis:
Two-Sample T-Test and CI: Machine 1, Machine 2
(Assuming equal variance)

Two-sample T for Machine 1 vs Machine 2

N            Mean  StDev      SE   Mean
Machine 1      5 5.2334 0.0143 0.0064
Machine 2      5 5.2208 0.0179 0.0080

Difference = μ (Machine 1) - μ (Machine 2)
Estimate for difference: 0.0126
95% CI for difference: (-0.0111, 0.0363)
T-Test of difference = 0 (vs ≠): T-Value = 1.23 P-Value = 0.255 DF = 8
Both use Pooled StDev = 0.0162
```

Since the *p*-value is > 0.05, we cannot reject the null hypothesis and conclude that the two machines are producing parts with equal means.

Two-Mean, Unequal Variance *t*-Test

In a two-mean, unequal variance *t*-test, the tests are between two sample means ($\bar{X}_1$ versus $\bar{X}_2$), and σ_1 and σ_2 are unknown but are not considered equal.

$$H_0: \mu_1 = \mu_2,\ H_1: \mu_1 \neq \mu_1$$

$$t = \frac{\bar{X}_1 - \bar{X}_2}{\sqrt{\frac{s_1^2}{n_1} + \frac{s_2^2}{n_2}}}$$

$$\text{df} = \frac{1}{\left(\frac{\frac{\frac{s_1^2}{n_1}}{\frac{s_1^2}{n_1}+\frac{s_2^2}{n_2}}}{n_1 - 1}\right) + \left(\frac{\frac{\frac{s_2^2}{n_2}}{\frac{s_1^2}{n_1}+\frac{s_2^2}{n_2}}}{n_2 - 1}\right)}$$

EXAMPLE

Use data from the prior example to perform an unequal variance test: df = 7.627 ≈ 8. Round off df to increase the confidence level rather than reduce it:

$$t = \frac{5.2334 - 5.2208}{\sqrt{\frac{(.0143)^2}{5} + \frac{(.0179)^2}{5}}} = 1.230$$

The critical value for $t_{.025,8} = 2.306$ (two-sided test for $\alpha = 0.05$). We fail to reject null hypothesis H_0.

```
Statistical Software Analysis:
Two-Sample T-Test and CI: Machine 1, Machine 2
(assuming unequal variances)*

Two-sample T for Machine 1 vs Machine 2

N            Mean   StDev      SE    Mean
Machine 1       5 5.2334 0.0143 0.0064
Machine 2       5 5.2208 0.0179 0.0080

Difference = µ (Machine 1) - µ (Machine 2)
Estimate for difference: 0.0126
95% CI for difference: (-0.0117, 0.0369)
T-Test of difference = 0 (vs ≠): T-Value = 1.23 P-Value = 0.260 DF = 7
```

Since the *p*-value > 0.5, we conclude that the parts manufactured from the two machines have equal variances, and fail to reject the null hypothesis.

*Green Belts are encouraged to first test for variances and apply the right test as to whether the data set has equal or unequal variances.

Paired *t*-Test

In these tests, subjects are matched in pairs and the outcomes are compared within each matched pair:

$$t = \frac{\bar{d}}{\frac{s_d}{\sqrt{n}}}$$

In general, the paired *t*-test is a more sensitive test than the comparison of two independent samples.

Note: The paired *t*-test is applied when we have a before-and-after scenario or there is a dependency between the two sets of measurements. For example, measurement of samples before and after heat treatment, measurement before and after calibration of equipment, or pre- and post-training skill assessment of an employee. It is important to know when to apply a paired *t*-test.

Procedure for Paired t-Test

1. Find the differences between each pair of data by subtracting one from the other.
2. Use these data and calculate the mean $\bar{d}$ and the standard deviation s of all the differences.
3. Let n be the number of paired differences.
4. Calculate the standard error $\frac{s}{\sqrt{n}}$.
5. Divide the mean $\bar{d}$ by the standard error $\frac{s}{\sqrt{n}}$ from step 4.

EXAMPLE

An operator is measuring the same set of five parts in a piece of test equipment and wants to compare the difference between the two sample means before and after calibration. Measurements are taken and their differences are calculated. A paired *t*-test is used, where

$$H_0 : \mu_1 = \mu_2, H_1 : \mu_1 \neq \mu_2$$

	Before	After	Difference (d)
1	5.257	5.243	0.014
2	5.220	5.193	0.027
3	5.235	5.225	0.010
4	5.230	5.220	0.010
5	5.225	5.223	0.002
			$\bar{d} = 0.0126$
			$S_d = 0.00915$

$$df = n - 1 = 4$$

$$t = \frac{\bar{d}}{\frac{S_d}{\sqrt{n}}} = 3.080$$

$t_{.025,\,4} = 2.776$ ($\alpha = 0.05$ for a two-sided test; a paired test is always a two-tailed test). The null hypothesis H_0 is rejected.

Continued

Continued

```
Statistical Software Analysis:

N               Mean     StDev       SE      Mean
Before            5 5.23340 0.01433 0.00641
After             5 5.22080 0.01795 0.00803
Difference        5 0.01260 0.00915 0.00409

95% CI for mean difference: (0.00123, 0.02397)
T-Test of mean difference = 0 (vs ≠ 0): T-Value = 3.08 P-Value =
0.037
```

Since the 95% CI for mean difference did not include "0" (hypothesized difference), we can conclude that there is a statistically significant difference between before and after calibration of the equipment. This is also reflected by the p-value < 0.05.

F-Test

The *F*-statistic is the ratio of two sample variances (also called two *chi-square distributions*) and is represented as

$$F = \frac{(s_1)^2}{(s_2)^2}$$

where

s_1^2 and s_2^2 = Sample variance of the two samples 1 and 2 under comparison

Procedure for Calculating Two-Sample-Variance F-Test

1. Set the conditions: populations are normal and samples are independent.
2. The hypothesis tests for comparing two population variances σ_1^2 and σ_2^2 are given by

$$H_0 : \sigma_1^2 = \sigma_2^2,\ H_0 : \sigma_1^2 \leq \sigma_2^2,\ H_0 : \sigma_1^2 \geq \sigma_2^2$$

$$H_1 : \sigma_1^2 \neq \sigma_2^2,\ H_1 : \sigma_1^2 > \sigma_2^2,\ H_1 : \sigma_1^2 < \sigma_2^2$$

   where H_0 represents the null hypothesis and H_1 represents the alternative hypothesis. The *F*-distribution has a nonsymmetrical shape and depends on the number of degrees of freedom associated with s_1^2 and s_2^2. Number of degrees of freedom are represented by v_1 and v_2.

3. Find the critical values in an *F* table.

4. Calculate the test statistic using

$$F = \frac{(s_1)^2}{(s_2)^2}$$

5. Reject the test hypothesis H_0 if it is in the reject region, otherwise do not reject H_0.

6. State the conclusion.

EXAMPLE

A pickle factory is studying the effect of aging on its product. They want to know if there is an improvement in consistency of crispness (strength) after aging for one month. The data collected are reported below (assume a 95 percent confidence level).

	Initial reading	After one month
Number of tests	8	6
Standard deviation	1800	600

Here

$H_0 : \sigma_1^2 \leq \sigma_2^2$ and $H_1 : \sigma_1^2 > \sigma_2^2$

$v_1 = 7$ and $\sigma_2 = 5$

Since this is concerned with an improvement in variation, a one-tail test with α risk in the right tail could be used. Using the *F* table (Appendix G), the critical value of *F* is 4.88. The null hypothesis rejection area is equal to or greater than 4.88.

$$F = \frac{(s_1)^2}{(s_2)^2} = \frac{(1800)^2}{(600)^2} = 9$$

The null hypothesis is rejected in this case as the calculated *F* value is in the critical region. There is enough evidence to prove reduced variance and higher consistency of crispness after aging for one month.

SINGLE-FACTOR ANALYSIS OF VARIANCE (ANOVA)

One-Way ANOVA

ANOVA is a statistical method for comparing several population means. We draw a *simple random sample* (SRS) from each population and use the data to test the null hypothesis that the population means are all equal.

A *factor* in ANOVA describes the cause of the variation in the data. When only one factor is being considered, the procedure is known as *one-way ANOVA*. This type of ANOVA has two parts: the variation between treatment means and the variation within treatments. To use one-way ANOVA, the following conditions must be present:

a. The populations of interest must be normally distributed.

b. The samples must be independent of each other.

c. Each population must have the same variance.

The basic idea here is to determine whether the variation caused by the factor is a sufficiently large multiple of the experimental error to reject the null hypothesis. The *F*-statistic measures that multiple. The experimental error is measured as the within-treatment variation.

Procedure for Calculating One-Way ANOVA

1. $H_0: \mu_1 = \mu_2 = \mu_3 = \ldots = \mu_k$, H_1 : not all the means are equal; this is a right-tailed test.

2. Determine the α value. This is similar to the use of α in confidence intervals. In hypothesis testing jargon, the value of α is referred to as the *significance level*.

3. Construct the ANOVA table:

Source of variation	Sum of squares	Degrees of freedom	Mean squares	F-statistic
Between treatment	SS_B	$k-1$	$MS_B = SS_B/(k-1)$	$F = MS_B/MS_W$
Within treatment	SS_W	$N-k$	$MS_W = SS_W/(N-k)$	
Total	SS_T	$N-1$		

A fundamental property of this table is that the total row is the total of the values of the entries above it in the sum of squares column and the degrees of freedom column, where

N = Number of readings

n = Number of readings per level (or treatment)

k = Number of levels (or treatments)

T = Grand total of readings $\Sigma y_i = \Sigma T_i$

C = Correction factor = T^2/N

y_i's = Individual measurements

SS_T = Sum of squares total = $\Sigma y_i^2 - C$

SS_B = Sum of squares between treatments = $\Sigma T_i^2 /n - C$

SS_W = Sum of squares within treatment = $SS_T - SS_B$

4. The test statistic is the F-value as defined in the table.
5. Find the critical value in an F table using $k - 1$ as the numerator degrees of freedom and $k(n - 1)$ as the denominator degrees of freedom.
6. Determine whether the null hypothesis should be rejected. Since this is a right-tailed test, if the value of the test statistic is ≥ the critical value, then the null hypothesis is rejected and the alternative hypothesis is accepted. If the value of the test statistic is < the critical value, the null hypothesis is not rejected.
7. State the conclusion in terms of the original problem.

EXAMPLE

A polyurethane casting process can be run at 200 °F, 220 °F, or 240 °F. Does the temperature significantly affect the moisture content? Use $\alpha = 0.05$.

To answer the question, four batches were run at each of the temperatures. The twelve runs were executed in random order. The temperature results are:

240 °F	220 °F	200 °F
10.8	11.4	14.3
10.4	11.9	12.6
11.2	11.6	13.0
9.9	12.0	14.2

The entries in the table are individual measurements, referred to as y-values, and are moisture content values in percentage H_2O.

The hypothesis test:

1. Assume that the conditions have been tested and are satisfied.
2. $H_0 : \mu_1 = \mu_2 = \mu_3$, H_1 : not all the means are equal, this is a right-tailed test.
3. $\alpha = .05$.
4. Construct the ANOVA table.
5. Calculate the average in each column.

240 °F	220 °F	200 °F
10.8	11.4	14.3
10.4	11.9	12.6
11.2	11.6	13.0
9.9	12.0	14.2
$T_1 = 42.3$	$T_2 = 46.9$	$T_3 = 54.1$
$\bar{y}_1 = 10.575$	$\bar{y}_2 = 11.725$	$\bar{y}_3 = 13.525$

Continued

Continued

where

number of readings $N = 12$

number of readings per level (or treatment) $n = 4$

number of levels (or treatments) $k = 3$

$$\Sigma y_i^2 = 10.8^2 + 10.4^2 + 11.2^2 + 9.9^2 + 11.4^2 + 11.9^2 + 11.6^2 + 12.0^2 + 14.3^2 + 12.6^2 + 13.0^2 + 14.2^2 = 1732.27$$

Grand total of readings: $T = \Sigma y_i = \Sigma T_i = 42.3 + 46.9 + 54.1 = 143.3$

Correction factor: $C = T^2 \div N = 143.3^2 \div 12 = 1711.24$

Sum of squares total $= SS_T = \Sigma y^2 - C = 1732.27 - 1711.24 = 21.03$

Sum of squares between treatments $= SS_B = \Sigma T_i^2/n - C$
$= 42.3^2/4 + 46.9^2/4 + 54.1^2/4 - 1711.24$
$= 17.69$

Sum of squares within treatment $= SS_W = SS_T - SS_B = 21.03 - 17.69 = 3.34$

These values fit into an ANOVA table as follows:

Source of variation	Sum of squares	Degrees of freedom	Mean square	F-statistic
Between treatment	$SS_B = 17.69$	$k - 1 = 2$	$17.69/2 = 8.85$	23.92
Within treatment	$SS_W = 3.3$	$N - k = 9$	$3.34/9 = 0.37$	
Total	$SS_T = 21.03$	$N - 1 = 11$		

6. The test statistic is defined as $F = MS_B/MS_W = 8.85/0.37 = 23.92$.
7. The critical values are in the F table located in Appendix G. The table is indexed by the degrees of freedom associated with the numerator and denominator of the fraction used to calculate F. In this case, the numerator MS_B has 2 degrees of freedom and the denominator MS_W has 9 degrees of freedom. From the $F_{.05}$ table in column 2 and row 9, $F = 4.26$. Here there is only one rejection region, which is the shaded area on the right tail of the distribution. This is a right-tailed test. The test statistic 23.92 exceeds the critical value 4.26, so the null hypothesis is rejected.
8. Thus, the conclusion is that at the .05 significance level the data indicate that temperature does have an impact on moisture content.

Statistical Software Analysis:

One-way ANOVA: 240 F, 220 F, 200 F

```
Method

Null hypothesis All means are equal
Alternative hypothesis At least one mean is different
Significance level α = 0.05
```

Continued

Continued

```
Equal variances were assumed for the analysis.

Factor Information

Factor Levels Values
Factor 3 240 F, 220 F, 200 F

Analysis of Variance

Source DF Adj SS Adj MS F-Value P-Value
Factor  2 17.687 8.8433   23.81    0.000
Error   9  3.342 0.3714
Total  11 21.029

Pooled StDev = 0.609417
```

Since the null hypothesis is rejected, we can conclude based on the alternate hypothesis that at least one of the population means is different from the others.

Using Spreadsheet FINV Function

One can generate critical *F*-statistics using a spreadsheet's FINV function, which is represented as

FINV(probability, deg_freedom1, deg_freedom2)

where

probability = The level of significance

deg_freedom1 = $\nu_1 = k - 1$

deg_freedom2 = $\nu_2 = N - k$

Figure 17.4 illustrates the FINV function used to determine the critical *F*-statistic with $\alpha = 0.05$, $\nu_1 = 3 - 1 = 2$, and $\nu_2 = 12 - 3 = 9$ from our previous example.

Cell A1 contains the spreadsheet formula =FINV(0.05,2,9), with the result being 4.2564. This value is very close to the one in the *F*-distribution table (Appendix G), 4.26.

Using Spreadsheets to Perform One-Way ANOVA

One can perform ANOVA using spreadsheets; here are the steps:

1. Enter the data in each column of a blank spreadsheet.
2. Select Data Analysis from the Tools menu bar.
3. Select ANOVA: Single Factor from the Data Analysis dialog box and click OK (see Figure 17.5).

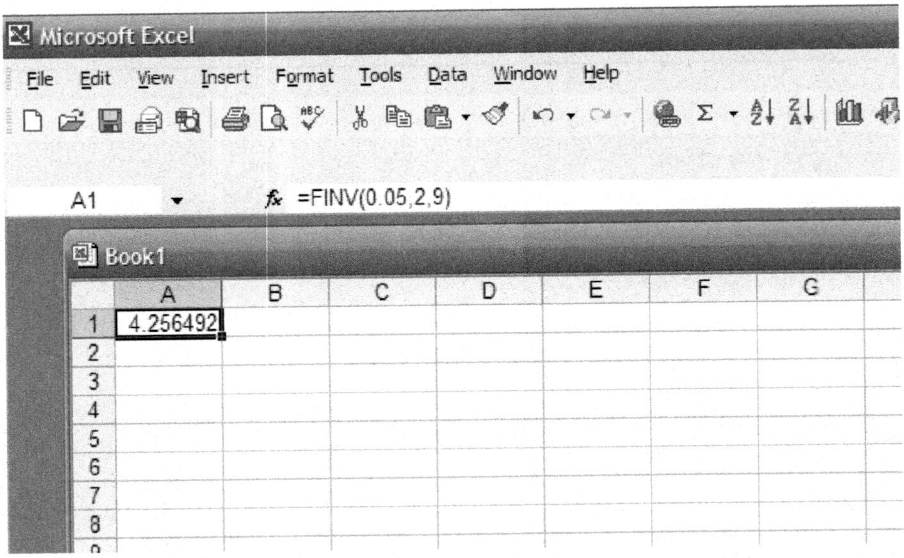

Figure 17.4 Spreadsheet FINV function.

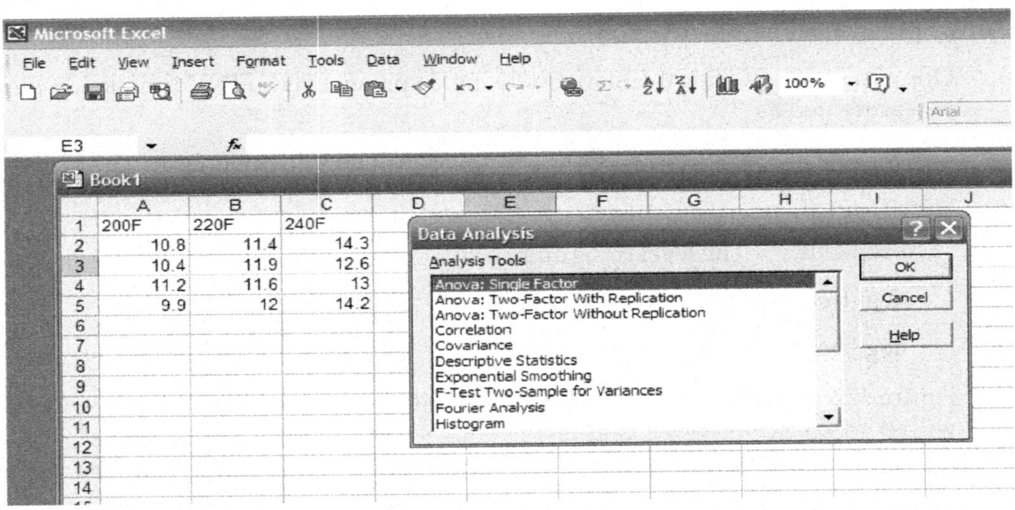

Figure 17.5 Setting up a one-way ANOVA in spreadsheet.

4. Set up the ANOVA: Single Factor dialog box as shown in Figure 17.6.
5. Click OK; Figure 17.7 shows the final ANOVA results.

These results are consistent with what we found doing it the hard way in the previous sections. Notice that the p-value = 0.000254 for this test, meaning we can reject H_0 as this p-value $\leq \alpha$, which is 0.05.

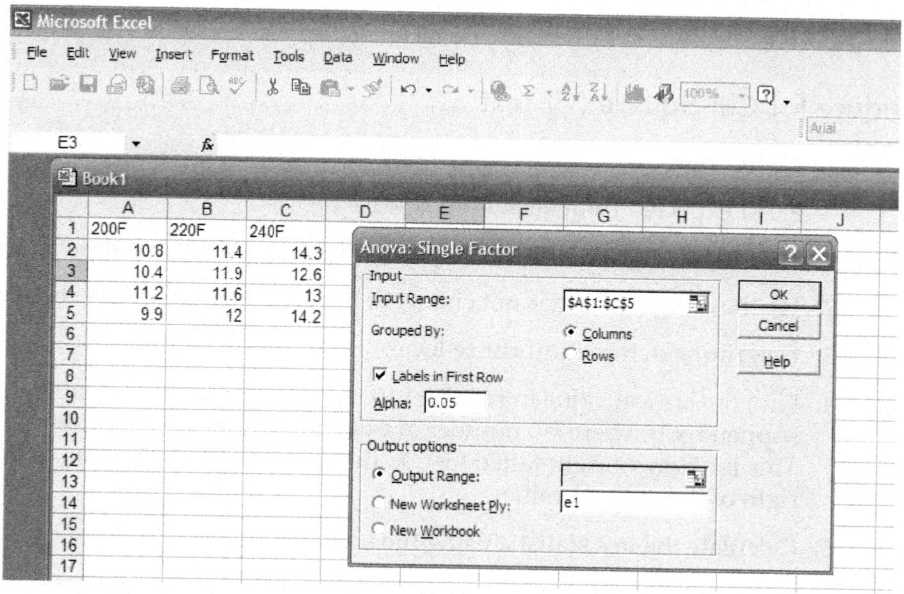

Figure 17.6 The ANOVA: Single Factor dialog box.

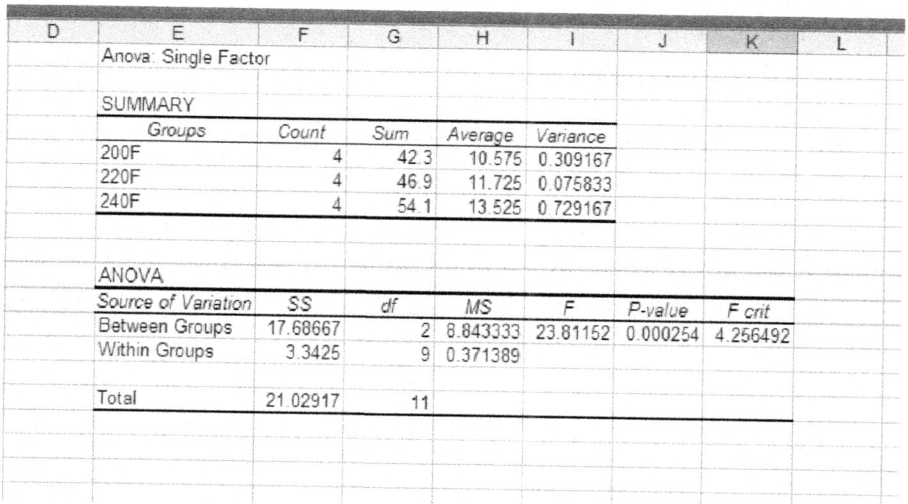

Anova: Single Factor

SUMMARY

Groups	Count	Sum	Average	Variance
200F	4	42.3	10.575	0.309167
220F	4	46.9	11.725	0.075833
240F	4	54.1	13.525	0.729167

ANOVA

Source of Variation	SS	df	MS	F	P-value	F crit
Between Groups	17.68667	2	8.843333	23.81152	0.000254	4.256492
Within Groups	3.3425	9	0.371389			
Total	21.02917	11				

Figure 17.7 Final results of one-way ANOVA in spreadsheet.

CHI SQUARE

Procedure for Chi-Square (χ^2) Test

1. Conditions:
   - All expected frequencies are at least 1
   - At most, 20 percent of the expected frequencies are less than 5
2. H_0: the distribution has not changed; H_1: the distribution has changed.
3. Determine α, the significance level.
4. Find the critical value in row $k - 1$ in the χ^2_α column of the χ^2 table, (Appendix I) where k = number of categories in the distribution. This is always a right-tailed test, so the reject region is the area to the right of this critical value.
5. Calculate the test statistic using the formula

   $\chi^2 = \Sigma[(O - E)^2/E]$ (the sum of the last column in a χ^2 table)

   where O is the observed frequency and E is the expected frequency.
6. Reject H_0 if the test statistic is in the reject region. Otherwise do not reject.
7. State the conclusion.

EXAMPLE

Suppose that all rejected products have exactly one of four types of defects and that historically they have been distributed as follows:

Paint run	16%
Paint blister	28%
Decal crooked	42%
Door cracked	14%
Total	100%

Data on rejected parts for a randomly selected week in the current year are:

Paint run	27
Paint blister	60
Decal crooked	100
Door cracked	21

Continued

Continued

The question that one needs to answer is, is the distribution of defect/deformity types different from the historical distribution? This is often called the χ^2 *goodness-of-fit test* (pronounced "chi square"). To get a feel for this test, construct a table that displays the number of defects that would be expected in each category if the sample exactly followed the historical percentages:

	Probability	Observed frequency	Expected frequency
Paint run	0.16	27	33.28
Paint blister	0.28	60	58.24
Decal crooked	0.42	100	87.36
Door cracked	0.14	21	29.12
Total			208.00

The expected frequency for paint run is found by calculating 16 percent of 208, and so on. It remains to be seen whether the difference between the expected frequencies and observed frequencies (Figure 17.8) is sufficiently large to conclude that the sample comes from a population that has a different distribution. Test this at the .05 significance level.

The test statistic is obtained by the following equation:

$$\chi^2 = \Sigma \frac{(O-E)^2}{E}$$

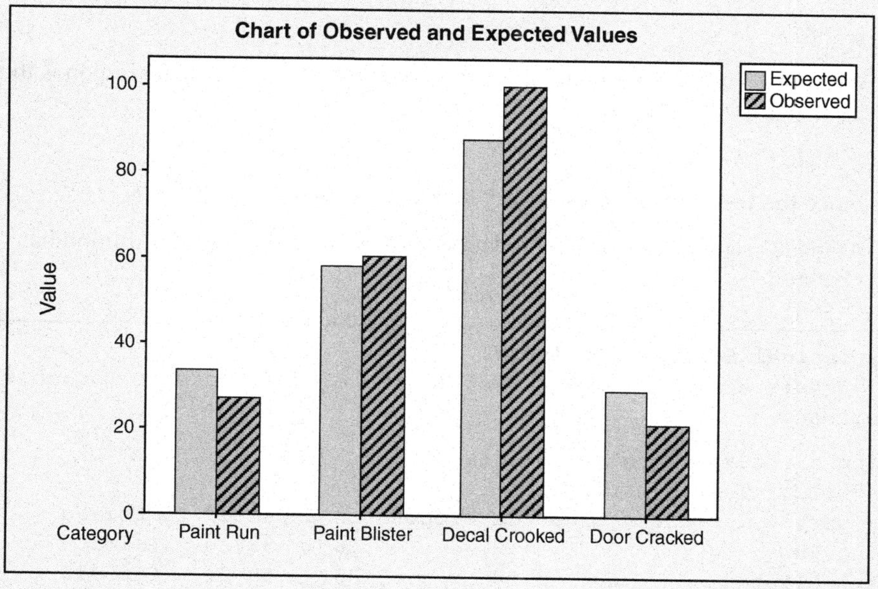

Figure 17.8 Expected versus observed chart.

Continued

Continued

The null hypothesis is that the distribution hasn't changed. This hypothesis will be rejected if the total of the last column is too large. The results are as follows:

	Probability	Observed frequency (O)	Expected frequency (E)	(O – E)	$\frac{(O-E)^2}{E}$
Paint run	0.16	27	33.28	–6.28	1.185
Paint blister	0.28	60	58.24	6.76	0.053
Decal crooked	0.42	100	87.36	7.64	1.829
Door cracked	0.14	21	29.12	–8.12	2.264
Total			208.00		5.33

EXAMPLE

Using the data from the previous example:

1. The conditions are met.
2. H_0: The distribution of defect types has not changed.

   H_1: The distribution of defect types has changed.
3. $\alpha = .05$
4. From row 3 of the $\chi^2_{.05}$ column, the critical value is 7.815. The reject region is the area to the right of 7.815.
5. $\chi^2 = \Sigma[(O - E)^2/E] = 4.9$
6. Since the test statistic does not fall in the reject region, do not reject H_0.
7. At the .05 significance level, the data do not indicate that the distribution has changed.

```
Statistical Software Analysis:
Chi-Square Goodness-of-Fit Test for Observed Counts in Variable:
Frequency

Using category names in Defects
Historical Test Contribution
Category        Observed Counts Proportion Expected to Chi-Sq
Paint Run             27    0.16          0.16     33.28    1.18505
Paint Blister         60    0.28          0.28     58.24    0.05319
Decal Crooked        100    0.42          0.42     87.36    1.82886
Door Cracked          21    0.14          0.14     29.12    2.26423
```

Continued

Continued

```
N    DF Chi-Sq  P-Value
208 3   5.33133 0.149
```

A statistical analysis with the same data provides a *p*-value of 0.149. Since the assumed alpha risk is 0.05, we fail to reject the null hypothesis, meaning there is no difference in the observed frequency compared to historical defective proportion defectives.

Using Spreadsheet to Perform Chi-Square Distribution

One can generate critical chi-square scores using spreadsheet CHIINV function, which has the following characteristics:

CHIINV(probability, deg_freedom)

where

probability = The level of significance α

deg_freedom = The number of degrees of freedom

For example, to determine the critical chi-square score for $\alpha = .05$ and df = 3 from the previous example, the CHIINV function is illustrated in Figure 17.9.

Cell A1 contains the spreadsheet formula =CHIINV(0.05,3) with the result being 7.815. This is similar to the value in the chi-square distribution table (Appendix I).

Contingency Tables

Contingency tables are two-dimensional classification tables with rows and columns containing original frequencies or count data that can be analyzed to determine whether the two variables are independent or have significant association. When the totals of rows and columns are analyzed in a certain way, the chi-square procedure will test whether there is dependency between the two classifications.

Figure 17.9 Spreadsheet CHIINV function.

Also, a *contingency coefficient* can be calculated. If the chi-square test shows a significant dependency, the contingency coefficient will show the strength of the correlation.

Parametric and Nonparametric Tests

Parametric test implies a descriptive measure calculated using population data. Usually, an assumption is made when performing a hypothesis test that the data are a sample from a certain distribution, commonly referred to as the *normal distribution*. *Nonparametric test* implies that there is no assumption of a specific distribution for the population.

Nonparametric techniques of hypothesis testing are applicable for many quality engineering problems and projects. These tests are also called *distribution-free* as they make no assumption regarding the population distribution. They can be applied to *ranking tests*, in which data are not specific in any continuous data or attribute sense, but are simply ranks. Three powerful nonparametric techniques commonly used are the Kendall coefficient of concordance, Spearman rank correlation coefficient, and Kruskal-Wallis one-way analysis of variance. We will not go into detailed descriptions of these techniques as they are outside the scope of the Green Belt BoK.

Chapter 18

Additional Analysis Methods

GAP ANALYSIS

> Analyze scenarios to identify performance gaps and compare current and future states using predefined metrics. (Analyze)
>
> **Body of Knowledge IV.C.1**

Baselining the current state and identifying performance gaps prior to the improvements is important to document and communicate to our key stakeholders. Without knowing where we started, it is unlikely to know where we want to land. Let us explore some of the reasons why this is important.

1. Identify appropriate stakeholders required for a cross-functional effort.
2. Identify resources required to accomplish the goals.
3. Make a leadership commitment.
4. Measure progress made post improvement.
5. Articulate the improvements made as financial and economic benefits.

Some performance gaps are apparent at the "define" stage of DMAIC implementation. Some are more subtle and are not apparent until an analysis is performed. For example, customer satisfaction level may be apparent at the "define" stage of the implementation. However, by the analysis stage, gaps in customer service will be more apparent as we peel the onion.

There are a number of scenarios to identify performance gaps. When an organization wants to identify opportunities for improvement (proactive and reactive), the impact of changes made to the organization must be considered.

Scenarios include but are not limited to;

A. Proactive

- Customer perception prior to a product/service launch
- Operational efficiency prior to rolling out an organizational change

B. Reactive

- Customer satisfaction after having launched a product/service
- Measuring operational efficiency after the organizational change

We need to start with the objective. If the objective is to understand the user experience after a feature change to a product or service, then identifying the attributes (predefined metrics) that will clearly show difference as a result of the change should be identified and measured upfront.

Examples of gap analysis may also include items like energy use, energy loss, engineering changes, etc. (Ref: ISO 50001:2018).

If we are planning to improve a process through Value Stream Mapping (VSM), then identifying and baselining the current state's predefined metrics are logical steps.

Predefined metrics for this example may be:

Predefined metrics	**Current state**	**Future state**	**Improvement**
Lead time (days)	21	7	14
Cycle time (minutes)	1800	600	1200
Inventory ($K)	500	50	450
People (#)	10	7	3
Wait time (minutes)	3000	500	2500
Transportation (meters)	1000	500	500
Rolled Throughput Yield (%)	90	95	5
Process Risk Priority Number(RPN) > 100	10	0	10
Ratio of Value add to lead time (Process Cycle Efficiency) %	3	10	7
OEE %	60	75	15
Cost of Poor Quality as % of Operational Expenses	25	10	15
Customer Satisfaction %	90	95	5

Once the current state is developed and the predefined metrics are baselined, it will be beneficial in analyzing the improvements made once the final state is achieved.

Gap analysis can be performed using many Lean Six Sigma tools. Audits, assessments, value stream analysis, Failure Mode Effects Analysis, hypothesis testing, etc. are effectively used by practitioners to perform this analysis.

5 WHY ANALYSIS (WHY-WHY ANALYSIS)

> Use cause and effect diagrams, relational matrices, 5 Whys, fault tree analysis, and other problem-solving tools to identify the true cause of a problem. (Analyze)
>
> **Body of Knowledge IV.C.2**

5 Why analysis is an easy-to-use approach to arriving at a root cause. This is a widely used approach due to its simplicity. No special training or skill is required. All the user needs is the persistence to ask "why" every time you get an answer and keep going until you reach the root cause.

If you have observed your five-year-old child, you will recognize this very common inquisitiveness. At this age, a child wants to know more by asking why. There are times we as adults run out of answers! So, if we all started young by asking questions, when did we stop? As we age, influenced by our experiences and the environment in which we grow up, we make up our mind on every question and continue to live with these assumptions. We hardly ever validate or challenge our assumptions.

The 5 Why analysis forces us to keep asking questions. Many Quality professionals get hung up on the number 5 in "5 Why" and literally force to ask why 5 times. However, depending on the complexity of issue, one might arrive at the root cause at the third "why" or the seventh "why" or even the ninth "why." *5 Why is merely a guideline and not literal.* To avoid this misunderstanding, nowadays organizations call this tool "Why-Why Analysis." There is always a question from the users as to when to stop asking why. One could keep going. If you stop too soon at the second or third "why," the cause may still be shallow. If you go way past the actionable root cause several whys deeper than required, the cause can get very broad, like "organizational culture" or "market conditions," etc. These are not quite actionable items for the RCA team. The ideal point for when to stop is where the root cause is specific and actionable, and the problem solver or the organization does not have any control to act on the subsequent "why." Also, it is important to verify that the root cause addresses a process or system-level breakdown to prevent recurrence.

Let us take an example from Toyota:

1. "Why did the robot stop?" The circuit has overloaded, causing a fuse to blow.

2. "Why is the circuit overloaded?" There was insufficient lubrication on the bearings, so they locked up.

3. "Why was there insufficient lubrication on the bearings?" The oil pump on the robot is not circulating sufficient oil.

4. "Why is the pump not circulating sufficient oil?" The pump intake is clogged with metal shavings.
5. "Why is the intake clogged with metal shavings?" Because there is no filter on the pump.

Now, if the manufacturing quality practitioner stopped at the first "why," the organization would have thought that the fuse is the root cause and kept an inventory of fuses as a maintenance spare part inventory! Or if they had stopped at the second "why," the operator would have been advised to keep pouring more lubrication. This would have resulted in spillage on the floor and factory accidents. Is the fifth "why," identifying the filter as the root cause, sufficient? In this case it is sufficient, as ensuring a filter and periodic replacement of the filter during preventive maintenance program will prevent the robot from stopping. However, there may be a missing washer in a different piece of equipment, or even worse, a missing safety-related equipment part. Since the above 5 Whys did not go far enough to address the process, or system-level breakdown, we must continue to ask why to arrive at the system-level root cause.

1. "Why is there no filter on the pump?" The maintenance technician forgot to replace it during the preventive maintenance.
2. "Why did the maintenance technician forget to replace it?" There was no error proofing and a process for preventive maintenance.

One should stop at this point to ensure there is defined, documented, and fully-deployed effective maintenance process that not only prevents a missing filter but also prevents any errors in the future. One should also provide a closed-loop feedback to the equipment vendor or retrofit the equipment with error proofing to be assembled only when filter is present.

Once a root cause has been identified in the 5 Why (or Why-Why) analysis, it is important to verify if the root cause can indeed address the problem effect. If you have the ability to turn ON and OFF the root cause to verify the problem prevention and problem re-creation, this is an effective way to confirm the root cause. However, this is not always practical. Some root causes are not easy to recreate. Even if we were able to re-create the root cause, it can create significant undesirable effects on our customer. In these scenarios, practitioners are advised to be creative in verification—for example, using a software simulation or equivalent methods. Finally, verify recurrence of the events. Complaints from the customer are a reactive approach to verify if the root cause and corresponding corrective action has been effective.

Fault Tree Analysis

Fault tree analysis (FTA) is a top-down failure analysis in which a failure of a system is analyzed using a combined series of lower-level events. This analysis method is mainly used in reliability engineering to understand how systems can fail. This also helps designers and engineers identify efficient ways to reduce risk and to determine overall reliability of a system.

Fault Tree Analysis can be used to:

- Understand the AND/OR logic leading to the failure of a system.
- Identify the significant contributors to the system-level failure and prioritize improvements.
- Optimize resources to achieve highest reliability of an overall system.
- Help maintenance technicians and those responsible for troubleshooting functions to identify root causes of the system-level failure.

Fault tree analysis is not only applicable to technical applications. Once effectively utilized, this tool for implementing programs can:

- Identify a top-level program objective.
- Identify the next-level objectives.
- Develop a logical relationship between the top-level and next-level objectives (AND/OR).
- Similarly identify the subsequent next-tier objectives.

Other analysis approaches assume one root cause. Approaches involve a series of probing from the top level to a root cause a few layers down. Even though similar in top-down structure, fault tree analysis takes into consideration multiple causes for each cause and assigns a logical relationship between the causes. An effect is caused by a combination of various causes with varying probability for each cause. These causes at each layer may occur in combination or independently. The probabilities are thus calculated for causes in series or in parallel. All the probabilities sum up to the top-level cause.

Part V

Improve Phase

Part V is an overview of the *improve* phase, including summaries of those Six Sigma methods and practices designed and intended to determine and prioritize improvements to products, processes, and organizations. It covers approximately 15 of the 100 questions that will be asked on the ASQ CSSGB exam. The investigative capabilities within design of experiments are supported by a practical overview of effective ways to analyze root causes.

The BoK was slightly reorganized for Part V, and now includes information on lean, with additional emphasis on overall waste elimination, cycle time reduction, kaizen, and kaizen blitz activities.

Chapter 19

Design of Experiments (DOE)

BASIC TERMS

> Define and describe terms such as independent and dependent variables, factors and levels, responses, treatments, errors, repetition, blocks, randomization, effects, and replication. (Understand)
>
> **Body of Knowledge V.A.1**

Design of experiments (DOE) can be summarized as a set of planned experiments performed by an experimenter. The experimenter adjusts a set of variables under controlled conditions in order to collect the outcomes needed to determine whether a significant outcome is realized from those adjustments. The basic terms provide context and clarification on the different elements and attributes characteristic of DOE initiatives.

Factor

A *factor* is the variable controlled by the experimenter and could be viewed as a stimulus: one of the controlled or uncontrolled variables whose influence on a response is studied in an experiment.

Levels

A *level* refers to the settings or possible values of a factor in an experimental design throughout the progress of the experiment. The "levels" of a factor could be quantitative measures (that is, three different temperatures) or qualitative (that is, on or off, high–medium–low, one of four different operators).

Treatment

A *treatment* is a single level assigned to a single factor or experimental unit during the experimental run, for example, pressure at 200 psi. A treatment is also a specific combination of factor levels whose effect is to be compared with other treatments. A *treatment combination* is the set or series of levels for all factors in a given experimental run, for example, pressure–200 psi, temperature–70 °F, feed–high.

Block

A *block* is a portion of the experimental material or environment that is common to itself and distinct from other portions (for example, samples from the same batch). Blocking will be explained later as an experimental design method.

Experimental Design

The *experimental design* or *pattern* is the formal experiment plan that includes the responses, factors, levels, blocks, and treatments, and the use of planned grouping, randomization, and replication. The approach used by Taguchi emphasizes design for robustness to external factors or component variation and minimization of overall variation.

Experimental Error

The variation in the response variable when levels and factors are held constant. Experimental error must be subtracted to determine the true effect of an experiment.

Planned Grouping

Planned grouping is a practice done to promote uniformity within blocks and minimize the effect of unwanted variables. This will make the experiment more effective in determining assignable causes.

Randomization

Randomization organizes the experiment to have treatment combinations done in a chance manner, improving statistical validity.

Replication

Replication repeats observations or measurements to increase precision, reduce measurement errors, and balance unknown factors. It is the repetition of the set of all the treatment combinations to be compared in an experiment. Each of the repetitions is called a *replicate* or a *replication* and is done to increase reliability.

Repetition

It is important to note the difference between replication and repetition in the context of DOE. Replication is a process of running the experimental trials in a random manner. In contrast, *repetition* is a process of running the experimental trials under the same setup of machine parameters. In other words, the variation due to machine setup can't be captured using repetition. Replication requires resetting of each trial condition; therefore, the cost of the experiment and also the time taken to complete the experiment may be increased to some extent.

Variables

Dependent variables are variables dependent on another variable. In simple terms, the independent variable is said to cause an apparent change in, or simply affect, the dependent variable. In analysis, researchers usually want to explain why the dependent variable has a given value. In research, the values of a dependent variable at different settings are usually compared. It is important to remember that the dependent variable does not change unless the independent variable on which it relies also changes.

Independent variables are variables presumed to affect or determine a dependent variable. They can be changed as required, and their values do not represent a problem requiring explanation in an analysis but are simply taken as given. The independent variable in an experiment is most commonly an input and does not have interactions with other variables.

Response variables are variables that show the observed results of an experimental treatment. The response is the outcome of the experiment as a result of controlling the levels, interactions, and number of factors.

Effects

An experimental design uses *effects* to determine whether or not setting a factor at a particular level has a significant impact on the process. This should allow the assignable causes to be traceable and capable of further analysis or action.

Main effects are defined as an estimate of the effect of a factor independent of any other means. The first step in calculating main effects, sometimes called *average main effects,* is to average the results for each level of each factor. This is accomplished by averaging the results of the runs for that level.

Interactions occur when the effect of one input factor on the output depends on the level of another input factor. The preferred DOE approach screens a large number of factors with highly fractional experiments. Once suspected factors have been reduced, interactions are explored or additional levels are examined.

DOE Overview—Planning and Organizing Experiments

History of DOE. Design of experiments is a structured, organized method that is used to determine the relationship between the different factors (*X*'s) affecting a process and the output of that process (*Y*). Sir Ronald A. Fisher, the renowned mathematician and geneticist, first developed this method in the 1920s and 1930s. Fisher started working as a statistician in 1919 at the Rothamsted Experimental Station. This research center in England was conducting a series of experiments to measure the effects of different fertilizers on various crops.

Key Point: The use of DOE in agriculture is one of the primary reasons that the United States maintains such a large lead in what farmers can produce compared to other parts of the world. Also, you may note that some of the terms that we still use in DOE have agricultural connotations.[1]

Purpose. The objective of a designed experiment is to generate knowledge about a product or process. The experiment seeks to find the effect that a set of independent variables has on a set of dependent variables. The independent variables that the experimenter controls are called *control factors* or *signal factors*, or sometimes, just *factors*. The other factors are called *noise factors*.

In the terminology of experimental design, each outcome (observation of the variable of interest, which is the expression level in our case) is measured in correspondence with a set of levels of different factors that may influence it. The experimenter specifically wants to study how the outcome depends on their variation. Some other effects correspond to the inevitable variability in the experimental setting, and the researcher mainly wants to control for their presence in interpreting the results.

Design of experiments involves designing a set of experiments to identify optimal conditions, the factors that most influence the results, and details such as the existence of interactions and synergies between factors. A *data matrix* is a table organizing the data into columns for analysis. The columns of the matrix represent factors, and the rows are the different experiments. Within the individual squares, the levels of each factor are shown, corresponding to the effect of that level.

DOE methods require well-structured data matrices. Analysis of variance (ANOVA) delivers accurate results even when the matrix that is analyzed is quite small. Experimental design is a strategy that can be applied to investigate a phenomenon in order to gain understanding or improve performance.

Building a design means carefully choosing a small number of experiments that are to be performed under controlled conditions. There are four interrelated steps in building a design:

1. Define an objective.
2. Define the variables that will be controlled.
3. Define the variables that will be measured to describe the outcome of the experimental runs (response variables) and examine their precision.
4. Choose the design that is compatible with the objective, number of design variables, and precision of measurements.

Planning Test Programs. Experiments should have carefully defined objectives, separation of the effects of factors, freedom from bias, and precision. Based on the background information, choose the factors and design the experimental program. Define the experiment by number of experimental factors, structure of experimental design, and the kind of information the experiment is primarily intended to provide.[2]

Classification of Experimental Designs

- *Completely randomized* experiments are appropriate when only one factor is analyzed.

- *Factorials* are appropriate when investigating several factors at two or more levels, and interaction is necessary.
- *Blocked factorials* reduce the number of runs and use blocking to run the experiment in subsets.
- *Fractional factorials* reduce the combinations of factors and levels required to be run, while resulting in a close estimate to the full factorial.
- *Randomized blocks* investigate a single factor when material or environment can be blocked.
- *Balanced incomplete blocks* are appropriate when all the treatments cannot be accommodated within a block.
- *Partially balanced incomplete blocks* expand on the balanced incomplete blocks if the experiment requires a larger number of blocks.
- *Latin square* and *Youden square* experiments are used to investigate a primary factor while controlling the interactions and effects of other variables. In Latin square, the number of rows, columns, and treatments must all be equal, and there must be no interactions between the row, the column, and the studied factor. For this reason, the Latin square is not a factorial design, which allows interactions between the several factors comprising the design. While Youden squares also have the same number of columns and treatments, a fairly wide choice in the number of rows is possible.
- *Nested designs* are appropriate when studying relative variability instead of mean effects.
- *Response surface* provides contour diagrams or maps of how the factors influence the response and direction for optimization of variable settings.
- *Mixture designs* are factorial experiments that express factor levels in percentages adding up to 100 percent. With mixtures, the property of interest depends on the proportions of the mixture components and not on the amounts.
- *Independent variable* is a factor that is or could be controlled.
- *Dependent variable* is a factor that is affected by a change in the independent variables.
- *Confounding* exists when there is an external variable that correlates directly with either an independent or dependent variable. This affects the analysis and conclusions obtained from the data as it "confounds" the traceability of the causes.[3]
- *Response variable* refers to the variables influenced by the other factors, for which experimental objectives are created.

Experimental Objectives. When preparing to conduct an experiment, the first consideration is, "What question are we seeking to answer?" Here are examples of experimental objectives.

Find the:

Inspection procedure that provides optimum precision.

Combination of mail and media ads that produces the most sales.

Cake recipe that produces the most consistent taste in the presence of oven temperature variation.

Combination of valve dimensions that produces the most linear output.

Sometimes, the design of experiments objective derives from a question. The objective must be related to the enterprise goals and objectives. The objective must also be measurable, and the measurement system must be reasonably simple and easy to operate. Once the DOE objective and a measurement system have been determined, the factors and levels are selected as the things (factors) the experimenters would change and the various values (levels) they would recommend. From these recommendations, the list of factors and the levels for each factor are determined. A practical guideline for selecting levels is "Be bold but not foolish."

The next step is to choose the appropriate design given affordability and time available. Establish a budget of time and other resources. It is usually best to begin with more modest screening designs whose purpose is to determine the variables and levels that need further study.

DOE Design Principles

Randomization. Randomization is a method of designing and organizing the experiment that attempts to lessen the effects of special cause variation by randomly assigning when the test will run.

Suppose there are eight treatments with five replications per treatment that produces 40 tests. The purpose of randomization is to spread out the variation caused by noise variables. The 40 tests may be randomized in several ways.

Number the tests from 1 to 40 and randomize those numbers to obtain the order in which the tests are performed. This type of numbering is referred to as a *completely randomized design.*

Suppose time of day is a noise factor such that products made before noon are different from those made after noon. With the completely randomized design, each run will likely have parts from both morning and afternoon.

Blocking. Blocking attempts to mitigate the effect of variables that we are trying to eliminate or avoid. Blocking is a method of designing and organizing the experiment that attempts to lessen the effects of special cause variation by grouping the experiments in batches of tests or runs known as a *randomized block* design.

For example, random selection might put runs 1, 4, 5, and 8 during first shift, and 2, 3, 6, and 7 during second shift. Another method that may be used to nullify the impact of the shift change would be to do the first three replicates of each run during the first shift and the remaining two replicates of each run during the second shift.

Replication. This is the repetition of the set of all the treatment combinations to be compared in an experiment in a random order.

Other Principal Items

- Sample size is determined by the type of experiment chosen.
- Interaction occurs when one or more factors have an influence on the behavior of another factor.
- Confounding occurs when a factor interaction cannot be separately determined from a major factor in an experiment.
- Screening experiments can be used to screen out insignificant factors.

Experimentation. A planned experiment (often called design of experiments, or DOE) can be very useful in understanding the variation, through testing and optimizing, of a process. The purpose of running a DOE is to determine better ways of doing things or understand the other factors in the process. There is a basic process that these designed experiments tend to follow.

A planned experiment on your operation can be very useful. During DOE, a list of exact activities will be defined for each trial. Confirmation tests determine if the experiment has successfully identified better process conditions.

The following actions support successful experimentation:

- Map the current process
- Brainstorm the causes of variation
- Use a cause-and-effect diagram to list and categorize sources of variation
- Brainstorm the key factors causing variation
- Determine variation levels usable in the process
- Define the experiment
- Run the experiment by operating the process using the various factors
- Collect samples from each run of the experiment
- Calculate which factors actually affect the process
- Run the process to confirm improvements
- Update operation sheets to show the new parameters of operation

Getting good results from DOE involves a number of steps. It is important to set objectives and select process variables and an experimental design appropriate for the objectives. After executing the design, a quick confirmation can check that the data are consistent with the experimental assumptions. The final stage is to analyze and interpret the results, and present the results for further actions or decisions.

There are important practical steps to consider when conducting design of experiments:

- Check performance of gages and measurement devices
- Keep experiments as simple as possible
- Check that all planned runs are feasible
- Watch for process drifts and shifts
- Avoid unplanned changes
- Allow some time and backup material for unexpected events
- Obtain buy-in from all parties involved
- Maintain effective ownership of each step in the experimental plan
- Preserve all raw data—not just summary averages
- Record everything that happens
- Reset equipment to original state after experiment

Experimental objectives can be summarized under four categories:

- *Comparative.* Conclude if a factor is significant
- *Screening.* Select a few important main effects
- *Response surface.* Find optimal settings and weak points of processes
- *Mixture.* Determine the best proportions of a mixture

Once the objectives are set, the experimentation plan can be developed.

A typical design of experiments checklist includes the following actions:

- Define the objective of the experiment
- Learn facts about the process
- Brainstorm a list of dependent and independent variables
- Debug equipment with test experiments that are not counted toward the statistical calculations
- Assign levels to each independent variable
- Select or develop a DOE plan
- Run experiments in random order and analyze periodically
- Draw conclusions and verify with replication

It is recommended to apply an iterative approach to DOE. Rather than limiting observations to a single experiment, it is common to perform two, three, or more experiments to get the final answer. Since each experiment provides a different answer, the experimenter can logically move through stages of experimentation.

There are additional assumptions that must be in place in order for the experiment to be valid and successful:

a. Measurement system capability
   - Confirm the measurement system before embarking on the experiment.
   - Confirm the measurement system throughout the life of the experiment.

b. Process stability
   - The experiment should start and end at standard process set points under identifiable operating conditions.
   - A baseline needs to be established to reveal whether the process drifted during experimentation.

c. Residuals
   - Residuals are estimates of experimental error (observed response–predicted response).
   - Should be normally and independently distributed with a mean of zero and constant variance.

Application of DOE. The operator will work with engineering, supervisors, and Six Sigma practitioners (those specially trained using Six Sigma problem-solving techniques) to look for variation. The operators know the machines best and will be able to give valuable insights on what might happen if different settings are used on the machines. During the DOE, a list of exact settings will be designed for each trial. There are typically eight or more test runs (trials) made for each DOE, and the operator will be asked to run the machine at the designed settings and take random parts from the machine for testing. Once the sample parts are tested and the measures recorded, the person who designed the DOE will probably use a computer to calculate the optimal setting based on the measurements that came from the samples.

A confirmation test should then be run using the new settings to see if the experiment has identified a better working condition for the machine. If proven out, the new setting should become part of the operation, and the operator will then need to update the documentation in his or her area to reflect the new process settings and parameters.

Another variation on this same theme is the SDCA (standardize, do, check, act) cycle. This is most commonly used once a process has been improved to update control plans and process sheets to lock in the improvements and standardize the changes throughout the organization. SDCA is the outcome of applying sound recommendations based on the correct interpretation of valid effects obtained from proper design of experiments.

Simulation Studies. As simulation software increases the versatility and usability of computerized simulation and experimentation, additional experiments can be conducted to show potential outcomes and interactions. A consistent set of steps should be regularly applied for any DOE initiative, whether directly performed or simulated:

- Specify the problem, questions, business needs, and expected outcomes or answers.
- Prepare a plan that includes the test data, data collection, alternative scenarios, milestones, and timelines.
- Collect the data, identifying and addressing potential or existing gaps.
- Build a model or framework connecting or relating the inputs, process variables, and outputs.
- Design and run the scenarios, distinguishing between fixed and variable parameters, and ensuring the reproducibility of results.
- Analyze and interpret the data, verifying statistical significance for meaningful effects.
- Map outcomes back to the original problems, questions, business needs, and expected outcomes to determine whether the intent of the experiment or simulation was fulfilled.

Key Point: DOE is more suitable for optimizing a process than for solving problems.

DOE GRAPHS AND PLOTS

> Interpret main effects analysis and interaction plots. (Apply)
>
> **Body of Knowledge V.A.2**

Main Effects

Main effects are defined as an estimate of the effect of a factor independent of any other means.

The first step in calculating main effects, sometimes called *average main effects*, is to average the results for each level of each factor. This is accomplished by averaging the results of the runs for that level.

For example, $F_{.01}$ (feed at the .01 in/min level) is calculated by averaging the results of four runs in which feed was set at the .01 level. Results of the four runs are 10, 4, 6, and 2.

Run	F	Response
1	–	10
2	–	4
3	–	6
4	–	2
5	+	7
6	+	6
7	+	6
8	+	3

These were runs 1, 2, 3, and 4, so

$$F_{.01} = (10 + 4 + 6 + 2) \div 4 = 5.5$$

Similarly,

$$F_{.04} = (7 + 6 + 6 + 3) \div 4 = 5.5$$

Runs numbered 1, 2, 5, and 6 had S at 1300 rev/min, so

$$S_{1300} = (10 + 4 + 7 + 6) \div 4 = 6.75$$

and

$$S_{1800} = (6 + 2 + 6 + 3) \div 4 = 4.25$$

$$C_{100} = (10 + 6 + 7 + 6) \div 4 = 7.25$$

$$C_{140} = (4 + 2 + 6 + 3) \div 4 = 3.75$$

The main effects may be graphed as shown in Figure 19.1.

Since the better surface finish (the quality characteristic of interest in this case) has the lowest "score," the team would choose the level of each factor that produces the lowest result. The team would suggest using a speed of 1800 rev/ min and coolant temp of 140 °F. What feed rate should be recommended? Since both $F_{.01}$ and $F_{.04}$ are 5.5, the feed rate doesn't impact surface finish in this range. The team would recommend a feed rate of .04 since it will result in a faster operation.

Factors with the greater difference between the "high" and "low" results are the factors with the greatest impact on the quality characteristic of interest. Most

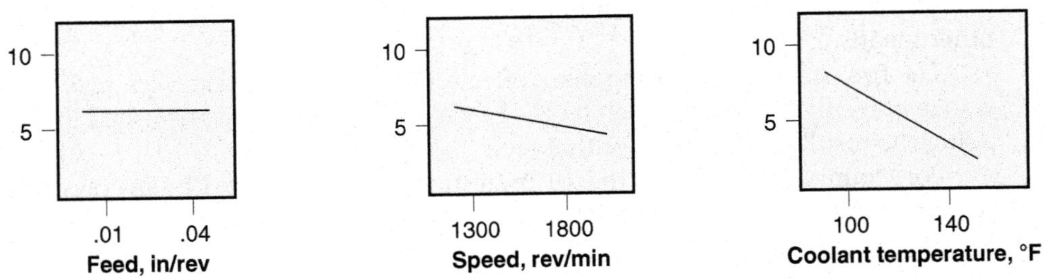

Figure 19.1 Main effects graphs.

authors refer to the main effect as the "high level" result minus the "low level" result for the factor. For example:

$$\text{Main effect of factor } F = F_{.04} - F_{.01} = 5.5 - 5.5 = 0$$

$$\text{Similarly, main effect of } S = S_{1800} - S_{1300} = 4.25 - 6.75 = -2.50$$

$$\text{and } C = C_{140} - C_{100} = 3.75 - 7.25 = -3.50$$

Using this definition of main effect, the larger the absolute value of the main effect, the more influence that factor has on the quality characteristic. It is possible that the perceived difference between "high" and "low" results is not statistically significant. This would occur if the experimental error is so large that it would be impossible to determine whether the difference between the high and low values is due to a real difference in the dependent variable or due to experimental error. This may be determined by using ANOVA procedures.

To clarify, the α-risk (alpha) is the probability that the analysis will show that there is a significant difference when there is not. The β-risk (beta) is the probability that the analysis will show that there is no significant difference when there is. The power of the experiment is defined as $1 - \beta$, so the higher the power of the experiment, the lower the β-risk. In general, a higher number of replications or a larger sample size provides a more precise estimate of experimental error, which in turn reduces the β-risk.

Interaction Effects

Interactions occur when the effect of one input factor on the output depends on the level of another input factor. The preferred DOE approach screens a large number of factors with highly fractional experiments. Once suspected factors have been reduced, interactions are explored or additional levels are examined.

To assess the interaction effects, return to the original experimental design matrix, replacing each high level with "+" and each low level with "–" as shown in Table 19.1.

Table 19.1 A 2^3 full-factorial design using + and – format.

Run	F	S	C	F × S	F × C	S × C	F × S × C
1	–	–	–				
2	–	–	+				
3	–	+	–				
4	–	+	+				
5	+	–	–				
6	+	–	+				
7	+	+	–				
8	+	+	+				

To find an entry in the column labeled "F × S," multiply the entries in the F and S columns, using the multiplication rule "If the signs are the same, the result is positive; otherwise, the result is negative." Fill in the other interaction columns the same way. To fill in the F × S × C column, multiply the F × S column by the C column (see Table 19.2).

The calculation methodology requires the values in the response column to be traced back to either a positive or negative sign.

To calculate the effect of the interaction between factors F and S, first find $F \times S_+$ by averaging the results of the runs that have a "+" in the F × S column:

$$F \times S_+ = (10 + 4 + 6 + 3) \div 4 = 5.75$$

Similarly,

$$F \times S_- = (6 + 2 + 7 + 6) \div 4 = 5.25$$

The calculated effects are compared between the average positive (+) and negative (–) interactions. The difference between the two effects is determined in order to find the effect of the F × S interaction, which is

$$5.75\ (+\ \text{Main effect}) - 5.25\ (-\ \text{Main effect}) = 0.50\ (\text{Overall main effect})$$

Similar calculations show that

$$F \times C = 1.50,\ S \times C = 0,\ \text{and}\ F \times S \times C = -1$$

The presence of interactions indicates that the main effects aren't additive.

The design shown in Table 19.3 uses only four of the eight possible runs; therefore, the experiment itself will consume only half the resources as the one shown in Table 19.2. It still has three factors at two levels each. It is traditional to call this a 2^{3-1} design because it has two levels and three factors, but only 2^{3-1} = 22 = 4 runs. It is also called a *half fraction of the full factorial* because it has half the number of runs as in the 2^3 full factorial design.

Table 19.2 A 2^3 full-factorial design showing interaction columns.

Run	F	S	C	F × S	F × C	S × C	F × S × C	Response
1	–	–	–	+	+	+	–	10
2	–	–	+	+	–	–	+	4
3	–	+	–	–	+	–	+	6
4	–	+	+	–	–	+	–	2
5	+	–	–	–	–	+	+	7
6	+	–	+	–	+	–	–	6
7	+	+	–	+	–	–	–	6
8	+	+	+	+	+	+	+	3

Table 19.3 Half fraction of 2^3 (also called a 2^{3-1} design).

Run #	A	B	C
1	–	–	+
2	–	+	–
3	+	–	–
4	+	+	+

Balanced Designs

An experimental design is called *balanced* when each setting of each factor appears the same number of times with each setting of every other factor.

The logical next question is, "Why use a full factorial design when a fractional design uses a fraction of the resources?" To see the answer, review the following example.

EXAMPLE

Columns have been added to the design in Table 19.3 with two- and three-level interactions using the multiplication rule as shown in Table 19.4.

Table 19.4 Half fraction of 2^3 design with interaction columns filled in.

Run #	A	B	C	A × B	A × C	B × C	A × B × C
1	–	–	+	+	–	–	+
2	–	+	–	–	+	–	+
3	+	–	–	–	–	+	+
4	+	+	+	+	+	+	+

Effects and Confounding

An experimental design uses effects to determine whether or not setting a factor at a particular level has a significant impact on the process. This should allow the assignable causes to be traceable and capable of further analysis or action.

Note that the A × B interaction column has the same configuration as the C column (see Table 19.4). Isn't that scary? This means that when the C main effect is calculated, it is not clear whether the effect is due to factor C or the interaction between A × B or, more likely, a combination of these two causes. Statisticians say that the main effect C is *confounded* with the interaction effect A × B.

Suppose the team has completed a number of full-factorial designs and determined that factors A, B, and C do not interact significantly in the ranges involved. Then there would be no significant confounding and the fractional factorial would be an appropriate design.

Design and Analysis of One-Factor Experiments

One-factor experiments are completely randomized when no tests are omitted and the order is completely random. A randomized block experiment involves taking a factor in each treatment and making exactly one measurement. Results are analyzed and evaluated with ANOVA, and significant values exceed the *F*-statistic derived from the samples.

Examples of one-factor experiments are Latin square and Graeco-Latin squares, which do not incorporate interactions among factors.

EXAMPLE

A process can be run at 180 °F, 200 °F, or 220 °F. Does the temperature significantly affect the moisture content?

Produce four batches at each of the temperatures.

Temperature, °F		
180	200	220
#1	#5	#9
#2	#6	#10
#3	#7	#11
#4	#8	#12

The 12 tests could be completely randomized. A completely randomized design would have a chart like the following to show the testing order, where test #3 is done first, and so on:

Temperature, °F		
180	200	220
#3	#11	#8
#7	#5	#1
#12	#9	#2
#6	#4	#10

If the team decided to produce one batch at each temperature each day for four days, they would randomize the order of the temperatures each day, thus using a randomized block design. The test order chart would then look like the following:

Continued

Continued

	Temperature, °F		
Day	**180**	**200**	**220**
1	#3	#1	#2
2	#1	#3	#2
3	#1	#2	#3
4	#2	#1	#3

The team might decide to block for two noise variables: the day the test was performed and the machine the test was performed on. In this case, a Latin square design could be used. However, these designs require that the number of levels of each of the noise factors is equal to the number of treatments. Since they have decided to test at three temperatures, they must use three days and three machines. This design is shown in Table 19.5.

Assume that the team decides on the completely randomized design and runs the 12 tests with the following results:

Temperature, °F		
180	**200**	**220**
10.8	11.4	14.3
10.4	11.9	12.6
11.2	11.6	13.0
9.9	12.0	14.2

The averages of the three columns are 10.6, 11.7, and 13.5, respectively. A dot plot of these data is shown in Figure 19.2.

Figure 19.3 suggests that an increase in temperature does cause an increase in moisture. How much spread is significant? That question is best answered by using analysis of variance (ANOVA) procedures, where the main effect is compared to the *F*-statistic.

Table 19.5 Latin square design.

Day	Machine #1	Machine #2	Machine #3
1	180	200	220
2	200	220	180
3	220	180	200

Continued

Continued

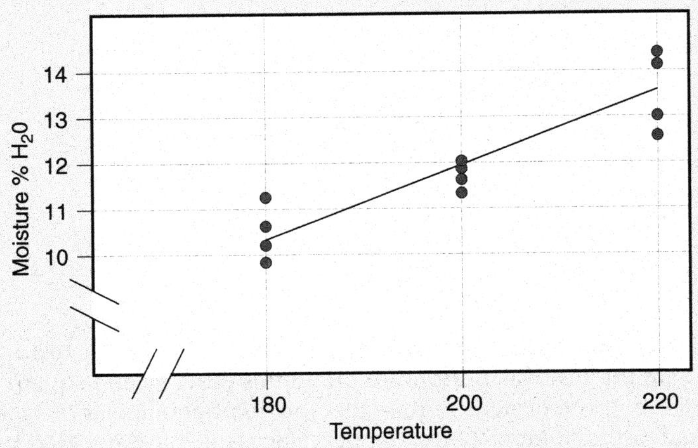

Figure 19.2 Graph of temperature and moisture content.

Design and Analysis of Full Factorial Experiments

Full factorial experiments look at every possible combination in order to complete a full study of interactions. Full factorial results can be retained and converted into different experimental designs.

EXAMPLE

A 2^2 full-factorial completely randomized experiment is conducted, with the results shown in Table 19.6.

The first step is to find the mean response for each run and calculate the interaction column as shown in Table 19.7.

The main effect of factor A is

$$(24.7 + 37.3) \div 2 - (28.4 + 33) \div 2 = 0.3$$

Table 19.6 A 2^2 full factorial completely randomized experiment with results.

Run #	A	B	Response, y			$\bar{y}$
1	–	–	28.3	28.6	28.2	28.4
2	–	+	33.5	32.7	32.9	33.0
3	+	–	24.6	24.6	24.8	24.7
4	+	+	37.2	37.6	37.0	37.3

Continued

Continued

Table 19.7 A 2^2 full factorial completely randomized experiment with main and interaction effects.

Run #	A	B	A x B	Response, y			$\bar{y}$
1	–	–	+	28.3	28.6	28.2	28.4
2	–	+	–	33.5	32.7	32.9	33.0
3	+	–	–	24.6	24.6	24.8	24.7
4	+	+	+	37.2	37.6	37.0	37.3

The main effect of factor B is

$$(33.0 + 37.3) \div 2 - (28.4 + 24.7) \div 2 = 8.6$$

The interaction effect A × B is

$$(28.4 + 37.3) \div 2 - (33.0 + 24.7) \div 2 = 4.0$$

The next issue is whether these effects are statistically significant or merely the result of experimental error. The larger the effect, the more likely that the effect is significant. The definitive answer to the question can be found by conducting a two-way ANOVA on the data. The calculations are quite cumbersome, so software packages are often employed.

Using a hypothesis testing model, the null hypothesis would be that the source of variation is not statistically significant. In this case, the p-values for the sources of variation are small enough to reject the null hypothesis and declare these factors—B and the A × B interaction—to be statistically significant at the significance level. What the ANOVA test really does is compare the between-treatment variation with the within-treatment variation.

Full versus fractional factorial must be understood to ensure that the experiment is properly designed. Full factorial is an experimental design that contains all levels of all factors. No possible treatments are omitted. Fractional factorial is a balanced experimental design that contains fewer than all combinations of all levels of all factors. The following examples will show how the same experiment can be displayed as either a full factorial or a fractional factorial.

FULL FACTORIAL EXAMPLE

Suppose you are cooking steak for 100 people, and the current approval rating is 75 percent acceptable. You want to know the effect of different cooking methods and approaches to see how the overall approval or "yield" is affected.

Continued

Continued

Assuming:

Cooking method: (+) is grill and (–) is fry

Meat: (+) is sirloin and (–) is rib eye

Marinade: (+) is red wine and rosemary and (–) is soya sauce and garlic

Experiment	Cooking method	Meat	Marinade	% approval
1	–	–	–	66
2	+	–	–	88
3	–	+	–	58
4	+	+	–	84
5	–	–	+	67
6	+	–	+	91
7	–	+	+	63
8	+	+	+	84
			Average	75.125

Next, we find the effect of setting the different factors to their respective levels:

Cooking method: [(88 + 84 + 91 + 84) – (66 + 58 + 67 + 63)]/4 = 23.25

Meat: [(58 + 84 + 63 + 84) – (66 + 88 + 67 + 91)]/4 = –5.75

Marinade: [(67 + 91 + 63 + 84) – (66 + 88 + 58 + 84)]/4 = 2.25

When the steak is grilled instead of fried, we gain 23 percent approval. The interaction effects can be checked with interaction columns.

Exp.	Cook	Meat	Mar.	Cook × Meat	Meat × Mar.	Cook × Mar.	Cook × Meat × Mar.	% approval
1	–	–	–	+	+	+	–	66
2	+	–	–	–	+	–	+	88
3	–	+	–	–	–	+	+	58
4	+	+	–	+	–	–	–	84
5	–	–	+	+	–	–	+	67
6	+	–	+	–	–	+	–	91
7	–	+	+	–	+	–	–	63
8	+	+	+	+	+	+	+	84

Continued

Continued

Cook × Meat interactions: [(66 + 84 + 67 + 84) – (88 + 58 + 91 + 63)]/4 = 1.25

Meat × Mar. interactions: [(66 + 88 + 63 + 84) – (58 + 84 + 67 + 91)]/4 = 0.25

Cook × Mar. interactions: [(66 + 58 + 91 + 84) – (88 + 84 + 67 + 63)]/4 = – 0.75

Cook × Meat × Mar. interactions: [(88 + 58 + 67 + 84) – (66 + 84 + 91 + 63)]/4 = –1.75

In this example, the interactions have minimal effects on approval or yield. The conclusion is that the cooking method (grilling the steak or frying) has greater effect on approval than the other considerations. The selection of the meat and marinade do not indicate that one choice is significantly superior to another.

To summarize, a full factorial experiment assesses all of the factors and levels, and ensures that main effects and interaction effects of an experiment are accurately revealed.

Design and Analysis of Two-Level Fractional Factorial Experiments

Fractional factorial experiments save time and money by not examining every possible combination. This method is used for quick exploratory tests, when interactions are insignificant, and many tests are needed rapidly.

In contrast, full factorial experiments require a large number of runs, especially if several factors or several levels are involved. Recall that the formula for the number of runs in a full factorial experiment is

$$\text{Number of runs} = L^F$$

where

L = Number of levels

F = Number of factors

If runs are replicated, the number of tests will be multiples of these values. Because of the extensive resource requirements of full factorial experiments, fractional factorial experimental designs were developed. The example below shows a full factorial experiment run with 50 percent of the runs in a balanced design so that the factors could be shown in their different levels equally.

FRACTIONAL FACTORIAL EXAMPLE

This design can identify a similar outcome with fewer experiments.

Experiment	Cooking method	Meat	Marinade	% approval
2	+	–	–	88
3	–	+	–	58
5	–	–	+	67
8	+	+	+	84
			Average	76.67

Cooking method: [(88 + 84) – (58 + 67)]/2 = 23.5

Meat: [(58 + 84) – (88 + 67)]/2 = – 6.5

Marinade: [(67 + 84) – (88 + 58)]/2 = 2.5

Although the results are not the same as a full factorial, similar conclusions can be derived.

Since we already know from the full factorial that the interaction effects are not significant, we can draw valid conclusions from the main effects of a fractional factorial. However, when the interaction effects are present, so are the risks of confounding.

Two-Level Fractional Factorial Experiment Procedure

The procedure can be summarized into three basic stages:

1. Select a process; identify the output factors of concern and the input factors and levels to be investigated
2. Select a design, conduct the experiment under predetermined conditions, and analyze the data
3. Analyze the data and draw conclusions

EXAMPLE

Step 1: Select a process; identify outputs and input factors.

a. Development and delivery of computer software to a customer.

b. Success is based on deployment without critical issues requiring high-priority resolutions.

c. Study the effect of seven variables at two levels (can be both variable-quantitative or attribute-qualitative).

Input factors	Level 1 (–)	Level 2 (+)
A. Requirements	20 high-level requirements	Defined 400+ business, functional, technical, and user requirements
B. Risk analysis	Minimal checklist	Formal
C. Architecture	Agile	Structured
D. Design and coding	Prototyping	Staged delivery
E. System integration	Freeware and shareware	Prequalified components
F. Product testing	Exploratory and random checks	Unit–integration–system–user acceptance tests
G. Quality assurance	Final inspection only	Series of five progressive go/no-go gates requiring approval

EXAMPLE

Step 2: Experiment.

Assuming that no interactions are evaluated or considered, the seven factors can be evaluated at two levels. *Pass* refers to the condition of having no critical or major issues obtained after product release, and *score* refers to a level of customer satisfaction with the software.

	Input factors							Outputs (results)	
Test	A	B	C	D	E	F	G	Pass	Score
1	+	+	–	+	–	–	+	Yes +	85
2	+	+	–	–	+	+	–	No –	57
3	+	–	+	+	–	+	–	No –	82
4	+	–	+	–	+	–	+	Yes +	100
5	–	+	+	+	+	–	–	No –	69
6	–	+	+	–	–	+	+	Yes +	87
7	–	–	–	+	+	+	+	No –	72
8	–	–	–	–	–	–	–	No –	44
							Average		74.5

EXAMPLE

Step 3: Analyze the data and draw conclusions.

	Input factors							Outputs (results)	
Test	A	B	C	D	E	F	G	Pass	Score
1	+85	+85	−85	+85	−85	−85	+85	Yes +	85
2	+57	+57	−57	−57	+57	+57	−57	No −	57
3	+82	−82	+82	+82	−82	+82	−82	No −	82
4	+100	−100	+100	−100	+100	−100	+100	Yes +	100
5	−69	+69	+69	+69	+69	−69	−69	No −	69
6	−87	+87	+87	−87	−87	+87	+87	Yes +	87
7	−72	−72	−72	+72	+72	+72	+72	No −	72
8	−44	−44	−44	−44	−44	−44	−44	No −	44
							Average		74.5

Find the differences and divide by 4 to find the effect.

Difference	52	0	80	20	0	0	92
Effect	13	0	20	5	0	0	23

For brevity, focus on C and G, which are the largest contributors.

C (architecture): structured architecture will improve satisfaction by 20%

G (quality assurance): quality assurance measures will improve satisfaction by 23%

Two-Level Fractional Factorial Experiment Conclusions

ANOVA is the appropriate statistical method for assessing the significance of the experimental effects. The ANOVA outcomes summarized in Table 19.8 indicate that at certain levels of confidence, changing the levels of two of the factors significantly affects the process outcomes.

Based on the *F*-value for significance, only factors G and C are significant effects influencing the overall process. The *F*-value is compared to the critical value found in the tables. If the test statistic for F is greater than the critical value, then the null hypothesis can be rejected at the alpha level of significance. In this example, the *F*-values for factors G and C exceeded the critical value, as shown in Table 19.8.

Through these examples, fractional factorial experiments have been shown to be a cost-effective alternative to full factorial experiments. They can also be specifically designed in their own right as a way to quickly interpret the main effects of an experiment.

Table 19.8 ANOVA outcomes.

Factor	Effect	Sum of squares	*F*-value significance
G	23	66.1	85.29 < 85 95% confidence
C	20	50	65.42 < 57 90% confidence
A	13	21.2	27.22 < 57 Not important
D	5	3.1	Error term value

Testing Main Effects and Interactions

The statistical significance of main effects and interactions needs to be tested. For any estimated main effect or interaction, if the absolute value of an estimated effect is greater than the standard error, the experimenter can conclude that the difference is statistically significant.

Design of Experiments Considerations

Along with statistical practices, some strategic approaches are needed to ensure a successful DOE initiative. Recommended steps include:

- Work with subject matter experts to determine which factors should be included in the experiment.
- Evaluate the process history to specify the levels for each factor.
- Determine representative factors and levels that reflect actual events and capabilities of the process.
- Use historical records to review existing data and determine how much additional data need to be collected in order to obtain statistical confidence.
- Review the possible treatment combinations to assess whether a condensed fractional factorial design would be adequate or if a full factorial design is required.

Taguchi recognized that quality is primarily determined at the design stage and should be sufficiently robust to withstand expected fluctuations in production or delivery. Without adequate design controls, the effort to ensure that workers and equipment remain within specifications will be much greater.[4]

Chapter 20

Implementation Planning

> Apply implementation planning by using proof of concepts, try-storming simulations, and pilot tests. (Apply)
>
> **Body of Knowledge V.B**

Implementation planning is a critical concept that enables the Improvement Phase for a Six Sigma project. These terms will be summarized below, with elaborations and additional examples throughout the chapter:

- *Proof of Concepts.* This technique is an early-stage design validation technique that confirms the feasibility of the initiative and presents it to the participants and stakeholders for the purpose of eliciting feedback and detailed inputs.
- *Try-storming simulations.* This extends traditional brainstorming by incorporating a simulated or prototype solution. The purpose is exploratory in pursuit of incrementally building the concept toward completion and implementation.
- *Pilot Tests.* This refers to the tests that are performed in the latter stages of the project. The purpose is to determine the readiness of the developed project for broader implementation, and to identify any outstanding critical areas requiring resolution, prior to continuous and regular use.

PROOF OF CONCEPT: IDEATION TOWARD FEASIBILITY

A proof of concept is used to demonstrate feasibility to show:

- Suitability of the proposed solution to address the problem or opportunity statements.
- Potential fulfillment of the needs analysis, within business costs and constraints.

- Success criteria can be fulfilled, to the standards and expectations of key stakeholders.
- Foundational justification and rationale for acquiring resources, estimating schedules, and securing commitments.

After determining this foundation, the ideation can be manifested into more specific and detailed solutions and completion items.

TRY-STORMING: RAPID ACTIONS FROM BRAINSTORMING

In this activity, the benefits from the quality practice of brainstorming are immediately applied. The most feasible ideas from the brainstorming sessions are quickly simulated or developed for demonstration. This demonstration is exploratory, which enables participants to apply their own incremental creative contributions.

Try-storming is also referenced as building prototypes or conducting role-playing exercises to simulate the process revisions. It is actually a very effective tool or technique for training Six Sigma participants. Try-storming is also used when information technology is applied to adjust or modify how a company receives, processes, and manages information. The user interfaces of internal websites can be rapidly adjusted to elicit and create more permanent solutions.

PILOT TESTS: READINESS FOR DEPLOYMENT

The logic behind pilot tests is that prior to the completion, there are opportunities to "sample" the solution. If this is a process change, the implementation could be tested during off-peak periods. The reduced scope and tight control of implementation activities performs several purposes:

- Verifies the effectiveness of the Six Sigma solution.
- Validates the implementation plan.
- Establishes that the participants and stakeholders can deploy the change.
- Confirms that the Six Sigma solution, as properly implemented, will provide the desired results.
- Identifies any unintended risks or side effects from the implementation.

For example, consider a Six Sigma project that speeds up a subscription renewal process by having the customer's credit card automatically renew, which helps with customer retention, but the removal of a personal contact eliminates an opportunity to upgrade or upsell the customer for greater revenue opportunities. By resolving the problem of missed payments, the project introduces a new problem of growth-oriented customer engagement touchpoints.

IMPLEMENTATION PLANNING WITH INTEGRATED APPROACHES

The timing and sequence of proof of concept, try-storming, and pilot tests are shown in a serial manner, but they can also be concurrent and interdependent. Ideally, the proof of concept should represent a project milestone or "gate," enabling broader commitments to the Six Sigma project. The gate of try-storming operates as a filter to quickly vet and discard approaches that are not feasible for further development.

The gate of pilot tests is based on control of scope and scale. If the pilot test reveals substantial risks and deficiencies, the measures taken to control and roll back changes are relatively within bounds. If the pilot test gate does not restrict problematic solutions, then the benefits of the Six Sigma project will be offset by the new risks, defects, and maladies introduced from the suboptimal implementation approach.

By incorporating these items as sequential gates, the overall progress of the Six Sigma implementation can be tracked as ongoing metrics. The efficiency of a project can be measured by time and resources between implementation steps. For example, if weeks and months go by between proof of concept and try-storming, then that would suggest a slippage or even abandonment of the project. In contrast, a rapid progression to pilot tests without adequate try-storming suggests that ideas could be lacking.

Chapter 21

Lean Tools

WASTE ELIMINATION

> Select and apply tools and techniques for eliminating or preventing waste, including pull systems, kanban, 5S, standard work, and poka-yoke. (Apply)
>
> Body of Knowledge V.C.1

Relevance of Lean to Six Sigma

All organizations must improve continuously on all fronts: quality, cost, delivery, customer satisfaction, and predictability. The objective is to achieve and sustain performance improvements. A holistic approach that blends the benefits of lean and Six Sigma will more effectively address:

- Key measures of performance, and identification of performance gaps
- Multiple aspects of process management, including design, improvement, and control
- Improvement across multiple functional areas beyond manufacturing to include service, transactional, and administrative processes
- Incorporating financial goals, budgets, and reviews within the improvement framework

As part of project selection, the appropriate approach should be selected to reflect organizational goals and performance gaps. As a prerequisite, value stream mapping should be applied to uncover the problems. Depending on the simplicity or complexity of the items revealed, the Six Sigma Green Belt (SSGB) can generate potential savings from the application of lean and Six Sigma techniques (see Figure 21.1).

As the SSGB gains proficiency in lean practices, the convergence of lean approaches and Six Sigma methods will more effectively deliver the intended benefits to cost, quality, delivery, customer satisfaction, and predictability (see Figure 21.2). By integrating lean within the Six Sigma system, the overall improvements

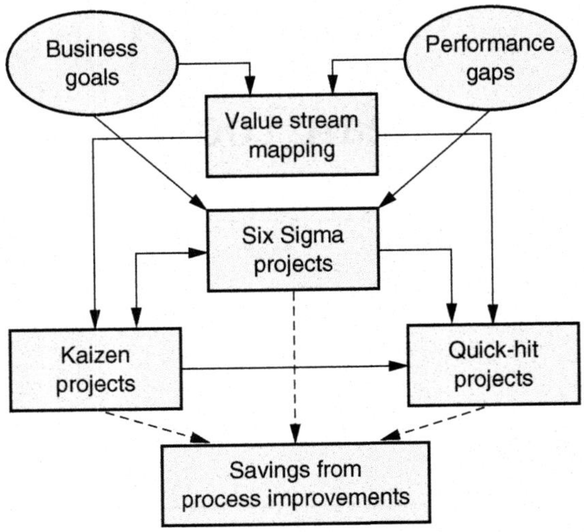

Figure 21.1 Project framework.

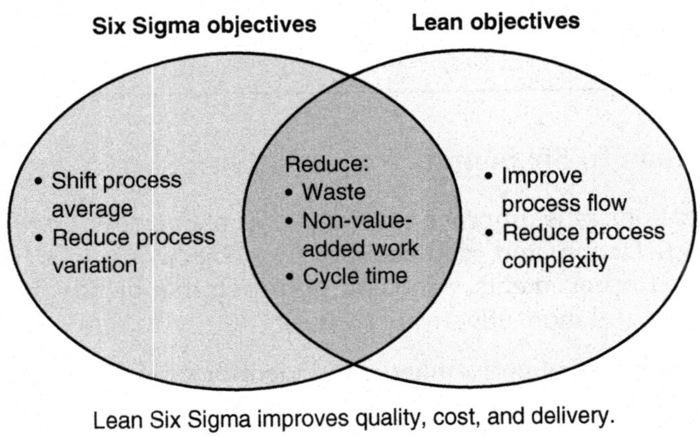

Figure 21.2 Lean Six Sigma objectives.

will be sustainable and continually reinforced by the combined methods and practices.[1]

Comparison of Lean to Traditional Systems

The benefits of lean and lean tools can be best appreciated when contrasted with the approach and characteristics of traditional systems. The following aspects of lean can be expected to transform an organization:

- *Product/service design.* Lean emphasizes standardization and incremental changes to yield the benefits of continuity.

- *Capacity.* Lean increases flexibility and overall utilization to reduce the extreme spikes associated with shifts in demand.
- *System layout.* Lean emphasizes the work flow and incorporates multiple work cells as opposed to having large work areas.
- *Workforce.* Lean demands and promotes team-based cooperation of empowered practitioners who are cross-trained with diverse skills. Leadership and change are derived by consensus and collaboration.
- *Scheduling.* Lean reduces setup times and increases concurrent production of different models or solutions due to standardization and improved practices.
- *Inventories.* Lean reduces the quantity of *work-in-process* (WIP) inventory, which reduces the need for large, segregated inventory areas, reducing carrying costs.
- *Suppliers.* Lean improves the partnership mentality and interdependency throughout the supply chain, reducing the quantity of vendors. Deliveries are made directly to the point of application rather than to a central dispatch area for further transportation.
- *Operations.* By operating with simple visual indicators, taking a continuous and preventive approach to quality and maintenance, and operating at the logical pace of progress instead of an artificially accelerated tempo, lean makes operations management more anticipatory and responsive to the demands and conditions of the organization.

Examples of Lean Tools

One of the principal problems in any organization is the control of its processes. In addition to the techniques of *statistical process control* (SPC), the Green Belt may employ tools grouped under the name "lean thinking."

It is important to specify value in terms of each specific product:

- Identify the value stream for each product
- Make value flow without interruption or delay
- Let the customer pull value through the process steps
- Pursue perfection and prevent rework

Lean is the application of tools for the removal of waste and variation in a process, which ultimately allows for outcomes that will be delivered more efficiently and be closer to the specifications and expectations set by the customer.[2] Lean tools for control include:

- *5S*, which stands for sort (*seiri*), straighten (*seiton*), shine (*seiso*), standardize (*seiketsu*), and sustain (*shitsuke*), is used to create order and organization.

- *Visual factory* applies visual displays and visual controls to allow anyone to promptly know the status of the process and interpret whether the process is operating properly.
- *Kaizen* pursues low-cost gradual improvement of processes either on an ongoing basis or as a dedicated endeavor known as a *kaizen blitz.*
- *Kanban* is a signal for a particular action including "obtain inventory," "produce part," or "move material from the upstream point to the downstream process." Kanban also ensures that no defective pieces are allowed to pass or be transferred.
- *Poka-yoke* is mistake-proofing by design and can include sizing devices, limit switches, color signals, and alarms.
- *Total productive maintenance* manages operations with preventive maintenance, load balancing, and streamlined flow control.
- *Standard work* applies capacity resource planning to determine the most efficient combinations of operations.
- *Pull systems* refers to production scheduling techniques where materials are staged at the point of consumption or application, and restocking is limited to only the amount that has been consumed by the process, avoiding the accumulation of excess inventory.

Pull Systems

Pull systems manage scheduling and material flow so that overproduction and excess inventories are eliminated. Unlike traditional systems where products are stocked in inventory and "pushed" through the various stages of completion to customer delivery, pull systems are driven by customer demand, and incorporate process metrics (for example, setup time, takt time) to determine proper timing and quantities. Pull systems operate consistently with *Little's law,* which defines a relationship between lead time, throughput, and work-in-process (WIP). By reducing WIP inventory levels throughout the enterprise, pull systems reduce lead times, resulting in more rapid delivery and satisfaction of customer demand.

Pull systems are enhanced and improved by the discovery and application of particular efficiencies including:

- Simplification
- Focus on minimizing time for setup and changeover
- Reduction of steps and handoffs
- Elimination of unnecessary loops or rework
- Managing constraints and "bottlenecks"
- Positioning inventory for more rapid deployment and application
- Reducing rework or redundancy

Mistake-Proofing (Error-Proofing or Poka-Yoke)

The goal of *poka-yoke* is to augment a work activity in a manner that prevents errors from being committed in the completion of a process step. Through this design or equipment change, the process step can only be completed if the material is placed or inserted in the correct position, or the working steps are fully and properly completed. Poka-yoke activities, by devising methods that make erroneous events impossible, further enable process control by automatically eliminating another source of variation. For example, a newly assigned press operator stacked finished parts on a pallet with the incorrect orientation. The next operator didn't notice the difference, resulting in several hundred spoiled products. A fixture for the pallet now makes it impossible to stack misoriented parts.

Mistake-proofing (poka-yoke) is achieved by limiting or restricting ways to complete a task to ensure accuracy and compliance (for example, electric plug design intended to ensure correct connection).

Poka-yoke tries either to limit or restrict ways to perform any given task in a process or alternatively to reveal the mistake visually. These efforts help to reduce variation in the process and help prevent nonconformance issues from occurring downstream in the process or when a customer uses the product or service. Have you noticed that when you try to plug an electrical appliance into an electrical outlet that the plug can only go in one way? This is true for both two- and three-prong plugs! This is an example of mistake-proofing that allows the electronics industry to help ensure proper usage of appliances in a home (see Figure 21.3).

Consider whether potential mistakes are possible (for example, multiple ways to insert a part, process steps could be done in a different order than designed). Addressing these potential deviations will improve consistency and help reduce the variation in the process. Also, review any references describing the processes, the process sequence, or the work area.

The originator of this method is a Japanese engineer named Shigeo Shingo, whose legacy is commemorated with the Shingo Prize, which honors excellence among lean practitioners.

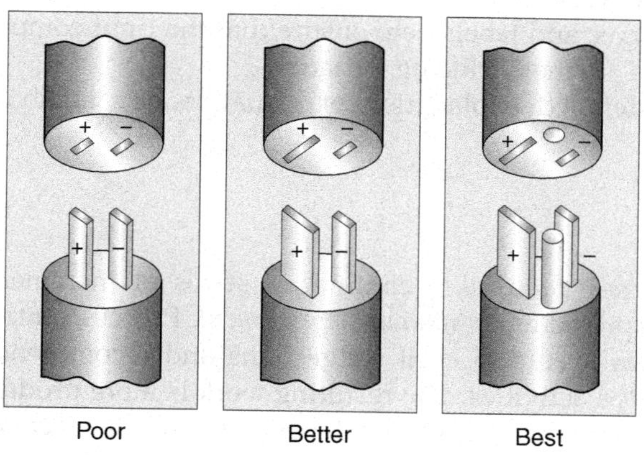

Figure 21.3 Poka-yoke example.

5S

Consider the cleanliness, lighting, and general housekeeping status of any area where measurements are conducted, since process control data are filtered through the measurement system. Applying the 5S steps (that is, sort [*seiri*], straighten [*seiton*], shine [*seiso*], standardize [*seiketsu*], and sustain [*shitsuke*]) will reduce problems arising from poor housekeeping or organization:

- *Sort* distinguishes the work that needs to be done and eliminates the unnecessary steps.
- *Straighten*, or *set in order*, follows the maxim, "A place for everything and everything in its place."
- *Shine*, or *scrub*, is to clean and remove the clutter from the workplace.
- *Standardize*, or *systemize*, refers to efforts to develop and implement standardized procedures. This will make the work environment more orderly.
- *Sustain* reinforces the prior four steps with repeated and continuous performance.

Visual Factory

The visual factory strives to make problems visible, notify employees of current operating conditions, and communicate process goals. Charts placed prominently in the workplace show trends displaying quality, delivery, downtime, productivity, and other measures. Production and schedule boards advise employees on current conditions. Similar steps can be made in non-factory environments (for example, services, education) to continuously keep all participants informed.

Accessible and clearly illustrated work instructions are critical to avoid negligence and deviations. This is especially true in situations where cross-trained personnel flex into various workstations, and mixed-model schedules are employed. Good lines, signs, and labels help ensure that the right component is at the right place and time, further reducing variation.

A more detailed explanation of visual factory attributes is provided in Chapter 23.

Kaizen

The lean journey requires a continuous series of incremental improvements. *Kaizen* is interpreted as *continuous improvement*. Process control can be enhanced through kaizen events to elicit suggestions and recommendations that reduce non-value-added activities. The resulting work is more productive and permits a closer process focus by all involved. Kaizen and its related concepts are elaborated in a specific section later in this chapter.

Kanban

A properly administered kanban system improves system control by assuring timely movement of products and information, ensuring material and information flow into and out of the process in a smooth and rational manner. If process inputs arrive before they are needed, unnecessary confusion, inventory, and costs generally occur. If process outputs are not synchronized with downstream processes, the results often are delays, disappointed customers, and associated costs.

Total Productive Maintenance

The effective use of equipment and infrastructure is essential to waste reduction. Managing these assets is the composite activity of the following initiatives:

- Avoiding reduced, idled, or stopped performance due to equipment breakdown.
- Reducing and minimizing the time spent on setup and changeover of equipment, which can otherwise idle machine operations and create bottlenecks.
- Avoiding stoppages arising from the processing or discovery of unacceptable products or services.
- Ensuring that processes and equipment are operating at the speed and pace for which they were designed. If the pace is slower or delayed, work to address and rectify the source of the delays.
- Increase the yield of acceptable material to reduce material waste, scrap, rework, and the need for material reviews.

Total productive maintenance (TPM) improves the maintenance practices for equipment and infrastructure and enables the prediction and/or prevention of anticipated failure. Through a coordinated effort integrating engineering, operations, and maintenance, a portion of the tasks can be shifted to the operating team members, who can perform maintenance as part of their ongoing process activities.

TPM aims to remove deficiencies from machines to minimize or eliminate defects and downtime. This extends beyond preventive maintenance to include management of people, processes, systems, and the environment. In any situation where mechanical devices are used, the working state of those devices has an impact on the control of the process. If equipment deteriorates even subtly, the process output may be affected, often in unsuspected ways.

Details on the specific attributes of *total productive maintenance* (TPM) are found in Chapter 23.

Standard Work

Standard work is a term describing the systematization of how a part is processed, and includes man–machine interactions and studies of human motion. Operations are safely carried out with all tasks organized in the best-known sequence and by using the most effective combination of resources. Finding better ways of producing an ever more consistent product is the essence of process control. Standard work contributes to this effort by assuring that product flowing into a process has minimal variation and that there is a reduction in the variation caused by the process.[3]

Key Point: There are many tools and techniques that can be used to help improve a given process or situation. Stay open-minded, take initial readings or measurements of the process, and use whatever is available to try to improve and monitor the process. The simple fact that you are focusing on an issue will go a long way toward highlighting to everyone involved that management is actually watching this area. Things will sometimes magically start improving when people are focused on the issues.

CYCLE-TIME REDUCTION

> Use various techniques to reduce cycle time [continuous flow, setup reduction, single minute exchange of dies (SMED)]. (Analyze)
>
> **Body of Knowledge V.C.2**

Standard Work

Standard work is a precise description of each work activity, specifying cycle time, takt time, the work sequence of specific tasks, and the minimum inventory of parts on hand needed to conduct the activity. All jobs are organized around human motion to create an efficient sequence without waste. The purpose of standard work is to provide a baseline for improvement. All processes possess waste. Standard work helps define the process, which is the steps and actions to reach a defined goal.

Standard work establishes the baseline for performing tasks and activities to obtain a desired result, linking to product and process quality, cost, schedule, execution, cycle time, and customer satisfaction. Improvements in quality and cycle time and reductions in waste (for example, elimination of excess movement, scrap, and rework) can be demonstrated relative to the baseline.

Standard work is established through analysis, observation, and employee involvement. Employees are involved because the people closest to the work understand it best. It should be noted that employees may be resistant to the establishment of standard work. Use of quality and management tools helps to assess the potential barriers through analysis and/or team involvement in the following sequence:

1. Communicate the change.
2. Identify the benefits.

3. Identify the barriers.
4. Identify how to capitalize on the benefits.
5. Identify the methods to mitigate the barriers.
6. Get buy-in/agreement to steps 4 and 5.
7. Implement the change.
8. Communicate the results.
9. Celebrate success.

Benefits of standard work include the following:

- Process stability—stability means repeatability
- Clear process definition
- Organizational learning
- Employee involvement
- Supporting baselines to enable lean practices (for example, poka-yoke, kaizen)

Three charts are utilized to create standard work: production capacity chart, standardized work combination table, and standardized work analysis chart.

Production Capacity Chart. The production capacity chart is used to determine the capacity of machines/humans in a process. Its purpose is to identify bottlenecks within the process. It is based on the capacity calculation:

- Capacity
- Process time
- Interval
- Setup time
- Operational time per shift

Standardized Work Combination Table. The standardized work combination table shows the work elements and their sequence. Each element is broken down into individual times, including operator and machine. The table may include the interactions between the operator and the machine or other operators and machines.

This table is commonly used to analyze the value-added times versus the non-value-added times during each process step. This provides another visual representation for assessing the process and identifying areas for improvement. It often highlights idle time of an operator that can be filled with another value-added work element.

Standardized Work Analysis Chart. The standardized work analysis chart can be used as rationalization of a process layout. It provides a visual aid to an employee in training. The chart should include the work/cell layout, process steps, and times. It is also a good chart for highlighting quality and safety items and defining standardized work-in-process.

By engaging the employees, communicating the benefits of standard work, understanding and addressing employee concerns, and involving the right people, success is easier to obtain.

Process Design

Processes in a lean environment are never developed in isolation; they require active involvement from suppliers and representatives from all affected areas of the organization. The focus throughout the design process is to meet or exceed customer needs without adding activities or costs that do not add value.

SIPOC

Suppliers-inputs-process-outputs-customers (SIPOC) charting offers process development based on understanding the customer's needs and studying how those needs are met. SIPOC highlights the need to serve internal customers in order to satisfy external customers. The SIPOC process identifies who receives or is affected by the outputs of the process. SIPOC aids in clarifying the process, including its specific requirements. The next step involves defining the inputs related to customer requirements, to be communicated to suppliers, both internal and external to the organization. By driving the process from the customer requirements, there is traceability linking the processes developed to the customer requirements fulfilled. For this reason, the customer requirements should be incorporated into the SIPOC scope.[4]

After analyzing and understanding the suppliers, inputs, outputs, and customers (including the requirements at each stage), the process is developed, identifying the steps needed to convert the inputs to outputs in an efficient and customer-focused manner. SIPOC, however, simplifies the process by providing a structured approach to increase knowledge of the company's overall system, showing the interrelationships and interdependencies between processes. Through the structured SIPOC method used to understand who serves whom, a SIPOC chart can greatly improve the design of processes.

Process Stability

Stable processes have predictable costs, quality, and cycle times. Alternatively, without stability, outcomes cannot be accurately determined, and planning becomes very difficult.

To achieve stability in a process:

- *Method.* Process steps are clearly defined and consistently followed.
- *People.* Minimum levels of training and qualification.
- *Machinery.* Designed and maintained in a consistent manner.

- *Measurement.* Equipment is properly used, maintained, and regularly calibrated.
- *Materials.* Consistent quality is achieved.

Process design helps to minimize the variation caused by each of these elements and achieve stability quickly. When the process is stable, however, this expected variation becomes predictable within limits determined by the process average and standard deviation. A predictable level of performance is a prerequisite to achieving a sustained level of quality, cost, and delivery.

Process Capability

Process capability is a measure that describes how well a process can meet a given target. In simple terms, process capability represents a comparison of the *voice of the customer* (VOC) with the *voice of the process* (VOP). The higher the result, the more capable the process is of meeting customer needs.

Process Capacity

A *process capacity sheet* is a common lean tool used to calculate and understand the capacity of the equipment in a particular work area based on the following example inputs:

- Machine time per piece
- Manual time per piece (operator time of activity performed while the machine is not producing product)
- Total cycle time per piece (calculated as machine time per piece + manual time per piece)
- Number of pieces before a tool change is needed
- Time required to perform a tool change
- Tool change time per piece (calculated as time to perform tool change / number of pieces before a tool change is needed)
- Available time per shift (reported in units similar to the other time-related measures)

Using this information, process capacity is calculated as:

$$\text{Capacity} = \frac{\text{Available time per shift}}{\text{Total cycle time per piece} + \text{Tool change time per piece}}$$

A particular process step that is not able to produce to its theoretical capacity identifies where problem-solving efforts should be focused. An additional benefit of understanding process capacity is related to takt time. If the capacity calculation shows that the process is not able to keep up with its takt time, something will need to change or customer satisfaction will suffer.

Other Factors

There are several other factors to keep in mind when applying lean thinking to the development or improvement of processes, including streamlined layout, one-piece flow, and quality at the source. Streamlining the layout of a process to improve the flow of materials and information is a critical focus of lean thinking. Lean thinking requires an intimate knowledge of the company's processes, systems, and customers in order to make the trade-off decisions necessary to serve customers well. Although some decisions can make an individual process operate more efficiently in isolation, lean thinking helps us understand whether they benefit the overall system as well.

Work Flow Analysis

In the context of lean, a *system* is anything that has interacting, interrelated, or interdependent processes. A *process* is a sequence of linked activities, or tasks, with the purpose of producing a product or performing a service for a customer (internal or external). However, since a system is a structure of repeatable processes, a system can be associated with flow—in other words, how well processes are linked together to achieve an expected outcome (product or service).

Work flow is the ability, or rate, at which work is processed through a system. It can be stated as the measure of how well the processes within a system are linked and work together. *Work flow analysis*, then, is the evaluation of flow in a system and how the efficiency and effectiveness of the flow can be improved.[5]

This section reviews four ways to evaluate work flow within an organization:

1. Flowcharts
2. Flow analysis charts
3. Value stream mapping
4. Takt time analysis

Flowcharting. A *flowchart* is a visual representation of a sequentially ordered set of tasks. A flowchart can be constructed from a list of tasks by writing each task on a small piece of paper and organizing them sequentially. The resulting collection of sequentially ordered activities is the foundation of a flowchart. A flowchart can include practically anything that is relevant to the process, for example, various process inputs and outputs, decision points, personnel, times, and metrics. The versatility of the flowchart allows for its use on any process.

Flow Analysis Charts. There are five distinct process elements that characterize flow:

1. *Processing.* Physical change in the material or the quality of a work item.
2. *Inspection.* Comparison to a standard.

3. *Transportation.* Movement of a work item between locations.

Delay is a period of time in which no processing, inspection, or transportation occurs. Shingo defined two delay elements:

4. *Process delay.* Delay of an entire production lot while the previous lot is processed, inspected, or moved (the three previous process elements).

5. *Lot delay.* The lot delay element is used to represent the situation where one piece is processed while the other pieces of the lot wait to be inspected or transported.

These five elements can be documented in a visual map of the process. A typical *flow analysis chart,* or *analytical process chart,* employs the sequential arrangement of these five elements that communicate the work flow through the process.

Value Stream Mapping. Value stream mapping illustrates sequential activities that take place within a value stream, showing where value is being created and where potential wastes exist. Value stream maps identify activities that add value and those that create waste. Value stream maps permit users to visualize:

- Flow of products, services, and information
- Waste and its sources
- Context for all levels and functions

This creates the foundation to support:

- Fact-based decisions about the flow of the process
- An improvement strategy and implementation plan for the value stream
- Relationships between information and product flows

Current State. The value stream map is a big picture of the flow of a given product/ service throughout the entire organization. Before beginning to map the process, it is important to plan how the process will be mapped. Begin at the shipping end of the value stream, closest to the customer. This helps keep a customer focus throughout the entire value stream. Obtain all information directly from the value stream through observations, time study, inventory counting, and so on. Draw the value stream manually during the mapping exercise. Gather pertinent information, which might include:

- Cycle time
- Changeover time
- Number of people
- Uptime/downtime
- Work-in-process
- Inventory

- Packaging size
- Scrap/defect rate
- Total operating time (minus breaks)
- Value-added time versus non-value-added time
- Lead time
- Number of changeovers

Future State. The intent is to brainstorm ideas for improving the current-state value stream and then reconstruct the value stream with the improvements in place (that is, the future-state value stream map).

Establish and Implement a Plan. Create an action plan (that is, countermeasures) to reach the future-state value stream goal. As a rule of thumb, a new current-state value stream map should be created when about 80% of the improvements (countermeasures) have been implemented.

Value stream mapping is simply a tool to visually show the current flow of product/service and information in the organization and to guide everyone in the organization through the analysis of the process to improve the flows and design improved value streams.

Takt Time Analysis. *Takt* is the German word for "pace" or "rhythm," like the beat an orchestra conductor uses to regulate the speed of the musicians' playing. *Takt time,* often referred to by lean practitioners as the heartbeat of the process, is a measure of customer demand. This can be defined as the time between the completion of process units (for example, items painted), the time calculated for each process step or position, or the pace of demand.

Takt time is calculated as

$$\text{Takt time} = \frac{\text{Available work time}}{\text{Customer demand}} \text{ (over a given period of time)}$$

If work time is measured in capacity metrics (for example, minutes/day) and demand is given in quantity metrics (for example, units/day), the takt time will measure the flow for each item.

Example: An 8-hour day has 480 minutes. A furniture factory has a painting process that requires 160 pieces to be painted in a day to meet customer demand. The average takt time is 3 minutes per unit. Therefore, the flow rate per step should be less than or equal to 3 minutes per process step in order to avoid a backlog.

Awareness of the takt time by the operators will help to retain and preserve the pace by identifying any delays or potential backlogs prior to process completion.

The primary purpose of takt time is to serve as a management tool to indicate, at a glance, whether the value stream or process is meeting customer demand at any given time. It also serves as a tool to align upstream processes and downstream processes in a value stream with the customer demand requirements.

The intent of takt time analysis is to match the value stream's process time (cycle time) with takt time (or customer demand) as closely as possible. Takt times of individual products, services, and processes can be compared to identify issues with production leveling, load leveling, and various forms of waste.

Line balancing is used to ensure continuous work flow through the process. As lean grew beyond manufacturing, the use of the line balancing was applied to service processes. Although line balancing can be performed independently, combining it with other lean tools, like takt time analysis, process flow, and value stream mapping, provides the best benefit. Since total time includes waste (*muda*) like waiting, as noted earlier, the partnering of load leveling with other lean tools eliminates waste within the overall process and within each individual process step.

The ideal state for a level loaded process is one in which all processes are as close to takt time as possible, with little or no waste time, and stacked in a sequence where the downstream process is slightly faster than the process preceding it. Line balancing analysis helps improve work flow. Line balancing analysis is performed through observation, motion studies, or process flow analysis. While a powerful tool on its own, load leveling can be combined with other lean tools to have a greater impact on reducing waste.

Countermeasure Activities

Countermeasures are the actions taken to reduce or eliminate the root causes of problems that are preventing you from reaching your goals. Countermeasures are typically identified by team members for the purpose of eliminating waste (*muda*), unevenness (*mura*), and/or unreasonableness (*muri*).[6]

Here are some examples of countermeasures that support lean objectives:

- *Mistake- and error-proofing (poka-yoke).* Mistake-proofing is the use of process design features to facilitate correct actions, prevent simple errors, or mitigate the negative impact of errors. *Poka-yoke* is Japanese slang for "mistake-proofing," a term coined by Shigeo Shingo.
- *Quick changeover/setup reduction.* Quick changeover or setup reduction is a process for changing over production equipment from one part to another in as little time as possible. The overall goal is to reduce the time devoted to setup/changeover so that more time is available to produce a wider variety of products.
- *One-piece flow.* One-piece flow refers to the concept of moving one workpiece at a time between operations within a work cell. The goals of one-piece flow are to make one part at a time, to make that part correctly all the time, and to produce without unplanned interruptions and lengthy queue times.
- *Right-sized equipment.* Right-sized equipment is highly capable, easy to maintain (therefore available to produce whenever needed), quick to change over, easy to move, and designed to be installed in small increments of capacity to facilitate capital and labor linearity.

- *Cellular flow.* Cellular flow manufacturing is the linking of manual and machine operations into the most efficient combination of resources to maximize value-added content while minimizing waste. The most efficient combination implies the concept of line balancing. When processes are balanced, the product flows continuously, parts movement is minimized, wait time between operations is reduced, inventory is reduced, and productivity increases.
- *Sensible automation.* Sensible automation, or *jidoka,* as it is referred to in lean, gives equipment the ability to distinguish good parts from bad parts autonomously, without being monitored by an operator. An essential component of jidoka is the stoppage of work to fix problems as they occur. This leads to improvements in the process that build in quality by eliminating the root causes of defects.
- *Material signals (kanban). Kanban* is a Japanese word that means "sign" or "signboard." At its core, kanban is a replenishment system that orders new material only after consumption of existing material.
- *Source inspection.* Source inspection is a quality inspection that a buyer performs at a vendor's location before material is received or shipped. Source inspection means the work is reviewed before it is passed on to the next station. It is used to identify, correct, and contain a problem before it enters the value stream.

Quick Changeover/Setup Reduction

Quick changeover, also referred to as *setup reduction* or *single minute exchange of die* (SMED), is a technique that can significantly reduce the time taken to switch from producing one product to producing another. A closely related concept is *one-touch exchange of die* (OTED).

The concept of quick changeover/SMED is important because long changeovers create bottlenecks, drive up lot sizes, and reduce the flexibility to provide a broad selection of products to customers. Traditionally, companies adhering to *economic order quantity* (EOQ) formulas endorsed making large amounts to amortize the cost of long setups. Quick changeover/SMED mitigates the hidden costs of producing and transporting large amounts of inventory.[7]

Setup time is the key metric for quick changeover/SMED. *Setup time* is the total elapsed time from the *last* piece of one product made from one process to the first *good* piece of the next product produced by another process. The overall goal of quick changeover/SMED is to reduce the time devoted to setup/changeover so that more time is available to set up and produce a wider variety of products/services.

After documenting and standardizing the setup method, train workers in the new method and hold them accountable for following it. Be on the lookout for additional ways to cut setup time and perform problem solving whenever the standard cannot be followed, costs are rising, quality problems are exposed, or customer conditions change.

A key goal for setups and changeovers is to reduce or eliminate all the activities that require the process to be stopped. Often, first changeover improvements are simple workplace organization efforts. 5S is a great place to start quick changeover/SMED, making sure all the tools, machines, materials, supplies, setup sheets, and information needed are accessible and well organized. Other early SMED improvement efforts focus on identifying steps that can be done while the prior job is still running, such as getting materials and documentation and presetting tooling.

Another key to reducing internal setup time is standardization. The following are some things to look into when focusing on improving internal setup times:

- Eliminate adjustments
- Standardize tools
- Standardize tool positions
- Standardize programs (for example, machine programming such as for CNCs, injection molding)
- Use setup teams

Once a standard for quick changeover/SMED is established, it should be documented like any other standardized work. This will spell out the conditions necessary for success and help ensure that time, quality, safety, and cost expectations are consistent for changeovers. By continuing to apply the PDCA process to changeover activities, operators experienced in setups and skilled in problem identification and problem solving will be able to point to more ways to reduce setups.

Cycle Time Reduction Through Constraints Management

The *theory of constraints* (TOC) approach popularized by Goldratt requires the identification of bottlenecks in the process and balancing of work flows within the system. This analysis tracks the flow of materials and products through the different transition points within the process.[8] Since the overall pace of the process is only as fast as the bottleneck, efforts should be directed toward reducing the impact or effect of that bottleneck with the following activities:

- Review the process to reduce the demands on that process step where it is discretionary, prioritizing it for essential items first
- Increase the resources and equipment capacity or speed
- Improve the skill level of the resources at the bottleneck
- Automate or incorporate technology solutions to reduce waiting times
- Simplify the process steps at the bottleneck
- Transfer some material preparation or adjustment to earlier in the process

In the same way that the critical path of a project defines the duration, the bottlenecks will affect the process duration such that a delay at the bottleneck will absorb any slack time and extend the duration of the process. For this reason, the overall inventories at each process point should be regulated and limited to the quantity that can be processed at the identified bottlenecks. This applies within both manufacturing and service scenarios, where a service bottleneck could be, for example, the applications desk at a university registration office.

The impact of a *capacity constrained resource* (CCR) can be significant if not properly managed. The CCR is a limiting factor that may already be or is at risk of becoming a bottleneck. If the capacity of this resource type is neglected, the throughput of the organization can be compromised. Unlike a bottleneck, a CCR can affect processes even if the operations are not running at full capacity. CCR examples include personnel, equipment, suppliers, restrictive policies, and other impediments to optimal work flow.

KAIZEN AND KAIZEN BLITZ

> Define and distinguish between these two methods and apply them in various situations. (Apply)
>
> **Body of Knowledge V.C.3**

Incremental Continuous Improvement (Kaizen)

Kaizen is a Japanese word synonymous with "continuous improvement." Kaizen thinking promotes small, easy, low-risk, and low-cost process improvements. Quick and simple experimentation also means the velocity with which ideas are generated, tested, and implemented will increase.[9]

Kaizen permits quick progression through the idea generation, testing, and evaluation phases. This results in an organization populated with people experiencing many, many learning cycles on a daily or weekly basis. This contrasts with the traditional approach, which misses the benefits of the small incremental improvements made over time that are immediately realized with kaizen. Through aggregated productivity improvement, large long-term projects are replaced by smaller, more frequent activities that can be rapidly implemented and the benefits immediately realized.

Kaizen Blitz

A kaizen approach depends on a continuous series of incremental improvements, the span of which can be measured in years. To accelerate the pace and reduce the duration, a hybrid activity known as a *kaizen blitz* can be organized. This can incorporate workshops and events and involves a cross-functional team completing a continuous project within a week. The prerequisite is that all participants receive training in kaizen and lean activities. The training is immediately reinforced with practical actions by completing a continuous improvement project, collecting and analyzing the data, and immediately implementing the recommended improvements. The targets can be similar to those of a kaizen effort, except the pace and timing are considerably more aggressive.

Breakthrough Continuous Improvement (Kaikaku)

Kaikaku, translated as "innovation," complements kaizen by emphasizing major redesigns of product, part manufacturing, and facility layout or business processes. Since kaizen improvements are small, incremental modifications to an established process, kaizen can be described as *evolutionary*. Due to the major rethinking related to kaikaku changes, kaikaku can be described as *revolutionary*. Kaikaku commonly refers to a focused kaizen action designed to address a particular issue over the course of a week, referred to as a *kaizen blitz* or *kaizen event*.

Without kaizen, kaikaku improvements will deteriorate over time, as no one is responsible for constantly looking for and resolving small problems, and process management and improvement are left to those who redesigned the process. With both kaizen and kaikaku, organizations will optimize their continuous improvement efforts.

Kaikaku is typically initiated by management since the scope of the change and the anticipated result will impact the overall business. Kaikaku can be focused on introducing a new knowledge set, new marketing or production strategies, new approaches, new production techniques, or new equipment. Kaikaku may be triggered by external factors, such as changing market conditions, new technology, or competitors' actions. Kaikaku projects often result in improvements at a higher level and can also provide a new base level for continued kaizen.

Gemba is defined as the real place where work happens. For effectiveness, the kaizen effort is conducted at the gemba by those operators and individuals who know it best. As a tool for carrying out *gemba kaizen*, team members use a checklist of economy of motion that includes such points as the following:

1. Eliminate unnecessary movement:
   a. Movement involved in looking for or selecting something
   b. Need for judgments and extra attention
   c. Transferring the workpiece from one hand to the other
2. Reduce eye movement:
   a. Confirm by listening instead of looking
   b. Use lamps
   c. Place items within operator's field of vision
   d. Use different coloring
   e. Use transparent containers

3. Combine operations:
    a. Process while carrying workpiece
    b. Inspect while carrying workpiece
4. Improve the workplace:
    a. Place materials and tools in a given area in front of the operator
    b. Place materials and tools in the same sequence as the work
5. Improve tools, jigs, machines:
    a. Use containers that make it easier for picking parts
    b. Combine two or more tools into one
    c. Replace levers and handles with a button for single-motion machine operation

The gemba kaizen guidelines include the aims of the project, the schedule, and the major activities, covering the following:

1. Set the target
2. Select leaders
3. Check the operating line
4. Confirm the inventory
5. Explain the purpose of the project
6. Prepare tools
7. Select kaizen plans
8. Instruct the operators
9. Prepare standards
10. Prepare the summary report

There are a few keys to successful gemba kaizen workshops:

- Set challenging but well-defined targets
- Form cross-functional teams to solve problems
- Take action with speed, on the spot, at the gemba
- Invest time in preparation, communication, and planning the workshop
- Make learning and skill transfer a kaizen objective

Since 85 percent of total costs, as well as the conditions for quality and delivery, are determined in the design planning stages, improvement in upstream management is the key to achieving successful quality, cost, and delivery. Gemba kaizen is but a starting point for bringing kaizen to upstream processes such as design, planning, and marketing.

Organizing for Kaizen

There are several levels of improvement for kaizen activities, of different scope and depth:

- *Individual (point kaizen).* At the individual workstation level, there are always opportunities to reduce waste—workplace organization, inventory and tool location, work sequence, ergonomics, poka-yoke, and on and on.
- *Work teams (mini kaizen).* Work teams or groups undertake improvement projects affecting their collective work area. Examples include work flows, cell layout, line balancing, 5S, and quality improvements.
- *Flow kaizen.* Flow kaizen teams typically work across a full value stream, led by a project manager, often assisted by a champion, and sometimes mentored by consultants. The team comprises multidisciplinary and cross-functional members. Flow kaizen projects usually address process issues, system issues, and organizational issues. Flow kaizen is about value stream improvement and getting flow going and is performed on the material and information flow of an overall value stream.[10]
- *Process kaizen.* Process kaizen is about the elimination of waste, with focus on a specific process or subprocess. Kaizen events focus on internal processes and on solving the customer's problems or improving the customer's effectiveness.
- *Supply chain kaizen.* Supply chain project teams involve participating companies within the value stream in focusing on optimizing the supply chain improvements. These teams usually have a project manager, typically from the *original equipment manufacturer* (OEM) company, and are supported by champions and consultants.

Kaizen Events

While the term *kaizen blitz* refers to a specific team approach to quickly tear down and rebuild a process to function more efficiently, the broader *"kaizen"* term is commonly used for all types of continuous improvement.

Kaizen events have a dual role—to make improvements and to teach and communicate. The kaizen principles from a lean perspective comprise the following:

- Define value as perceived by the customer
- Identify the value stream
- Eliminate waste

- Create flow
- Establish pull where flow is not possible
- Pursue perfection

Kaizen principles, kaizen thinking, or a kaizen mindset, consists of the following elements:

- Kaizen is everyone's job
- "Go to the gemba," observe, and document reality
- Emphasis on problem awareness
- Use of problem-solving tools
- Bias for action
- Standardization once improvement has been achieved
- Focus on improving both process and results
- Applies to any aspect of the work
- Continual improvement to achieve higher standards by involving and engaging the workforce

For a kaizen event, following are typical attributes contributing to kaizen success:

- *Charter.* A charter establishes the framework, determines the problem statement, relevant background information, time frame, team members, resources involved, and how the improvement will be measured.
- *Identification of critical success factors.* The elements that are critical to the process need to be identified, defined, and measured. This allows for verification of the effectiveness of action items and countermeasures.
- *Scope.* The amount tackled within the kaizen event is important for driving efforts to completion.
- *Link kaizen event to business plan (organizational strategy).* Align kaizen efforts to meet an organization's goals and get the best use of allocated resources.
- *Team selection.* People's capabilities and skill sets, individuals' expertise or knowledge, and subject matter experts should be considered.
- *Follow-through.* Good follow-through is necessary to ensure that the improvements that are made are maintained and prevent backsliding.
- *Presenting results.* Results promote learning throughout the organization, as another area with a similar situation could benefit from benchmarking examples and knowledge transfer.

- *Visibility.* Those who are not directly involved need to be aware of the improvements the team is making. If they are made aware, they will be more inclined to support the improvement.
- *Management commitment.* Management must support and actively participate in kaizen initiatives. Management must ensure that the team has everything it needs to be successful and recognize the accomplishments of the team.[11]

The following lean concepts have been included within this chapter: value stream mapping, kaizen, total productive maintenance (TPM), standard work, cycle time reduction, and quick changeover.

Part VI

Control Phase

Part VI is an overview of the Six Sigma *control* phase, emphasizing those practices that are intended to sustain and entrench the improvements made in the prior phases.

The BoK was slightly reorganized for Part VI, and now includes additional information on lean with respect to its application to process control. The statistical process control and control plan content from the previous editions were retained, along with examples and templates that can be applied in the workplace.

Chapter 22

Statistical Process Control (SPC)

SPC BASICS

> Describe the theory and objectives of SPC, including measuring and monitoring process performance for both continuous and discrete data. Define and distinguish between common and special cause variation and how these conditions can be deduced from control chart analysis. (Analyze)
>
> **Body of Knowledge VI.A.1**

Statistical Process Control

The basis of the *control* portion of a Six Sigma program is the correct application and interpretation of control charts, which is generally categorized as *statistical process control* (SPC).

The purpose of SPC is not specifically to create arrays and files of control charts for their own sake, but to apply this information constructively toward building and adapting processes, documenting procedures, and maintaining gains realized from successful improvement initiatives. Through these methods and techniques, the necessary process controls can be entrenched as a new baseline on which future improvements can be made.

Key Definitions

control process—A feedback loop through which we measure actual performance, compare it with a standard, and act on the difference.

statistical process control (SPC)—The application of statistical techniques for measuring and analyzing (and hopefully reducing) the variation in processes.

statistical quality control (SQC)—The application of statistical techniques for measuring and improving the quality of processes.

statistical process display (SPD)—The misapplication of statistical techniques to pretend that information is being used to run a process, when in reality it is only for show (usually for a customer) instead of an actual application to improve a process.

Supplemental Definitions

These are added to support the overall understanding of SPC, and reinforce the key points. These are elaborated in Chapter 17.

assignable cause—The observed change is caused by the behavior of a major variable in the process, or reflective of a new major variable.

chance or unassignable cause—The observed change is caused by the interplay of multiple minor variables in the process.

hypothesis—This is an assertion concerning a numerical property of the population, based on a tested sample. The statistical tests can either reject the hypothesis (that is, a significant increase in defects for a particular process is explained within statistically expected variances) or accept the hypothesis (there has been a significant change to the process resulting in higher defect levels).

type 1 errors—Rejection of the hypothesis when it is true; reflected by alpha (α).

type 2 errors—Acceptance of the hypothesis when it is false; reflected by beta (β).

SPC Theory

SPC originated in the 1920s, when Dr. Walter Shewhart of Bell Telephone Laboratories discovered that variation in manufacturing could be attributed to inherent (random) variation and intermittent (assignable) variation. SPC has evolved since that time to become a core competency among inspectors, quality specialists, and Six Sigma practitioners.

Tactics

- $\bar{X}$ and R charts are the most common type of control chart, and are calculated for each subgroup (generally four to 10 samples) and plotted in order of production on separate charts.
- After an assignable cause of variation is discovered and removed, new control limits calculated from 25 new subgroup averages and ranges often give a substantially narrower process capability, becoming the economic limit to improvement.
- In 1953, Rath and Strong developed pre-control for IBM, which is a simple algorithm. The tolerance band is divided into a target zone bounded by two cautionary zones. A pair of individual samples is measured periodically, and if both fall within the cautionary zone or either falls outside tolerance, the process is adjusted immediately. Pre-control requires no calculations or charting.

- Factorial experiments and multi-vari analysis are other innovations in diagnostic techniques.

Objectives and Benefits

The goal of any quality activity is to meet the needs of the customer. Statistical process control consists of a set of tools and activities that contribute to this goal through the following objectives:

- Monitoring processes in real time
- Identifying whether processes are operating as expected
- Identifying whether processes have changed and corrective action may be required
- Making statistically valid decisions
- Centering the process
- Determining when and when not to take action on the process
- Determining the type of action to take (that is, actions to eliminate special causes, actions to improve the overall process)
- Quantifying and reducing variation
- Improving understanding of products and processes
- Improving product and process design
- Monitoring continual improvement and confirming that changes were effective

The SPC tools achieve these objectives by collecting and analyzing data.

Statistical Process Control: Errors

- Controlling process performance involves sampling coordinated activity and modifying the process behavior.
- Control charts are used to separate assignable causes from random variation.
- Type I errors occur when a behavior treated as a special cause has no effect in the process.
- Type II errors occur when special causes affecting the process are not addressed.
- Tampering or impulsively modifying the process to reduce variation will actually contribute to increasing variation.

To use SPC effectively, create a data collection plan and collect all the data possible. Historical data may be available, but used with caution. One approach is to place the collected data on a histogram and calculate the mean and standard deviation.

The histogram provides a visual picture of the variation and center of the process, while the mean and standard deviation provide numerical values for comparison.

Process improvement is not limited to the factory floor or manufacturing facility. Any process (machinery, office, sports team, household, and so on) can be monitored using basic SPC techniques. Once control of the process is established, you can then make changes to see if these either alter the variation (range) or move the target (average).

Process Capability: Special versus Common Causes

In the 1920s, Shewhart developed control charts to distinguish between assignable variation—characteristic of systems out of control—and chance variations within controlled systems. The assignable variation that causes the process to go out of control should be detectable with the appropriate control chart.

Every process has variation. Process improvement requires reducing the amount of variation that is currently present. Variation can be physical or mechanical (that is, tool, machine, maintenance, equipment, environment) or procedural (operator, accuracy, legibility, workload). The specification limits should reflect the voice of the customer. The process variation reflects the voice of the process. This is expanded in Chapter 15.

Process variation has two main categories: special and common. Variation must be traceable to its sources, making it necessary to distinguish between common and special causes.

The *common causes* of variation are those that are inherent to the process and generally are not controllable by process operators. Common cause variation is also known as *natural variation* and refers to the many sources of variation within a process. Common causes reside in processes within statistical control, and can be characterized by location (process average), spread (piece-to-piece variability), and shape (distribution) for predictability.

Special causes of variation include unusual events that the operator, when properly alerted, can usually remove or adjust. Special causes are sometimes called *assignable causes.* Unless all the special causes of variation are identified and mitigated, the process output will be unpredictably influenced, with random results.

The principal purpose of control charts is to recognize the presence of special causes so that appropriate action can be taken. While both special and common causes can be detected with statistical techniques, common causes are more difficult to isolate and remove. A process is considered to be in statistical control when only common causes remain after special causes have been removed.

Tactics

A principal problem is the separation of special and common causes. If you adjust a process in response to common cause variation, the result is usually more variation rather than less. This is sometimes called *overadjustment* or *overcontrol.* If you fail to respond to the presence of a special cause of variation, this cause is likely to produce additional process variation. This is referred to as *underadjustment* or *undercontrol.*

Consider trying this test: Using a stopwatch, start the timer as you turn on the ignition of your primary means of transportation going to work or school. Turn off the timer once you arrive at your designation. Record your times for a period of time and identify what is causing the variation in times. What are the common causes and what are the special causes?[1]

Special Cause Examples

Figures 22.1 through 22.3 represent the effects of special causes on a process, showing the disruptions to average and variation when these are not stabilized.

Common Cause Example

Figure 22.4 represents the effects of common causes on a process, which reflect movement within stable process behavior.

Process Data: Discrete versus Continuous

In quality, there are generally two types of inspection. The first type (type 1) refers to a variable that can be measured or scaled. The second type (type 2) determines whether a characteristic exists and refers to attributes that can include the choices of being acceptable or defective, timely or late, or other characteristics that can be tabulated without measurement.

The appropriate SPC method must be selected based on whether the data are discrete or continuous. Control charts for variables data require measurement on a continuous scale (for example, length, weight, resistance), while charts for attributes data address discrete "either/or" outcomes.

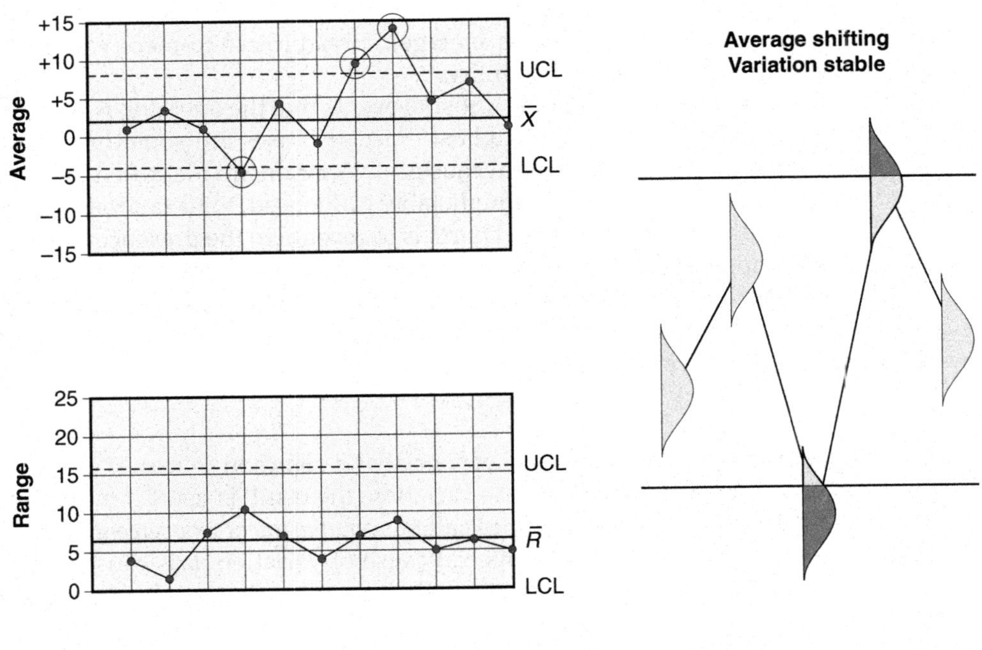

Figure 22.1 Average shifting, variation stable.

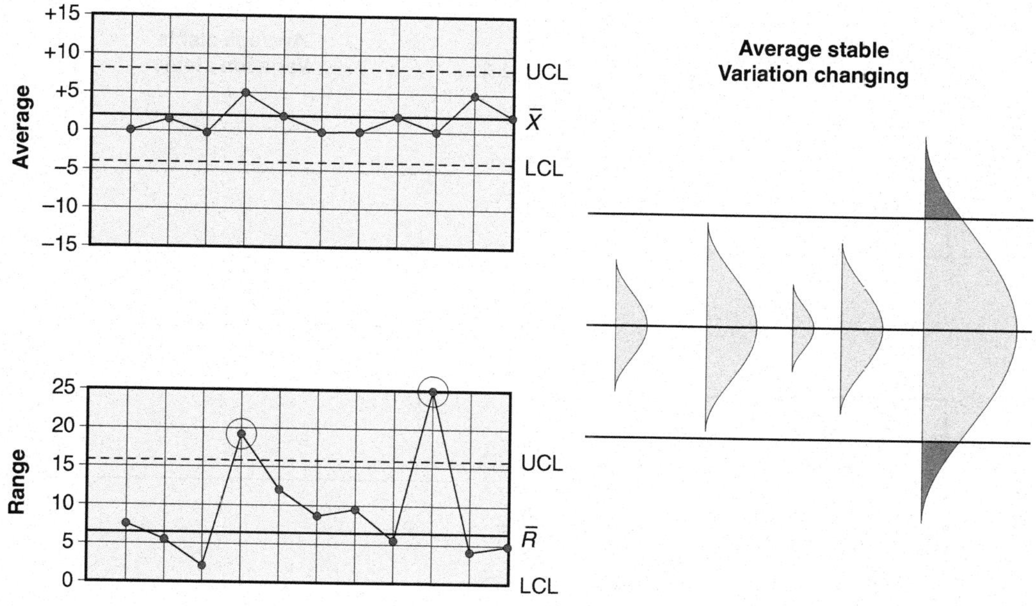

Figure 22.2 Average stable, variation changing.

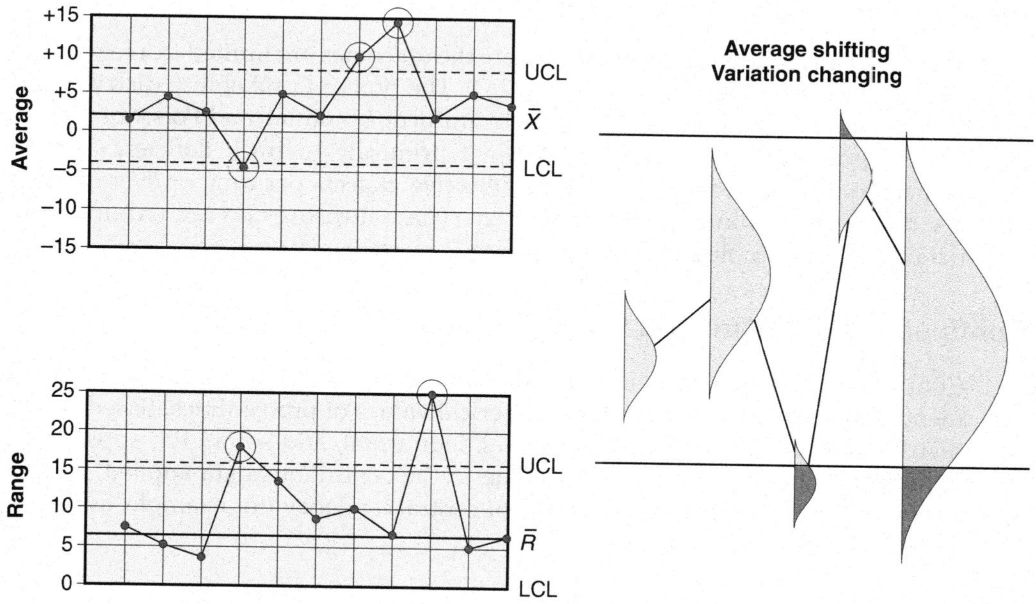

Figure 22.3 Average shifting, variation changing.

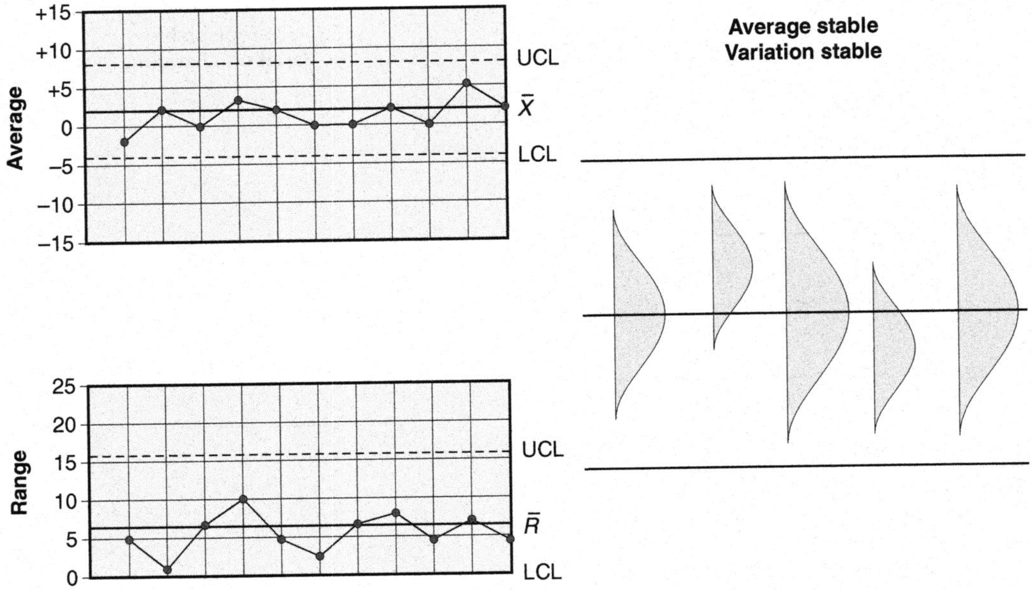

Figure 22.4 Average stable, variation stable.

Discrete Data (Attributes Control Charts)

Discrete data are used in situations where the outcomes are limited to a few values (for example, pass/fail, true/false; 0, 1, 2, 3). The types of probability distributions generated from this kind of data include uniform, binomial, and Poisson.

Control charts have been established primarily to track deficiencies with respect to units defective, percentage defective, defects per unit, or overall number of defects within a given sample size. These measures do not require input details beyond simple indicators as assigned by the operator.

Continuous Data (Variables Control Charts)

Continuous data are used in situations where a continuous range of measurements can be tracked, resulting in numerical values of probabilities. Based on the distribution of the data (such as normal, lognormal, and so on) the appropriate distribution model can be selected. Due to the continuous data applied, control charts can incorporate measurements of central tendency (for example, averages) or variation (for example, range).

Process Behavior Charts (Control Charts)

Key Point: The idea behind changing the name to *process behavior charts* is to emphasize that we want to study the process in question, not the people! How is the process behaving, and is there something that we should or can do about it?

Walter Shewhart, known as the "father of modern-day quality control," laid the foundations for process behavior charts (i.e., control charts) in the late 1920s. Today, there are over 30 different charts that can be used; however, we typically only use six or seven on a regular basis. These charts display the process variation while work is being done, which allows the operator to ensure that the process is stable and continues to operate within the process boundaries (not necessarily specification limits) that have been established for that process. If something does start to deteriorate or change in the process, the process behavior chart will give the operator an early warning indicator that something needs to be adjusted or changed to bring the process back into control.[2]

Process Behavior Charts (Statistical Process Control)

- Typically, only six or seven types are used on a regular basis
- Create a picture of the process variation while work is being done
- Ensure that the process is stable and continuing to operate within the process boundaries established for that process
- Provide an early warning to adjust or change the process to bring it back into control
- Variables data are continuous and come from scales or measures
- Attributes data are discrete and come from indicators
- Ensure that the measurements from the process are recorded, calculated, and plotted appropriately
- Refer to an upper control limit or lower control limit value from a table
- Being in statistical control refers to being between upper and lower control limits

Control charts are used to attain a state of statistical control, monitor a process, and determine process capability. Control charts can monitor process stability and achieve parts per million defect levels. Reduction of variation is achieved through other techniques, referencing control charts.

A state of *statistical control* means that only random causes are present in the process. It does not mean that products meet specifications. Conversely, a process not in statistical control may still produce product conforming to specifications:

- Control charts attain statistical control using 25 subgroups, a log of process changes during data collection, trial control limits computed from the data, charting the data for each subgroup, eliminating assignable causes of excessive variation, and continuing the control measures.
- The mean and standard deviation are estimated based on the sample data. Averages are more sensitive to change than individual readings.
- Control limits based on the statistical variation of the process can be established at ±3 standard deviations from the mean.

The *operating characteristic* (OC) curve is a plot of the true value of a process parameter against the probability that a single sample will fall within the control limits (see Figure 22.5). It shows the ability of the chart to detect process changes. The OC curve plots the probability of accepting the original hypothesis as reflecting the true value of the population. This highlights the risk of type II (β) error in the sample. This is applicable for one or two-tail tests.

Setting Up Control Charts

- Choose the characteristic to be charted based on what is defective and controllable or adjustable by the worker.

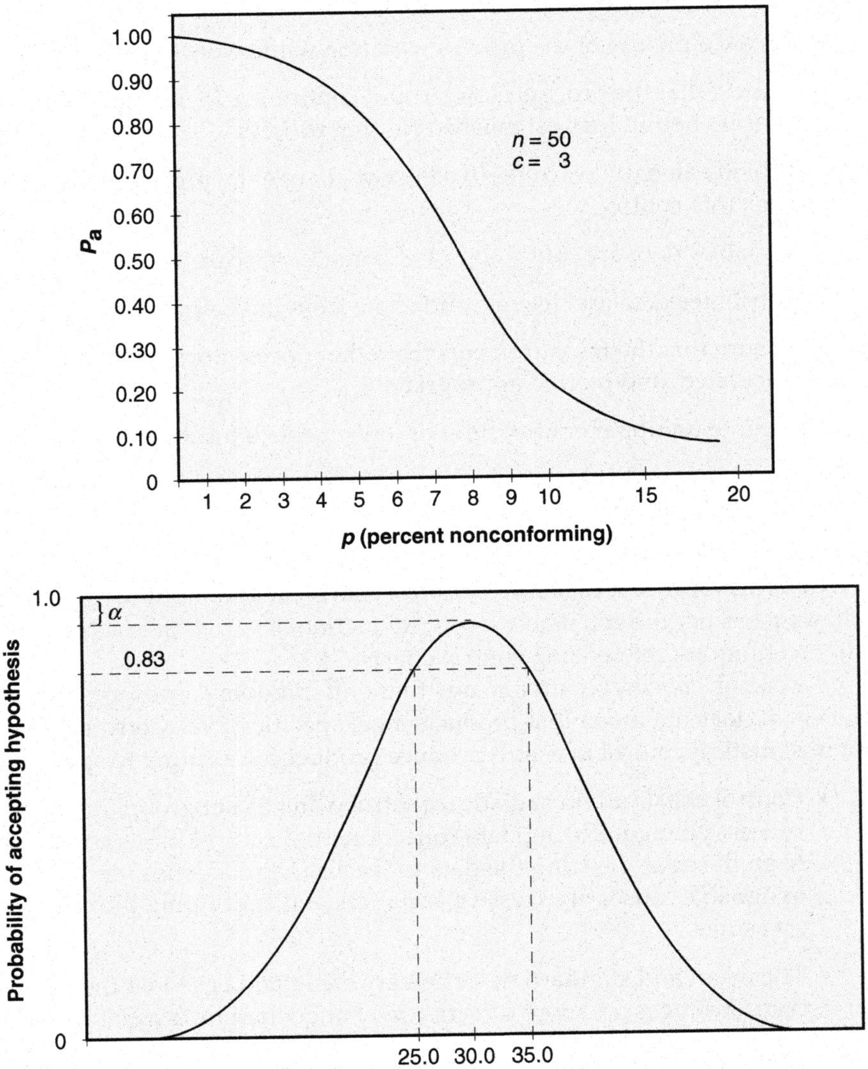

Figure 22.5 An operating characteristic (OC) curve.

- Identify the process variables and conditions contributing to product outcomes and characteristics.
- Consider attributes data (that is, percentage defective) and variables data (that is, numerical measurements) to diagnose causes and determine action. Charts for attributes require discrete measurements (that is, pass/fail, counts) and will be useful provided that the defective rate is high enough to show on a chart with a reasonable subgroup size. Variables charts require measurements on a continuous scale (that is, length, weight, and so on.)
- Determine the earliest point in the process where testing can be done to get information on assignable causes. The earlier the cause can be identified, the more likely the consequences can be effectively contained and mitigated.
- Choose the type of control chart used. An example of an $\bar{X}$ and R control chart is shown in Figure 22.6.

$\bar{X}$ and R Control Chart Machine_____ Process_____

Product/part name and number: *mp plate w239* Gage: *64e* Specification limits: 7.125+.010

Date/operator: *3/17 G. Turner*

Time		7am	8	9	10	11	Noon	1pm	2	3	4		
	1	7.127	7.125	7.123	7.127	7.128	7.125	7.126	7.126	7.127	7.128		
	2	7.123	7.126	7.129	7.127	7.125	7.125	7.123	7.126	7.129	7.123		
	3	7.123	7.121	7.129	7.124	7.126	7.127	7.123	7.127	7.128	7.122		
	4	7.126	7.122	7.124	7.125	7.127	7.128	7.125	7 .128	7.129	7.124		
	5												
Ave, $\bar{X}$		7.125	7.124	7.126	7.126	7.127	7.126	7.124	7.127	7.128	7.124		
Range, R		.004	.005	.006	.003	.003	.003	.003	.002	.002	.006		

Notes:

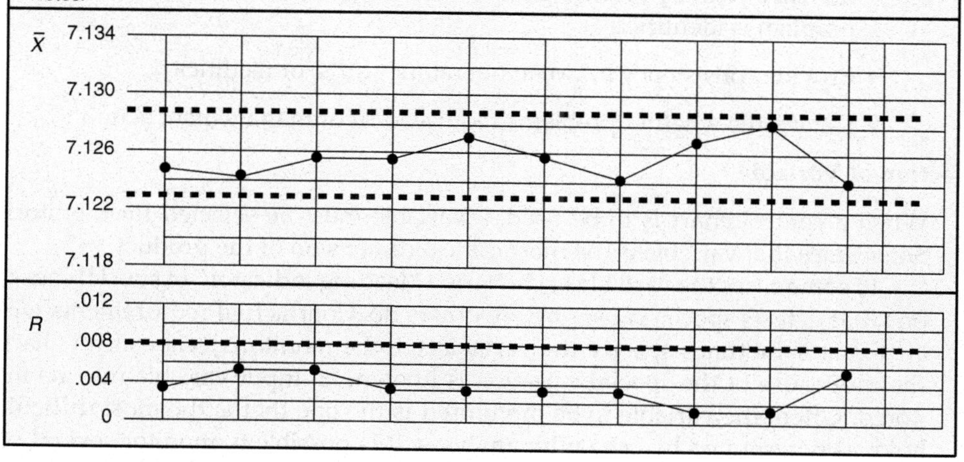

Figure 22.6 $\bar{X}$ and R control chart example with data plotted.

- Determine the central line and the basis for calculating control limits.
- Choose the rational subgroup and the appropriate strategy (subgroup frequency, size, and so on).
- Provide the system for collecting data.
- Calculate the control limits and provide specific instructions on interpretation of results and actions to be taken.

By identifying and resolving the special causes, the Six Sigma Green Belt can facilitate bringing the process into statistical control. At this point, the process results are predictable, and the suitability of the process to achieve customer specifications is revealed. From here out, continual improvement can be realized.

Control charts can also be used for maintaining statistical control. After the expected process level and dispersion are attained, the expected range of variation will become the standard against which subsequent samples are compared to detect the presence of significant causes of variation. Use of the expected range of variation (sometimes referred to as "natural control limits") as a standard will help detect significant changes in the process.

Control charts maintain statistical control and provide traceable evidence through three key actions:

- *Collection*. Run the process and collect the data for plotting on a graph or chart.
- *Control*. Calculate control limits to support analysis and establish process variability.
- *Capability*. Establish the ability of the process to meet customer specifications.

Advantages of Control Charts

- Data able to be collected at the process by the operator
- Increase yield by revealing and containing problems at the stage the problem is identified
- Provide consistency between operators, shifts, or facilities
- Determine whether problems require local or management action

Selection of Variable

When a control chart is to be used, a variable must be selected for monitoring. Sometimes that variable is the most critical dimension of the product.

In some cases the variable of choice is a "leading indicator" of special causes—one that detects special causes before others do. Contractual requirements with a customer sometimes specify the variable(s) to be monitored via control chart. If the root cause of the special variation is known, an input variable may be monitored. Often, the variable to be monitored is the one that is the most difficult to hold, as determined by capability analyses. It is possible to monitor several variables on separate control charts, especially if computerized charting is employed. The selection of the variable to be charted depends on experience and judgment.[3]

Selection of Variables

a. *Key process input variables* (KPIVs) may be analyzed to determine their effect on a process.

b. *Key process output variables* (KPOVs) determine process capability and process monitoring using control charting.

c. *Design of experiments* (DOE) and *analysis of variance* (ANOVA) methods may also identify variables significant to process control.

Variables that are critical to quality should be selected for control charts based on:

- Importance to customer perception
- Objectivity (counted or measured)
- Clear indicators to suggest whether quality is being achieved
- Quality function deployment, which identifies the customers' needs and wants

Characteristics for Analysis

- Concentrate on characteristics that are most promising for process improvement
- Consider the needs of the customer or end user
- Address current and potential problem areas
- For mature systems, investigate relationships, interactions, and correlations

RATIONAL SUBGROUPING

> Define and describe how rational subgrouping is used. (Understand)
>
> Body of Knowledge VI.A.2

In order to properly set up a control chart, several steps must have been completed first:

- The process variables and conditions have been identified
- The testing point has been determined
- The type of control chart has been selected
- The central line to be used for calculating process limits has been determined

For process control, rational subgroups should be chosen so that the units within the subgroup have very close alignment, and units between subgroups show a difference. This will enable each point on a control chart to reflect or represent several units of the product under measurement.

Rational subgrouping will make data collection and interpretation easier to manage and control, and will make the determination of assignable causes easier to detect on the control charts on which they are used.

Rational Subgrouping Method

The method used to select samples for a control chart must be logical, or "rational." Processes must not be out of control (otherwise use a simple run chart) so that the samples used will be valid. A rational subgroup is a sample set that is sufficient to determine common-cause scenarios. Normally the average of a subgroup is used.

In rational subgrouping:

- The division of observations into rational subgroups is key.
- Success of control charting depends on the selection of subgroups.
- Selection should result in groups as homogeneous as possible.
- The first subgroup should reflect product all produced at one time as much as possible.
- One subgroup should be representative of all of the production over a given period of time.
- More useful information is derived from smaller groups (that is, five subgroups of five rather than one subgroup of 25). Larger subgroups provide too much opportunity for process change within the subgroup.
- Attributes control charts are based on Poisson or binomial distributions and require 50 or more samples within subgroups.

Choosing the rational subgroup requires care to make sure the same process is producing each item.

In the case of the $\bar{X}$ and R chart, the $\bar{X}$ chart should detect any process shift, while the R chart should capture only common cause variation. That means there should be a high probability of variation between successive samples, while the variation within the sample is kept small. Therefore, samples frequently consist of parts that are produced successively by the same process to minimize the within-sample variation. The next sample is chosen some time later so that any process shifts that have occurred will be displayed on the chart as between-sample variation.

The choice of sample size depends to some extent on the resources available to do the measuring. The traditional charts have room for five readings per sample, but fewer of the spaces may be used if necessary. In general, the larger the sample size, the more sensitive the chart. Sensitivity also depends on the type of charting technique, with variables charts being more sensitive to special causes than attributes charts.

For variables charts, data are reported from a particular characteristic of a process output in small subgroups of two to five sequential pieces taken methodically (for example, every 20 minutes, three times per shift).

For attributes charts, larger subgroup sizes (that is, 50 to 200) are required in order to observe detectable process shifts. Subgroup sizes need not be exactly repeated, but should not vary by more than 20 percent. In the event that sample sizes change, control limits should be recalculated. Quantity and frequency of collection is dependent on the process, and should support the sufficient data collection necessary to view the full range of process variation.

Control charts can be sensitive and susceptible to false alarms. One of the main advantages of a control chart is that it detects changes in a process that would otherwise go unnoticed. However, if a process is adjusted when it has not statistically changed, that "oversteering" could contribute to the variation. The risk of false alarms leads to overadjustment, which can contribute to additional losses. Appropriate sample sizes help to offset false alarms and reveal the true nature of the process.[4]

Rational Subgrouping Approach

- Select the size, frequency, and number of subgroups based on the control chart using the groups.
- Choose subgroups to minimize piece-to-piece variation within the subgroup.
- Ensure that sample sizes remain constant for all subgroups under review.
- Collect subgroups frequently enough to reveal potential opportunities for variation due to shifts, operators, materials, timing, and so on.
- Major sources of variation should have an opportunity to appear through the subgroup selection.
- A quantity of 25 or more subgroups containing 100 or more individual readings should provide stability and indicate process location and spread for further analysis.
- Where possible, apply existing data that were recently collected and that are consistent with the subgroup approach.

Sampling without considering rational subgrouping will incorporate the variations from the different streams (see Figure 22.7).

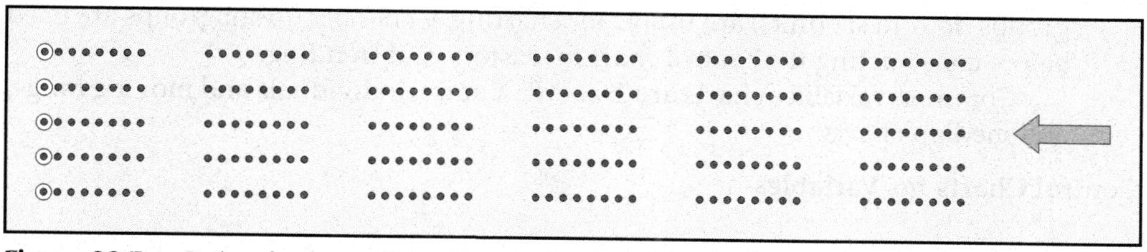

Figure 22.7 Rational subgrouping.

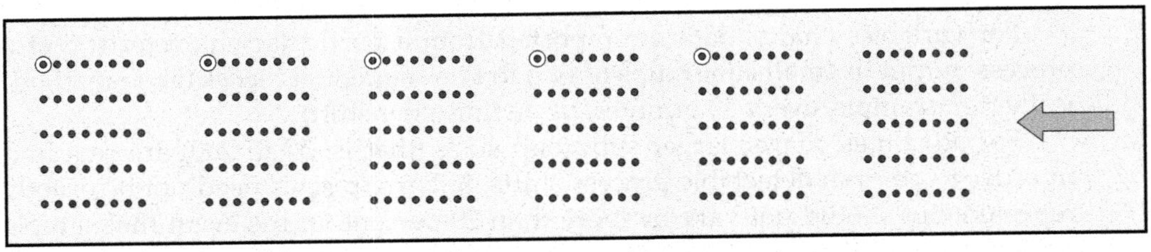

Figure 22.8 Rational subgrouping with randomness.

Rational subgrouping can be applied to enhance randomness and reduce "piece-to-piece" variation (see Figure 22.8).

CONTROL CHARTS

> Identify, select, construct, and use control charts: $\bar{X}$–R, $\bar{X}$–s, individual and moving range (ImR or XmR), median, p, np, c, and u. (Apply)
>
> **Body of Knowledge VI.A.3**

Selection and Application of Control Charts

Key Point: There are many other forms of control charts/process behavior charts that have been developed over the years for specific applications. The ones mentioned here are the more common ones.

Statistical process control (SPC) is a data collection and analysis method used to support process decisions with objective evidence. The values in control charts may be precise and accurate continuous measurements (variables) or counts of occurrences (attributes).

Variables Charts

The *variables chart* is so named because the data to be plotted result from measurement on a variable or continuous scale. This type of measurement occurs when for each pair of values there are an infinite number of possible values between them. Variables data are always quantitative and continuous, therefore smaller sample groups (four to six units) are usable for charting. Generally, 25 subgroups are used before constructing the control chart for history and trending.

Common variables charts are $\bar{X}$ and R, $\bar{X}$ and s, individuals and moving range, and median charts.

Control Charts for Variables

- Plot specific measurements of process characteristics

- $\bar{X}$ and R charts (data are readily available)
- Run charts (single-point data)
- $\overline{MX}$–MR charts (limited data, moving average, moving range)
- XmR charts (limited data, individual moving range)
- $\bar{X}$ and s charts (sigma is available)
- Median charts
- Short-run charts

Best results are obtained from the following preparation steps:

- Ensure a responsive environment with candid and forthcoming people committed to quality and improvement.
- Define the process elements (that is, people, equipment, material, methods, environment), relationships (upstream, downstream, serial, parallel), and documentation/records.
- Define the measurement system.

Benefits of Analysis of Variables

- Focus on measurable characteristics
- Measurement value is more powerful than a "yes/no" or "pass/fail" statement
- Reduce total inspection costs with better data
- Reduce turnaround time for corrective actions
- Processes can be analyzed even within specification limits

Control Limits

Control limits are calculated based on data from the process. Formulas for control limits are given in Appendix D. Several constants are needed in the formulas. The values of these constants can be found in Appendix E.

When calculating control limits, it is prudent to collect as much data as practical. Many authorities specify at least 25 samples. It is very important that sample size be held constant.

$\bar{X}$ and R Control Charts

This control chart shows how the process average changes, along with corresponding changes in process variation. Both average and variation must be controlled. Control charts start with data collected during a base period to establish an overall mean and range, against which subsequent results can be compared to determine process behavior.

Constructing $\bar{X}$ and R charts:

- Determine sample size and frequency
- Calculate average and range, and the averages of both measures
- Calculate the control limits based on the subgroup sample size (Appendix D)
- Plot the data and analyze the chart

Control limits for $\bar{X}$ and R control charts are given by the following formulas:

Upper control limit for the averages chart: $UCL_{\bar{X}} = \bar{\bar{X}} + A_2\bar{R}$

Lower control limit for the averages chart: $LCL_{\bar{X}} = \bar{\bar{X}} - A_2\bar{R}$

Upper control limit for the range chart: $UCL_R = D_4\bar{R}$

Lower control limit for the range chart: $LCL_R = D_3\bar{R}$

where

$\bar{\bar{X}}$ = averages of the sample averages (the process average)

$\bar{R}$ = average of the ranges

A_2, D_3, and D_4 are constants depending on sample size from Appendix D.

EXAMPLE

Data are collected from a face-and-plunge operation done on a lathe. The dimension being measured is the groove inside diameter (ID), which has a tolerance of 7.125 ± .010. Four parts are measured every hour. These values have been entered in Figure 21.9.

The next step is to calculate the average ($\bar{X}$) and range (R) for each sample. These values have been entered in Figure 22.9. Next, calculate the average of the averages ($\bar{\bar{X}}$) and the average range ($\bar{R}$). These values are 7.126 and .0037, respectively. Following are the control limit calculations:

$UCL_{\bar{X}} = \bar{\bar{X}} + A_2\bar{R} = 7.126 + .729 \times .0037 \approx 7.129$

$LCL_{\bar{X}} = \bar{\bar{X}} - A_2\bar{R} = 7.126 - .729 \times .0037 \approx 7.123$

$UCL_R = D_4\bar{R} = 2.282 \times .0037 \approx .008$

There is no lower control limit for the R chart because D_3 is undefined for this sample size.

The row for subgroup size 4 is used because each hourly sample has four readings. The next step is to choose a scale on the average and range charts that includes the control limits. The control limits are then drawn, usually with a dashed line, and the average lines are drawn on each chart, usually with solid lines. Finally, the points are plotted and connected with a broken line. The final chart is shown in Figure 22.10.

Continued

Continued

$\bar{X}$ and R Control Chart **Machine_____** **Process_____**

Product/part name and number: *mp plate w239* Gage: *64e* Specification limits: 7.125±.010

Date/operator: *3/17 G. Turner*

Time		7am	8	9	10	11	Noon	1pm	2	3	4		
	1	7.127	7.125	7.123	7.127	7.128	7.125	7.126	7.126	7.127	7.128		
	2	7.123	7.126	7.129	7.127	7.125	7.125	7.123	7.126	7.129	7.123		
	3	7.123	7.121	7.129	7.124	7.126	7.127	7.123	7.127	7.128	7.122		
	4	7.126	7.122	7.124	7.125	7.127	7.128	7.125	7 .128	7.129	7.124		
	5												
Ave, $\bar{X}$		7.125	7.124	7.126	7.126	7.127	7.126	7.124	7.127	7.128	7.124		
Range, R		.004	.005	.006	.003	.003	.003	.003	.002	.002	.006		

Notes:

$\bar{X}$

R

Figure 22.9 Measurement data entered in an $\bar{X}$ and R control chart.

Continued

Continued

$\bar{X}$ and *R* Control Chart **Machine_____** **Process_____**

Product/part name and number: *mp plate w239* Gage: *64e* Specification limits: 7.125+.010

Date/operator: *3/17 G. Turner*

Time		7am	8	9	10	11	Noon	1pm	2	3	4		
	1	7.127	7.125	7.123	7.127	7.128	7.125	7.126	7.126	7.127	7.128		
	2	7.123	7.126	7.129	7.127	7.125	7.125	7.123	7.126	7.129	7.123		
	3	7.123	7.121	7.129	7.124	7.126	7.127	7.123	7.127	7.128	7.122		
	4	7.126	7.122	7.124	7.125	7.127	7.128	7.125	7 .128	7.129	7.124		
	5												
Ave, $\bar{X}$		7.125	7.124	7.126	7.126	7.127	7.126	7.124	7.127	7.128	7.124		
Range, *R*		.004	.005	.006	.003	.003	.003	.003	.002	.002	.006		

Notes:

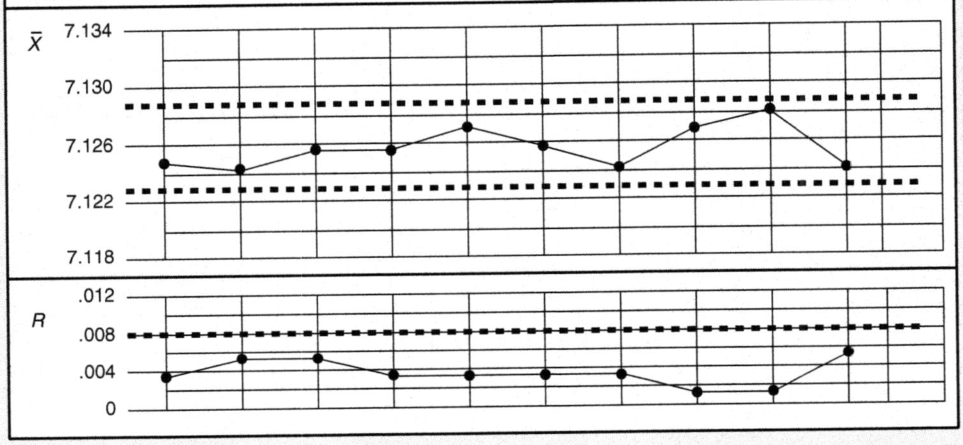

Figure 22.10 Completed $\bar{X}$ and *R* control chart.

$\bar{X}$ and *s* Control Charts

This chart shows the average and the sigma for the process values. Unlike the range, which is calculated by subtracting the smallest value in the subgroup from the largest, the *sigma,* or *standard deviation,* value is calculated from all values in the subgroup. When the subgroup is smaller (that is, five or fewer), the range and sigma charts yield similar results. The increase or decrease in sigma provides a more precise indicator of process change requiring further investigation.

The $\bar{X}$ and *s* control chart is very similar to the $\bar{X}$ and *R* chart except that each value in the range row is replaced by the sample standard deviation *s*. Calculation of control limits also is very similar. Instead of using $\bar{R}$ these formulas use $\bar{s}$ and the appropriate constants from Appendix E.

Upper control limit for the averages chart: $UCL_{\bar{X}} = \bar{\bar{X}} + A_3\bar{s}$

Lower control limit for the averages chart: $LCL_{\bar{X}} = \bar{\bar{X}} + A_3\bar{s}$

Upper control limit for the standard deviation chart: $UCL_s = B_4\bar{s}$

Lower control limit for the standard deviation chart: $LCL_s = B_3\bar{s}$

An example of an $\bar{X}$ and s control chart is shown in Figure 22.11, using the same data that were used in Figure 22.10.
The formula calculations:

$$UCL_{\bar{X}} = \bar{\bar{X}} + A_3\bar{s} = 7.126 + (1.628)(.0020) \approx 7.129$$

$$LCL_{\bar{X}} = \bar{\bar{X}} + A_3\bar{s} = 7.126 - (1.628)(.0020) \approx 7.123$$

Control limits for the standard deviation chart:

$$UCL_s = B_4\bar{s} = (2.266)(.0020) \approx .005$$

$$LCL_s = B_3\bar{s} = 0(.0020) = 0$$

$\bar{X}$ and *s* Control Chart **Machine_____** **Process_____**

Product/part name and number: *mp plate w239* Gage: *64e* Specification limits: 7.125±.010

Date/operator: *3/17 Martha Jane*

Time		7am	8	9	10	11	Noon	1pm	2	3	4		
	1	7.127	7.125	7.123	7.127	7.128	7.125	7.126	7.126	7.127	7.128		
	2	7.123	7.126	7.129	7.127	7.125	7.125	7.123	7.126	7.129	7.123		
	3	7.123	7.121	7.129	7.124	7.126	7.127	7.123	7.127	7.128	7.122		
	4	7.126	7.122	7.124	7.125	7.127	7.128	7.125	7 .128	7.129	7.124		
	5												
Ave, $\bar{X}$		7.125	7.124	7.126	7.126	7.127	7.126	7.124	7.127	7.128	7.124		
Std dev, *s*		.002	.002	.003	.002	.002	.002	.002	.001	.001	.003		

Notes:

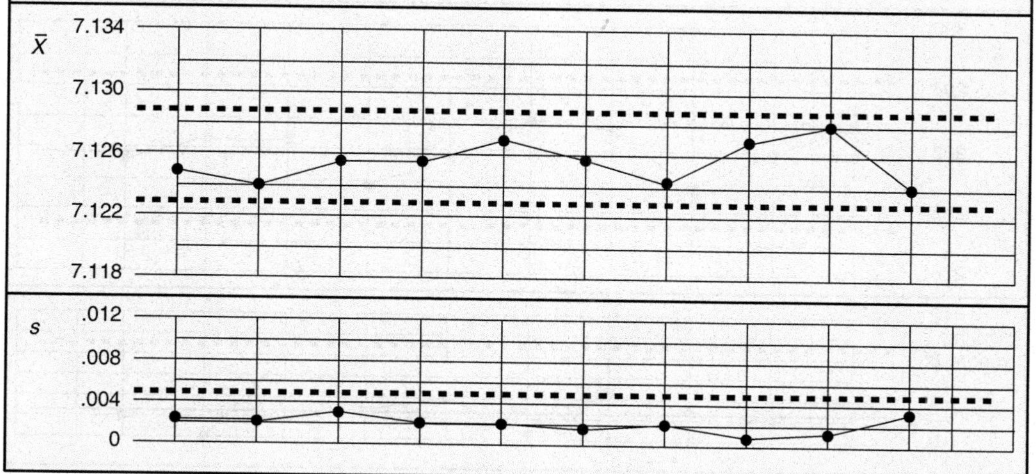

Figure 22.11 Example of an $\bar{X}$ and s control chart.

Individuals and Moving Range Control Charts

Recall that larger sample sizes produce more sensitive charts. In some situations, however, a sample size of one must be used. If the sample size is one, an individuals and moving range (also known as ImR or XmR) chart is appropriate. An example of an ImR chart is shown in Figure 22.12. The data are entered in the row numbered 1. The moving range is calculated by taking the absolute value of the difference between each measurement and the previous one. This value is entered in the row labeled "Moving *R*."

The control limit formulas for the ImR control chart are given in Appendix D and repeated here:

$$UCL_X = \bar{X} + E_2\bar{R}$$

$$LCL_X = \bar{X} - E_2\bar{R}$$

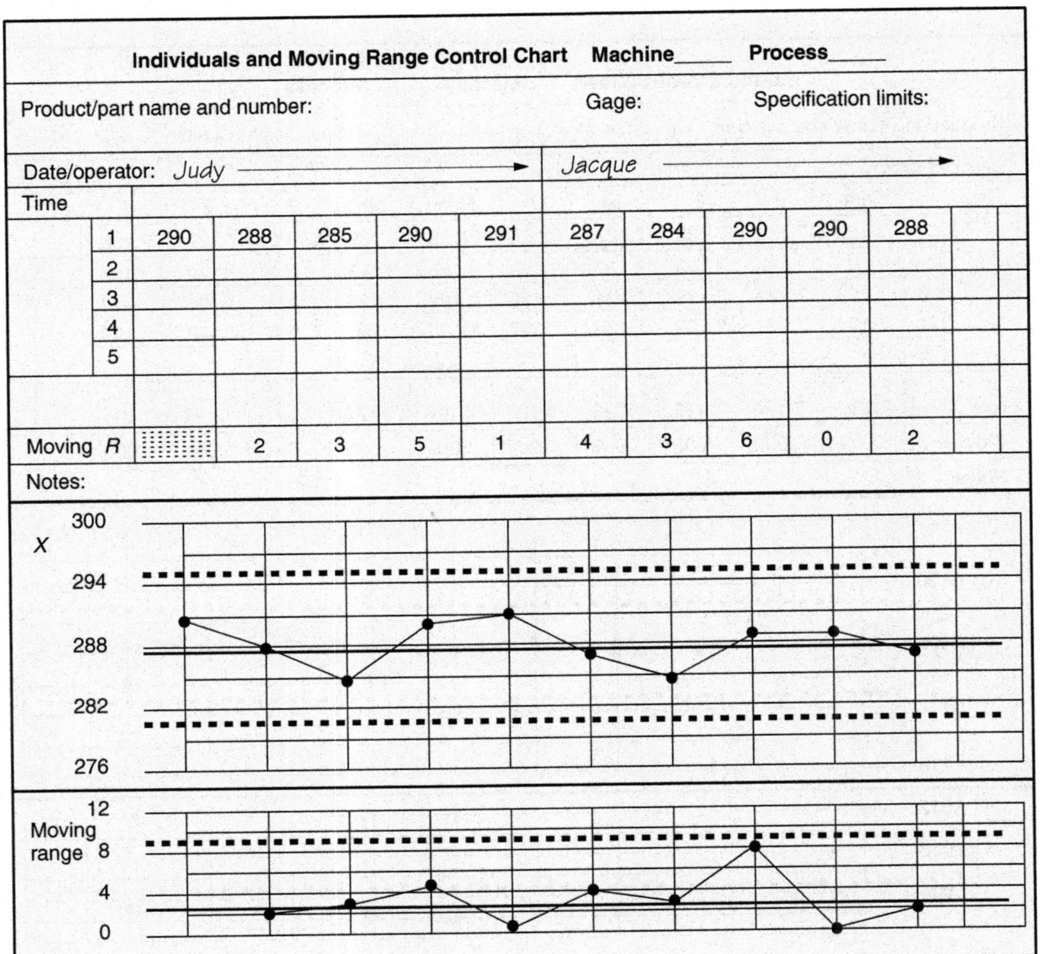

Figure 22.12 Example of an individuals and moving range control chart.

$$UCL_R = D_4\bar{R}$$

$$LCL_R = D_3\bar{R}$$

The values of E_2, D_3, and D_4 are also found in Appendix E. The sample size is two because two subsequent subgroups are compared in each moving range.

The measurements in Figure 22.12 have an average of 288.3 and an average range of 2.889.

$$UCL_x = 288.3 + (2.660)(2.889) \approx 296$$

$$LCL_x = 288.3 - (2.660)(2.889) \approx 280.6$$

$$UCL_r = 3.267\ (2.889) \approx 9.439$$

$$LCL_r = 0\ (2.889) = 0$$

These control limits are drawn, and the measurements and ranges are then plotted.

Median Control Charts

Median charts plot the median of the sample rather than the average. This chart is often used when outliers are expected. All data points in the sample are plotted, and the user connects the middle point in successive samples. Figure 22.13 illustrates a median chart. To detect ranges that are outside range control limits, a paper or plastic gage is constructed that has a length equal to the distance between range control limits. The user of the chart places the gage over the plotted points for a particular sample. If the gage can't cover all the points, the range exceeds control limits. Trends and other anomalies in the range are difficult to detect using the median chart.

The control limits for the median chart are calculated in the usual way:

$$UCL = \bar{X}' + A'_2\bar{R} = 7.125 + (1.880)(.0036) \approx 7.132$$

$$LCL = \bar{X}' - A'_2\bar{R} = 7.125 - (1.880)(.0036) \approx 7.118$$

$$UCL_r = D_4\bar{R} = (2.574)(.0036) \approx .009$$

$$LCL_r = D_3\bar{R} = 0$$

where

A'_2 is the special A_2 constant for use with median charts

x' is the sample median

Attributes Control Charts

Quality can also be measured in counts, fractions, or percentages of defective items. Attributes control charts require only a count of defective parts or services, rather than precise and careful measurements. The simplicity of this approach is offset by the requirement for more sample data needed for process control.

Attributes charts are used for count data. For attributes control charts, if every item is in one of two categories, such as good or bad, "defectives" are counted.

Median Control Chart Machine_____ Process_____

Product/part name and number: *mp plate w239* Gage: *64e* Specification limits: 7.125±.010

Date/operator: *3/17 Deb H.*

Time		7am	8	9	10	11	Noon	1pm	2	3	4		
	1	7.127	7.125	7.123	7.122	7.128	7.125	7.126	7.126	7.127	7.128		
	2	7.123	7.126	7.129	7.122	7.125	7.125	7.123	7.126	7.129	7.123		
	3	7.122	7.121	7.129	7.124	7.126	7.121	7.123	7.127	7.128	7.122		
	4												
	5												

Notes:

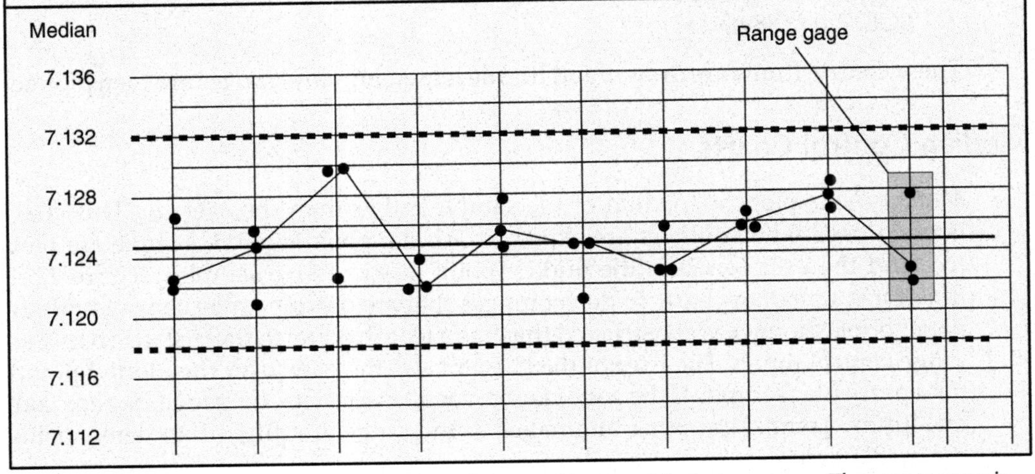

Figure 22.13 Example of a median control chart with associated range gage. The range gage is also shown in position over the points plotted for the 4 pm sample, showing that they are within the range control limits.

If each item may have several flaws, "defects" are counted. Attributes data are generally qualitative, but can be counted, recorded, and analyzed. Examples include nonconformities, nonconforming units, and percentage nonconforming. Larger sample sizes are required for analysis (50 to 200) to fit binomial or Poisson distributions.

One of the benefits of attributes control charts is that they illustrate the number of defects, which supports process tracking and defect investigation. This is particularly helpful if the probability of defect occurrence is relatively low.

Control Charts for Attributes

- *p*-charts (Defectives—sample size varies)
- *np*-charts (Defectives—sample size fixed)
- *c*-charts (Defects—sample size fixed)
- *u*-charts (Defects—sample size varies)
- Short-run charts for *p*, *np*, *c*, *u*

Constructing attributes charts follows a process similar to that for variables charts, except for the use of the much larger sample size:

- Follow trends and cycles to evaluate process changes
- Use subgroup size greater than 50
- Calculate the control limits using the formulas
- Plot the data and analyze the chart

p Control Charts

The *p*-chart measures the proportion of defective parts or pieces within the group under review. This could be for a single characteristic or multiple characteristics. Pieces either conform or are rejected. The rejected portion is expressed as a decimal fraction of the sample size.

The *p*-chart is used to chart binary data where each item is in one of two categories. This would be the appropriate chart for plotting numbers of defectives, for instance. In the following example, each blood sample is in one of two categories, so the *p*-chart is appropriate, although neither category is defective.

EXAMPLE

A test for the presence of the Rh factor in 12 samples of donated blood yields the data shown on the *p*-chart in Figure 21.14.

Control limits for the *p*-chart are given by the following formulas:

$$UCL_p = \bar{p} + 3\sqrt{\frac{\bar{p}(1-\bar{p})}{\bar{n}}}$$

$$LCL_p = \bar{p} - 3\sqrt{\frac{\bar{p}(1-\bar{p})}{\bar{n}}}$$

where

$$\bar{n} = \frac{\text{Sum of the sample sizes}}{\text{Number of samples}}$$

$$\bar{p} = \frac{\text{Sum of the discrepancies}}{\text{Sum of the sample sizes}} = \frac{\sum \text{discrepancies}}{\sum n}$$

Note: when the formula for LCL produces a negative number, no LCL is used.

The control limit formulas use the average sample size. Some software packages recompute control limits each time the sample size changes. While technically correct, this is somewhat difficult when charts are being plotted by hand. As a compromise, the Automotive Industry Action Group (AIAG) recommends recalculating control limits whenever the sample is more than 25 percent above or below the average sample size. In the example in Figure 22.14:

Continued

Continued

$$\bar{n} = \frac{\text{Sum of the sample sizes}}{\text{Number of samples}} = 1496 \div 12 = 124.7$$

$$\bar{p} = \frac{\text{Sum of the discrepancies}}{\text{Sum of the sample sizes}} = \frac{\sum \text{discrepancies}}{\sum n} = 177 \div 1496 = 0.118$$

$$\sqrt{\frac{\bar{p}(1-\bar{p})}{\bar{n}}} = \sqrt{\frac{.118(.882)}{124.7}} \approx .029$$

$$\text{UCL} = \bar{p} + 3\sqrt{\frac{\bar{p}(1-\bar{p})}{\bar{n}}} = 0.118 + 3 \times 0.029 \approx 0.205$$

$$\text{LCL} = \bar{p} - 3\sqrt{\frac{\bar{p}(1-\bar{p})}{\bar{n}}} = 0.118 - 3 \times 0.029 \approx 0.030$$

p-Chart

Machine/process: *Blood analysis*

Product: *Donated blood* | Defective = *Rh negative*

Date: 2015 *Sept. 8 8 8 8 9 9 12 12 12 12 12 12 12*

Operator: *Emily* ---------------- *Dana* ----------------

Defectives, *np*	# 14	18	13	17	15	15	16	11	14	13	14	17			
Sample size:	125	111	133	120	118	137	108	110	124	128	144	138			
Fraction, *p*	.11	.16	.10	.14	.13	.11	.15	.10	.11	.10	.10	.12			

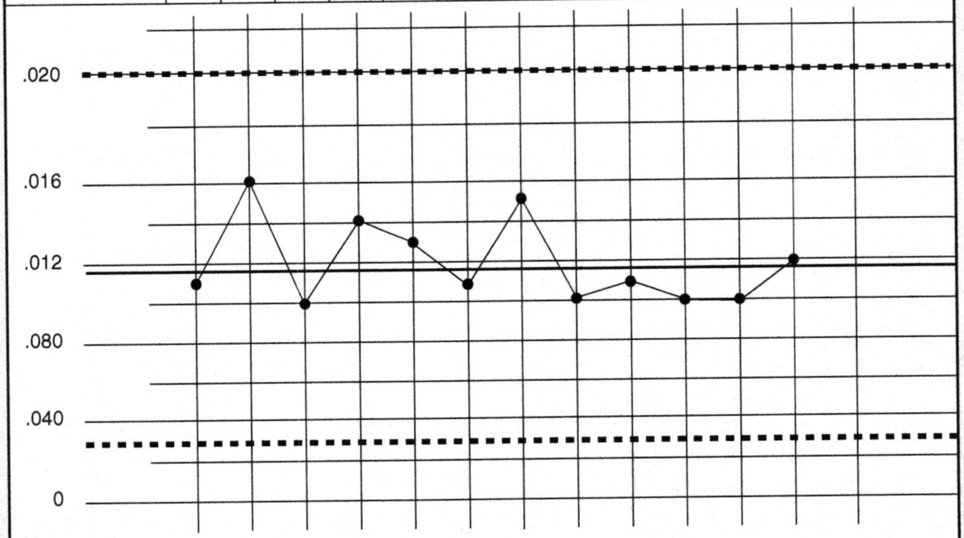

Notes:

Figure 22.14 Example of a *p*-chart.

np Control Charts

The *np*-chart measures the number of rejected items in a sample with an integer rather than a proportion. The *np*-chart is most useful when sample sizes are constant and the integer number is more meaningful and relevant than the proportional decimal amount.

If defectives are being counted and the sample size remains constant, the *np*-chart can be used instead of the *p*-chart.

EXAMPLE

Packages containing 1000 light bulbs are randomly selected and all 1000 bulbs are light-tested. The data have been entered in the *np*-chart shown in Figure 22.15. Note that this chart is slightly simpler to use than the *p*-chart because the number of defectives is plotted rather than the fraction of the sample.

The formulas for control limits:

$$UCL_{np} = \overline{np} + 3\sqrt{\overline{np}\left(1 - \frac{\overline{np}}{n}\right)}$$

$$LCL_{np} = \overline{np} - 3\sqrt{\overline{np}\left(1 - \frac{\overline{np}}{n}\right)}$$

where

$\overline{np}$ = Average number of defectives

n = Sample size

Note: When the formula for LCL produces a negative number, no LCL is used.

For the example shown in Figure 21.15.

$$\overline{np} = 120\sqrt{13} \approx 9.23 \text{ and } n = 1000.$$

$$UCL_{np} = 9.23 + 3(3.02) \approx 18.29$$

$$LCL_{np} = 9.23 - 3(3.02) \approx 0.17$$

Continued

Continued

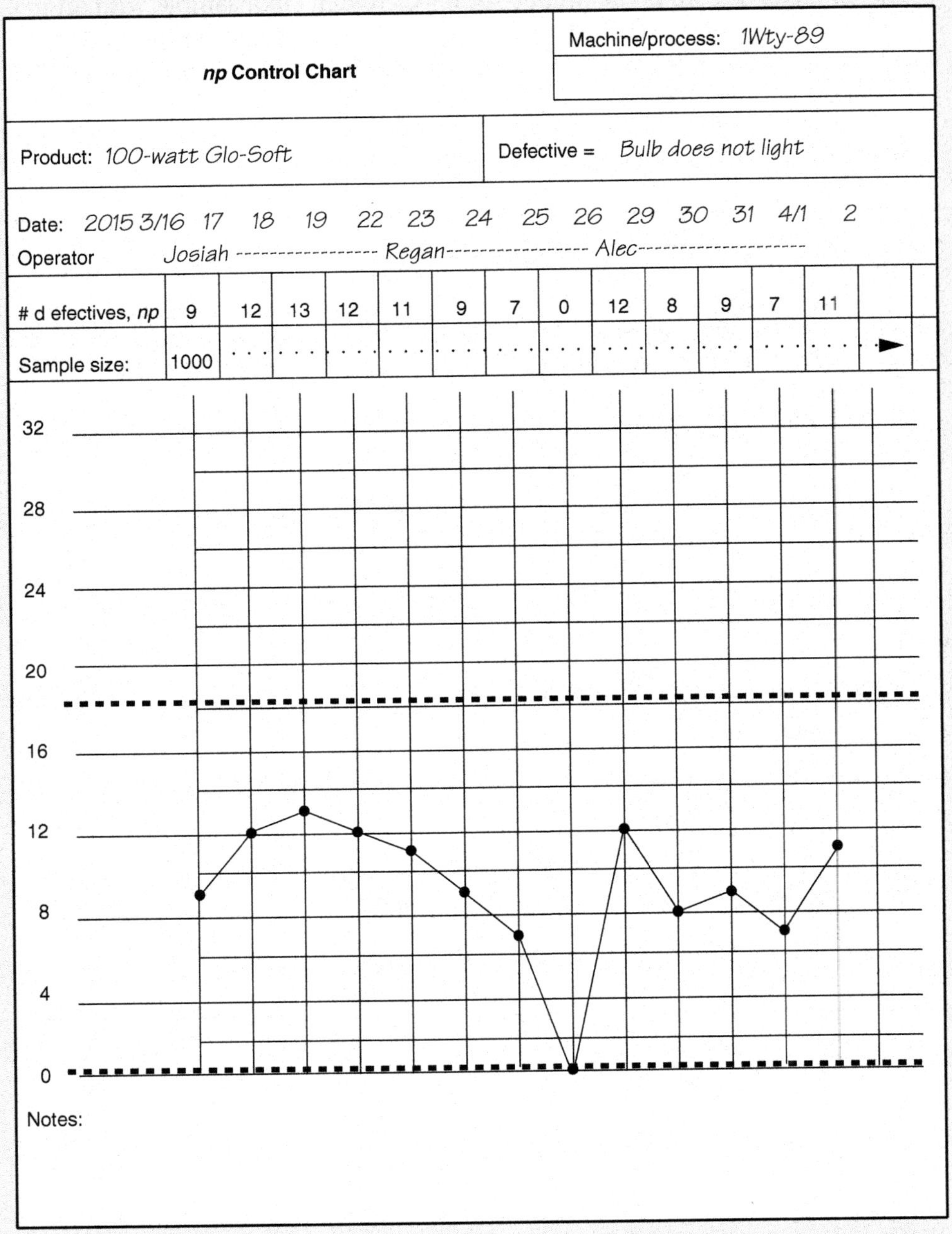

Figure 22.15 Example of an *np*-chart.

u Control Chart

The *u*-chart is appropriate to use when defects rather than defectives are counted. The *u*-chart measures the number of defects or problems on a per unit basis. The example in Figure 22.16 shows the results of inspecting panes of glass in which defects include bubbles, scratches, chips, inclusions, waves, and dips. The number of defects is counted and recorded for each sample, and the fraction (# defects ÷ sample size) is calculated and plotted.

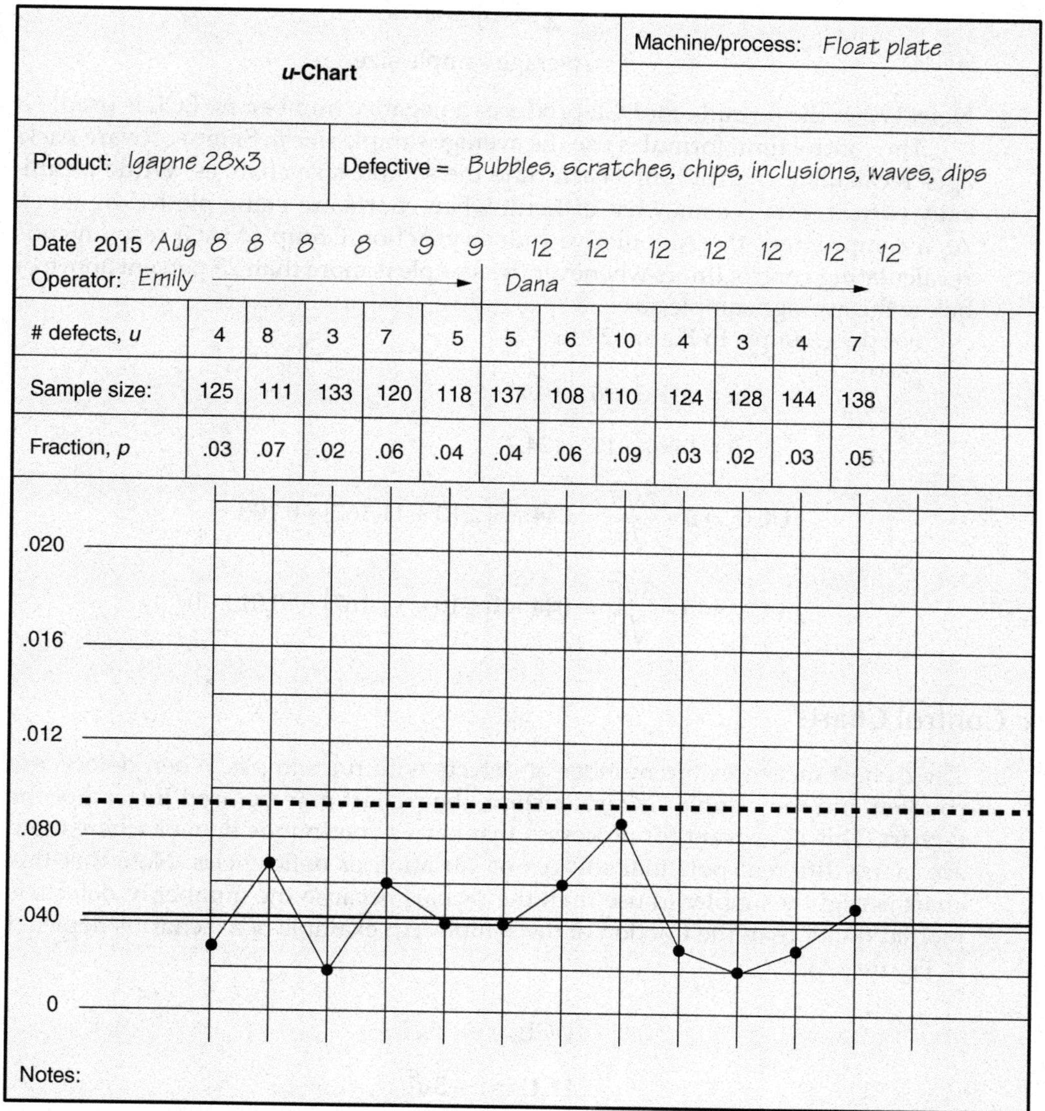

Figure 22.16 Example of a *u*-chart.

$$\text{UCL}_u = \bar{u} + \frac{3\sqrt{\bar{u}}}{\sqrt{\bar{n}}}$$

$$\text{LCL}_u = \bar{u} - \frac{3\sqrt{\bar{u}}}{\sqrt{\bar{n}}}$$

where

$$\bar{u} = \frac{\sum \text{defects}}{\sum \text{sample sizes}}$$

$$\bar{n} = \text{Average sample size}$$

Note: When the formula for LCL produces a negative number, no LCL is used.

The control limit formulas use the average sample size $\bar{n}$. Some software packages recalculate control limits each time the sample size changes. While technically correct, this is somewhat difficult when charts are being plotted by hand. As a compromise, the Automotive Industry Action Group (AIAG) recommends recalculating control limits whenever the sample is more than 25 percent above or below the average sample size.

For the example in Figure 22.16:

$$\bar{u} = 66 \div 1496 \approx .044$$

$$\bar{n} = 1496 \div 12 \approx 124.7$$

$$\text{UCL}_u = \bar{u} + \frac{3\sqrt{\bar{u}}}{\sqrt{\bar{n}}} = .044 + 3(.210 \div 11.167) \approx 0.100$$

$$\text{LCL}_u = \bar{u} - \frac{3\sqrt{\bar{u}}}{\sqrt{\bar{n}}} = .044 - 3(.210 \div 11.167) \approx -0.012 \approx 0$$

c Control Charts

The *c*-chart measures the number of defects within a sample. When defects are counted and the sample size is constant, the *c*-chart may be used instead of the *u*-chart. This is relevant for processes that have a continuous flow or where there are many different potential sources of variation or deficiencies. Note that this chart is slightly simpler to use than the *u*-chart because the number of defects is plotted rather than the fraction of the sample. An example of a *c*-chart is depicted in Figure 22.17.

$$\text{UCL}_c = \bar{c} + 3\sqrt{\bar{c}}$$

$$\text{LCL}_c = \bar{c} - 3\sqrt{\bar{c}}$$

where

$\bar{c}$ = Average number of defects

Note: When the formula for LCL produces a negative number, no LCL is used.

For the example in Figure 22.17:

$$\bar{c} = 217 \div 13 \approx 16.7$$

$$\text{UCL}_c = 16.7 + (3 \times 4.1) = 29$$

$$\text{LCL}_c = 16.7 - (3 \times 4.1) = 4.4$$

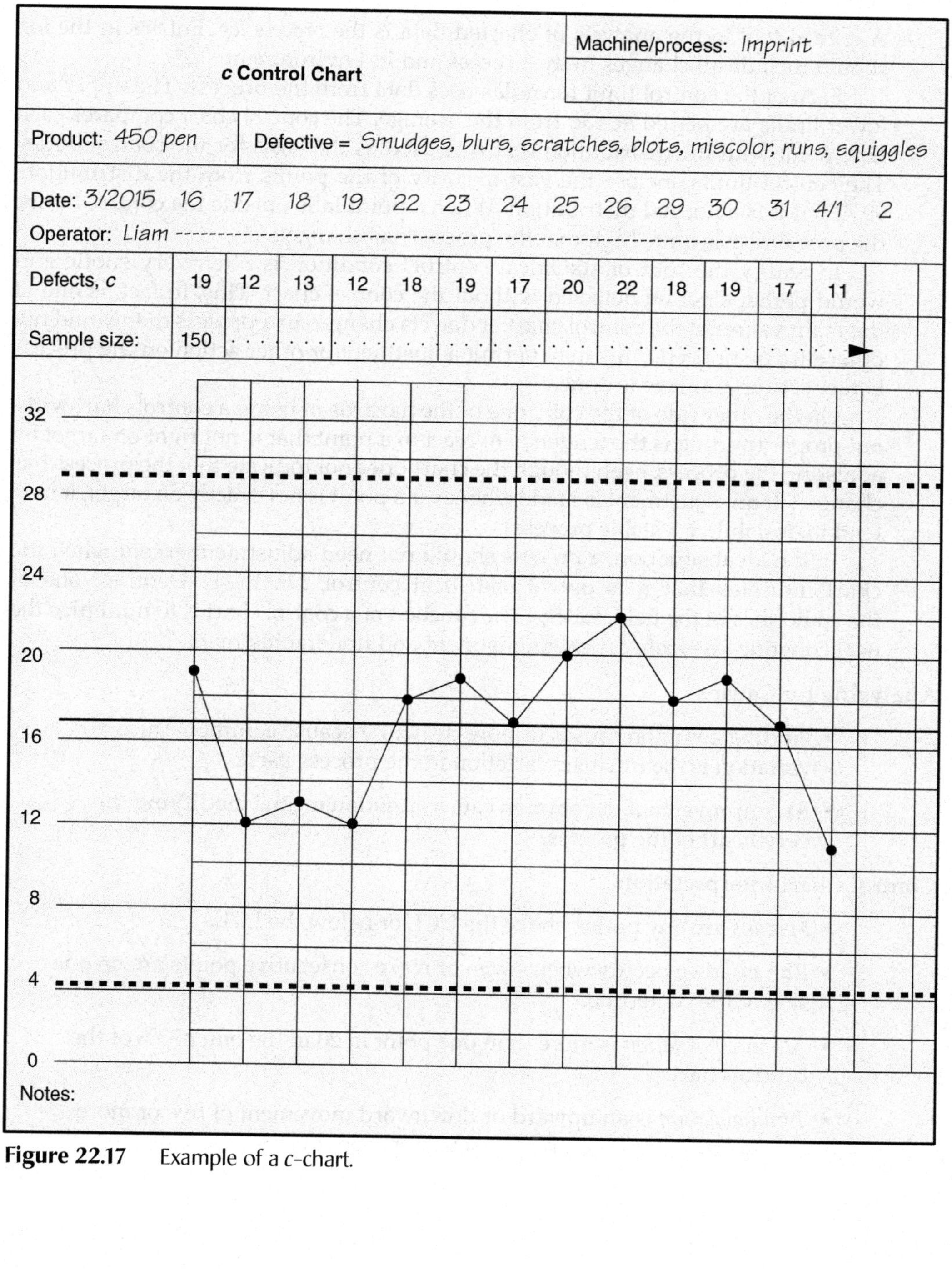

Figure 22.17 Example of a *c*-chart.

Key Point: The use of short-run techniques involves setting a target value for the particular item being run on the machinery and measuring/charting the plus and minus values from the target value. By doing this, you can develop a true process behavior chart for the process (for example, the way a particular machine is operating) instead of a particular component.

Analysis of Control Charts

A critical tool in the analysis of charted data is the *process log*. Entries in the log should include all changes in the process and its environment.

Each of the control limit formulas uses data from the process. The upper and lower limits are placed at $\pm 3\sigma$ from the average. The control chart compares each new point with the distribution that was used as the basis for the control limits. The control limits enclose the vast majority of the points from the distribution, 99.72% if it is a normal distribution. When a point falls outside the control limits, the probability is quite high that the process has changed.

In reality the "out of statistical control" condition is often very subtle and would perhaps not be detected without the control chart. This, in fact, is one of the main values of the control chart: it detects changes in a process that would not otherwise be noticed. This may permit adjustment or other action on the process before serious damage is done.

On the other side of the coin, one of the hazards of using a control chart without proper training is the tendency to react to a point that is not right on target by adjusting the process, even though the chart does not indicate that the process has changed. If an adjustment is made whenever a point is not exactly on target, it may tend to destabilize a stable process.

In the ideal situation, a process should not need adjustment except when the chart indicates that it is out of statistical control. Dr. W. E. Deming, one of the authorities in the field, states, "The function of a control chart is to minimize the net economic loss from . . . overadjustment and underadjustment."[5]

Analyzing for Causes

- Finding common causes is more difficult because common cause variation is the intrinsic variation in the process itself.
- An improvement in common cause variation means modifying the very heart of the process.

Control Chart Interpretation

- *Specials* are any points above the UCL or below the LCL.
- *Run violation* occurs when seven or more consecutive points are on one side of the centerline.
- A *1 in 20 violation* is more than one point in 20 in the outer 33% of the control chart.
- *Trend violation* is an upward or downward movement of five or more consecutive points or drifts of seven or more points.

Process Control

- A process is in control when both the average and variation are stable.
- Trends can be corrected with equipment repair, rotation, or replacement.
- Jumps reflect abrupt changes in material, method, or performance, and can be corrected with consistency.
- Recurring cycles reflect wear and fatigue and can be overcome with tighter controls and reduced cycle times.
- Points near or outside limits could indicate overadjustment or material variation, and can be corrected with test and inspection and restriction of operator controls.
- Lack of variability suggests that control limits are too loose or there may be a measurement issue (that is, fudging, tampering).

Process change can be determined by studying charts and identifying process shifts. Statistical indicators of process change are available. Two of the most widely used are Minitab and the Automotive Industry Action Group (AIAG) *SPC Manual*. The eight rules used by the software package Minitab are:

- One point more than 3σ from the centerline (either side)
- Nine points in a row on the same side of the centerline
- Six points in a row all increasing or all decreasing
- Fourteen points in a row alternating up and down
- Two out of three points more than 2σ from the centerline (same side)
- Four out of five points more than 1σ from the centerline (same side)
- Fifteen points in a row within 1σ of the centerline (either side)
- Eight points in a row more than 1σ from the centerline (either side)

The six rules listed by the AIAG in its *SPC Manual* are:

- Points beyond the control limits
- Seven points in a row on one side of the average
- Seven points in a row that are consistently increasing (equal to or greater than the preceding points) or consistently decreasing
- Over 90 percent of the plotted points are in the middle third of the control limit region (for 25 or more subgroups)
- Fewer than 40 percent of the plotted points are in the middle third of the control limit region (for 25 or more subgroups)
- Obvious nonrandom patterns such as cycles

Each of these rules is illustrated in Figures 22.18 through 22.31. Any points that violate the rule are circled.

If, for instance, an increase in values represents a safety hazard, it would not be necessary to wait for the specified number of successively increasing points to take action. Control limits are somewhat arbitrary and could conceivably be adjusted based on the economic trade-off between the costs of not taking action when an out-of-control condition occurs and the costs of taking action when an out-of-control condition has not occurred. Deming stated (in a private conversation in October 1985), however, that moving the control limits up and down can be a source of additional problems, and it would be better in most cases to put that energy into reducing variation.

Ensure that each point is calculated and plotted correctly. For variables charts, the range section should be analyzed first. Increases in the range values represent

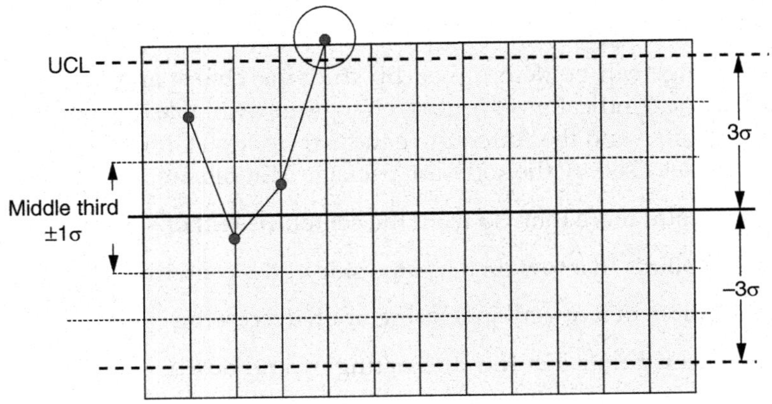

Figure 22.18 One point more than 3σ from the centerline (either side).

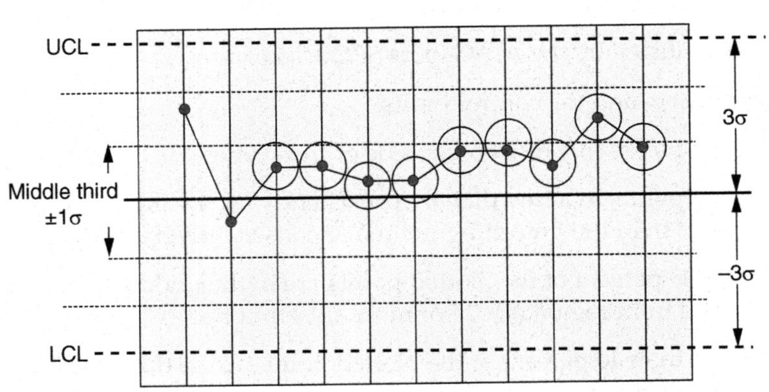

Figure 22.19 Nine points in a row on the same side of the centerline.

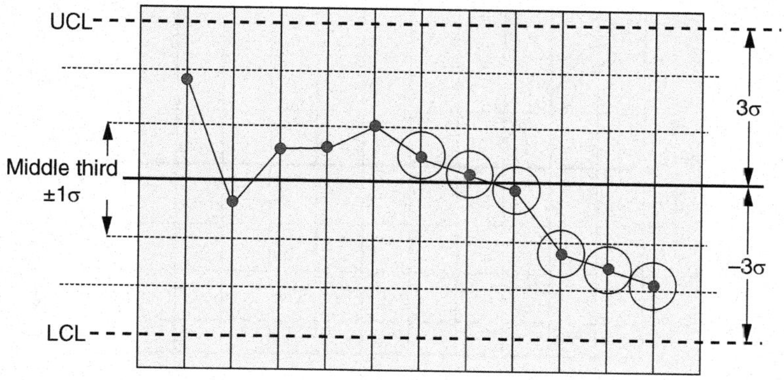

Figure 22.20 Six points in a row all increasing or decreasing.

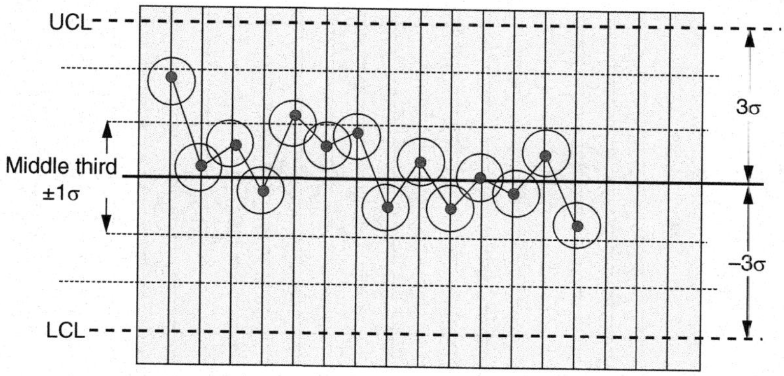

Figure 22.21 Fourteen points in a row alternating up and down.

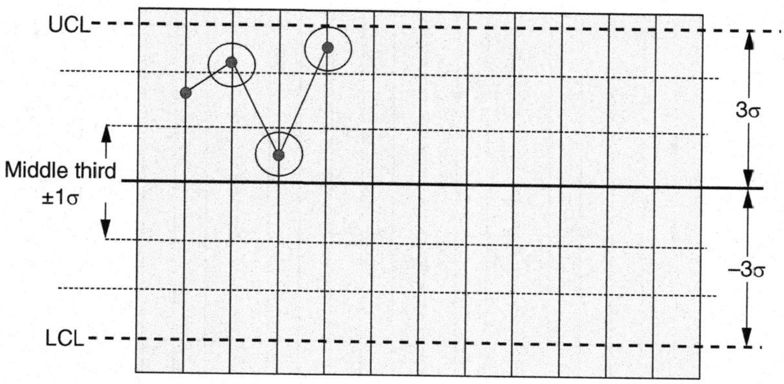

Figure 22.22 Two out of three points more than 2σ from the centerline (same side).

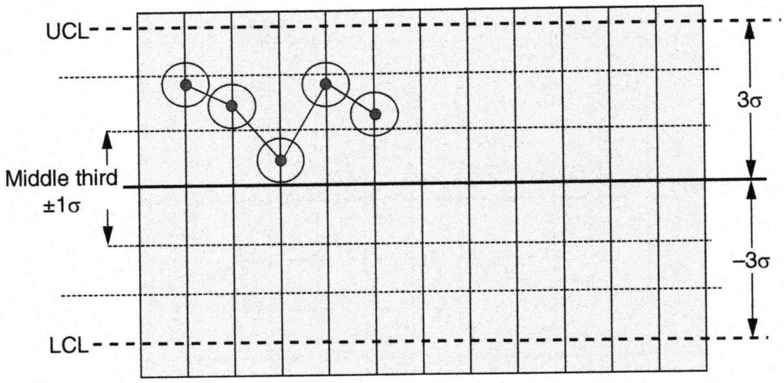

Figure 22.23 Four out of five points more than 1σ from the centerline (same side).

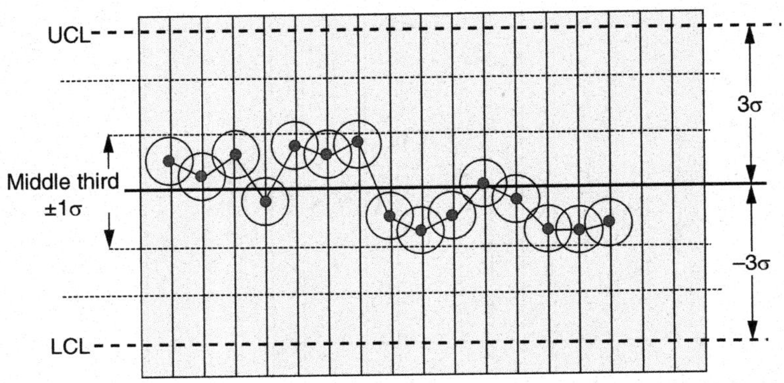

Figure 22.24 Fifteen points in a row within 1σ of the centerline (either side).

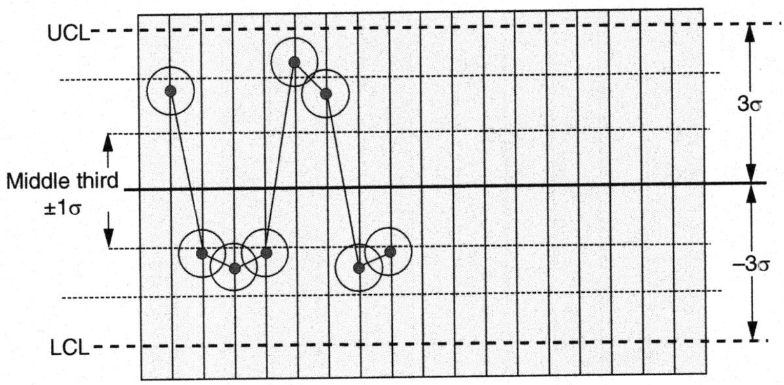

Figure 22.25 Eight points in a row more than 1σ from the centerline (either side).

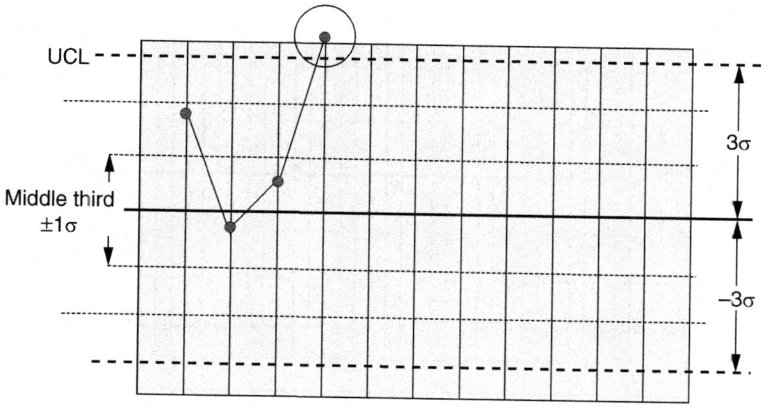

Figure 22.26 Points beyond the control limits.

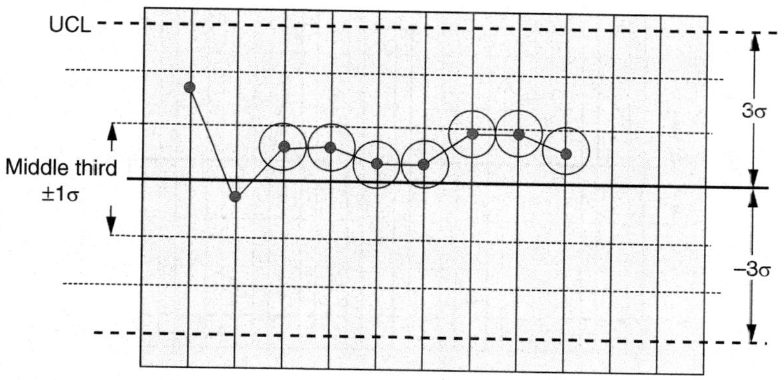

Figure 22.27 Seven points in a row more on one side of the average.

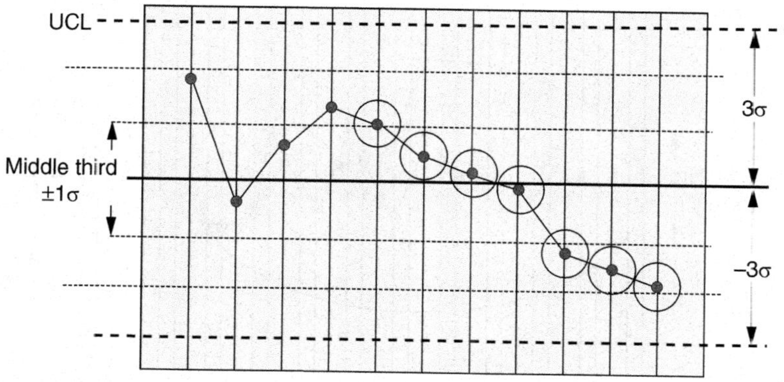

Figure 22.28 Seven points in a row that are consistently increasing or consistently decreasing.

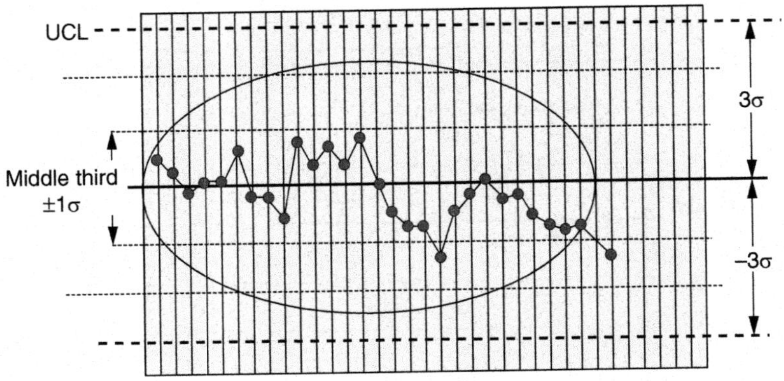

Figure 22.29 Over 90 percent of the plotted points are in the middle third of the control limit region (for 25 or more subgroups).

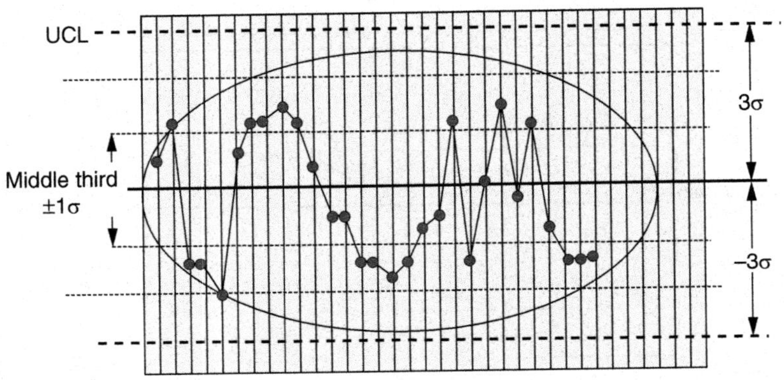

Figure 22.30 Fewer than 40 percent of the plotted points are in the middle third of the control limit region (for 25 or more subgroups).

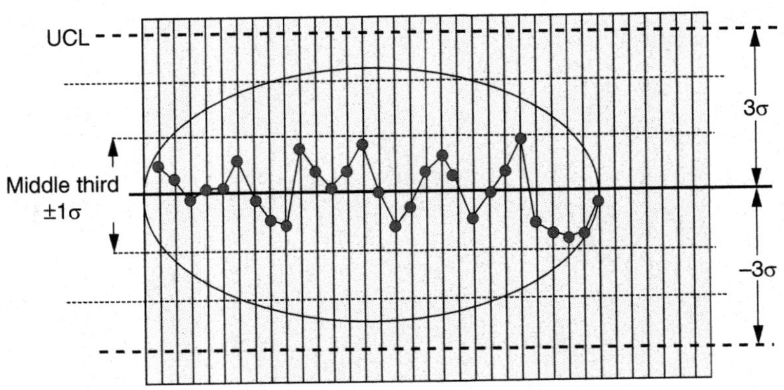

Figure 22.31 Obvious nonrandom patterns such as cycles.

increased variation between the readings within an individual sample. Possible causes include bearings, tooling, or fixtures. Changes in the averages chart represent some sort of shift in the process. Frequent causes include tool wear, changes in raw materials, and changes in measurement systems or process parameters such as machine settings, voltages, pneumatic pressure, or other settings.

It will be more useful if we can relate the violation to possible causes. Only when the causes of the deviation are assignable can continual improvement occur. It is useful to construct a list of things to check when certain chart characteristics occur. Such a list can come from a discussion among experienced personnel as well as from data from a process log.

In most cases, "out of control" conditions reflect process readings at or beyond the control limits. These can be traced to several causes, including inaccurate control point calculations, declining process performance, or errors in the measurement system. Nonrandom patterns or trends likely indicate the presence of a special cause (for example, novice operator, worn machine). In this case, the points associated with the special causes should be excluded when recalculating control limits, which should consequently be more restrictive.

In some cases, the events on the "out of control" lists represent improved situations, particularly if the readings centralize to the mean and do not come close to reaching or exceeding control limits. The process should be investigated to determine what changed and to see whether this change can be perpetuated. If a log is maintained for the process, it may be possible to find changes that correspond to the time that the improvement occurred.

In the event of "false signals," a common cause may be incorrectly linked to a special cause, forcing a process modification. Modifying the process in this way will actually increase process variation and make the process less stable and more likely to fall out of control. Special causes should be confirmed and validated. Experience is the best teacher when it comes to chart interpretation, and efforts should be made to document a body of knowledge about each process. This is reflective of type II error (producer's risk), where the hypothesis that the process has changed is accepted even though the process is actually within statistical control. Excessive adjustments will contribute to even greater levels of variation, which may actually put the process further out of control.

The following sources of variability may contribute to process variation:

- Long-term variation (product spread or process spread)
- Lot-to-lot variation
- Stream-to-stream variation (assuming lots have multiple streams)
- Time-to-time variation
- Piece-to-piece variation
- Within-piece variation (assumes complexity of the piece)
- Inherent error of measurement
- Inherent process capability

The null hypothesis is that the process hasn't changed, and as each point is plotted, the chart is examined to determine whether there is sufficient evidence to reject the null hypothesis and conclude that the process has changed.

Key Point: When reading a process behavior chart, always start with the variables section. Variables data are usually more revealing of process behavior due to the trends and patterns that are more easily identified from the inputs. Focus on the range first to determine whether the process is changing internally. The range tracks the variation in the process. Following this review, look at the target value (averages) to understand other forces on the process. When these forces are significant and verifiable, they could be investigated as special causes, or determined to be common causes of the process.

Chapter 23

Sustain Improvements

CONTROL PLAN

> Assist in developing and implementing a control plan to document and monitor the process. (Apply)
>
> Body of Knowledge VI.B.1

Control plans make the operator aware of items in the system for controlling parts and processes during full production. The control plan is a document, updated as needed, that explains how to control the work flow in your process. It should have, as a minimum, the following basic information:

- A flowchart or other graphical representation of the process with the desired outcomes displayed
- Any special or safety characteristics clearly displayed on the control plan
- A description of the direct relationship between any highlighted characteristics and their controlling process settings or parameters
- Identification of any gages or test equipment needed for the operations
- Identification of appropriate sample sizes and frequencies of all testing
- Any reactions to failure mode and effects analysis (FMEA) conditions spelled out to prevent nonconforming products or out-of-control conditions
- Reaction plans that can be easily understood by the operators
- Criteria that can be used to verify accuracy and validate alignment with the next operation in the process

The purpose of a control plan is to *"hold the gains"* realized through the prior Six Sigma phases of design, measure, analyze, and improve. The following steps can be used to create a suitable control plan in order to ensure the correct monitoring and measurement of control subjects, and the appropriate actions to perform:

1. Specify those variables with direct or indirect impact to the remedy and the customer.
2. Establish the control limits and standards for when to take action.
3. Measure and set baselines for the different control variables.
4. Specify the timing and location for measurements, and determine the appropriate control chart.
5. Assign and delegate those who will review and analyze monitoring and measurement results to identify when processes are out of control, and diagnose the assignable causes.
6. Put troubleshooting and corrective actions in place to restore the process quickly, incorporating adaptations to the control plan as part of ongoing improvement.

The control plan should outline the steps to be followed during each phase of the process to ensure that all process outputs will be in a state of control. Operators need to feel comfortable working with the paperwork in their area.

Some elements and attributes that could be part of a control plan include the following items:

- *Control subject.* Those features of a product or process that are brought under monitoring and/or measurement as a way of tracking and identifying conditions where the process may be out of control.
- *Performance standard.* This enables the process operator to evaluate the current performance relative to the predetermined control limits.
- *Unit(s) of measure.* Consistency of measurement is imperative for objectivity and acceptance of control outcomes. Confusion in units (for example, inches versus centimeters, days versus hours) can skew control charts if not properly managed.
- *Measurement rules.* This includes the measurement device, frequency of measurement, sample size, and manner of recording.
- *Actions.* If the process measures fall outside of the control limits or standards, the control plan should specify the actions for troubleshooting, correction, and restoration.
- *Decision/deployment.* There should be a place to show who decides on and carries out the actions to be taken.
- *Record.* The control plan should specify where actions are recorded and archived for future reference or more extensive trend analysis.
- *Version.* This is necessary to communicate whether the control plan is the "latest and greatest" released and in use.[1]

Sample Control Plan Layout				
Plant	Operation	Date control limits calculated	Part number	Specification
Machine	Characteristic	Sample size/ frequency	Part name	Control item
Averages chart				Actions on special causes
Ranges chart				Action instructions
Readings				Subgroup size
Sum of readings				
Process Log				
Date/time	Material change	Methods/ equipment change	Operator change	Comments

Figure 23.1 A sample control plan template.

A simple plan based on the template in Figure 23.1 could be applied to track the necessary data on a single page or sheet. Examples are shown as Figures 23.2 and 23.3.

Deming was well known for his constant harping on managers to fix the process, not the people. Studies have shown that 80 percent of the time the real issue is something in the system. Operators need to use data collection tools to demonstrate that they are following the control plan so that any issue that may arise can be shown to be due to system operation.

Control plans provide a structured approach for the design, selection, and implementation of value-added control methods for the total system. The scope

Soft Start-Up Valve Control Plan

Control plan number: CP714				**Control plan revision level:** C		**Revision date:** 12/01/15			
Part/assembly number/rev: 714647-H & 714648-J				**Product line:** Soft start air dump valve		**Originator:** J. Hausner			
				Methods					**Reaction plan code**
						Sample			
Sta #	**Process description**	**Machine tools/ equipment**	**Print no.**	**Characteristic specification**	**Evaluation measurement equipment**	**Size**	**Freq.**	**Control method**	
14	Machine needle bleed port on cover	Drill press	714648	0.060" min diameter	0.60 (minus) gage pin S/N 15-50-2118	1	1 per hour	Check sheet	A
18	Pressure gage torque	Torque driver	714647 714648	20 +/– 5 IN LB	Torque gage S/N 15-50-2019	5	1 per shift	$\bar{X}$ chart	E, F
23	Body-cover screw torque	Torque driver	714647 714648	60 +/– 15 IN LB	Torque gage S/N 15-50-2120	3 per screw	2 per shift	Separate $\bar{X}$ charts	E, F
27	Solenoid assembly torque	Torque driver	209647 209648	14 +/– 7 IN LB	Torque gage S/N 15-50-2019	5	1 per shift	$\bar{X}$ chart	E, F
29	Final air test	Test tank	209647 209648	Functional test and leak check	Visual: ref. QA spec 203795 Functional: ref. assy instruction	1	100%	Go/no-go	A, B, C, D
All	All	All	209647 209648	Workmanship	Visual	1	100%	Go/no-go	See note 2

Note 1: At all times, quarantine one hour worth of product before releasing to shipping. In the event of a final test failure, the last hour of production should be set aside for possible retest. This should be done on all final test failures with the exception of porosity.

Note 2: Compare suspect unit to visual accept/reject standards. If unit is unacceptable, stop the line and follow standard four-step reaction plan: (A) contain suspect units; (B) diagnose the root cause and implement corrective action; (C) verify that the corrective action is effective; (D) disposition suspect material (sort, scrap, rework, use as-is).

Figure 23.2 An example control plan–first page.

Soft Start-Up Valve Control Plan

Control plan number: CP714	**Key contact:** J. Hausner	**Control plan revision level:** C	**Revision date:** 12/01/15
Part/assembly number/rev: 714647-H & 714648-J	**Part name/description:** Soft start air dump valve HG & HJ series	**Product line:** Airlogic control valve series	**Originator:** J. Hausner

Failure mode	**Reaction plan**	**Code**
Valve fails to open	Containment: Segregate nonconforming unit and previous hour of production for MRB. Disposition: Verify that wire leads and power supply are hooked up correctly. Verify needle port diameter > 0.060". If port diameter is under spec, switch to 100% inspection for the next 50 units and notify the product engineer (PE) if another failure is found. Replace drill bit if hole is not drilled through or burrs are present. Verify that piston ring is installed and free of nicks. Verify that needle valve is open at least one complete turn. Verify that the solenoid port resistor is installed. Try another solenoid. If other tests fail, check diameter of diaphragm. Contact the PE if additional diagnosis is required. Verification: Verify that corrective action eliminates problem. Disposition: Scrap nonconforming components. Rework assemblies as necessary and retest 100% of the previous hour of production.	
Valve fails to close	Containment: Segregate nonconforming product for MRB. Diagnosis: Verify that wire leads and power supply are hooked up correctly. Verify that flow control is open. Verify that diaphragm is installed correctly and check for voids in the seal bead. Verify that the dump hole is drilled completely through bonnet. Check that the fluid resistor is in place. Try another solenoid. If solenoid sticks open, quarantine current batch and switch to a new batch of solenoids. Contact PE if further diagnosis is required to determine cause. Verification: Verify that corrective action eliminates problem. Notify PE if another failure is found on the next 50 units. Disposition: Scrap nonconforming components. Rework assembly and retest.	
Body–bonnet leak	Containment: Segregate nonconforming product for MRB. Diagnosis: Verify torque. For torque adjustments, see Reaction Code "E" below. Ensure that diaphragm is installed correctly and that there are no voids present on the bead. Verify that the bead grooves on the bonnet and body are free of nicks or porosity and the diameters are within tolerance. Verify that the milled slot on the body is within tolerance. Contact PE if further diagnosis is required. Verification: Verify that corrective action eliminates problem. Disposition: Scrap nonconforming components. Rework assembly and retest. Contact line lead or PE if there are two or more consecutive failures or three failures within one hour.	
Leak at fittings	Containment: Segregate nonconforming product for MRB. Diagnosis: Verify that fittings are installed correctly and have the correct torque applied. Verify that the threads on the fitting and assembly are free of nicks or porosity. Contact PE if further diagnosis is required. Verification: Verify that corrective action eliminates problem. Notify PE if another failure is found on the next 50 units. Disposition: Scrap nonconforming components. Rework assembly and retest.	
Torque out of spec	Containment: Segregate nonconforming product for MRB. Diagnosis: Verify torque using another torque gage. For torque adjustments, take at least 10 samples and adjust torque gun if average is more than one standard deviation away from the nominal. Notify maintenance if average is close to nominal and there are any observations out of spec. Contact PE for further diagnosis. Verification: Measure a minimum of three subgroups and verify that the process is near nominal and in control. Disposition: If undertorqued, retorque assembly. If overtorqued, replace screw(s) and retorque.	
SPC out of control, but parts in spec	Refer to QA/SPC procedure 231573. Comply with SPC procedure requirements. Document the root cause and corrective action in a note on the control chart.	

Figure 23.3 An example control plan–second page.

of control plans includes dynamic control plans, quality process sheets, and standard operating procedures. In whatever form, control plans are dynamic documents that explain how to control the process work flow.

Additional considerations characteristic of successful control plans include:

- Confirmation with operators, engineers, internal customers, and those affected by the process under control
- Visual representation of controls with flowcharts or graphical symbols
- Display of special characteristics
- Descriptions of relationships between characteristics and controlled process settings or parameters
- Identification of gages, test equipment, sample sizes, and frequency of testing
- Specification of troubleshooting steps or responses to nonconforming products or out-of-control conditions, and traceability of actions back to the control plan

Another variation on this same theme is the SDCA (standardize, do, check, act) cycle. This is most commonly used once a process has been improved to update the control plans and process sheets to lock in the improvements and standardize the changes throughout the organization.[2]

Examples of Process Controls

Control plans can be used to manage not just the outputs of the process, but the overall process itself. Process controls allow the operators to support the decision of whether to start, continue, pause, or stop running the process. Process controls should consider the potential failures or hazards, the measurements or comparisons, and the feedback needed for operator decisions.

Examples can be applied throughout the phases from raw material acceptance to completion of finished goods or services. Visual inspections of process conditions, first pieces produced, and incoming materials reveal failure modes early, before expensive work is performed. Ongoing monitoring and control charts can reveal deviations or alarms requiring intervention. Acceptance inspections and audits can identify nonconformities and opportunities for corrective action. Supporting activities and devices (such as training, references, documentation, measuring tools, environmental controls) support successful process completion.

Changes to processes must be communicated and references must be updated in order to prevent the process controls from being obsolete or irrelevant. These adaptations must be reinforced with ongoing maintenance and sustainment of process control assets and resources.

Dynamic Control Planning

The *dynamic control plan* (DCP) combines necessary information into one document to help plan, monitor, control, study, and maintain your process. Some of the documents include: standard operating procedures (SOP), control plans, failure mode and effects analysis (FMEA), gage control plan, quality planning sheets (QPS), and others.

The DCP is often called a living document where the operators have the right and responsibility to update the DCP any time that things change; the documents need to be updated to communicate to others so they know that something is different in the process.

The basic DCP includes a matrix (sometimes referred to as the *DCP critical path*) of the following items:

- DCP launch
- Team structure
- Question log
- Support information
- Prelaunch or preliminary controls
- Process failure mode and effects analysis (PFMEA)
- Control plan
- Illustrations and instructions
- Implementation and maintenance

When starting a DCP process, everyone needs to know what is going to happen, and management must be committed to support the efforts. The team's focus for the DCP is to maintain the control plan and the control planning process. Teams should also remember to maintain a log to keep a process history that can be used in the lessons learned database as well as for detailed study of the process when designed experiments are to be used.

Supporting information includes any number of items, including but not limited to blueprints, engineering specifications, prototype plans, FMEAs (design and process), special or critical characteristic identification, process sheets, flowcharts, statistical information, and so on. All this information should be available to the operators involved prior to any new line being started in the plant or prior to launching a new product or process so that activities will work out better and more quickly to get things running smoothly. The process FMEA and the control plan (see Table 23.1) are the primary focuses of the DCP process and include any number of illustrations and instructions, with some of these being enlarged and posted on the job site to allow for ease of use in running the operation.

It is the responsibility of the operators and supervisors to ensure that, once started, this process is maintained and updated regularly to ensure quality of products and services. If problems occur, the records and documentation contribute to helping fix the process.

The dynamic control plan enhances the control plan with its inclusion of a team structure, support information, preliminary controls, and a process failure mode and effects summary. Dynamic control planning supports the control plans with illustrations and instructions and procedures for implementation and maintenance.

Table 23.1 Process FMEA and control plan.

Failure mode and effects analysis (FMEA)							Control plan	
Failure mode	**Failure effect**	**Critical**	**Priority**	**Ease to detect**	**Risk priority**	**Action**	**Mitigation**	**Validation**
Run-time error	System shutdown	5	5	2	50	Restore system	Graceful shutdown with frequent data saving	Validated in process
Bar code mismatch	Invalid sample accepted	4	4	4	64	Reject sample	Insert bar code checking safeguards	Validated in process
Timing error	Expiry of sample	4	4	3	48	Dispose sample	Set alarm	Validated in process
Power outage	System shutdown	5	5	1	25	Restore system	Graceful shutdown at 30 percent power, run on cable	Validated in process

In order to help ensure that the process runs smoothly, the dynamic control plan documents must be constantly updated, indicating how to plan, monitor, control, study, and maintain the process. Changes to the document must be communicated to all necessary parties.

Gage Control Plan

Gage R&R (repeatability and reproducibility) is used to measure the accuracy of the gauging system. The gage repeatability and reproducibility reports support the conclusions of a gage control plan. The parameters in a gage control plan can include proper storage and care of the gage or test equipment, calibration requirements, handling requirements, and indication of what parts or processes the gage or test equipment is used for. For more details, refer to Chapter 14.

Standard Operating Procedures

When the process is updated, confirmed, and validated as being suitable, *standard operating procedures* (SOPs) can be developed. Standard operating procedures are a step-by-step description of how to complete tasks. Documented evidence will go a long way in preventing finger-pointing or faultfinding and the operator being blamed for something out of their control. Standard operating procedures create consistency and establish the proper methods for completing a process.

Standardization is the process of locking in the gains made during the improvement process. SDCA refers to standardize, do, check, adjust. Following the SDCA process, standardization is done after control plans and process documentation have been updated.

Many different terms are used for SOPs, such as *work instructions, level three ISO 9001 documentation, operating guides, job aids,* or *standard job practices.*

SOPs should give the details, and address things such as:

- What is the job?
- Where does the SOP apply?
- When does the SOP apply?
- Who is responsible?

Operators are responsible for following the SOPs as written. If at any time deviations are taken, then the operator needs to document what was done and why. This will be a big help if at a later date a problem arises and an investigation is done.

The SOP should be a living document; if something changes in the system, then the operator should ensure that the SOP is updated. When something changes in the process and a new desirable level is achieved, the operator should update all documents relating to that process. These updates should be reviewed by other affected functional areas and technical experts (that is, engineering, purchasing, shipping and receiving, and so on) to validate their suitability and gain support for SOP deployment and implementation.

Continual Improvement

Continual improvement (CI) is a process of keeping an open mind and looking for ways to make the things that you do better, cheaper, or faster. As the Industrial Revolution progressed into the early 1900s, Frederick Taylor developed a method of work specialization that is still used by many organizations today. It was during this time that workers first stopped checking their own work, and specialized inspectors were employed in inspection teams. This process progressed and developed for several decades, and professional organizations developed around doing inspection better.

During the late 1920s, Walter Shewhart developed the first control chart, and statistical process control (SPC) was born (alternatively referred to as *process behavior charting*). Many organizations continued to rely on inspectors, but the use of charting that could bring operators back into looking at the quality of their work became a requirement in the United States during World War II. It was in 1951 when Armand Feigenbaum first published the book *Total Quality Control* and the total quality management (TQM) age started. This was an era where quality was first embraced by senior management as a strategic imperative, and deployed across the entire organization.

During the 1960s and 1970s, the use of quality circles and employee involvement became the next evolution in continual improvement. This was followed by a major resurgence of SPC during the 1980s. During the 1990s, the International

Organization for Standardization's (ISO) quality management system standard ISO 9001 and the Malcolm Baldrige National Quality Award became the preferred strategies for continual improvement.

Other terms that have been used of late include *value analysis/value engineering, lean manufacturing/lean office, kaizen, poka-yoke,* and others. Six Sigma has become the latest wave of the ongoing continual improvement movement and is bringing many fields of study back into the hands of the people doing the work.

Some people refer to these various methods as *continuous improvement* as they feel that we should always be advancing in everything we do. Unfortunately, nature and human beings do not work that way. Even in evolution, sometimes things have to step back or level off every now and then. As we learn new things, sometimes humans have to relearn old knowledge to gain new. Thus, Deming changed the term *continuous* to *continual.*

Within this change, Deming also developed the *system of profound knowledge.* This concept involves an appreciation for a system, knowledge about variation, theory of knowledge, and psychology. By using each of these concepts, continual improvement can and will become a reality in our organizations. Our goal is to always maintain and improve the quality of the products or services we provide to customers, both internal and external.

When the plan–do–study–act (PDSA) and standardize–do–check–adjust (SDCA) cycles are used together, the operator will see a complete system for identifying processes, improving those processes, and applying lessons learned. The two cycles working together with the other tools in this book will help the operator continually improve the work that is done, with an eye toward satisfying the customer.

Process Improvement

Process improvement is the act of making the system work better to satisfy customer wants and needs. It is a vital element in order for continual improvement to become a reality. We are looking at reducing overall *variability,* not just the variation. Variability is made up of three components: instability, variation, and off-target.

In dealing with variability, most practitioners have traditionally only dealt with the variation question. Variation is very important, and we use the tools covered in this book to help reduce variation as much as possible. The other two components are also very important. Without knowledge of these two, instability and off-target, we could miss some very important factors and even cause major problems in the shop.

Instability is the lack of a stable operating process. Common cause and special cause variation are unchecked and not responded to. Without a stable process, capability values are not worth calculating, and customers can, and do, see any number of issues come and go without rhyme or reason. The best method for monitoring a process is process behavior charts. Operators should play a big role in monitoring the processes to ensure that the jobs they perform are stable and in control.

Off-target is often the responsibility of the engineers who design the parts and production. The operator can only monitor whether the process is centered within the engineering specifications and/or control limits. Even though today we talk about the Taguchi loss function and how processes should be centered on the customers' wants and needs, many jobs we work in today were designed years ago when engineers put the target wherever it made the most economic sense for the company instead of the customer. So, the operator should monitor the process and be ready to give up-to-date information to engineers when processes are to be redesigned so that the new thinking becomes a reality on the shop floor.

DOCUMENT CONTROL

> Understand document control and its role in controlling and sustaining improvements. (Understand)
>
> **Body of Knowledge VI.B.2**

Documented information is necessary to establish that the activities described are actually performed, and that the affected processes and systems are compliant. This includes the templates supporting the work and the completed records and artifacts demonstrating the proper completion of the work.

The Six Sigma Improve phase is often characterized by changes to process inputs, process steps, and process outcomes or deliverables. These changes should prompt timely and commensurate revisions in the affected documentation. For example, if the Six Sigma improvement resulted in a more efficient material handling process, there should be corresponding documents and records associated with the changes. Otherwise, the continued use of obsolete procedures, control plans, and records would undermine the effectiveness of the process change.

Documents are controlled by having approved releases as of a particular date, by a recognized authority (e.g., Quality Director, Plant Manager). The documents should be traceable to the process or procedures involved and should match the requirements for the work needed. If operators make adjustments, these should be reviewed to remove non-essential portions while inserting necessary components into new versions.

Document controls make the operator aware of documents and records currently in circulation. These should have, as a minimum, the following basic information:

- An indicator of the purpose of the document, its intended distribution, and any restrictions (e.g., internal, confidential)
- References to the affected processes with the associated documents displayed
- Any special documents and records clearly displayed on the document control plan
- A description of the direct relationship between process documents and the parent references for traceability

- A statement warning that printed versions are not controlled, and to reference the online version for the most current revision

The document controls should outline the steps to be followed during each phase of the process to ensure that all process outputs will be in a state of control. Operators need to feel comfortable working with the paperwork in their area. For process records, the control, storage, and disposition of artifacts must be established to ensure proper archiving.

Impact of Document Controls on Improvements

Documents serve as both tools and artifacts of the improvement. Without documentation, their function or organization would be challenged to demonstrate that the Six Sigma improvement has been properly deployed and implemented.

Documents that represent templates (either in print or electronic form) are released to support the correct performance of the work. The completion of these templates will enable the processes to be completed. For this reason, the release date of such documents must align with and cross-reference the process improvements made.

Records that demonstrate proper process completion represent the accountability and performance (or evidence of non-performance) by the workers and their functions. If there are gaps, the Six Sigma team must revisit the project to understand why the improvement was not fully deployed and sustained. Details may have been overlooked that need to be corrected or mitigated.

From the perspective of a management system, document controls should extend from process controls, and provide traceability up to the core quality characteristics and objectives of the organization. This traceability is also demonstrated through practices like Enterprise Resource Planning, which link data and process steps across multiple functions within an organization.

TRAINING PLANS

> Develop training plans to implement and sustain improvements. (Apply)
>
> **Body of Knowledge VI.B.3**

Competence, awareness, and training are needed to demonstrate that people have been provided with the necessary orientation, skills, and resources. When organizations improve internal processes and systems, these improvements necessitate incremental and commensurate training plans to correspond to the changes.

Training is distinguished from general education by the expectation that the outcome of successful training will result in specific competencies. This should build upon existing knowledge and practices such that training is an incremental augmentation of ongoing capabilities. Since the scope of training is to enable long-term improvement, it complements the actions needed to sustain improvements within an organization.

Training also must be adapted to the learning styles and preferences of the trainees, along with the expected attention span and need for amusement and novelty. Training must reinforce learning in a way that is memorable and enjoyable. If this is done within a workplace, the expected workplace conduct and rules are also important.

Training plans are necessary to ensure coverage of the materials and sustain the improvements through targeted propagation of new methods and practices to the affected participants and stakeholders:

- Needs analysis and business case
- Strategic planning
- Definition of learning experiences
- Training documentation, facilitation, and provision
- Training validation and confirmation
- Reinforcement of concepts

Training Plans

Training plans should reflect the necessary course material, and provide a profile of the intended learners and the training objectives. In this context, the training is being deployed to impart new skills and enable the affected workers and interested parties to complete their portion of the Six Sigma improvements.

Details of the training plans should include the expectations, information, and schedule. This ensures that training will be completed according to the plans, and that alignment and accountability can be established from the training to subsequent work performance.

The level of training should extend beyond knowledge and comprehension to application; it is not sufficient to know and understand the new material, it must be correctly applied. The purpose and motivation for the training should be made clear to create the necessary interest and urgency in the subject material.

For technical subjects, guided learning and demonstrations must be planned and included. If these require equipment or infrastructure, the training plans should include the necessary logistical arrangements. If additional items (e.g., personal protective equipment) are required, the training plan should specify any prerequisites.

The training plans should outline the steps to be followed during the training portion of the Improve Phase to ensure that all process outputs will be in a state of control. Participants and stakeholders need to feel comfortable working with and adhering to the new controls in their respective areas.

Needs Analysis and Business Case

Training needs analysis is prompted by several conditions including organizational restructuring, changes to functional or departmental priorities and objectives, and adjustments to new technologies or process steps. The business case can determine the scope of training, the schedule when it must be performed, and the comparative value of timing training periods or methods with existing work commitments.

Cross-training should also be provided to those impacted by the improvements in order for them to know what is expected for prerequisite and downstream processes.

Needs analysis should address gaps in skills and knowledge, flawed environments and work incentives, and worker motivation. The business cases should highlight the incremental value derived from this training for the benefit of the organization and its affected stakeholders.

Consider the example of improving the process of handling third-party components that must be returned to the vendor. A practical training program should include the following elements:

- Reinforcement of existing practices and capabilities must be addressed. This can include existing practices for workplace safety, material handling, and documentation.
- The "Net-New" training resulting from the Six Sigma improvements must also be considered and incorporated into a training program. This can be a combination of explanations, demonstrations, monitored work, and a validation that the employees or participants are suitably trained.

Strategic Planning, Learning Experiences

The desired and appropriate learning experiences should drive the strategy, which should be used to support the creation of detailed training plans. For the context of the Six Sigma Improve Phase, the training should be a prescribed process that emphasizes clarity and specific outcomes.

The scope of training could extend beyond colleagues and employees at different organizational levels, to include suppliers, vendors, partners, customers, and other constituents and stakeholders. An initial effort is needed to establish the optimal learning experiences for these constituent groups. These learning experiences could include online or in-person learning seminars, webinars, simulations, demonstrations, and actual work performance.

Some elements and attributes that could be part of a training plan include the following items:

- Scope and context of expected training
- Background of trainees (specify different approaches for novice and expert groups)
- Training support systems and resources (includes content and infrastructure)
- Allocation of time and budget
- Learning mode and methods (e.g., in class, on site, online self-study)

- Opportunities to demonstrate and apply new knowledge

The training plan should be documented and applied to the overall Six Sigma project. Since the purpose of training is to sustain improvements, there should be a reconciliation between the scope of training and the process or project improvements made.

For the example of material handling, the training plan should address any gaps in skills and knowledge identified as impacting the implementation of the Six Sigma improvements within the organization.

Documentation, Facilitation, and Provision

Training objectives should be expressed as performance targets. There should be a clear mapping between the training activities and the desired performance. This should define the learning objectives for each module. Once the learning objectives are known, the documentation and content is structured in a way that fulfills the learning objectives, bringing the trainees closer to meeting the performance targets.

The facilitation of this training has to consider objectives beyond simply raising awareness of the process changes. The training is often competing with legacy practices and adherence to a status quo. Facilitators must go beyond subject expertise to be positive advocates for process improvement.

The provision of this training has to align with the desired learning experiences, as reflected within the training plans. For the material handling example, if the workers are more oriented toward tactile experiences, they will learn more from the physical movement of items than from taking notes in a classroom.

The performance objectives are such that training has to extend beyond the online webinar or classroom. Training is not concluded until the imparted skills have been successfully adopted and demonstrated. Provision of training extends to include ongoing monitoring and verification of successful work completion, process adoption, and sustained improvements.

Training Validation and Confirmation

The validation of the training should extend beyond whether the attendees were satisfied with the events or learning programs. There must be tangible demonstrations that learning has occurred and behavior has changed.

Validation of training can be achieved through oral and/or written examinations, demonstrations, ongoing monitoring, and the verification that completed work meets performance expectations.

- Training records must be kept to show who received training and what training was provided. This sustains the improvement by creating accountability and performance benchmarks for future reference.
- The effect of training on the trainees must be captured to know the value of the training methods, and the basis for future improvements.

- The results of training on sustaining the Six Sigma improvements should be reflected in the affected processes (ideally as the elimination of special causes).

Reinforcement of Concepts

Concepts are best reinforced when they are immediately applied in situations. For the example of material handling, the newly trained individuals should apply the improvements to material handling methods. The systems and infrastructure should be modified to remove the options of reverting back to old methods.

Periodic training and confirmation is helpful. Organizations do this for workplace safety or information security, by making people repeat the exam annually.

Training records can be updated as well to highlight the recency of training. This is relevant if processes are dependent on technology solutions that have frequent changes or upgrades.

Promoting Behavioral and Performance Changes

The reason for this training is to sustain the Six Sigma improvements and to realize the promises of performance optimization. The success of training is not measured by the count of certificates nor the time spent in classrooms or viewing online webinars. The results have to be manifested in sustainable changes to behavior, leading to performance improvements.

AUDITS

> Define first-, second-, and third-party audits. (Remember)
>
> **Body of Knowledge VI.B.4**

Audits are a type of quality assessment activity intended to reconcile the description of systems, processes, activities, and objective evidence with their actual outcomes. In the context of Six Sigma programs, the audits would primarily be within the framework of the organization's overall management system. The scope of audits can include:

- Document audits are preliminary or prerequisite activities needed to determine the scope and purpose of the audit. This stage establishes the acceptance criteria against which the audit observations would be compared in order to highlight observations, divergences, or nonconformities.
- Products and services, comparing the quality characteristics (i.e., functions, dimensions, attributes) as described with what is actually observed.

- Processes, linking the series of tasks and activities for consistent and repeatable outcomes. These audits start with a baseline of descriptions or references to set expectations, and then observation of the actions and outcomes to determine the extent to which such expectations are fulfilled.
- Systems, which are the comprehensive aggregation of documents, actions, and deliverables that could be combined toward the overall organization. This would go beyond compliance to highlight any misalignment between organizational functions (e.g., the just-in-time purchasing department supporting a build-to-stock manufacturing strategy).

The distinction of audit types relates to how the auditor and the auditees are connected.

First-, Second-, and Third-Party Audits

All of these audit types are valid, and may be integrated into a Six Sigma program. The Six Sigma project could use audits as a way to sustain its improvements by reinforcing those practices that contributed to improvement.

For example, consider a Six Sigma project that has determined that for safety reasons, it must use only electronic products certified for EMC and safety from a vendor with a CE Mark designation. In order for this improvement to be sustained, it has been documented within the management system as a required practice.

First-party audits are conducted within an organization, between functional areas. While there should be independence (to ensure freedom from bias), the audits should still be sufficiently rigorous and meaningful to generate actionable observations.

- *Sustain improvements.* The audits would be performed internally on the producers or providers of the products and services, for the benefit of the internal customers. This could also be done to show alignment between the commitments made and the actual deliverables.
- *Procurement example.* Since there is an explicit requirement to provide only certified electronic products, internal audits could ensure that Purchasing specifies this requirement on quotations and invoices, Shipping and Receiving knows to look for particular labels and reports, and that those individuals designing or producing with the electronic products will not approve any products lacking these certification requirements. The audits will look for the controls in place, and whether these controls are observed and respected.

Second-party audits are conducted by external providers (i.e., the direct customer), and are frequently a condition of continued purchases or preservation of the business relationship. The scope can extend beyond the products or services procured to the organization itself, depending upon the priorities of other external interested parties.

- *Sustain improvements.* The audits would be performed by the customer on the producers or providers of the products and services. This could be done as a prerequisite for an order, or to ensure that ongoing orders have controls needed to ensure this requirement.
- *Procurement example.* Since there is an explicit requirement to provide only certified electronic products, supplier audits could ensure that this requirement is prominent on sales orders, the product inspection teams know to look for particular labels and reports, and that those individuals releasing the electronic products will not approve any products lacking these certification requirements. The audits will look for the controls in place, and whether these controls are observed and respected.

Third-party audits are conducted by a representative of a regulatory or accreditation body. While there is no direct customer-supplier relationship, these audits are frequently necessary to overcome a barrier of entry to a particular market. The outcomes are to confirm that the criteria for an independent standard has been fulfilled; or alternatively there are some major or minor nonconformances that need to be overcome. This category also includes representatives of the government, who must audit as part of the regulatory controls for particular industries (e.g., medical, nuclear energy).

- *Sustain improvements.* The audits would be performed by the organization to demonstrate that their controls are needed to ensure this requirement adheres to and complies with the necessary regulations or standards. This external validation would be manifested with a certificate and public communication of acceptance.
- *Procurement example.* Since there is an explicit requirement to provide only certified electronic products, third-party audits could ensure that this requirement is prominent throughout the organization and its relevant supply chain. The audits will look for the controls in place, and whether these controls are observed and respected.

Audit Methods and Six Sigma

All of these audit types are valid, and may be integrated into a Six Sigma program. The Six Sigma project could use audits as a way to sustain its improvements by reinforcing those practices that contributed to improvement.

For example, the comparison of audit findings (objective evidence and observations) to audit criteria (expected references and behavior) will lead to the audit conclusions. The expectation is that audits will be performed in a timely and professional manner and conducted with integrity.

One impact is that the audit may reveal shortcomings or discrepancies in the improvement itself. The objectivity of the audit enables the auditees to freely impart their own perspectives. For this reason, the audits not only validate the compliance, but also the effectiveness of any processes or changes.

Through a combination of independence and confidentiality, the Six Sigma project can benefit from a disclosure of evidence and risks. This enables the Six Sigma team to enhance or modify the recommendations.

In the case of the procurement example, an audit may reveal that having certified electronic components would be undermined if those components are mishandled while in the organization's care. The potential exposure to hazards (e.g., static electricity) could offset the benefits from careful vendor selection and procurement.

PLAN-DO-CHECK-ACT (PDCA)

> Apply and distinguish between the steps of plan-do-check-act (PDCA). (Apply)
>
> **Body of Knowledge VI.B.5**

Plan-Do-Check-Act

PDCA is an improvement cycle methodology that evolved between 1939 and 1993 from Shewhart to Deming and improvisation of the model between Ishikawa, Japanese Quality Circle movement. There are variations of the improvement model.

The model is a simple circle divided into four quadrants named "Plan," "Do," "Check," and "Act." The improvement team brainstorms the activities that the four quadrants should encompass in the context of the problem at hand and fits them into the relevant quadrant. Quality gurus like Deming and Ishikawa have already provided a foundation to this model to build upon.

The initial PDCA cycle (Figures 23.4 and 23.5) was created by solving shop floor problems and preventing their recurrence. Seven Quality Tools were used, along with the PDCA model. The model is robust and applies to many applications, including product design and development, service offering, educational

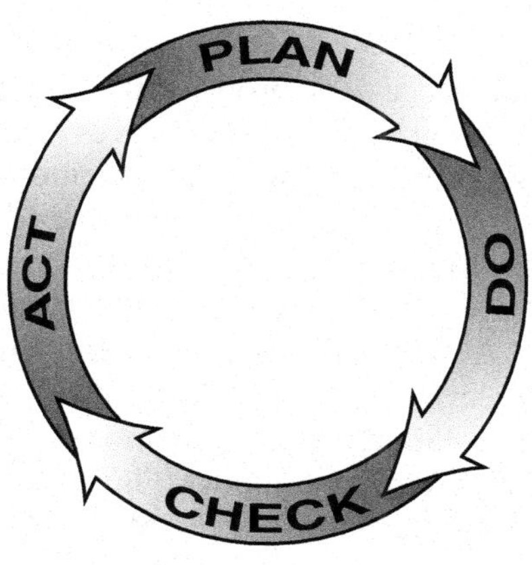

Figure 23.4 The Plan-Do-Check-Act cycle.

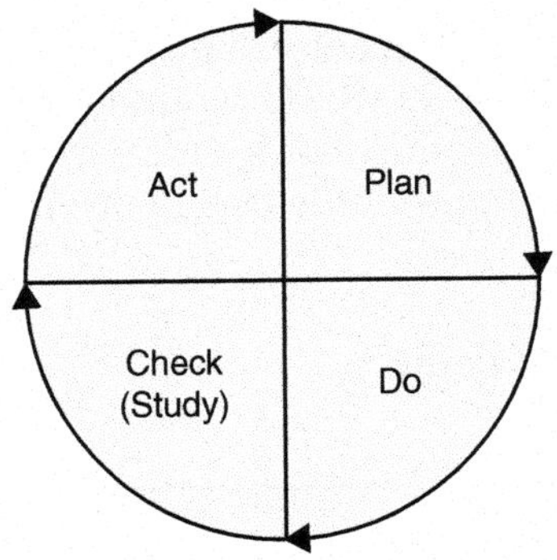

Figure 23.5 The Plan-Do-Check/Study-Act cycle.

curriculum, healthcare, etc. No matter what the application is, we can map the activities to P-D-C-A. This forces a disciplined approach to program management and provides the team members with a big picture of what needs to be accomplished and where they are in the quadrant.

Plan: Define a problem and hypothesize possible causes and solutions.

Do: Implement a solution.

Check: Evaluate the results.

Act: Return to the plan step if the results are unsatisfactory, or standardize the solution if the results are satisfactory.

Later the model was improved by Ishikawa by incorporating additions to the Plan and Do steps (Figure 23.6).

Plan: Determine goals and targets. Determine methods of reaching goals.

Do: Engage in education and training. Implement work.

Check: Check the effects of implementation.

Act: Take appropriate action.

At the "plan" stage, the process or problem that is selected for improvement is described. The improvement goals and objectives are defined. A current process that is resulting in the problem situation is described through process mapping. Quality tools are used to analyze the problem and identify the root causes. Use a cross-functional team to develop a solution, action plan. At the "do" stage, the team implements the solution. At the "check" stage, the team evaluates the effects of the implementation. At the "act" stage, the team takes appropriate actions as to whether to continue the cycle by going back to the "plan" stage or to standardize the learning to ensure sustainability of the actions.

Deming differentiates his PDSA model (Figure 23.7) from the Japanese PDCA as more of a plan for management. The "study" step of the PDSA compares the

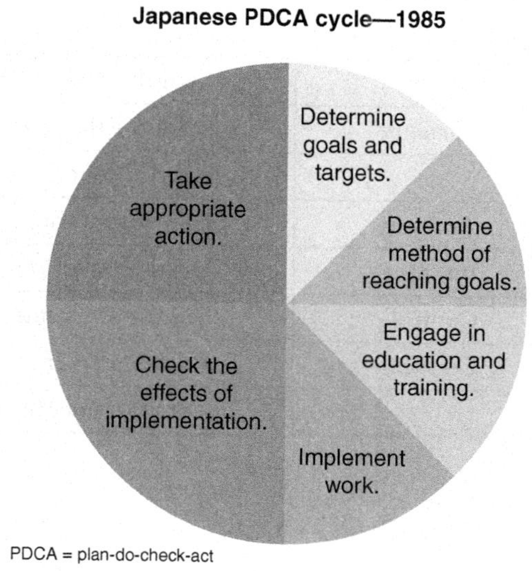

Figure 23.6 The improved PDCA model.

results with the initial predictions as process learning. The improvement team is encouraged to develop a checklist using the PDCA/PDSA model.

The PDCA, PDSA models can also be mapped to the Six Sigma DMAIC methodology and AIAG's 8 Discipline approach.

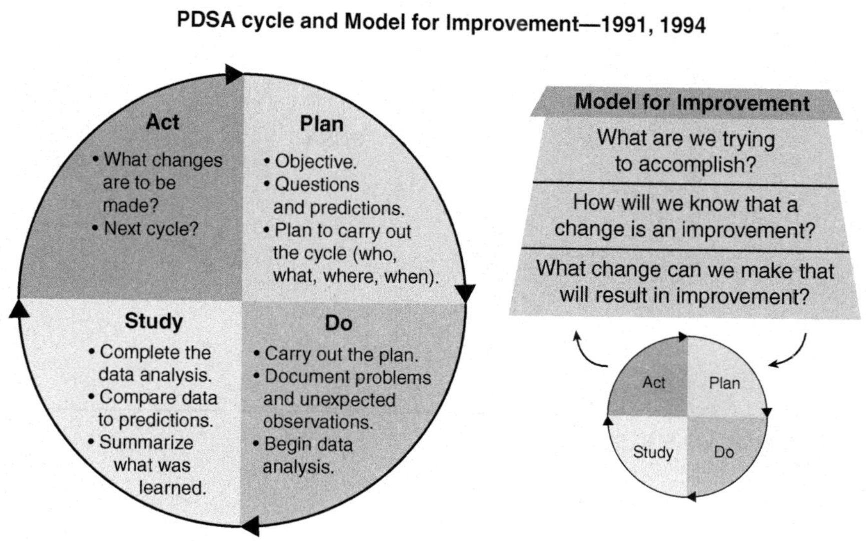

Figure 23.7 PDSA cycle.
Source: Quality Tool Box, Figures Ref: Circling Back- Quality Progress, November 2010.

Table 23.2 List of PDCA tasks.

Plan	Define	Identify the problem, use team approach
DO	Measure	Describe the problem
		Interim containment
CHECK/STUDY	Analyze	Define root cause
	Improve	Develop solutions
ACT	Control	Implement solutions
	Control	Prevent recurrence
		Congratulate the team

Sustaining Improvements with PDSA and SDSA

"Plan-do-study-act" (PDSA) and "standardize-do-check-act" (SDCA) are derivatives of Walter A. Shewhart's "plan-do-check-act" (PDCA) cycle (see Figure 23.8). PDSA should be used when reviewing a process for possible improvement ideas, and SDCA should be used to help ensure that the process is stable and maintaining the gains of previous improvement efforts. Skills/knowledge needed: An understanding of business and manufacturing processes and the fact that continual improvement is never finished."[1]

Variations include the "plan-do-study-act" cycle to emphasize improvements, with a "standardize-do-study-act" cycle to promote stability.

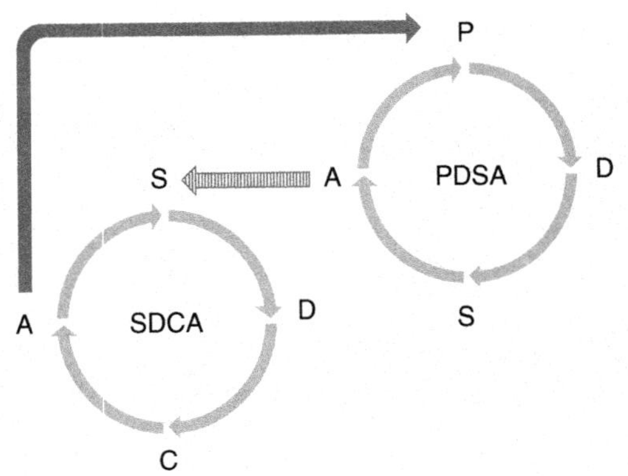

Figure 23.8 PDSA and SDCA.

Chapter 24

Total Productive Maintenance

TOTAL PRODUCTIVE MAINTENANCE (TPM)

> Define the elements of TPM, including use of predictive maintenance and describe how they can be used to control the improved process. (Understand)
>
> Body of Knowledge VI.C.1

Total productive maintenance (TPM) improves the maintenance practices for equipment and infrastructure and enables the prediction and/or prevention of anticipated failure. Through a coordinated effort integrating engineering, operations, and maintenance, a portion of the tasks can be shifted to the operating team members, who can perform maintenance as part of their ongoing process activities.[1]

The effective use of equipment and infrastructure is essential to waste reduction. Managing these assets is the composite activity of the following initiatives:

- Avoiding reduced, idled, or stopped performance due to equipment breakdown.
- Reducing and minimizing the time spent on setup and changeover of equipment, which can otherwise idle machine operations and create bottlenecks.
- Avoiding stoppages arising from the processing or discovery of unacceptable products or services.
- Ensuring that processes and equipment are operating at the speed and pace for which they were designed. If the pace is slower or delayed, work to address and rectify the source of the delays.
- Increasing the yield of acceptable material to reduce material waste, scrap, rework, and the need for material reviews.

TPM aims to remove deficiencies from machines to minimize or eliminate defects and downtime. This extends beyond preventive maintenance to include management of people, processes, systems, and the environment. In any situation where mechanical devices are used, the working state of those devices has an impact on the control of the process. If equipment deteriorates even subtly, the process output may be affected, often in unsuspected ways.

The benefits of effective TPM are twofold:

- TPM will reveal insights into problems that can be applied to other parts of the improvement cycle. This integrates lean back into the overall Six Sigma program by incorporating data collection, problem reporting, and continual improvement for integration into the products and projects.
- TPM will support the efforts taken to identify and address the various forms of waste, or muda, enabling lean systems to more effectively detect and remove waste. Using the *Tim Wood* acronym,[2] the wastes can be summarized as:
  - Transportation
  - Inventory
  - Motion
  - Waiting
  - Overproduction
  - Overprocessing
  - Defects

Total productive maintenance is a system used to ensure that every machine in a production process is able to perform its required tasks so that production is never interrupted. Uptime is maximized, along with machine performance and first-pass yield.[3]

TPM is a lean maintenance strategy for maximizing equipment reliability. TPM evolved from:

- *Breakdown maintenance.* The act of repairing a piece of equipment after it breaks down
- *Corrective maintenance.* Improving or modifying equipment to prevent breakdowns or make maintenance easier
- *Preventive maintenance.* Scheduled maintenance activities performed to prevent unforeseen breakdowns
- *Productive maintenance.* A combination of preventive maintenance, equipment reliability engineering, equipment maintainability engineering, and equipment engineering economics

TPM embraces various disciplines, as operators can inspect, clean, lubricate, adjust, and even perform simple calibrations on their respective equipment. This frees the technical workforce for higher-level preventive maintenance activities that require more of their technical expertise.

TPM relies on data concerning equipment uptime, utilization, and efficiency. Having zero breakdowns, maximum productivity, and zero defects are goals shared by everyone under TPM. TPM focuses on the entire equipment life cycle by coordinating all departments and involving all employees in group activities, from the design of a product, through production, to the end of its useful life. TPM fosters an environment where improvement efforts in safety, quality, cost, delivery, and creativity are encouraged through employee participation. TPM systems have certain elements, including:

- Maintenance infrastructure, goals, and objectives
- Data collection system (overall equipment effectiveness)
- Training and education
- Workflow and controls
- Operational involvement

The key metric for TPM is *overall equipment effectiveness* (OEE), explained further in the OEE section below. OEE is maximized by reducing equipment breakdowns, improving throughput and quality, reducing inventory, and reducing overall lead times, while lowering operating costs.[4] To understand the current situation and measure the effectiveness of any improvement activity, you need a baseline measurement of OEE:

$$\text{OEE} = \text{Availability} \times \text{Performance efficiency} \times \text{Quality rate}$$

Availability is sometimes referred to as *uptime* or *machine utilization*. Availability is used to track the unscheduled downtime losses with the machine. The availability rate is equal to the actual run time (equipment scheduled and operating) divided by the net operating time (scheduled run time). It is critical to note that scheduled run time is not based on calendar time (that is, it does not include planned downtime for activities such as preventive maintenance or when the equipment is not needed for production).

Performance efficiency reflects whether equipment is running at full capacity (or speed) for individual products. Performance is used to track the speed losses with the machine. The performance rate is the actual output divided by the targeted output. This can also be calculated based on the ideal time it takes to produce the actual number of parts made divided by the actual operating time.

OEE is broken down into different losses. The loss of availability is a downtime loss. The loss of performance is a speed loss. The loss of quality is a quality loss. A TPM program includes activities to help avoid the potential drain of productivity caused by the Six Big Losses: breakdowns, changeovers, idling and minor stoppages, reduced speed, scrap and rework, and start-up losses.

While downtime is a common measure, other types of data may be useful to collect:

- *Mean time to repair* (MTTR). The average time taken for repairs to correct a failure. This can be interpreted as a measure of overall maintenance process effectiveness with respect to response and corrections.

- *Mean time between failures* (MTBF). The average time between successive failures of repairable equipment, reflecting overall reliability. This can be interpreted as either the machine reliability or the effectiveness of the maintenance on the machine.

The early signs of a breakdown (that is, hidden failures, minor failures) are anomalies that do not cause any loss of functional capabilities but whose detection helps prevent breakdowns and improve our understanding of our equipment. TPM means listening and watching for anomalies, and taking action before the breakdown.

Loss Description

- *Breakdowns.* A mechanical failure that either stops production or causes production to be stopped while repairs are made.
- *Changeovers.* A tooling or setup change that causes production to be stopped. This category can include periodic tests that must be performed for quality or process capability. For calculating OEE, changeover spans the time between the end of the initial production run and the start of the subsequent production run. Another definition used for changeover time is the duration between the last acceptable product of the predecessor process to the first acceptable product or output following changeover.
- *Idling and minor stoppages. Idling* describes readiness but inactivity due to outside forces such as lack of raw material. *Minor stoppages* are machine faults that can be corrected in less than a minute.
- *Reduced speed.* Intentional slowness for one reason or another; slowing may also be due to problems with machine functionality that do not result in complete breakdown.
- *Scrap and rework.* Manufacturing product that does not conform to customer standards.
- *Start-up losses.* The machine is running but is not yet manufacturing salable product.

Analyzing the breakdown of the OEE figure and comparing it with industry and benchmark data may indicate that performance can readily be improved. Likely areas in which to focus improvement efforts include improving the performance rate by applying a TPM program to the press or improving the quality rate by applying mistake-proofing.

Alternatively, the availability rate for an eight-hour service period may be affected by downtime due to breakdowns and changeovers, similarly to a manufacturing environment. For a clerical operation, the breakdown may be due to equipment malfunctions with the computer system. A changeover could occur when clerks switch shifts.

The performance rate can also be monitored. Idling and minor stops can occur when important information is missing. Similarly, reduced speed losses occur due to lack of training or knowledge capability issues.

TPM can be enhanced by different participants as shown below.

Operators

- Prevent deterioration through regular inspections
- Correct deterioration through regular cleaning of equipment
- Measure deterioration through data collection
- Participate in team activities as equipment experts
- Assist in maintenance activities

Maintenance Personnel

- Planned maintenance activities to reduce disruption to production
- Preventive maintenance through scheduled checks and adjustments
- Training in TPM and OEE principles for operators and supervisors
- Design of equipment—helping eliminate design weaknesses
- Root cause analysis—helping understand root causes of downtime
- Participate in team activities as a maintenance resource

Engineers

- Early equipment management—designing for maintainability and life cycle
- Root cause analysis—helping understand root causes of downtime
- Participate in team activities as an engineering resource
- Prevent contamination at equipment design stage

Managers and Supervisors

- Root cause analysis—helping understand root causes of downtime
- Participate in team activities, offering leadership and direction
- Training—provide facilitation and development
- Accountability—to ensure that data collection and worksheets are completed

It is important to communicate the positive effects of a well-deployed TPM program, such as how it:

- Improves safety and quality
- Increases equipment productivity
- Reduces energy and maintenance costs

Overall Equipment Effectiveness (OEE)

Measurement of equipment efficiency has long been a traditional measure of productivity in manufacturing. Reliability measures have included failure rate, mean time to repair (MTTR), mean time between failures (MTBF), and availability. However, the significance of these measures is diminished without a true understanding of what was included or missed in the measurement.

OEE has three main factors: availability, performance, and quality. These measures can track the efficiency and effectiveness of processes and equipment. The idea of OEE is to assess when equipment is making good product and compare that duration with the total time possible that the equipment could theoretically make good product (the amount of product that passes through the process without requiring scrap or rework):

- *Potential available time* is the theoretical total time possible that the equipment, process, labor, and plant could make good product, or hours (time) of potential operation.
- *Planned downtime* includes all times that should be removed from the OEE calculation because they are times when the equipment is not planned to run production (for example, the facility is closed, scheduled downtime due to maintenance activities, lunch breaks, or when customer demand is met and the equipment is shut down so as to not create overproduction [muda]).
- *Planned available time* is the time that a process, equipment, labor, assembly line, plant, and so on, is planned to be in production making good product, which can be calculated as follows:

$$\text{Planned available time} = \text{Potential available time} - \text{Planned downtime}$$

The next step is to analyze the inefficiencies and productivity losses that occur during planned available time. Inefficiencies and productivity losses are broken down into three groups:

1. Unplanned downtime loss
2. Performance loss
3. Quality loss

Unplanned downtime loss is the sum of the losses of equipment availability due to unplanned stoppage of production over a period of time. The period of time should be the same as that used in the calculation of potential available time and planned available time. Examples may include running out of material/parts, changeovers, and equipment failures.

With unplanned downtime loss calculated, equipment actual operating time can be determined:

$$\text{Actual operating time} = \text{Planned available time} - \text{Sum of all unplanned downtime losses}$$

Availability, the first factor of OEE, is the percentage of time that equipment, the process, labor, or the production line is operating compared with the planned time of operation, or available time downtime losses.

Performance loss, the second group of productivity losses, is the sum of losses during equipment operation due to factors that cause the equipment to operate at less than the maximum designed efficiency over a period of time. Examples may include wear, operator inefficiency, material variations, part jams, and so on:

Actual performance time = Actual operating time –
Sum of all performance time losses

While *actual operating time* is the planned performance of equipment as designed (designed performance), *actual performance time* is the efficiency level at which the equipment actually performs during operation (actual performance).

Performance, the second factor of OEE, is the percentage of time of actual performance (net operating time) that equipment, the process, labor, or the production line runs during operation compared with the designed performance of operation. Performance can also be calculated based on equipment/process cycle times.

Quality loss, the third group of productivity losses, is the sum of quality losses during equipment operation due to defects and rework. This is simply the percentage of good production, or first-pass yield. With quality loss captured, equipment first-pass yield, or first-pass yield, can be determined.

Quality, the third and final factor of OEE, is first-pass yield. In simplest terms, OEE is a ratio of first-pass yield to planned available time.

The measure of OEE is the product of three factors (numerical examples are shown for illustrative purposes):

- *Availability.* Percentage of time equipment is available for production within the total working period
  - This reflects losses due to equipment failures, setup, and adjustments:

    Available hours: 1500

    Downtime from equipment deficiencies and delays: 250

    Availability: (1500 – 250) / 1500 = 83.33%
- *Performance efficiency.* A function of the cycle time, processing time, and equipment operating time
  - This reflects losses due to idling, stoppages, and slower pace:

    Cycle time: 0.5 hours/unit

    Quantity: 2000 units

    Operating time: 1250 hours

    Performance efficiency: (0.5 × 2000) / 1250 = 0.80
- *Rate of quality output.* Yield on the device, equipment, or infrastructure item

- This reflects losses due to process defects and reduced product yield:

  Quantity: 2000 units

  Acceptable units: 1960 units

  Rate of quality output: (1960 / 2000) = 98.0%

The formula and example measurement units for OEE can be summarized as:

OEE = Availability (Available hours) × Performance efficiency
(Proportion of production flow to operating time) ×
Rate of quality products (Percentage yield of acceptable products
to total units produced)

While OEE originated for use in manufacturing operations to monitor equipment, machines, and automated production, it can also be utilized to monitor labor-intensive operations (for example, manual assembly, administrative duties, and service functions). By modifying the definitions of each factor (availability, performance, and quality) for labor-intensive processes, OEE can be utilized as an effective measure for service and administrative functions. This can be called *overall labor effectiveness* (OLE).

VISUAL FACTORY

> Define the elements of a visual factory (Andon, Jidoka) and describe how they can be used to control the improved process. (Understand)
>
> **Body of Knowledge VI.C.2**

The *visual factory* strives to make problems visible, notify employees of current operating conditions, and communicate process goals. Charts placed prominently in the workplace display trends in quality, delivery, downtime, productivity, and other measures. Production and schedule boards advise employees on current conditions. Similar steps can be made in non-factory environments (for example, services, education) to continuously keep all participants informed.[5]

Accessible and clearly illustrated work instructions are critical to avoid negligence and deviations. This is especially true in situations where cross-trained personnel flex into various workstations and mixed-model schedules are employed. Good lines, signs, and labels help ensure that the right component is at the right place and time, further reducing variation.

Visual Management

Visual management uses techniques such as color coding, clear containers for materials and equipment, and improved signage and process indicators within a well-developed 5S (well-organized workplace) environment. Visual systems make vital information available to those who need to know it, when they need to know it, in a simple, straightforward manner. When abnormalities stand out, changes can be made quickly, and processes can get back on track.

Visual management aids are simple tools utilized as communication aids to show work standards (examples of good and defective product) and inspection methods. One form of visual management aids is the *single-point lesson,* used to show how a task should be performed, how a product or service should be inspected, and what to look for during the inspection. The single-point lesson relies heavily on visual displays and contains very little, if any, writing. The concept of the single-point lesson is to utilize a picture to guide the employee through the process and make it easier to identify defects before more value is added to the product or service.[6]

Visual Workplace

Visual management and the "visual workplace" is one of several continuous process improvement systems, and should be deployed in conjunction with other improvement solutions, which include:

1. Lot size reduction
2. Load leveling
3. 3P (production process preparation)
4. Total productive maintenance (TPM)
5. Standard work
6. Built-in feedback
7. Strategic business alignment
8. Continuous improvement process methodology
9. Quality systems
10. Corrective action system
11. Project management
12. Process design
13. Pull system
14. Knowledge transfer

With these systems in place, an organization will be able to see what and where improvements are needed. These improvements can then be reflected and communicated within the visual workplace.

The visual workplace is a compelling operational imperative, central to reducing waste, and crucial to meeting daily performance goals, vastly reduced lead times, and dramatically improved quality.

Visual workplace is defined as follows: a visual workplace is a self-ordering, self-explaining, self-regulating, and self-improving work environment. When a workplace gets visual, it functions differently—safer, better, faster, smoother.[7] Specifically, a visual workplace:

- Is in order
- Explains itself, sharing vital information about what to do/what not to do, how and when to do it, and how to respond if something goes wrong
- Is transparent and self-regulating through high-impact/low-cost visual devices
- Becomes self-improving because visual devices are constantly providing feedback on performance

Visual Devices: The Voice of Your Operations. A visual workplace is made up of many visual devices created by the workforce that needs them.

In a visual workplace, information is converted into simple, commonly understood visual devices, installed in the process of work itself—as close to the point of use as possible. The result is the transformation into a workplace that speaks clearly and precisely about how to perform error-free work safely, smoothly, reliably, and on time.

The Problem: Information Deficits. Workplace information can change quickly and often—production schedules, customer requirements, engineering specifications, operational methods, tooling and fixtures, material procurement, work-in-process, and the thousand other details on which the daily life of the enterprise depends. In any single day, literally thousands of informational transactions are required to keep work current, accurate, and timely.

Data can be found everywhere—in quality reports, SPC graphs, management briefings, in team meetings, and weekly and annual reports. Data flood the workplace. But without understanding the meaning of the data, we cannot make sound decisions and move the company and the people who work there forward.

Calculating the level of information deficits (missing answers) is the quickest way for you to diagnose the extent to which a visual work environment is both absent and needed.

Lean Tools in the Visual Factory. *Lean* is a composite of proven tools and techniques (shown in the "house of lean" diagram in Figure 24.1) intended to promote transparency, standardization, rapid repetition, and incremental improvement of operations and activities. The visual work environment, which may also be characterized as the *visual workplace* or *visual factory*, enhances the effectiveness of lean tools by:

- Placing all tooling, parts, production activities, and indicators in plain view to permit immediate communication and confirmation of system status

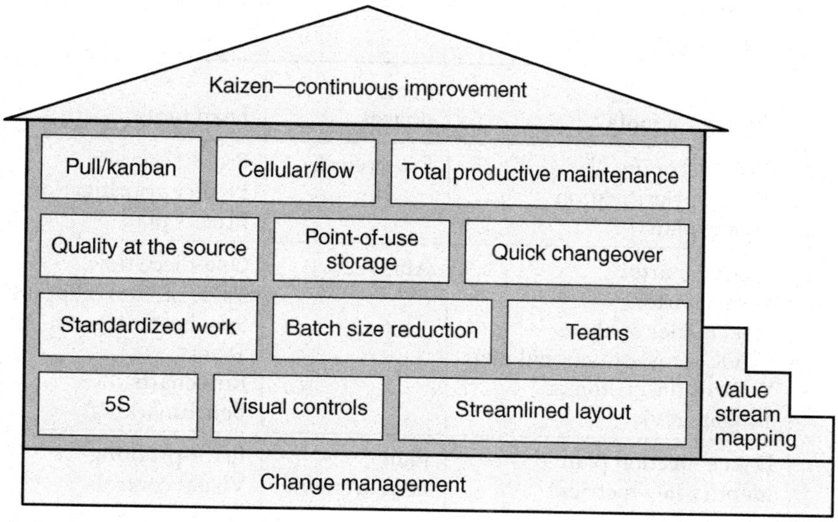

Figure 24.1 The house of lean.

- Labeling and posting of frequently performed activities, workstations, material storage resources, and corresponding diagrams and documents
- Streamlining and standardizing the workplace layout so that team members can obtain the necessary information without unnatural redirection or interruptions of normal work flow

Practices characteristic of both lean and Six Sigma make use of different types of visual indicators and controls as part of the visual factory. In fact, the practices are interdependent and only effective when they are available and displayed to the operators at the point of use. Table 24.1 depicts the lean practice of visual controls as mapping to the *control* phase of a Six Sigma initiative. Control charts and dashboards would be included as examples of visual controls within the visual factory.

As shown in the table, the visual factory first emerges in the earlier *measure* phase of the Lean Six Sigma initiative and continually evolves to incorporate the identified measures, metrics, improvements, process controls, and work flows.

A Gigantic Adherence Mechanism. Visual devices translate the thousands of informational transactions that occur every day at work into visible meaning. Visual devices can show status (on time, process running, help needed), share work priorities (as in a work priority display board), prevent defects (from simple signage reminders to complex mistake-proofing systems), provide order on the plant floor (through clear borders for WIP and deliveries, along with person-width borders for easy access), and of course much more.[8]

The key to a visual workplace is that the visual devices used are easily understood and accessed by all. The use of pictures, colors, shapes, and so forth, makes it easier and more efficient to quickly comprehend and react to events within a process than a spreadsheet filled with numbers, a written procedure, or no

Table 24.1 Lean Six Sigma tools.

DMAIC phases	Six Sigma tools	Lean phases	Lean tools
Pre-project	Project scoping Project prioritization Project plan	Pre-project	Project scoping Project prioritization Project plan
Define	Project charter Team charter Stakeholder analysis SIPOC, cross-functional map Voice of the customer Tollgate review	Analyze	One-piece flow Value stream mapping Spaghetti diagram Teams Run charts Benchmarking
Measure	Data collection plan Identify key metrics Gap analysis Process sigma calculation Capability study Control charts Tollgate review	Plan improvement	Error-proofing Visual controls Total productive maintenance Streamlined layout
Analyze	Pareto chart Ishikawa diagram Five whys Run charts Relations graph Correlation Regression analysis Hypothesis testing Tollgate review	Focus improvement	Visual display 5S Value stream mapping Root cause analysis Five whys
Improve	Brainstorming Mistake-proofing Design of experiments Pugh matrix House of quality Failure mode and effects analysis Tollgate review	Deliver performance	Kaizen Kanban Changeover reduction Point-of-use storage Standardized work Failure mode and effects analysis
Control	Control charts Process sigma Dashboards Balanced scorecards Storyboarding Tollgate review	Improve performance	Visual controls 5S Continous flow and cell design Quality at the source Balanced scorecards

information at all. Colors are used to quickly identify various situations of volumes, status, and so forth.

Visual workplace can be applied anywhere, from our daily life to the sophisticated manufacturing and service workplace. An employee photo ID badge is a visual aid to ensure that outsiders do not enter an organization's work area, mingle, and cause risk to employee and information security. Most documented standard work uses a pictorial work flow so that the operators can comprehend the process effectively. Incorporating a watermark feature in a document helps identify whether the document is controlled/uncontrolled, a draft, or obsolete.

Inspection tags in various colors interpret product as "good," "scrap," "on hold," and so forth, thus preventing the shipping of bad products to the customer and overprocessing an already defective product.

Perhaps the most well-known type of visual workplace is the Toyota *andon board*. This visual display notifies the group leader, supervisor, or maintenance with a light that pinpoints the process step and sometimes includes an audible alarm when a process is stopped. The andon board is a visual representation of production/process status. Some andon boards incorporate scrolling LED message boards that update the employee of the current production rate, total production, production variation, operational equipment efficiency, production goals, and takt time. With costs of technology dropping, large flat-screen monitors are being used more in production and service workplaces. The displays can be revised in real time on the screen. Marker boards and grease boards used in emergency rooms for tracking patients are also being replaced with flat-screen monitors.

Another form of visual workplace is the use of *stack lights*. This light is fitted to machines and has red, yellow, and green lights. *Green* indicates normal manufacturing with no interruption. *Yellow* indicates the operator needs help. *Red* indicates that manufacturing has stopped. The operator may signal yellow to draw attention and ask for help at any time. The line supervising personnel should provide assistance and record the incident to resolve the issue and prevent recurrence. As stated earlier, visual management is utilized in service organizations as well.

Material planning also uses various visual controls to display material levels of various parts in inventory. Using the visual indicators of green, yellow, and red, visual controls can be put in place to trigger restock of parts, material, or processes. Visual controls can be utilized for material planning on the production line, in the warehouse, and in office areas to control reorder points and minimize inventory levels.

When a pull system is created, a kanban card is used as an effective visual indicator to trigger production of parts and components of an assembly process when they are depleted from inventory. Kanban cards may be used to create another visual management system in which a subassembly process will be visually alerted when to produce parts.

In service industries, visual management can be utilized to indicate when additional resources are needed in the process. Systems like this may be utilized in service industries to trigger the addition of a bank teller, checkout clerk, counter associate, and so forth, when a line of customers reaches a given length.

Simple color-coding techniques are used in doctor's offices, hospitals, schools, and many other places as a file management tool to reduce the time to locate files. The same color-coding technique is also used in inventory management systems to allow employees to quickly locate specific parts and supplies in a warehouse or storage space.

A colored diagonal line can be used to quickly indicate whether a critical manual or documentation is not in its designated place. Methods like this can be deployed in maintenance areas or organizations that are regulated to maintain critical records in the event of an emergency or audit.

Equipment maintenance personnel use visual controls to identify the status of the equipment as "up" or "down" or under preventive maintenance. They also use *"lock out tag out"* (LOTO) tools in high-risk areas to prevent inadvertent use of equipment that, if used, could cause a safety issue to the operator and the equipment. Hazardous areas are designated with signs that alert an employee of work hazards.

There are many manufacturing applications for visual devices. Manufacturing control valves are pictured in the correct direction for "open" and displayed along the valve so that employees know if the valve is mistakenly closed by anyone. Another typical application is putting limits on analog gages. This is a cut piece of paper that identifies the sector of the circle in the gage that is the operating range. This will help the operator stay within the operating range or identify when equipment requires servicing. For digital gages, audio alarms and LEDs are available that provide a similar indication.

Standard work can display visual defects to help train new inspectors and also verify product during inspection. In training of employees, visual control employing different-colored tags is used to distinguish the trainees from the skilled experts. Visual hour-by-hour charts are used in areas where processes are expected to have a rate of output based on takt time. They are specifically applicable in processes where flow has been established. These charts consist of expected output per a given period of time, or *pitch*. This is a good tool for employees to quickly identify whether a process is operating as expected and to document reasons for deviations. Color coding (for example, green for output met expected, and red for production not met expected) can add a level of efficiency to highlight trouble areas.

The visual workplace is limited only by the imagination. The intent is to make the comparison of actual performance with expected performance easy and accessible by all employees. The purpose is to focus on the process and highlight areas in need of focus.

There are challenges to the visual workplace. Employees and management must respond to the visual signal. If someone fails to respond to an alarm light, if management does not address a lost production issue, if warehouse personnel do not replenish the depleted stock, if an operator ignores a visual defect and passes it on to the next process, or if employees choose to ignore a safety display and enter the danger zone without protection, then any good visual system will have no purpose.

Jidoka

Jidoka is sometimes termed "intelligent automation" or "automation with a human touch," and is otherwise referenced as "autonomation." The integration of automation with human intelligence reduces the need for direct monitoring and enables productivity gains by multi-process handling. Operators are alerted to the automated stoppages, enabling a response based on human judgment.

With this method, abnormalities (e.g., process malfunctions, product defects) can be quickly identified and corrected, improving the cycle times and reducing defective products or components. The objectives are continuous improvement and employee empowerment.

Characteristics of Jidoka include techniques such as poka-yoke, automatic stops, and in-station quality controls. The outcomes of this approach are demonstrated by:

- Harmonizing of humans and machines.
- Automated stoppages and responses to quality issues or irregularities.
- Detection of special cause variation.

The elements of total productive maintenance have been described and shown as a way to control the improved processes. To reiterate, these TPM elements include the seven wastes, overall equipment effectiveness, kanban, visual factory, and Jidoka. Through the deployment and application of these methods, the Six Sigma controls can be effectively applied and sustained.

Part VII

Appendices

Appendix A
ASQ Code of Ethics

INTRODUCTION

The purpose of the American Society for Quality (ASQ) Code of Ethics is to establish global standards of conduct and behavior for its members, certification holders, and anyone else who may represent or be perceived to represent ASQ. In addition to the code, all applicable ASQ policies and procedures should be followed. Violations to the Code of Ethics should be reported. Differences in work style or personalities should be first addressed directly with others before escalating to an ethics issue.

The ASQ Professional Ethics and Qualifications Committee, appointed annually by the ASQ Board of Directors, is responsible for interpreting this code and applying it to specific situations, which may or may not be specifically called out in the text. Disciplinary actions will be commensurate with the seriousness of the offense and may include permanent revocation of certifications and/or expulsion from the society.

FUNDAMENTAL PRINCIPLES

ASQ requires its representatives to be honest and transparent. Avoid conflicts of interest and plagiarism. Do not harm others. Treat them with respect, dignity, and fairness. Be professional and socially responsible. Advance the role and perception of the Quality professional.

EXPECTATIONS OF A QUALITY PROFESSIONAL

1. **Act with Integrity and Honesty**
    1. Strive to uphold and advance the integrity, honor, and dignity of the Quality profession.
    2. Be truthful and transparent in all professional interactions and activities.
    3. Execute professional responsibilities and make decisions in an objective, factual, and fully informed manner.

4. Accurately represent and do not mislead others regarding professional qualifications, including education, titles, affiliations, and certifications.

5. Offer services, provide advice, and undertake assignments only in your areas of competence, expertise, and training.

2. Demonstrate Responsibility, Respect, and Fairness

1. Hold paramount the safety, health, and welfare of individuals, the public, and the environment.

2. Avoid conduct that unjustly harms or threatens the reputation of the Society, its members, or the Quality profession.

3. Do not intentionally cause harm to others through words or deeds. Treat others fairly, courteously, with dignity, and without prejudice or discrimination.

4. Act and conduct business in a professional and socially responsible manner.

5. Allow diversity in the opinions and personal lives of others.

3. Safeguard Proprietary Information and Avoid Conflicts of Interest

1. Ensure the protection and integrity of confidential information.

2. Do not use confidential information for personal gain.

3. Fully disclose and avoid any real or perceived conflicts of interest that could reasonably impair objectivity or independence in the service of clients, customers, employers, or the Society.

4. Give credit where it is due.

5. Do not plagiarize. Do not use the intellectual property of others without permission. Document the permission as it is obtained.

Appendix B

Six Sigma Green Belt Body of Knowledge Map 2014–2022

The Certified Six Sigma Green Belt (CSSGB) Body of Knowledge (BoK) has been updated to ensure that the most current state of Six Sigma Green Belt practice is being tested in the examination. If you would like more information on how a BoK is updated, see a description of the process on the ASQ website (www.asq.org).

Part of the updating process is to conduct a job analysis survey to determine whether the topics in the 2014 BoK are still relevant to the job role of Six Sigma Green Belts and to identify any new topics that have emerged since that BoK was developed. The results of the CSSGB job analysis survey showed that nearly all of the topics that were in the 2014 BoK are still relevant to the job roles of Six Sigma Green Belts in 2022.

The 2021 Certified Six Sigma Green Belt (CSSGB) BoK was introduced at the August 2022 administration.

GENERAL COMMENTS ABOUT ASQ BODY OF KNOWLEDGE UPDATES

When the Body of Knowledge (BoK) is updated for an ASQ exam, the majority of the material covered in the BoK remains the same. There are very few programs that change dramatically over a 5–7 year period. One of the points that we make to all of the exam development committees is that ASQ certification exams need to reflect "the state of practice" not "the state of the art"—this helps to keep the programs grounded in what people currently do, rather than being driven by the latest hot-topic improvement idea or trend. Typically, the biggest change in any updated BoK is in how the content is organized. When a new BoK is announced and posted on the ASQ website, we also include a "BoK Map" that highlights the changes between the two bodies of knowledge, old and new. The BoK map also clearly identifies any new content that has been added to the exam, as well as any content that has been removed from the exam.

2014 BoK Section	2022 BoK Details	Notes
	I. Overview: Six Sigma and the Organization [11 Questions]	Number of questions changed from 13 to 11
	A. Six sigma and organizational goals	
I.A.1	1. Value of six sigma Recognize why organizations use six sigma, how they apply its philosophy and goals, and the evolution of six sigma from quality leaders such as Juran, Deming, Shewhart, Ishikawa, and others. (Understand)	
I.A.2	2. Organizational goals and six sigma projects Identify the linkages and supports that need to be established between a selected six sigma project and the organization's goals including SMART goals, and describe how process inputs, outputs, and feedback at all levels can influence the organization as a whole. (Understand)	Added SMART goals
I.A.3	3. Organizational drivers and metrics Recognize key business drivers (profit, market share, customer satisfaction, efficiency, product differentiation, key performance indicators (KPIs)) for all types of organizations. Understand how key metrics and scorecards are developed and how they impact the entire organization. (Understand)	Added key performance indicators (KPIs)
I.B	**B. Lean principles in the organization**	
I.B.1	1. Lean concepts Define and describe lean concepts such as theory of constraints, value chain, flow, takt time, just-in-time (JIT), Gemba, spaghetti diagrams, and perfection. (Apply)	Added takt time, just-in-time (JIT), Gemba and spaghetti diagrams
I.B.2	2. Value-stream mapping Use value-stream mapping to identify value-added processes and steps or processes that produce waste, including excess inventory, unused space, test inspection, rework, transportation, and storage. (Understand)	
I.C	**C. Design for six sigma (DFSS) methodologies**	
I.C.1	1. Road maps for DFSS Distinguish between DMADV (define, measure, analyze, design, verify) and IDOV (identify, design, optimize, verify), and recognize how they align with DMAIC. Describe how these methodologies are used for improving the end product or process during the design (DFSS) phase. Understand how verification and validation are used to compare results against stated goals. (Understand)	Added verification and validation
I.C.2	2. Basic failure mode and effects analysis (FMEA) Use FMEA to evaluate a process or product and determine what might cause it to fail and the effects that failure could have. Identify and use scale criteria, calculate the risk priority number (RPN), and analyze the results. (Analyze)	
I.C.3	3. Design FMEA and process FMEA Define and distinguish between these two uses of FMEA. (Apply)	

Continued

Continued

2014 BoK	2022 BoK Details	Notes
	II. Define Phase [20 Questions]	Number of questions changed from 23 to 20
	A. Project identification	
II.A.1	1. Project selection Describe the project selection process and what factors should be considered in deciding whether to use the six sigma DMAIC methodology or another problem-solving process. (Understand)	
II.A.2	2. Process elements Define and describe process components and boundaries. Recognize how processes cross various functional areas and the challenges that result for process improvement efforts. (Analyze)	
II.A.3	3. Benchmarking Understand various types of benchmarking, including competitive, collaborative and best practices. (Understand)	
II.A.4	4. Process inputs and outputs Identify process input and output variables and evaluate their relationships using the supplier, inputs, process, output, customer (SIPOC) model. (Analyze)	
II.A.5	5. Owners and stakeholders Identify the process owners and other stakeholders in a project. (Apply)	
	B. Voice of the customer (VOC)	
II.B.1	1. Customer identification Identify the internal and external customers of a project, and what effect the project will have on them. (Apply)	
II.B.2	2. Customer data Collect feedback from customers using surveys, focus groups, interviews, and various forms of observation. Identify the key elements that make these tools effective. Review data collection questions to eliminate vagueness, ambiguity, and any unintended bias. (Apply)	
II.B.3	3. Customer requirements Use quality function deployment (QFD), Critical to X (CTX when 'X' can be quality, cost, safety, etc.), Critical to Quality tree (CTQ), and Kano model to translate customer requirements statements into product features, performance measures, or opportunities for improvement. Use weighting methods as needed to amplify the importance and urgency of different kinds of input; telephone call vs. survey response; product complaint vs. expedited service request. (Apply)	Added CTX, CTQ, and Kano model
	C. Project management basics	
NEW	1. Project methodology Define and apply agile and top-down project management methods. (Apply)	
II.C.1	2. Project charter Define and describe elements of a project charter and develop a problem statement that includes baseline data or current status to be improved and the project's goals. (Apply)	
II.C.2	3. Project scope Help define the scope of the project using process maps, Pareto charts, and other quality tools. (Apply)	

Continued

Continued

2014 BoK	2022 BoK Details	Notes
II.C.3	4. Project metrics Help develop primary metrics (reduce defect levels by x-amount) and consequential metrics (the negative effects that making the planned improvement might cause). (Apply)	
II.C.4	5. Project planning tools Use work breakdown structures (WBS), Gantt charts, critical path method (CPM), program evaluation and review technique (PERT) charts, and toll-gate reviews to plan projects and monitor their progress. (Apply)	Added work breakdown structure (WBS), and toll-gate reviews
II.C.5	6. Project documentation Describe the types of data and input needed to document a project. Identify and help develop appropriate presentation tools (storyboards, spreadsheet summary of results) for phase reviews and management updates. (Apply)	
II.C.6	7. Project risk analysis and management Describe the elements of project risk analysis, including feasibility, potential impact, risk priority number (RPN), and risk management. Identify the potential effect risk can have on project goals and schedule, resources (materials and personnel), business continuity planning, costs and other financial measures, and stakeholders. (Understand)	Revised subtopic name; added risk management and business continuity planning
II.C.7	8. Project closure Review with team members and sponsors the project objectives achieved in relation to the charter and ensure that documentation is completed and stored appropriately. Identify lessons learned and inform other parts of the organization about opportunities for improvement. (Apply)	
II.D	**D. Management and planning tools** Define, select, and apply these tools: 1) affinity diagrams, 2) interrelationship digraphs, 3) tree diagrams, 4) prioritization matrices, 5) matrix diagrams, 6) process decision program charts (PDPC), 7) activity network diagrams, and 8) SWOT analysis. (Apply)	Added SWOT analysis
II.E	**E. Business results for projects**	
II.E.1	1. Process performance Calculate process performance metrics such as defects per unit (DPU), rolled throughput yield (RTY), cost of poor quality (COPQ), defects per million opportunities (DPMO), sigma levels, and process capability indices. Track process performance measures to drive project decisions. (Analyze)	
II.E.2	2. Communication Define and describe communication techniques used in organizations: top-down, bottom-up, and horizontal. (Apply)	
II.F	**F. Team dynamics and performance**	
II.F.1	1. Team stages and dynamics Define and describe the stages of team evolution, including forming, storming, norming, performing, adjourning, and recognition. Identify and help resolve negative dynamics such as overbearing, dominant, or reluctant participants, the unquestioned acceptance of opinions as facts, groupthink, feuding, floundering, the rush to accomplishment, attribution, discounts, digressions, and tangents. (Understand)	

Continued

Continued

2014 BoK	2022 BoK Details	Notes
II.F.2	2. Team roles and responsibilities Use tools, such as RACI, to describe and define the roles and responsibilities of participants on six sigma and other teams, including black belt, master black belt, green belt, champion, executive, coach, facilitator, team member, sponsor, and process owner. (Apply)	Added RACI
II.F.3	3. Team tools and decision-making concepts Define and apply team tools such as brainstorming, and decision-making concepts such as nominal group technique, and multi-voting. (Apply)	Revised subtopic name and added decision-making concepts to subtext
II.F.4	4. Team communication Identify and use appropriate communication methods (both within the team and from the team to various stakeholders) to report progress, conduct reviews, and support the overall success of the project. (Apply)	
	III. Measure Phase [20 Questions]	Number of questions changed from 23 to 20
III.A	**A. Process analysis and documentation** Develop process maps and review written procedures, work instructions, and flowcharts to identify any gaps or areas of the process that are misaligned. (Create)	
III.B	**B. Probability and statistics**	
III.B.1	1. Basic probability concepts Describe and interpret basic probability concepts: independent events, mutually exclusive events, multiplication rules, permutations, and combinations. (Understand)	
III.B.2	2. Central limit theorem Define the central limit theorem and describe its significance in relation to confidence intervals, hypothesis testing, and control charts. (Understand)	
III.C	**C. Statistical distributions** Define and describe various distributions as they apply to statistical process control and probability: normal, binomial, Poisson, chi square, Student's t, and F. (Understand)	
III.D	**D. Collecting and summarizing data**	
III.D.1	1. Types of data and measurement scales Identify and classify continuous (variables) and discrete (attributes) data. Describe and define nominal, ordinal, interval, and ratio measurement scales. (Analyze)	
III.D.2	2. Sampling and data collection plans and methods Define and apply various sampling methods (random and stratified) and data collection methods (check sheets and data coding). Prepare data collection plans that include gathering data and performing quality checks (e.g., minimum/maximum values, erroneous data, null values). (Apply)	Revised subtopic name and added data collection plans and quality checks
III.D.3	3. Descriptive statistics Define, calculate, and interpret measures of dispersion and central tendency. Develop and interpret frequency distributions and cumulative frequency distributions. (Evaluate)	

Continued

Continued

2014 BoK	2022 BoK Details	Notes
III.D.4	4. Graphical methods Construct and interpret diagrams and charts that are designed to communicate numerical analysis efficiently, including scatter diagrams, normal probability plots, histograms, stem-and-leaf plots, box-and-whisker plots. (Create)	
III.E	**E. Measurement system analysis (MSA)** Calculate, analyze, and interpret measurement system capability using gauge repeatability and reproducibility (GR&R) studies, measurement correlation, bias, linearity, percent agreement, and precision/tolerance (P/T). (Evaluate)	
III.F	**F. Process and performance capability**	
III.F.1	1. Process performance vs. process specifications Define and distinguish between natural process limits and specification limits, and calculate process performance metrics. (Evaluate)	
III.F.2	2. Process capability studies Define, describe, and conduct process capability studies, including identifying characteristics, specifications, and tolerances, and verifying stability and normality. (Evaluate)	
III.F.3	3. Process capability (C_p, C_{pk}) and process performance (P_p, P_{pk}) indices Describe the relationship between these types of indices. Define, select, and calculate process capability and process performance. Describe when C_{pm} measures can be used. Calculate the sigma level of a process. (Evaluate)	
III.F.4	4. Short-term vs. long-term capability and sigma shift Describe the assumptions and conventions that are appropriate to use when only short-term data are used. Identify and calculate the sigma shift that occurs when long- and short-term data are compared. (Evaluate)	
	IV. Analyze Phase [18 Questions]	Number of questions changed from 15 to 18
IV.A	**A. Exploratory data analysis**	
IV.A.1	1. Multi-vari studies Select appropriate sampling plans to create multi-vari study charts and interpret the results for positional, cyclical, and temporal variation. (Create)	
IV.A.2	2. Correlation and linear regression Describe the difference between correlation and causation. Calculate the correlation coefficient and linear regression and interpret the results in terms of statistical significance (p-value). Use regression models for estimation and prediction. (Evaluate)	
IV.B	**B. Hypothesis testing**	
IV.B.1	1. Basics Distinguish between statistical and practical significance. Determine appropriate sample sizes and develop tests for significance level, power, and type I and type II errors. (Apply)	

Continued

Continued

2014 BoK	2022 BoK Details	Notes
IV.B.2	2. Tests for means, variances, and proportions Conduct hypothesis tests to compare means, variances, and proportions (paired-comparison t-test, F-test, analysis of variance (ANOVA), chi square) and interpret the results. (Analyze)	
NEW	**C. Additional analysis methods**	
NEW	1. Gap analysis Analyze scenarios to identify performance gaps and compare current and future states using predefined metrics. (Analyze)	
V.B	2. Root cause analysis Use cause and effect diagrams, relational matrices, 5 Whys, fault tree analysis, and other problem-solving tools to identify the true cause of a problem. (Analyze)	
	V. Improve Phase [16 Questions]	Number of questions changed from 15 to 16
	A. Design of experiments (DOE)	
V.A.1	1. Basic terms Define and describe terms such as independent and dependent variables, factors and levels, responses, treatments, errors, repetition, blocks, randomization, effects, and replication. (Understand)	
V.A.2	2. DOE graphs and plots Interpret main effects analysis and interaction plots. (Apply)	
NEW	**B. Implementation planning** Apply implementation planning by using proof of concept, try-storming, simulations, and conducting pilot tests. (Apply)	
	C. Lean tools	
V.C.1	1. Waste elimination Select and apply tools and techniques for eliminating or preventing waste, including pull systems, kanban, 5S, standard work, and poka-yoke. (Apply)	
V.C.2	2. Cycle-time reduction Use various techniques to reduce cycle time (continuous flow, setup reduction), single-minute exchange of dies (SMED). (Analyze)	Added SMED
V.C.3	3. Kaizen and kaizen blitz Define and distinguish between these two methods and apply them in various situations. (Apply)	
	VI. Control Phase [15 Questions]	Number of questions changed from 11 to 15
VI.A	**A. Statistical process control (SPC)**	
VI.A.1	1. SPC Basics Describe the theory and objectives of SPC, including measuring and monitoring process performance for both continuous and discrete data. Define and distinguish between common and special cause variation and how these conditions can be deduced from control chart analysis. (Analyze)	
VI.A.2	2. Rational subgrouping Define and describe how rational subgrouping is used. (Understand)	

Continued

Continued

2014 BoK	2022 BoK Details	Notes
VI.A.3	3. Control charts Identify, select, construct, and use control charts: $\bar{X}-R$, $\bar{X}-s$, individual and moving range (ImR or XmR), median, p, np, c, and u. (Apply)	
NEW	**B. Sustain improvements**	
VI.B	1. Control plan Assist in developing and implementing a control plan to document and monitor the process. (Apply)	Removed "maintain the improvements"
NEW	2. Document control Understand document control and its role in controlling and sustaining improvements. (Understand)	
NEW	3. Training plans Develop training plans to implement and sustain improvements. (Apply)	
NEW	4. Audits Define first-, second-, and third-party audits. (Remember)	
NEW	5. Plan-do-check-act (PDCA) Apply and distinguish between the steps of plan-do-check-act (PDCA). (Apply)	
VI.C	**C. Lean tools for process control**	
VI.C.1	1. Total productive maintenance (TPM) Define the elements of TPM, including use of predictive maintenance and describe how they can be used to control the improved process. (Understand)	Added predictive maintenance
VI.C.2	2. Visual factory Define the elements of a visual factory (Andon, Jidoka) and describe how they can be used to control the improved process. (Understand)	Added Andon and Jidoka as examples

Appendix C

ASQ Certified Six Sigma Green Belt (CSSGB) Body of Knowledge (2022)

Included in this body of knowledge (BOK) are explanations (subtext) and cognitive levels for each topic or subtopic in the test. These details will be used by the Examination Development Committee as guidelines for writing test questions and are designed to help candidates prepare for the exam by identifying specific content within each topic that can be tested. Except where specified, the subtext is not intended to limit the subject or be all-inclusive of what might be covered in an exam but is intended to clarify how topics are related to the role of the Certified Six Sigma Green Belt (SSGB). The descriptor in parentheses at the end of each subtext entry refers to the highest cognitive level at which the topic will be tested. A complete description of cognitive levels is provided at the end of this document.

I. Overview: Six Sigma and the Organization (11 Questions)

A. Six sigma and organizational goals

1. Value of six sigma

Recognize why organizations use six sigma, how they apply its philosophy and goals, and the evolution of six sigma from quality leaders such as Juran, Deming, Shewhart, Ishikawa, and others. (Understand)

2. Organizational goals and six sigma projects

Identify the linkages and supports that need to be established between a selected six sigma project and the organization's goals including SMART goals, and describe how process inputs, outputs, and feedback at all levels can influence the organization as a whole. (Understand)

3. Organizational drivers and metrics

Recognize key business drivers (profit, market share, customer satisfaction, efficiency, product differentiation, key performance indicators (KPIs)) for all types of organizations. Understand how key metrics and scorecards are developed and how they impact the entire organization. (Understand)

B. **Lean principles in the organization**

1. **Lean concepts**

   Define and describe lean concepts such as theory of constraints, value chain, flow, takt time, just-in-time (JIT), Gemba, spaghetti diagrams, and perfection. (Apply)

2. **Value-streaming mapping**

   Use value-stream mapping to identify value-added processes and steps or processes that produce waste, including excess inventory, unused space, test inspection, rework, transportation, and storage. (Understand)

C. **Design for six sigma (DFSS) methodologies**

1. **Road maps for DFSS**

   Distinguish between DMADV (define, measure, analyze, design, verify) and IDOV (identify, design, optimize, verify), and recognize how they align with DMAIC. Describe how these methodologies are used for improving the end product or process during the design (DFSS) phase. Understand how verification and validation are used to compare results against stated goals. (Understand)

2. **Basic failure mode and effects analysis (FMEA)**

   Use FMEA to evaluate a process or product and determine what might cause it to fail and the effects that failure could have. Identify and use scale criteria, calculate the risk priority number (RPN), and analyze the results. (Analyze)

3. **Design FMEA and process FMEA**

   Define and distinguish between these two uses of FMEA. (Apply)

II. **Define Phase (20 Questions)**

A. **Project identification**

1. **Project selection**

   Describe the project selection process and what factors should be considered in deciding whether to use the six sigma DMAIC methodology or another problem-solving process. (Understand)

2. **Process elements**

   Define and describe process components and boundaries. Recognize how processes cross various functional areas and the challenges that result for process improvement efforts. (Analyze)

3. **Benchmarking**

   Understand various types of benchmarking, including competitive, collaborative, and best practices. (Understand)

4. Process inputs and outputs

Identify process input and output variables and evaluate their relationships using the supplier, inputs, process, output, customer (SIPOC) model. (Analyze)

5. Owners and stakeholders

Identify the process owners and other stakeholders in a project. (Apply)

B. Voice of the customer (VOC)

1. Customer identification

Identify the internal and external customers of a project, and what effect the project will have on them. (Apply)

2. Customer data

Collect feedback from customers using surveys, focus groups, interviews, and various forms of observation. Identify the key elements that make these tools effective. Review data collection questions to eliminate vagueness, ambiguity, and any unintended bias. (Apply)

3. Customer requirements

Use quality function deployment (QFD), Critical to X (CTX when 'X' can be quality, cost, safety, etc.), Critical to Quality tree (CTQ), and Kano model to translate customer requirements statements into product features, performance measures, or opportunities for improvement. Use weighting methods as needed to amplify the importance and urgency of different kinds of input; telephone call vs. survey response; product complaint vs. expedited service request. (Apply)

C. Project management basics

1. Project methodology

Define and apply agile and top-down project management methods. (Apply)

2. Project charter

Define and describe elements of a project charter and develop a problem statement that includes baseline data or current status to be improved and the project's goals. (Apply)

3. Project scope

Help define the scope of the project using process maps, Pareto charts, and other quality tools. (Apply)

4. Project metrics

Help develop primary metrics (reduce defect levels by x-amount) and consequential metrics (the negative effects that making the planned improvement might cause). (Apply)

5. Project planning tools

Use work breakdown structures (WBS), Gantt charts, critical path method (CPM), program evaluation and review technique (PERT) charts, and toll-gate reviews to plan projects and monitor their progress. (Apply)

6. Project documentation

Describe the types of data and input needed to document a project. Identify and help develop appropriate presentation tools (storyboards, spreadsheet summary of results) for phase reviews and management updates. (Apply)

7. Project risk analysis and management

Describe the elements of project risk analysis, including feasibility, potential impact, risk priority number (RPN), and risk management. Identify the potential effect risk can have on project goals and schedule, resources (materials and personnel), business continuity planning, costs and other financial measures, and stakeholders. (Understand)

8. Project closure

Review with team members and sponsors the project objectives achieved in relation to the charter and ensure that documentation is completed and stored appropriately. Identify lessons learned and inform other parts of the organization about opportunities for improvement. (Apply)

D. Management and planning tools

Define, select, and apply these tools: 1) affinity diagrams, 2) interrelationship digraphs, 3) tree diagrams, 4) prioritization matrices, 5) matrix diagrams, 6) process decision program charts (PDPC), 7) activity network diagrams, and 8) SWOT analysis. (Apply)

E. Business results for projects

1. Process performance

Calculate process performance metrics such as defects per unit (DPU), rolled throughput yield (RTY), cost of poor quality (COPQ), defects per million opportunities (DPMO), sigma levels, and process capability indices. Track process performance measures to drive project decisions. (Analyze)

2. Communication

Define and describe communication techniques used in organizations: top-down, bottom-up, and horizontal. (Apply)

F. Team dynamics and performance

1. Team stages and dynamics

Define and describe the stages of team evolution, including forming, storming, norming, performing, adjourning, and recognition. Identify and help resolve negative dynamics such as overbearing, dominant, or reluctant participants, the unquestioned acceptance of opinions as facts, groupthink, feuding, floundering, the rush to accomplishment, attribution, discounts, digressions, and tangents. (Understand)

2. Team roles and responsibilities

Use tools, such as RACI, to describe and define the roles and responsibilities of participants on six sigma and other teams, including black belt, master black belt, green belt, champion, executive, coach, facilitator, team member, sponsor, and process owner. (Apply)

3. Team tools and decision-making concepts

Define and apply team tools such as brainstorming, and decision-making concepts such as nominal group technique, and multi-voting. (Apply)

4. Team communication

Identify and use appropriate communication methods (both within the team and from the team to various stakeholders) to report progress, conduct reviews, and support the overall success of the project. (Apply)

III. Measure Phase (20 Questions)

A. Process analysis and documentation

Develop process maps and review written procedures, work instructions, and flowcharts to identify any gaps or areas of the process that are misaligned. (Create)

B. Probability and statistics

1. Basic probability concepts

Identify and use basic probability concepts: independent events, mutually exclusive events, multiplication rules, permutations, and combinations. (Apply)

2. **Central limit theorem**

   Define the central limit theorem and describe its significance in relation to confidence intervals, hypothesis testing, and control charts. (Understand)

C. **Statistical distributions**

Define and describe various distributions as they apply to statistical process control and probability: normal, binomial, Poisson, chi square, Student's t, and F. (Understand)

D. **Collecting and summarizing data**

1. **Types of data and measurement scales**

   Identify and classify continuous (variables) and discrete (attributes) data. Describe and define nominal, ordinal, interval, and ratio measurement scales. (Analyze)

2. **Sampling and data collection plans and methods**

   Define and apply various sampling methods (random and stratified) and data collection methods (check sheets and data coding). Prepare data collection plans that include gathering data and performing quality checks (e.g., minimum/maximum values, erroneous data, null values). (Apply)

3. **Descriptive statistics**

   Define, calculate, and interpret measures of dispersion and central tendency. Develop and interpret frequency distributions and cumulative frequency distributions. (Evaluate)

4. **Graphical methods**

   Construct and interpret diagrams and charts that are designed to communicate numerical analysis efficiently, including scatter diagrams, normal probability plots, histograms, stem-and-leaf plots, box-and-whisker plots. (Create)

E. **Measurement system analysis (MSA)**

Calculate, analyze, and interpret measurement system capability using gauge repeatability and reproducibility (GR&R) studies, measurement correlation, bias, linearity, percent agreement, and precision/tolerance (P/T). (Evaluate)

F. **Process and performance capability**

1. **Process performance vs. process specifications**

   Define and distinguish between natural process limits and specification limits, and calculate process performance metrics. (Evaluate)

2. **Process capability studies**

   Define, describe, and conduct process capability studies, including identifying characteristics, specifications, and tolerances, and verifying stability and normality. (Evaluate)

3. **Process capability (C_p, C_{pk}) and process performance (P_p, P_{pk}) indices**

   Describe the relationship between these types of indices. Define, select, and calculate process capability and process performance. Describe when C_{pm} measures can be used. Calculate the sigma level of a process. (Evaluate)

4. **Short-term vs. long-term capability and sigma shift**

   Describe the assumptions and conventions that are appropriate to use when only short-term data are used. Identify and calculate the sigma shift that occurs when long- and short-term data are compared. (Evaluate)

IV. Analyze Phase (18 Questions)

A. Exploratory data analysis

1. **Multi-vari studies**

   Select appropriate sampling plans to create multi-vari study charts and interpret the results for positional, cyclical, and temporal variation. (Create)

2. **Correlation and linear regression**

   Describe the difference between correlation and causation. Calculate the correlation coefficient and linear regression and interpret the results in terms of statistical significance (p-value). Use regression models for estimation and prediction. (Evaluate)

B. Hypothesis testing

1. **Basics**

   Distinguish between statistical and practical significance. Determine appropriate sample sizes and develop tests for significance level, power, and type I and type II errors. (Apply)

2. **Tests for means, variances, and proportions**

   Conduct hypothesis tests to compare means, variances, and proportions (paired-comparison t-test, F-test, analysis of variance (ANOVA), chi square) and interpret the results. (Analyze)

C. **Additional analysis methods**

1. **Gap analysis**

Analyze scenarios to identify performance gaps and compare current and future states using predefined metrics. (Analyze)

2. **Root cause analysis**

Use cause and effect diagrams, relational matrices, 5 Whys, fault tree analysis, and other problem-solving tools to identify the true cause of a problem. (Analyze)

V. **Improve Phase (16 Questions)**

A. **Design of experiments (DOE)**

1. **Basic terms**

Define and describe terms such as independent and dependent variables, factors and levels, responses, treatments, errors, repetition, blocks, randomization, effects, and replication. (Understand)

2. **DOE graphs and plots**

Interpret main effects analysis and interaction plots. (Apply)

B. **Implementation planning**

Apply implementation planning by using proof of concepts, try-storming simulations, and pilot tests. (Apply)

C. **Lean tools**

1. **Waste elimination**

Select and apply tools and techniques for eliminating or preventing waste, including pull systems, kanban, 5S, standard work, and poka-yoke. (Apply)

2. **Cycle-time reduction**

Use various techniques to reduce cycle time (continuous flow, setup reduction, single-minute exchange of dies (SMED)). (Analyze)

3. **Kaizen and kaizen blitz**

Define and distinguish between these two methods and apply them in various situations. (Apply)

VI. **Control Phase (15 Questions)**

A. **Statistical process control (SPC)**

1. **SPC Basics**

Describe the theory and objectives of SPC, including measuring and monitoring process performance for both continuous and discrete data. Define and distinguish between common and special cause variation and how these conditions can be deduced from control chart analysis. (Analyze)

2. **Rational subgrouping**

   Define and describe how rational subgrouping is used. (Understand)

3. **Control charts**

   Identify, select, construct, and use control charts: $\bar{X}$–R, $\bar{X}$–s, individual and moving range (ImR or XmR), median, p, np, c, and u. (Apply)

B. **Sustain improvements**

1. **Control plan**

   Assist in developing and implementing a control plan to document and monitor the process. (Apply)

2. **Document control**

   Understand document control and its role in controlling and sustaining improvements. (Understand)

3. **Training plans**

   Develop training plans to implement and sustain improvements. (Apply)

4. **Audits**

   Define first-, second-, and third-party audits. (Remember)

5. **Plan-do-check-act (PDCA)**

   Apply and distinguish between the steps of plan-do-check-act (PDCA). (Apply)

C. **Lean tools for process control**

1. **Total productive maintenance (TPM)**

   Define the elements of TPM, including use of predictive maintenance and describe how they can be used to control the improved process. (Understand)

2. **Visual factory**

   Define the elements of a visual factory (Andon, Jidoka) and describe how they can be used to control the improved process. (Understand)

LEVELS OF COGNITION BASED ON BLOOM'S TAXONOMY—REVISED (2001)

In addition to *content* specifics, the subtext detail also indicates the intended *complexity level* of the test questions for that topic. These levels are based on the Revised "Levels of Cognition" (from Bloom's Taxonomy, 2001) and are presented below in rank order, from least complex to most complex.

Remember

Be able to remember or recognize terminology, definitions, facts, ideas, materials, patterns, sequences, methodologies, principles, etc. (Also commonly referred to as recognition, recall, or rote knowledge)

Understand

Be able to read and understand descriptions, communications, reports, tables, diagrams, directions, regulations, etc.

Apply

Be able to apply ideas, procedures, methods, formulas, principles, theories, etc., in job-related situations.

Analyze

Be able to break down information into its constituent parts and recognize the parts' relationship to one another and how they are organized; identify sublevel factors or salient data from a complex scenario.

Evaluate

Be able to make judgments regarding the value of proposed ideas, solutions, methodologies, etc., by using appropriate criteria or standards to estimate accuracy, effectiveness, economic benefits, etc.

Create

Be able to put parts or elements together in such a way as to show a pattern or structure not clearly there before; able to identify which data or information from a complex set is appropriate to examine further or from which supported conclusions can be drawn.

Appendix D

Control Limit Formulas

VARIABLES CHARTS

$\bar{x}$ and R chart:

$Averages\ chart: \bar{\bar{x}} \pm A_2\bar{R}$ $\quad$ $Range\ chart: LCL = D_3\bar{R} \quad UCL = D_4\bar{R}$

$\bar{x}$ and s chart:

$Averages\ chart: \bar{\bar{x}} \pm A_3\bar{s}$ $\quad$ $Standard\ deviation\ chart: LCL = B_3\bar{s} \quad UCL = B_4\bar{s}$

Individuals and moving range chart (two-value moving window):

$Individuals\ chart: \bar{x} \pm 2.66\bar{R}$ $\quad$ $Moving\ range: UCL = 3.267\bar{R}$

Moving average and moving range (two-value moving window):

$Moving\ average: \bar{\bar{x}} \pm 1.88\bar{R}$ $\quad$ $Moving\ range: UCL = 3.267\bar{R}$

Median chart:

$Median\ chart: \bar{x}' \pm A_2'\bar{R}$ $\quad$ $Range\ chart: LCL = D_3\bar{R} \quad UCL = D_4\bar{R}$

ATTRIBUTES CHARTS

Variable sample size:		Constant sample size:	
p chart:	$\bar{p} \pm 3\sqrt{\frac{\bar{p}(1-\bar{p})}{\bar{n}}}$	np chart:	$n\bar{p} \pm 3\sqrt{n\bar{p}(1-\bar{p})}$
u chart:	$\bar{u} \pm 3\sqrt{\frac{\bar{u}}{\bar{n}}}$	c chart:	$\bar{c} \pm 3\sqrt{\bar{c}}$
D chart:	$\bar{D} \pm 3\sigma_D$	U chart:	$\bar{U} \pm 3\sigma_U$

Appendix E
Constants for Control Charts

Subgroup size N	A_2	d_2	D_3	D_4	A_3	c_4	B_3	B_4	E_2	A_2 for median charts	A_4
2	1.880	1.128	–	3.267	2.659	0.798	–	3.267	2.660	1.880	2.224
3	1.023	1.693	–	2.574	1.954	0.886	–	2.568	1.772	1.187	1.091
4	0.729	2.059	–	2.282	1.628	0.921	–	2.266	1.457	0.796	0.758
5	0.577	2.326	–	2.114	1.427	0.940	–	2.089	1.290	0.691	0.594
6	0.483	2.534	–	2.004	1.287	0.952	0.030	1.970	1.184	0.548	0.495
7	0.419	2.704	0.076	1.924	1.182	0.959	0.118	1.882	1.109	0.508	0.429
8	0.373	2.847	0.136	1.864	1.099	0.965	0.185	1.815	1.054	0.433	0.380
9	0.337	2.970	0.184	1.816	1.032	0.969	0.239	1.761	1.010	0.412	0.343
10	0.308	3.078	0.223	1.777	0.975	0.973	0.284	1.716	0.975	0.362	0.314

Appendix F

Areas Under Standard Normal Curve

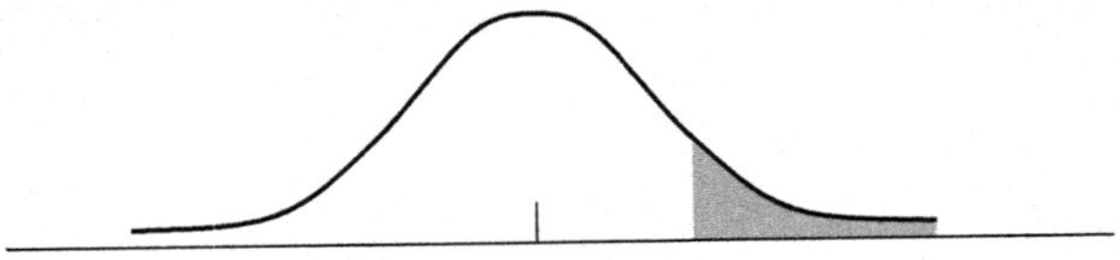

z	0.00	0.01	0.02	0.03	0.04
0.0	0.50000000000	0.49601064369	0.49202168628	0.48803352659	0.48404656315
0.1	0.46017216272	0.45620468746	0.45224157398	0.44828321335	0.44432999519
0.2	0.42074029056	0.41683383652	0.41293557735	0.40904588486	0.40516512830
0.3	0.38208857781	0.37828047818	0.37448416528	0.37069998106	0.36692826396
0.4	0.34457825839	0.34090297377	0.33724272685	0.33359782060	0.32996855366
0.5	0.30853753873	0.30502573090	0.30153178755	0.29805596539	0.29459851622
0.6	0.27425311775	0.27093090378	0.26762889347	0.26434729212	0.26108629969
0.7	0.24196365222	0.23885206809	0.23576249778	0.23269509230	0.22964999716
0.8	0.21185539858	0.20897008787	0.20610805359	0.20326939183	0.20045419326
0.9	0.18406012535	0.18141125489	0.17878637961	0.17618554225	0.17360878034
1.0	0.15865525393	0.15624764502	0.15386423037	0.15150500279	0.14916995033
1.1	0.13566606095	0.13349951324	0.13135688104	0.12923811224	0.12714315056
1.2	0.11506967022	0.11313944644	0.11123243745	0.10934855243	0.10748769707
1.3	0.09680048459	0.09509791780	0.09341750899	0.09175913565	0.09012267246
1.4	0.08075665923	0.07926984145	0.07780384053	0.07635850954	0.07493369953
1.5	0.06680720127	0.06552171209	0.06425548782	0.06300836446	0.06178017671
1.6	0.05479929170	0.05369892815	0.05261613845	0.05155074849	0.05050258347
1.7	0.04456546276	0.04363293652	0.04271622079	0.04181513761	0.04092950898
1.8	0.03593031911	0.03514789358	0.03437950245	0.03362496942	0.03288411866
1.9	0.02871655982	0.02806660666	0.02742894970	0.02680341888	0.02618984494
2.0	0.02275013195	0.02221559443	0.02169169377	0.02117826964	0.02067516287
2.1	0.01786442056	0.01742917794	0.01700302265	0.01658580668	0.01617738337
2.2	0.01390344751	0.01355258115	0.01320938381	0.01287372144	0.01254546144
2.3	0.01072411002	0.01044407706	0.01017043867	0.00990307556	0.00964186995
2.4	0.00819753592	0.00797626026	0.00776025355	0.00754941142	0.00734363096
2.5	0.00620966533	0.00603655808	0.00586774172	0.00570312633	0.00554262344
2.6	0.00466118802	0.00452711113	0.00439648835	0.00426924341	0.00414530136
2.7	0.00346697380	0.00336416041	0.00326409582	0.00316671628	0.00307195922
2.8	0.00255513033	0.00247707500	0.00240118247	0.00232740021	0.00225567669
2.9	0.00186581330	0.00180714378	0.00175015693	0.00169481002	0.00164106123

Continued

Continued

z	0.00	0.01	0.02	0.03	0.04
3.0	0.00134989803	0.00130623845	0.00126387343	0.00122276869	0.00118289074
3.1	0.00096760321	0.00093543672	0.00090425520	0.00087403152	0.00084473917
3.2	0.00068713794	0.00066367486	0.00064095298	0.00061895109	0.00059764850
3.3	0.00048342414	0.00046647986	0.00045008724	0.00043422992	0.00041889195
3.4	0.00033692927	0.00032481440	0.00031310568	0.00030179062	0.00029085709
3.5	0.00023262908	0.00022405335	0.00021577340	0.00020777983	0.00020006352
3.6	0.00015910859	0.00015309850	0.00014730151	0.00014171061	0.00013631902
3.7	0.00010779973	0.00010362962	0.00009961139	0.00009573989	0.00009201013
3.8	0.00007234804	0.00006948340	0.00006672584	0.00006407163	0.00006151716
3.9	0.00004809634	0.00004614806	0.00004427448	0.00004247293	0.00004074080
4.0	0.00003167124	0.00003035937	0.00002909907	0.00002788843	0.00002672560
4.1	0.00002065751	0.00001978296	0.00001894362	0.00001813816	0.00001736529
4.2	0.00001334575	0.00001276853	0.00001221512	0.00001168457	0.00001117599
4.3	0.00000853991	0.00000816273	0.00000780146	0.00000745547	0.00000712414
4.4	0.00000541254	0.00000516853	0.00000493505	0.00000471165	0.00000449794
4.5	0.00000339767	0.00000324138	0.00000309198	0.00000294918	0.00000281271
4.6	0.00000211245	0.00000201334	0.00000191870	0.00000182833	0.00000174205
4.7	0.00000130081	0.00000123858	0.00000117922	0.00000112260	0.00000106859
4.8	0.00000079333	0.00000075465	0.00000071779	0.00000068267	0.00000064920
4.9	0.00000047918	0.00000045538	0.00000043272	0.00000041115	0.00000039061
5.0	0.00000028665	0.00000027215	0.00000025836	0.00000024524	0.00000023277
5.1	0.00000016983	0.00000016108	0.00000015277	0.00000014487	0.00000013737
5.2	0.00000009964	0.00000009442	0.00000008946	0.00000008476	0.00000008029
5.3	0.00000005790	0.00000005481	0.00000005188	0.00000004911	0.00000004647
5.4	0.00000003332	0.00000003151	0.00000002980	0.00000002818	0.00000002664
5.5	0.00000001899	0.00000001794	0.00000001695	0.00000001601	0.00000001512
5.6	0.00000001072	0.00000001012	0.00000000955	0.00000000901	0.00000000850
5.7	0.00000000599	0.00000000565	0.00000000533	0.00000000502	0.00000000473
5.8	0.00000000332	0.00000000312	0.00000000294	0.00000000277	0.00000000261
5.9	0.00000000182	0.00000000171	0.00000000161	0.00000000151	0.00000000143
6.0	0.00000000099	0.00000000093	0.00000000087	0.00000000082	0.00000000077

Continued

Continued

z	0.05	0.06	0.07	0.08	0.09
0.0	0.48006119416	0.47607781735	0.47209682982	0.46811862799	0.46414360741
0.1	0.44038230763	0.43644053711	0.43250506832	0.42857628410	0.42465456527
0.2	0.40129367432	0.39743188680	0.39358012680	0.38973875244	0.38590811880
0.3	0.36316934882	0.35942356678	0.35569124520	0.35197270758	0.34826827346
0.4	0.32635522029	0.32275811025	0.31917750878	0.31561369652	0.31206694942
0.5	0.29115968679	0.28773971885	0.28433884905	0.28095730890	0.27759532475
0.6	0.25784611081	0.25462691467	0.25142889510	0.24825223045	0.24509709367
0.7	0.22662735238	0.22362729244	0.22064994634	0.21769543759	0.21476388416
0.8	0.19766254312	0.19489452125	0.19215020210	0.18942965478	0.18673294304
0.9	0.17105612631	0.16852760747	0.16602324606	0.16354305933	0.16108705951
1.0	0.14685905638	0.14457229966	0.14230965436	0.14007109009	0.13785657203
1.1	0.12507193564	0.12302440305	0.12100048442	0.11900010746	0.11702319602
1.2	0.10564977367	0.10383468112	0.10204231507	0.10027256795	0.09852532905
1.3	0.08850799144	0.08691496195	0.08534345082	0.08379332242	0.08226443868
1.4	0.07352925961	0.07214503697	0.07078087699	0.06943662333	0.06811211797
1.5	0.06057075800	0.05937994059	0.05820755564	0.05705343324	0.05591740252
1.6	0.04947146803	0.04845722627	0.04745968180	0.04647865786	0.04551397732
1.7	0.04005915686	0.03920390329	0.03836357036	0.03753798035	0.03672695570
1.8	0.03215677480	0.03144276298	0.03074190893	0.03005403896	0.02937898004
1.9	0.02558805952	0.02499789515	0.02441918528	0.02385176434	0.02329546775
2.0	0.02018221541	0.01969927041	0.01922617223	0.01876276643	0.01830889985
2.1	0.01577760739	0.01538633478	0.01500342297	0.01462873078	0.01426211841
2.2	0.01222447266	0.01191062542	0.01160379152	0.01130384424	0.01101065832
2.3	0.00938670553	0.00913746753	0.00889404263	0.00865631903	0.00842418640
2.4	0.00714281074	0.00694685079	0.00675565261	0.00656911914	0.00638715476
2.5	0.00538614595	0.00523360816	0.00508492575	0.00494001576	0.00479879660
2.6	0.00402458854	0.00390703257	0.00379256235	0.00368110801	0.00357260095
2.7	0.00297976324	0.00289006808	0.00280281463	0.00271794492	0.00263540208
2.8	0.00218596145	0.00211820504	0.00205235899	0.00198837585	0.00192620913
2.9	0.00158886965	0.00153819521	0.00148899875	0.00144124192	0.00139488724

Continued

Continued

z	0.05	0.06	0.07	0.08	0.09
3.0	0.00114420683	0.00110668496	0.00107029385	0.00103500297	0.00100078248
3.1	0.00081635231	0.00078884569	0.00076219469	0.00073637526	0.00071136397
3.2	0.00057702504	0.00055706107	0.00053773742	0.00051903543	0.00050093691
3.3	0.00040405780	0.00038971236	0.00037584092	0.00036242915	0.00034946312
3.4	0.00028029328	0.00027008769	0.00026022918	0.00025070689	0.00024151027
3.5	0.00019261558	0.00018542740	0.00017849061	0.00017179710	0.00016533898
3.6	0.00013112015	0.00012610762	0.00012127523	0.00011661698	0.00011212703
3.7	0.00008841729	0.00008495668	0.00008162377	0.00007841418	0.00007532364
3.8	0.00005905891	0.00005669351	0.00005441768	0.00005222823	0.00005012211
3.9	0.00003907560	0.00003747488	0.00003593632	0.00003445763	0.00003303665
4.0	0.00002560882	0.00002453636	0.00002350657	0.00002251785	0.00002156866
4.1	0.00001662376	0.00001591238	0.00001522998	0.00001457545	0.00001394772
4.2	0.00001068853	0.00001022135	0.00000977365	0.0000934467	0.0000893366
4.3	0.00000680688	0.00000650312	0.00000621233	0.00000593397	0.00000566753
4.4	0.00000429351	0.00000409798	0.00000391098	0.00000373215	0.00000356116
4.5	0.00000268230	0.00000255768	0.00000243862	0.00000232488	0.00000221623
4.6	0.00000165968	0.00000158105	0.00000150600	0.00000143437	0.00000136603
4.7	0.00000101708	0.00000096796	0.00000092113	0.00000087648	0.00000083391
4.8	0.00000061731	0.00000058693	0.00000055799	0.00000053043	0.00000050418
4.9	0.00000037107	0.00000035247	0.00000033476	0.00000031792	0.00000030190
5.0	0.00000022091	0.00000020963	0.00000019891	0.00000018872	0.00000017903
5.1	0.00000013024	0.00000012347	0.00000011705	0.00000011094	0.00000010515
5.2	0.00000007605	0.00000007203	0.00000006821	0.00000006459	0.00000006116
5.3	0.00000004398	0.00000004161	0.00000003937	0.00000003724	0.00000003523
5.4	0.00000002518	0.00000002381	0.00000002250	0.00000002127	0.00000002010
5.5	0.00000001428	0.00000001349	0.00000001274	0.00000001203	0.00000001135
5.6	0.00000000802	0.00000000757	0.00000000714	0.00000000673	0.00000000635
5.7	0.00000000446	0.00000000421	0.00000000396	0.00000000374	0.00000000352
5.8	0.00000000246	0.00000000231	0.00000000218	0.00000000205	0.00000000193
5.9	0.00000000134	0.00000000126	0.00000000119	0.00000000112	0.00000000105
6.0	0.00000000072	0.00000000068	0.00000000064	0.00000000060	0.00000000056

Appendix G

F Distributions

$F_{0.1}$

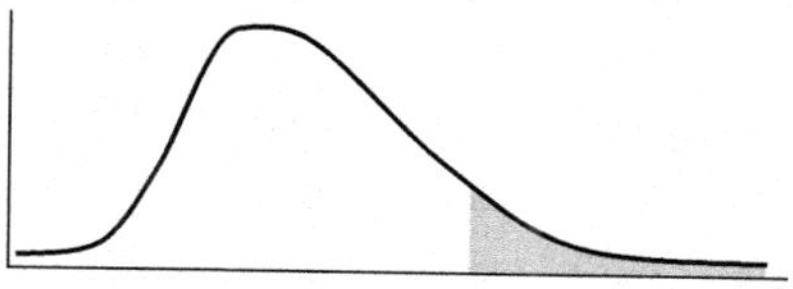

F distribution $F_{0.1}$

Denominator degrees of freedom	Numerator degrees of freedom										
	1	2	3	4	5	6	7	8	9	10	11
1	39.86	49.50	53.59	55.83	57.24	58.20	58.91	59.44	59.86	60.19	60.47
2	8.53	9.00	9.16	9.24	9.29	9.33	9.35	9.37	9.38	9.39	9.40
3	5.54	5.46	5.39	5.34	5.31	5.28	5.27	5.25	5.24	5.23	5.22
4	4.54	4.32	4.19	4.11	4.05	4.01	3.98	3.95	3.94	3.92	3.91
5	4.06	3.78	3.62	3.52	3.45	3.40	3.37	3.34	3.32	3.30	3.28
6	3.78	3.46	3.29	3.18	3.11	3.05	3.01	2.98	2.96	2.94	2.92
7	3.59	3.26	3.07	2.96	2.88	2.83	2.78	2.75	2.72	2.70	2.68
8	3.46	3.11	2.92	2.81	2.73	2.67	2.62	2.59	2.56	2.54	2.52
9	3.36	3.01	2.81	2.69	2.61	2.55	2.51	2.47	2.44	2.42	2.40
10	3.29	2.92	2.73	2.61	2.52	2.46	2.41	2.38	2.35	2.32	2.30
11	3.23	2.86	2.66	2.54	2.45	2.39	2.34	2.30	2.27	2.25	2.23
12	3.18	2.81	2.61	2.48	2.39	2.33	2.28	2.24	2.21	2.19	2.17
13	3.14	2.76	2.56	2.43	2.35	2.28	2.23	2.20	2.16	2.14	2.12
14	3.10	2.73	2.52	2.39	2.31	2.24	2.19	2.15	2.12	2.10	2.07
15	3.07	2.70	2.49	2.36	2.27	2.21	2.16	2.12	2.09	2.06	2.04
16	3.05	2.67	2.46	2.33	2.24	2.18	2.13	2.09	2.06	2.03	2.01
17	3.03	2.64	2.44	2.31	2.22	2.15	2.10	2.06	2.03	2.00	1.98
18	3.01	2.62	2.42	2.29	2.20	2.13	2.08	2.04	2.00	1.98	1.95
19	2.99	2.61	2.40	2.27	2.18	2.11	2.06	2.02	1.98	1.96	1.93
20	2.97	2.59	2.38	2.25	2.16	2.09	2.04	2.00	1.96	1.94	1.91
21	2.96	2.57	2.36	2.23	2.14	2.08	2.02	1.98	1.95	1.92	1.90
22	2.95	2.56	2.35	2.22	2.13	2.06	2.01	1.97	1.93	1.90	1.88
23	2.94	2.55	2.34	2.21	2.11	2.05	1.99	1.95	1.92	1.89	1.87
24	2.93	2.54	2.33	2.19	2.10	2.04	1.98	1.94	1.91	1.88	1.85
25	2.92	2.53	2.32	2.18	2.09	2.02	1.97	1.93	1.89	1.87	1.84
26	2.91	2.52	2.31	2.17	2.08	2.01	1.96	1.92	1.88	1.86	1.83
27	2.90	2.51	2.30	2.17	2.07	2.00	1.95	1.91	1.87	1.85	1.82
28	2.89	2.50	2.29	2.16	2.06	2.00	1.94	1.90	1.87	1.84	1.81
29	2.89	2.50	2.28	2.15	2.06	1.99	1.93	1.89	1.86	1.83	1.80
30	2.88	2.49	2.28	2.14	2.05	1.98	1.93	1.88	1.85	1.82	1.79
40	2.84	2.44	2.23	2.09	2.00	1.93	1.87	1.83	1.79	1.76	1.74
60	2.79	2.39	2.18	2.04	1.95	1.87	1.82	1.77	1.74	1.71	1.68
100	2.76	2.36	2.14	2.00	1.91	1.83	1.78	1.73	1.69	1.66	1.64

Continued

F distribution $F_{0.1}$ *(continued)*

Denominator degrees of freedom	Numerator degrees of freedom										
	12	**13**	**14**	**15**	**16**	**17**	**18**	**19**	**20**	**21**	**22**
1	60.71	60.90	61.07	61.22	61.35	61.46	61.57	61.66	61.74	61.81	61.88
2	9.41	9.41	9.42	9.42	9.43	9.43	9.44	9.44	9.44	9.44	9.45
3	5.22	5.21	5.20	5.20	5.20	5.19	5.19	5.19	5.18	5.18	5.18
4	3.90	3.89	3.88	3.87	3.86	3.86	3.85	3.85	3.84	3.84	3.84
5	3.27	3.26	3.25	3.24	3.23	3.22	3.22	3.21	3.21	3.20	3.20
6	2.90	2.89	2.88	2.87	2.86	2.85	2.85	2.84	2.84	2.83	2.83
7	2.67	2.65	2.64	2.63	2.62	2.61	2.61	2.60	2.59	2.59	2.58
8	2.50	2.49	2.48	2.46	2.45	2.45	2.44	2.43	2.42	2.42	2.41
9	2.38	2.36	2.35	2.34	2.33	2.32	2.31	2.30	2.30	2.29	2.29
10	2.28	2.27	2.26	2.24	2.23	2.22	2.22	2.21	2.20	2.19	2.19
11	2.21	2.19	2.18	2.17	2.16	2.15	2.14	2.13	2.12	2.12	2.11
12	2.15	2.13	2.12	2.10	2.09	2.08	2.08	2.07	2.06	2.05	2.05
13	2.10	2.08	2.07	2.05	2.04	2.03	2.02	2.01	2.01	2.00	1.99
14	2.05	2.04	2.02	2.01	2.00	1.99	1.98	1.97	1.96	1.96	1.95
15	2.02	2.00	1.99	1.97	1.96	1.95	1.94	1.93	1.92	1.92	1.91
16	1.99	1.97	1.95	1.94	1.93	1.92	1.91	1.90	1.89	1.88	1.88
17	1.96	1.94	1.93	1.91	1.90	1.89	1.88	1.87	1.86	1.86	1.85
18	1.93	1.92	1.90	1.89	1.87	1.86	1.85	1.84	1.84	1.83	1.82
19	1.91	1.89	1.88	1.86	1.85	1.84	1.83	1.82	1.81	1.81	1.80
20	1.89	1.87	1.86	1.84	1.83	1.82	1.81	1.80	1.79	1.79	1.78
21	1.87	1.86	1.84	1.83	1.81	1.80	1.79	1.78	1.78	1.77	1.76
22	1.86	1.84	1.83	1.81	1.80	1.79	1.78	1.77	1.76	1.75	1.74
23	1.84	1.83	1.81	1.80	1.78	1.77	1.76	1.75	1.74	1.74	1.73
24	1.83	1.81	1.80	1.78	1.77	1.76	1.75	1.74	1.73	1.72	1.71
25	1.82	1.80	1.79	1.77	1.76	1.75	1.74	1.73	1.72	1.71	1.70
26	1.81	1.79	1.77	1.76	1.75	1.73	1.72	1.71	1.71	1.70	1.69
27	1.80	1.78	1.76	1.75	1.74	1.72	1.71	1.70	1.70	1.69	1.68
28	1.79	1.77	1.75	1.74	1.73	1.71	1.70	1.69	1.69	1.68	1.67
29	1.78	1.76	1.75	1.73	1.72	1.71	1.69	1.68	1.68	1.67	1.66
30	1.77	1.75	1.74	1.72	1.71	1.70	1.69	1.68	1.67	1.66	1.65
40	1.71	1.70	1.68	1.66	1.65	1.64	1.62	1.61	1.61	1.60	1.59
60	1.66	1.64	1.62	1.60	1.59	1.58	1.56	1.55	1.54	1.53	1.53
100	1.61	1.59	1.57	1.56	1.54	1.53	1.52	1.50	1.49	1.48	1.48

Continued

F distribution $F_{0.1}$ *(continued)*

	Numerator degrees of freedom										
Denominator degrees of freedom	**23**	**24**	**25**	**26**	**27**	**28**	**29**	**30**	**40**	**60**	**100**
1	61.94	62.00	62.05	62.10	62.15	62.19	62.23	62.26	62.53	62.79	63.01
2	9.45	9.45	9.45	9.45	9.45	9.46	9.46	9.46	9.47	9.47	9.48
3	5.18	5.18	5.17	5.17	5.17	5.17	5.17	5.17	5.16	5.15	5.14
4	3.83	3.83	3.83	3.83	3.82	3.82	3.82	3.82	3.80	3.79	3.78
5	3.19	3.19	3.19	3.18	3.18	3.18	3.18	3.17	3.16	3.14	3.13
6	2.82	2.82	2.81	2.81	2.81	2.81	2.80	2.80	2.78	2.76	2.75
7	2.58	2.58	2.57	2.57	2.56	2.56	2.56	2.56	2.54	2.51	2.50
8	2.41	2.40	2.40	2.40	2.39	2.39	2.39	2.38	2.36	2.34	2.32
9	2.28	2.28	2.27	2.27	2.26	2.26	2.26	2.25	2.23	2.21	2.19
10	2.18	2.18	2.17	2.17	2.17	2.16	2.16	2.16	2.13	2.11	2.09
11	2.11	2.10	2.10	2.09	2.09	2.08	2.08	2.08	2.05	2.03	2.01
12	2.04	2.04	2.03	2.03	2.02	2.02	2.01	2.01	1.99	1.96	1.94
13	1.99	1.98	1.98	1.97	1.97	1.96	1.96	1.96	1.93	1.90	1.88
14	1.94	1.94	1.93	1.93	1.92	1.92	1.92	1.91	1.89	1.86	1.83
15	1.90	1.90	1.89	1.89	1.88	1.88	1.88	1.87	1.85	1.82	1.79
16	1.87	1.87	1.86	1.86	1.85	1.85	1.84	1.84	1.81	1.78	1.76
17	1.84	1.84	1.83	1.83	1.82	1.82	1.81	1.81	1.78	1.75	1.73
18	1.82	1.81	1.80	1.80	1.80	1.79	1.79	1.78	1.75	1.72	1.70
19	1.79	1.79	1.78	1.78	1.77	1.77	1.76	1.76	1.73	1.70	1.67
20	1.77	1.77	1.76	1.76	1.75	1.75	1.74	1.74	1.71	1.68	1.65
21	1.75	1.75	1.74	1.74	1.73	1.73	1.72	1.72	1.69	1.66	1.63
22	1.74	1.73	1.73	1.72	1.72	1.71	1.71	1.70	1.67	1.64	1.61
23	1.72	1.72	1.71	1.70	1.70	1.69	1.69	1.69	1.66	1.62	1.59
24	1.71	1.70	1.70	1.69	1.69	1.68	1.68	1.67	1.64	1.61	1.58
25	1.70	1.69	1.68	1.68	1.67	1.67	1.66	1.66	1.63	1.59	1.56
26	1.68	1.68	1.67	1.67	1.66	1.66	1.65	1.65	1.61	1.58	1.55
27	1.67	1.67	1.66	1.65	1.65	1.64	1.64	1.64	1.60	1.57	1.54
28	1.66	1.66	1.65	1.64	1.64	1.63	1.63	1.63	1.59	1.56	1.53
29	1.65	1.65	1.64	1.63	1.63	1.62	1.62	1.62	1.58	1.55	1.52
30	1.64	1.64	1.63	1.63	1.62	1.62	1.61	1.61	1.57	1.54	1.51
40	1.58	1.57	1.57	1.56	1.56	1.55	1.55	1.54	1.51	1.47	1.43
60	1.52	1.51	1.50	1.50	1.49	1.49	1.48	1.48	1.44	1.40	1.36
100	1.47	1.46	1.45	1.45	1.44	1.43	1.43	1.42	1.38	1.34	1.29

$F_{0.05}$

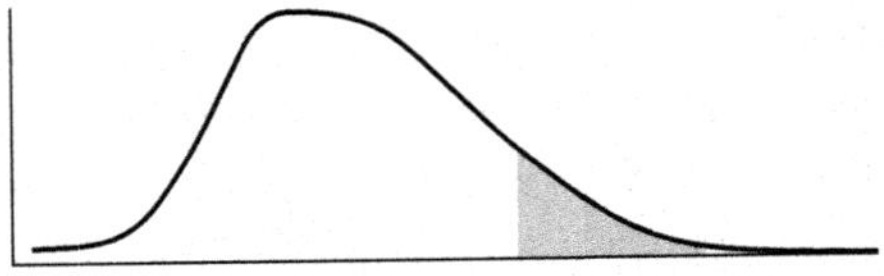

F distribution $F_{0.05}$

Denominator degrees of freedom	Numerator degrees of freedom										
	1	2	3	4	5	6	7	8	9	10	11
1	161.4	199.5	215.7	224.6	230.2	234.0	236.8	238.9	240.5	241.9	243.0
2	18.51	19.00	19.16	19.25	19.30	19.33	19.35	19.37	19.38	19.40	19.40
3	10.13	9.55	9.28	9.12	9.01	8.94	8.89	8.85	8.81	8.79	8.76
4	7.71	6.94	6.59	6.39	6.26	6.16	6.09	6.04	6.00	5.96	5.94
5	6.61	5.79	5.41	5.19	5.05	4.95	4.88	4.82	4.77	4.74	4.70
6	5.99	5.14	4.76	4.53	4.39	4.28	4.21	4.15	4.10	4.06	4.03
7	5.59	4.74	4.35	4.12	3.97	3.87	3.79	3.73	3.68	3.64	3.60
8	5.32	4.46	4.07	3.84	3.69	3.58	3.50	3.44	3.39	3.35	3.31
9	5.12	4.26	3.86	3.63	3.48	3.37	3.29	3.23	3.18	3.14	3.10
10	4.96	4.10	3.71	3.48	3.33	3.22	3.14	3.07	3.02	2.98	2.94
11	4.84	3.98	3.59	3.36	3.20	3.09	3.01	2.95	2.90	2.85	2.82
12	4.75	3.89	3.49	3.26	3.11	3.00	2.91	2.85	2.80	2.75	2.72
13	4.67	3.81	3.41	3.18	3.03	2.92	2.83	2.77	2.71	2.67	2.63
14	4.60	3.74	3.34	3.11	2.96	2.85	2.76	2.70	2.65	2.60	2.57
15	4.54	3.68	3.29	3.06	2.90	2.79	2.71	2.64	2.59	2.54	2.51
16	4.49	3.63	3.24	3.01	2.85	2.74	2.66	2.59	2.54	2.49	2.46
17	4.45	3.59	3.20	2.96	2.81	2.70	2.61	2.55	2.49	2.45	2.41
18	4.41	3.55	3.16	2.93	2.77	2.66	2.58	2.51	2.46	2.41	2.37
19	4.38	3.52	3.13	2.90	2.74	2.63	2.54	2.48	2.42	2.38	2.34
20	4.35	3.49	3.10	2.87	2.71	2.60	2.51	2.45	2.39	2.35	2.31
21	4.32	3.47	3.07	2.84	2.68	2.57	2.49	2.42	2.37	2.32	2.28
22	4.30	3.44	3.05	2.82	2.66	2.55	2.46	2.40	2.34	2.30	2.26
23	4.28	3.42	3.03	2.80	2.64	2.53	2.44	2.37	2.32	2.27	2.24
24	4.26	3.40	3.01	2.78	2.62	2.51	2.42	2.36	2.30	2.25	2.22
25	4.24	3.39	2.99	2.76	2.60	2.49	2.40	2.34	2.28	2.24	2.20
26	4.23	3.37	2.98	2.74	2.59	2.47	2.39	2.32	2.27	2.22	2.18
27	4.21	3.35	2.96	2.73	2.57	2.46	2.37	2.31	2.25	2.20	2.17
28	4.20	3.34	2.95	2.71	2.56	2.45	2.36	2.29	2.24	2.19	2.15
29	4.18	3.33	2.93	2.70	2.55	2.43	2.35	2.28	2.22	2.18	2.14
30	4.17	3.32	2.92	2.69	2.53	2.42	2.33	2.27	2.21	2.16	2.13
40	4.08	3.23	2.84	2.61	2.45	2.34	2.25	2.18	2.12	2.08	2.04
60	4.00	3.15	2.76	2.53	2.37	2.25	2.17	2.10	2.04	1.99	1.95
100	3.94	3.09	2.70	2.46	2.31	2.19	2.10	2.03	1.97	1.93	1.89

Continued

F distribution $F_{0.05}$ *(continued)*

Numerator degrees of freedom (columns); Denominator degrees of freedom (rows)

	12	13	14	15	16	17	18	19	20	21	22
1	243.9	244.7	245.4	245.9	246.5	246.9	247.3	247.7	248.0	248.3	248.6
2	19.41	19.42	19.42	19.43	19.43	19.44	19.44	19.44	19.45	19.45	19.45
3	8.74	8.73	8.71	8.70	8.69	8.68	8.67	8.67	8.66	8.65	8.65
4	5.91	5.89	5.87	5.86	5.84	5.83	5.82	5.81	5.80	5.79	5.79
5	4.68	4.66	4.64	4.62	4.60	4.59	4.58	4.57	4.56	4.55	4.54
6	4.00	3.98	3.96	3.94	3.92	3.91	3.90	3.88	3.87	3.86	3.86
7	3.57	3.55	3.53	3.51	3.49	3.48	3.47	3.46	3.44	3.43	3.43
8	3.28	3.26	3.24	3.22	3.20	3.19	3.17	3.16	3.15	3.14	3.13
9	3.07	3.05	3.03	3.01	2.99	2.97	2.96	2.95	2.94	2.93	2.92
10	2.91	2.89	2.86	2.85	2.83	2.81	2.80	2.79	2.77	2.76	2.75
11	2.79	2.76	2.74	2.72	2.70	2.69	2.67	2.66	2.65	2.64	2.63
12	2.69	2.66	2.64	2.62	2.60	2.58	2.57	2.56	2.54	2.53	2.52
13	2.60	2.58	2.55	2.53	2.51	2.50	2.48	2.47	2.46	2.45	2.44
14	2.53	2.51	2.48	2.46	2.44	2.43	2.41	2.40	2.39	2.38	2.37
15	2.48	2.45	2.42	2.40	2.38	2.37	2.35	2.34	2.33	2.32	2.31
16	2.42	2.40	2.37	2.35	2.33	2.32	2.30	2.29	2.28	2.26	2.25
17	2.38	2.35	2.33	2.31	2.29	2.27	2.26	2.24	2.23	2.22	2.21
18	2.34	2.31	2.29	2.27	2.25	2.23	2.22	2.20	2.19	2.18	2.17
19	2.31	2.28	2.26	2.23	2.21	2.20	2.18	2.17	2.16	2.14	2.13
20	2.28	2.25	2.22	2.20	2.18	2.17	2.15	2.14	2.12	2.11	2.10
21	2.25	2.22	2.20	2.18	2.16	2.14	2.12	2.11	2.10	2.08	2.07
22	2.23	2.20	2.17	2.15	2.13	2.11	2.10	2.08	2.07	2.06	2.05
23	2.20	2.18	2.15	2.13	2.11	2.09	2.08	2.06	2.05	2.04	2.02
24	2.18	2.15	2.13	2.11	2.09	2.07	2.05	2.04	2.03	2.01	2.00
25	2.16	2.14	2.11	2.09	2.07	2.05	2.04	2.02	2.01	2.00	1.98
26	2.15	2.12	2.09	2.07	2.05	2.03	2.02	2.00	1.99	1.98	1.97
27	2.13	2.10	2.08	2.06	2.04	2.02	2.00	1.99	1.97	1.96	1.95
28	2.12	2.09	2.06	2.04	2.02	2.00	1.99	1.97	1.96	1.95	1.93
29	2.10	2.08	2.05	2.03	2.01	1.99	1.97	1.96	1.94	1.93	1.92
30	2.09	2.06	2.04	2.01	1.99	1.98	1.96	1.95	1.93	1.92	1.91
40	2.00	1.97	1.95	1.92	1.90	1.89	1.87	1.85	1.84	1.83	1.81
60	1.92	1.89	1.86	1.84	1.82	1.80	1.78	1.76	1.75	1.73	1.72
100	1.85	1.82	1.79	1.77	1.75	1.73	1.71	1.69	1.68	1.66	1.65

Continued

F distribution $F_{0.05}$ *(continued)*

Denominator degrees of freedom	Numerator degrees of freedom 23	24	25	26	27	28	29	30	40	60	100
1	248.8	249.1	249.3	249.5	249.6	249.8	250.0	250.1	251.1	252.2	253.0
2	19.45	19.45	19.46	19.46	19.46	19.46	19.46	19.46	19.47	19.48	19.49
3	8.64	8.64	8.63	8.63	8.63	8.62	8.62	8.62	8.59	8.57	8.55
4	5.78	5.77	5.77	5.76	5.76	5.75	5.75	5.75	5.72	5.69	5.66
5	4.53	4.53	4.52	4.52	4.51	4.50	4.50	4.50	4.46	4.43	4.41
6	3.85	3.84	3.83	3.83	3.82	3.82	3.81	3.81	3.77	3.74	3.71
7	3.42	3.41	3.40	3.40	3.39	3.39	3.38	3.38	3.34	3.30	3.27
8	3.12	3.12	3.11	3.10	3.10	3.09	3.08	3.08	3.04	3.01	2.97
9	2.91	2.90	2.89	2.89	2.88	2.87	2.87	2.86	2.83	2.79	2.76
10	2.75	2.74	2.73	2.72	2.72	2.71	2.70	2.70	2.66	2.62	2.59
11	2.62	2.61	2.60	2.59	2.59	2.58	2.58	2.57	2.53	2.49	2.46
12	2.51	2.51	2.50	2.49	2.48	2.48	2.47	2.47	2.43	2.38	2.35
13	2.43	2.42	2.41	2.41	2.40	2.39	2.39	2.38	2.34	2.30	2.26
14	2.36	2.35	2.34	2.33	2.33	2.32	2.31	2.31	2.27	2.22	2.19
15	2.30	2.29	2.28	2.27	2.27	2.26	2.25	2.25	2.20	2.16	2.12
16	2.24	2.24	2.23	2.22	2.21	2.21	2.20	2.19	2.15	2.11	2.07
17	2.20	2.19	2.18	2.17	2.17	2.16	2.15	2.15	2.10	2.06	2.02
18	2.16	2.15	2.14	2.13	2.13	2.12	2.11	2.11	2.06	2.02	1.98
19	2.12	2.11	2.11	2.10	2.09	2.08	2.08	2.07	2.03	1.98	1.94
20	2.09	2.08	2.07	2.07	2.06	2.05	2.05	2.04	1.99	1.95	1.91
21	2.06	2.05	2.05	2.04	2.03	2.02	2.02	2.01	1.96	1.92	1.88
22	2.04	2.03	2.02	2.01	2.00	2.00	1.99	1.98	1.94	1.89	1.85
23	2.01	2.01	2.00	1.99	1.98	1.97	1.97	1.96	1.91	1.86	1.82
24	1.99	1.98	1.97	1.97	1.96	1.95	1.95	1.94	1.89	1.84	1.80
25	1.97	1.96	1.96	1.95	1.94	1.93	1.93	1.92	1.87	1.82	1.78
26	1.96	1.95	1.94	1.93	1.92	1.91	1.91	1.90	1.85	1.80	1.76
27	1.94	1.93	1.92	1.91	1.90	1.90	1.89	1.88	1.84	1.79	1.74
28	1.92	1.91	1.91	1.90	1.89	1.88	1.88	1.87	1.82	1.77	1.73
29	1.91	1.90	1.89	1.88	1.88	1.87	1.86	1.85	1.81	1.75	1.71
30	1.90	1.89	1.88	1.87	1.86	1.85	1.85	1.84	1.79	1.74	1.70
40	1.80	1.79	1.78	1.77	1.77	1.76	1.75	1.74	1.69	1.64	1.59
60	1.71	1.70	1.69	1.68	1.67	1.66	1.66	1.65	1.59	1.53	1.48
100	1.64	1.63	1.62	1.61	1.60	1.59	1.58	1.57	1.52	1.45	1.39

$F_{0.01}$

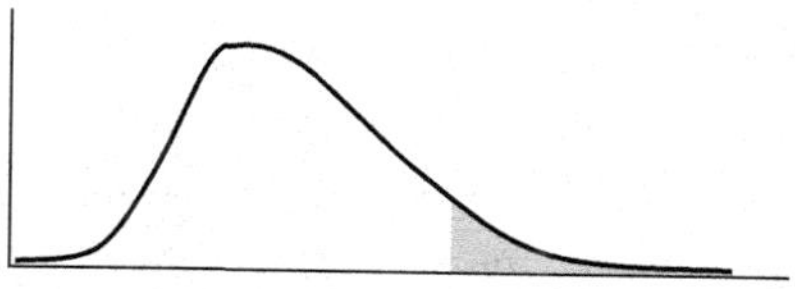

F distribution $F_{0.01}$

	Numerator degrees of freedom										
Denominator degrees of freedom	1	2	3	4	5	6	7	8	9	10	11
1	4052	4999	5404	5624	5764	5859	5928	5981	6022	6056	6083
2	98.5	99	99.16	99.25	99.3	99.33	99.36	99.38	99.39	99.4	99.41
3	34.12	30.82	29.46	28.71	28.24	27.91	27.67	27.49	27.34	27.23	27.13
4	21.2	18	16.69	15.98	15.52	15.21	14.98	14.8	14.66	14.55	14.45
5	16.26	13.27	12.06	11.39	10.97	10.67	10.46	10.29	10.16	10.05	9.963
6	13.75	10.92	9.78	9.148	8.746	8.466	8.26	8.102	7.976	7.874	7.79
7	12.25	9.547	8.451	7.847	7.46	7.191	6.993	6.84	6.719	6.62	6.538
8	11.26	8.649	7.591	7.006	6.632	6.371	6.178	6.029	5.911	5.814	5.734
9	10.56	8.022	6.992	6.422	6.057	5.802	5.613	5.467	5.351	5.257	5.178
10	10.04	7.559	6.552	5.994	5.636	5.386	5.2	5.057	4.942	4.849	4.772
11	9.646	7.206	6.217	5.668	5.316	5.069	4.886	4.744	4.632	4.539	4.462
12	9.33	6.927	5.953	5.412	5.064	4.821	4.64	4.499	4.388	4.296	4.22
13	9.074	6.701	5.739	5.205	4.862	4.62	4.441	4.302	4.191	4.1	4.025
14	8.862	6.515	5.564	5.035	4.695	4.456	4.278	4.14	4.03	3.939	3.864
15	8.683	6.359	5.417	4.893	4.556	4.318	4.142	4.004	3.895	3.805	3.73
16	8.531	6.226	5.292	4.773	4.437	4.202	4.026	3.89	3.78	3.691	3.616
17	8.4	6.112	5.185	4.669	4.336	4.101	3.927	3.791	3.682	3.593	3.518
18	8.285	6.013	5.092	4.579	4.248	4.015	3.841	3.705	3.597	3.508	3.434
19	8.185	5.926	5.01	4.5	4.171	3.939	3.765	3.631	3.523	3.434	3.36
20	8.096	5.849	4.938	4.431	4.103	3.871	3.699	3.564	3.457	3.368	3.294
21	8.017	5.78	4.874	4.369	4.042	3.812	3.64	3.506	3.398	3.31	3.236
22	7.945	5.719	4.817	4.313	3.988	3.758	3.587	3.453	3.346	3.258	3.184
23	7.881	5.664	4.765	4.264	3.939	3.71	3.539	3.406	3.299	3.211	3.137
24	7.823	5.614	4.718	4.218	3.895	3.667	3.496	3.363	3.256	3.168	3.094
25	7.77	5.568	4.675	4.177	3.855	3.627	3.457	3.324	3.217	3.129	3.056
26	7.721	5.526	4.637	4.14	3.818	3.591	3.421	3.288	3.182	3.094	3.021
27	7.677	5.488	4.601	4.106	3.785	3.558	3.388	3.256	3.149	3.062	2.988
28	7.636	5.453	4.568	4.074	3.754	3.528	3.358	3.226	3.12	3.032	2.959
29	7.598	5.42	4.538	4.045	3.725	3.499	3.33	3.198	3.092	3.005	2.931
30	7.562	5.39	4.51	4.018	3.699	3.473	3.305	3.173	3.067	2.979	2.906
40	7.314	5.178	4.313	3.828	3.514	3.291	3.124	2.993	2.888	2.801	2.727
60	7.077	4.977	4.126	3.649	3.339	3.119	2.953	2.823	2.718	2.632	2.559
100	6.895	4.824	3.984	3.513	3.206	2.988	2.823	2.694	2.59	2.503	2.43

Continued

F* distribution $F_{0.01}$ *(continued)

Denominator degrees of freedom	Numerator degrees of freedom										
	12	**13**	**14**	**15**	**16**	**17**	**18**	**19**	**20**	**21**	**22**
1	6107	6126	6143	6157	6170	6181	6191	6201	6208.7	6216.1	6223.1
2	99.42	99.42	99.43	99.43	99.44	99.44	99.44	99.45	99.448	99.451	99.455
3	27.05	26.98	26.92	26.87	26.83	26.79	26.75	26.72	26.69	26.664	26.639
4	14.37	14.31	14.25	14.2	14.15	14.11	14.08	14.05	14.019	13.994	13.97
5	9.888	9.825	9.77	9.722	9.68	9.643	9.609	9.58	9.5527	9.5281	9.5058
6	7.718	7.657	7.605	7.559	7.519	7.483	7.451	7.422	7.3958	7.3721	7.3506
7	6.469	6.41	6.359	6.314	6.275	6.24	6.209	6.181	6.1555	6.1324	6.1113
8	5.667	5.609	5.559	5.515	5.477	5.442	5.412	5.384	5.3591	5.3365	5.3157
9	5.111	5.055	5.005	4.962	4.924	4.89	4.86	4.833	4.808	4.7855	4.7651
10	4.706	4.65	4.601	4.558	4.52	4.487	4.457	4.43	4.4054	4.3831	4.3628
11	4.397	4.342	4.293	4.251	4.213	4.18	4.15	4.123	4.099	4.0769	4.0566
12	4.155	4.1	4.052	4.01	3.972	3.939	3.91	3.883	3.8584	3.8363	3.8161
13	3.96	3.905	3.857	3.815	3.778	3.745	3.716	3.689	3.6646	3.6425	3.6223
14	3.8	3.745	3.698	3.656	3.619	3.586	3.556	3.529	3.5052	3.4832	3.463
15	3.666	3.612	3.564	3.522	3.485	3.452	3.423	3.396	3.3719	3.3498	3.3297
16	3.553	3.498	3.451	3.409	3.372	3.339	3.31	3.283	3.2587	3.2367	3.2165
17	3.455	3.401	3.353	3.312	3.275	3.242	3.212	3.186	3.1615	3.1394	3.1192
18	3.371	3.316	3.269	3.227	3.19	3.158	3.128	3.101	3.0771	3.055	3.0348
19	3.297	3.242	3.195	3.153	3.116	3.084	3.054	3.027	3.0031	2.981	2.9607
20	3.231	3.177	3.13	3.088	3.051	3.018	2.989	2.962	2.9377	2.9156	2.8953
21	3.173	3.119	3.072	3.03	2.993	2.96	2.931	2.904	2.8795	2.8574	2.837
22	3.121	3.067	3.019	2.978	2.941	2.908	2.879	2.852	2.8274	2.8052	2.7849
23	3.074	3.02	2.973	2.931	2.894	2.861	2.832	2.805	2.7805	2.7582	2.7378
24	3.032	2.977	2.93	2.889	2.852	2.819	2.789	2.762	2.738	2.7157	2.6953
25	2.993	2.939	2.892	2.85	2.813	2.78	2.751	2.724	2.6993	2.677	2.6565
26	2.958	2.904	2.857	2.815	2.778	2.745	2.715	2.688	2.664	2.6416	2.6211
27	2.926	2.872	2.824	2.783	2.746	2.713	2.683	2.656	2.6316	2.609	2.5886
28	2.896	2.842	2.795	2.753	2.716	2.683	2.653	2.626	2.6018	2.5793	2.5587
29	2.868	2.814	2.767	2.726	2.689	2.656	2.626	2.599	2.5742	2.5517	2.5311
30	2.843	2.789	2.742	2.7	2.663	2.63	2.6	2.573	2.5487	2.5262	2.5055
40	2.665	2.611	2.563	2.522	2.484	2.451	2.421	2.394	2.3689	2.3461	2.3252
60	2.496	2.442	2.394	2.352	2.315	2.281	2.251	2.223	2.1978	2.1747	2.1533
10	2.368	2.313	2.265	2.223	2.185	2.151	2.12	2.092	2.0666	2.0431	2.0214

Continued

F distribution $F_{0.01}$ *(continued)*

Denominator degrees of freedom	Numerator degrees of freedom										
	23	24	25	26	27	28	29	30	40	60	100
1	6228.7	6234.3	6239.9	6244.5	6249.2	6252.9	6257.1	6260.4	6286.4	6313	6333.9
2	99.455	99.455	99.459	99.462	99.462	99.462	99.462	99.466	99.477	99.484	99.491
3	26.617	26.597	26.579	26.562	26.546	26.531	26.517	26.504	26.411	26.316	26.241
4	13.949	13.929	13.911	13.894	13.878	13.864	13.85	13.838	13.745	13.652	13.577
5	9.4853	9.4665	9.4492	9.4331	9.4183	9.4044	9.3914	9.3794	9.2912	9.202	9.13
6	7.3309	7.3128	7.296	7.2805	7.2661	7.2528	7.2403	7.2286	7.1432	7.0568	6.9867
7	6.092	6.0743	6.0579	6.0428	6.0287	6.0156	6.0035	5.992	5.9084	5.8236	5.7546
8	5.2967	5.2793	5.2631	5.2482	5.2344	5.2214	5.2094	5.1981	5.1156	5.0316	4.9633
9	4.7463	4.729	4.713	4.6982	4.6845	4.6717	4.6598	4.6486	4.5667	4.4831	4.415
10	4.3441	4.3269	4.3111	4.2963	4.2827	4.27	4.2582	4.2469	4.1653	4.0819	4.0137
11	4.038	4.0209	4.0051	3.9904	3.9768	3.9641	3.9522	3.9411	3.8596	3.7761	3.7077
12	3.7976	3.7805	3.7647	3.7501	3.7364	3.7238	3.7119	3.7008	3.6192	3.5355	3.4668
13	3.6038	3.5868	3.571	3.5563	3.5427	3.53	3.5182	3.507	3.4253	3.3413	3.2723
14	3.4445	3.4274	3.4116	3.3969	3.3833	3.3706	3.3587	3.3476	3.2657	3.1813	3.1118
15	3.3111	3.294	3.2782	3.2636	3.2499	3.2372	3.2253	3.2141	3.1319	3.0471	2.9772
16	3.1979	3.1808	3.165	3.1503	3.1366	3.1238	3.1119	3.1007	3.0182	2.933	2.8627
17	3.1006	3.0835	3.0676	3.0529	3.0392	3.0264	3.0145	3.0032	2.9204	2.8348	2.7639
18	3.0161	2.999	2.9831	2.9683	2.9546	2.9418	2.9298	2.9185	2.8354	2.7493	2.6779
19	2.9421	2.9249	2.9089	2.8942	2.8804	2.8675	2.8555	2.8442	2.7608	2.6742	2.6023
20	2.8766	2.8594	2.8434	2.8286	2.8148	2.8019	2.7898	2.7785	2.6947	2.6077	2.5353
21	2.8183	2.801	2.785	2.7702	2.7563	2.7434	2.7313	2.72	2.6359	2.5484	2.4755
22	2.7661	2.7488	2.7328	2.7179	2.704	2.691	2.6789	2.6675	2.5831	2.4951	2.4218
23	2.7191	2.7017	2.6857	2.6707	2.6568	2.6438	2.6316	2.6202	2.5355	2.4471	2.3732
24	2.6764	2.6591	2.643	2.628	2.614	2.601	2.5888	2.5773	2.4923	2.4035	2.3291
25	2.6377	2.6203	2.6041	2.5891	2.5751	2.562	2.5498	2.5383	2.453	2.3637	2.2888
26	2.6022	2.5848	2.5686	2.5535	2.5395	2.5264	2.5142	2.5026	2.417	2.3273	2.2519
27	2.5697	2.5522	2.536	2.5209	2.5069	2.4937	2.4814	2.4699	2.384	2.2938	2.218
28	2.5398	2.5223	2.506	2.4909	2.4768	2.4636	2.4513	2.4397	2.3535	2.2629	2.1867
29	2.5121	2.4946	2.4783	2.4631	2.449	2.4358	2.4234	2.4118	2.3253	2.2344	2.1577
30	2.4865	2.4689	2.4526	2.4374	2.4233	2.41	2.3976	2.386	2.2992	2.2079	2.1307
40	2.3059	2.288	2.2714	2.2559	2.2415	2.228	2.2153	2.2034	2.1142	2.0194	1.9383
60	2.1336	2.1154	2.0984	2.0825	2.0677	2.0538	2.0408	2.0285	1.936	1.8363	1.7493
100	2.0012	1.9826	1.9651	1.9489	1.9337	1.9194	1.9059	1.8933	1.7972	1.6918	1.5977

Appendix H

Binomial Distribution

Probability of x or fewer occurrences in a sample of size n

Binomial distribution

		p																	
n	*x*	0.01	0.02	0.03	0.04	0.05	0.06	0.07	0.08	0.09	0.10	0.15	0.20	0.25	0.30	0.35	0.40	0.45	0.50
2	0	0.980	0.960	0.941	0.922	0.903	0.884	0.865	0.846	0.828	0.810	0.723	0.640	0.563	0.490	0.423	0.360	0.303	0.250
2	1	1.000	1.000	0.999	0.998	0.998	0.996	0.995	0.994	0.992	0.990	0.978	0.960	0.938	0.910	0.878	0.840	0.798	0.750
3	0	0.970	0.941	0.913	0.885	0.857	0.831	0.804	0.779	0.754	0.729	0.614	0.512	0.422	0.343	0.275	0.216	0.166	0.125
3	1	1.000	0.999	0.997	0.995	0.993	0.990	0.986	0.982	0.977	0.972	0.939	0.896	0.844	0.784	0.718	0.648	0.575	0.500
3	2	1.000	1.000	1.000	1.000	1.000	1.000	1.000	0.999	0.999	0.999	0.997	0.992	0.984	0.973	0.957	0.936	0.909	0.875
4	0	0.961	0.922	0.885	0.849	0.815	0.781	0.748	0.716	0.686	0.656	0.522	0.410	0.316	0.240	0.179	0.130	0.092	0.063
4	1	0.999	0.998	0.995	0.991	0.986	0.980	0.973	0.966	0.957	0.948	0.890	0.819	0.738	0.652	0.563	0.475	0.391	0.313
4	2	1.000	1.000	1.000	1.000	1.000	0.999	0.999	0.998	0.997	0.996	0.988	0.973	0.949	0.916	0.874	0.821	0.759	0.688
4	3	1.000	1.000	1.000	1.000	1.000	1.000	1.000	1.000	1.000	1.000	0.999	0.998	0.996	0.992	0.985	0.974	0.959	0.938
5	0	0.951	0.904	0.859	0.815	0.774	0.734	0.696	0.659	0.624	0.590	0.444	0.328	0.237	0.168	0.116	0.078	0.050	0.031
5	1	0.999	0.996	0.992	0.985	0.977	0.968	0.958	0.946	0.933	0.919	0.835	0.737	0.633	0.528	0.428	0.337	0.256	0.188
5	2	1.000	1.000	1.000	0.999	0.999	0.998	0.997	0.995	0.994	0.991	0.973	0.942	0.896	0.837	0.765	0.683	0.593	0.500
5	3	1.000	1.000	1.000	1.000	1.000	1.000	1.000	1.000	1.000	1.000	0.998	0.993	0.984	0.969	0.946	0.913	0.869	0.813
5	4	1.000	1.000	1.000	1.000	1.000	1.000	1.000	1.000	1.000	1.000	1.000	1.000	0.999	0.998	0.995	0.990	0.982	0.969
6	0	0.941	0.886	0.833	0.783	0.735	0.690	0.647	0.606	0.568	0.531	0.377	0.262	0.178	0.118	0.075	0.047	0.028	0.016
6	1	0.999	0.994	0.988	0.978	0.967	0.954	0.939	0.923	0.905	0.886	0.776	0.655	0.534	0.420	0.319	0.233	0.164	0.109
6	2	1.000	1.000	0.999	0.999	0.998	0.996	0.994	0.991	0.988	0.984	0.953	0.901	0.831	0.744	0.647	0.544	0.442	0.344
6	3	1.000	1.000	1.000	1.000	1.000	1.000	1.000	0.999	0.999	0.999	0.994	0.983	0.962	0.930	0.883	0.821	0.745	0.656
6	4	1.000	1.000	1.000	1.000	1.000	1.000	1.000	1.000	1.000	1.000	1.000	0.998	0.995	0.989	0.978	0.959	0.931	0.891
6	5	1.000	1.000	1.000	1.000	1.000	1.000	1.000	1.000	1.000	1.000	1.000	1.000	1.000	0.999	0.998	0.996	0.992	0.984
7	0	0.932	0.868	0.808	0.751	0.698	0.648	0.602	0.558	0.517	0.478	0.321	0.210	0.133	0.082	0.049	0.028	0.015	0.008
7	1	0.998	0.992	0.983	0.971	0.956	0.938	0.919	0.897	0.875	0.850	0.717	0.577	0.445	0.329	0.234	0.159	0.102	0.063
7	2	1.000	1.000	0.999	0.998	0.996	0.994	0.990	0.986	0.981	0.974	0.926	0.852	0.756	0.647	0.532	0.420	0.316	0.227
7	3	1.000	1.000	1.000	1.000	1.000	1.000	0.999	0.999	0.998	0.997	0.988	0.967	0.929	0.874	0.800	0.710	0.608	0.500
7	4	1.000	1.000	1.000	1.000	1.000	1.000	1.000	1.000	1.000	1.000	0.999	0.995	0.987	0.971	0.944	0.904	0.847	0.773
7	5	1.000	1.000	1.000	1.000	1.000	1.000	1.000	1.000	1.000	1.000	1.000	1.000	0.999	0.996	0.991	0.981	0.964	0.938
7	6	1.000	1.000	1.000	1.000	1.000	1.000	1.000	1.000	1.000	1.000	1.000	1.000	1.000	1.000	0.999	0.998	0.996	0.992

Continued

Binomial distribution *(continued)*

		p																	
n	***x***	**0.01**	**0.02**	**0.03**	**0.04**	**0.05**	**0.06**	**0.07**	**0.08**	**0.09**	**0.10**	**0.15**	**0.20**	**0.25**	**0.30**	**0.35**	**0.40**	**0.45**	**0.50**
8	**0**	0.923	0.851	0.784	0.721	0.663	0.610	0.560	0.513	0.470	0.430	0.272	0.168	0.100	0.058	0.032	0.017	0.008	0.004
8	**1**	0.997	0.990	0.978	0.962	0.943	0.921	0.897	0.870	0.842	0.813	0.657	0.503	0.367	0.255	0.169	0.106	0.063	0.035
8	**2**	1.000	1.000	0.999	0.997	0.994	0.990	0.985	0.979	0.971	0.962	0.895	0.797	0.679	0.552	0.428	0.315	0.220	0.145
8	**3**	1.000	1.000	1.000	1.000	1.000	0.999	0.999	0.998	0.997	0.995	0.979	0.944	0.886	0.806	0.706	0.594	0.477	0.363
8	**4**	1.000	1.000	1.000	1.000	1.000	1.000	1.000	1.000	1.000	1.000	0.997	0.990	0.973	0.942	0.894	0.826	0.740	0.637
8	**5**	1.000	1.000	1.000	1.000	1.000	1.000	1.000	1.000	1.000	1.000	1.000	0.999	0.996	0.989	0.975	0.950	0.912	0.855
8	**6**	1.000	1.000	1.000	1.000	1.000	1.000	1.000	1.000	1.000	1.000	1.000	1.000	1.000	0.999	0.996	0.991	0.982	0.965
8	**7**	1.000	1.000	1.000	1.000	1.000	1.000	1.000	1.000	1.000	1.000	1.000	1.000	1.000	1.000	1.000	0.999	0.998	0.996
9	**0**	0.914	0.834	0.760	0.693	0.630	0.573	0.520	0.472	0.428	0.387	0.232	0.134	0.075	0.040	0.021	0.010	0.005	0.002
9	**1**	0.997	0.987	0.972	0.952	0.929	0.902	0.873	0.842	0.809	0.775	0.599	0.436	0.300	0.196	0.121	0.071	0.039	0.020
9	**2**	1.000	0.999	0.998	0.996	0.992	0.986	0.979	0.970	0.960	0.947	0.859	0.738	0.601	0.463	0.337	0.232	0.150	0.090
9	**3**	1.000	1.000	1.000	1.000	0.999	0.999	0.998	0.996	0.994	0.992	0.966	0.914	0.834	0.730	0.609	0.483	0.361	0.254
9	**4**	1.000	1.000	1.000	1.000	1.000	1.000	1.000	1.000	0.999	0.999	0.994	0.980	0.951	0.901	0.828	0.733	0.621	0.500
9	**5**	1.000	1.000	1.000	1.000	1.000	1.000	1.000	1.000	1.000	1.000	0.999	0.997	0.990	0.975	0.946	0.901	0.834	0.746
9	**6**	1.000	1.000	1.000	1.000	1.000	1.000	1.000	1.000	1.000	1.000	1.000	1.000	0.999	0.996	0.989	0.975	0.950	0.910
9	**7**	1.000	1.000	1.000	1.000	1.000	1.000	1.000	1.000	1.000	1.000	1.000	1.000	1.000	1.000	0.999	0.996	0.991	0.980
9	**8**	1.000	1.000	1.000	1.000	1.000	1.000	1.000	1.000	1.000	1.000	1.000	1.000	1.000	1.000	1.000	1.000	0.999	0.998
10	**0**	0.904	0.817	0.737	0.665	0.599	0.539	0.484	0.434	0.389	0.349	0.197	0.107	0.056	0.028	0.013	0.006	0.003	0.001
10	**1**	0.996	0.984	0.965	0.942	0.914	0.882	0.848	0.812	0.775	0.736	0.544	0.376	0.244	0.149	0.086	0.046	0.023	0.011
10	**2**	1.000	0.999	0.997	0.994	0.988	0.981	0.972	0.960	0.946	0.930	0.820	0.678	0.526	0.383	0.262	0.167	0.100	0.055
10	**3**	1.000	1.000	1.000	1.000	0.999	0.998	0.996	0.994	0.991	0.987	0.950	0.879	0.776	0.650	0.514	0.382	0.266	0.172
10	**4**	1.000	1.000	1.000	1.000	1.000	1.000	1.000	0.999	0.999	0.998	0.990	0.967	0.922	0.850	0.751	0.633	0.504	0.377
10	**5**	1.000	1.000	1.000	1.000	1.000	1.000	1.000	1.000	1.000	1.000	0.999	0.994	0.980	0.953	0.905	0.834	0.738	0.623

Appendix I

Chi-Square Distribution

Chi-square distribution

df	$\chi^2_{0.995}$	$\chi^2_{0.99}$	$\chi^2_{0.975}$	$\chi^2_{0.95}$	$\chi^2_{0.90}$	$\chi^2_{0.10}$	$\chi^2_{0.05}$	$\chi^2_{0.025}$	$\chi^2_{0.01}$	$\chi^2_{0.005}$
1	0.000	0.000	0.001	0.004	0.016	2.706	3.841	5.024	6.635	7.879
2	0.010	0.020	0.051	0.103	0.211	4.605	5.991	7.378	9.210	10.597
3	0.072	0.115	0.216	0.352	0.584	6.251	7.815	9.348	11.345	12.838
4	0.207	0.297	0.484	0.711	1.064	7.779	9.488	11.143	13.277	14.860
5	0.412	0.554	0.831	1.145	1.610	9.236	11.070	12.832	15.086	16.750
6	0.676	0.872	1.237	1.635	2.204	10.645	12.592	14.449	16.812	18.548
7	0.989	1.239	1.690	2.167	2.833	12.017	14.067	16.013	18.475	20.278
8	1.344	1.647	2.180	2.733	3.490	13.362	15.507	17.535	20.090	21.955
9	1.735	2.088	2.700	3.325	4.168	14.684	16.919	19.023	21.666	23.589
10	2.156	2.558	3.247	3.940	4.865	15.987	18.307	20.483	23.209	25.188
11	2.603	3.053	3.816	4.575	5.578	17.275	19.675	21.920	24.725	26.757
12	3.074	3.571	4.404	5.226	6.304	18.549	21.026	23.337	26.217	28.300
13	3.565	4.107	5.009	5.892	7.041	19.812	22.362	24.736	27.688	29.819
14	4.075	4.660	5.629	6.571	7.790	21.064	23.685	26.119	29.141	31.319
15	4.601	5.229	6.262	7.261	8.547	22.307	24.996	27.488	30.578	32.801
16	5.142	5.812	6.908	7.962	9.312	23.542	26.296	28.845	32.000	34.267
17	5.697	6.408	7.564	8.672	10.085	24.769	27.587	30.191	33.409	35.718
18	6.265	7.015	8.231	9.390	10.865	25.989	28.869	31.526	34.805	37.156
19	6.844	7.633	8.907	10.117	11.651	27.204	30.144	32.852	36.191	38.582
20	7.434	8.260	9.591	10.851	12.443	28.412	31.410	34.170	37.566	39.997
21	8.034	8.897	10.283	11.591	13.240	29.615	32.671	35.479	38.932	41.401
22	8.643	9.542	10.982	12.338	14.041	30.813	33.924	36.781	40.289	42.796
23	9.260	10.196	11.689	13.091	14.848	32.007	35.172	38.076	41.638	44.181
24	9.886	10.856	12.401	13.848	15.659	33.196	36.415	39.364	42.980	45.558
25	10.520	11.524	13.120	14.611	16.473	34.382	37.652	40.646	44.314	46.928
26	11.160	12.198	13.844	15.379	17.292	35.563	38.885	41.923	45.642	48.290
27	11.808	12.878	14.573	16.151	18.114	36.741	40.113	43.195	46.963	49.645
28	12.461	13.565	15.308	16.928	18.939	37.916	41.337	44.461	48.278	50.994

Continued

Chi-square distribution *(continued)*

df	$\chi^2_{0.995}$	$\chi^2_{0.99}$	$\chi^2_{0.975}$	$\chi^2_{0.95}$	$\chi^2_{0.90}$	$\chi^2_{0.10}$	$\chi^2_{0.05}$	$\chi^2_{0.025}$	$\chi^2_{0.01}$	$\chi^2_{0.005}$
29	13.121	14.256	16.047	17.708	19.768	39.087	42.557	45.722	49.588	52.335
30	13.787	14.953	16.791	18.493	20.599	40.256	43.773	46.979	50.892	53.672
31	14.458	15.655	17.539	19.281	21.434	41.422	44.985	48.232	52.191	55.002
32	15.134	16.362	18.291	20.072	22.271	42.585	46.194	49.480	53.486	56.328
33	15.815	17.073	19.047	20.867	23.110	43.745	47.400	50.725	54.775	57.648
34	16.501	17.789	19.806	21.664	23.952	44.903	48.602	51.966	56.061	58.964
35	17.192	18.509	20.569	22.465	24.797	46.059	49.802	53.203	57.342	60.275
40	20.707	22.164	24.433	26.509	29.051	51.805	55.758	59.342	63.691	66.766
45	24.311	25.901	28.366	30.612	33.350	57.505	61.656	65.410	69.957	73.166
50	27.991	29.707	32.357	34.764	37.689	63.167	67.505	71.420	76.154	79.490
55	31.735	33.571	36.398	38.958	42.060	68.796	73.311	77.380	82.292	85.749
60	35.534	37.485	40.482	43.188	46.459	74.397	79.082	83.298	88.379	91.952
65	39.383	41.444	44.603	47.450	50.883	79.973	84.821	89.177	94.422	98.105
70	43.275	45.442	48.758	51.739	55.329	85.527	90.531	95.023	100.425	104.215
75	47.206	49.475	52.942	56.054	59.795	91.061	96.217	100.839	106.393	110.285
80	51.172	53.540	57.153	60.391	64.278	96.578	101.879	106.629	112.329	116.321
85	55.170	57.634	61.389	64.749	68.777	102.079	107.522	112.393	118.236	122.324
90	59.196	61.754	65.647	69.126	73.291	107.565	113.145	118.136	124.116	128.299
95	63.250	65.898	69.925	73.520	77.818	113.038	118.752	123.858	129.973	134.247
100	67.328	70.065	74.222	77.929	82.358	118.498	124.342	129.561	135.807	140.170

Appendix J

Exponential Distribution

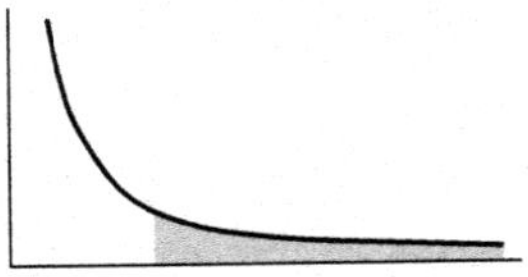

Exponential distribution

X	Area to left of *X*	Area to right of *X*
0	0.00000	1.00000
0.1	0.09516	0.90484
0.2	0.18127	0.81873
0.3	0.25918	0.74082
0.4	0.32968	0.67032
0.5	0.39347	0.60653
0.6	0.45119	0.54881
0.7	0.50341	0.49659
0.8	0.55067	0.44933
0.9	0.59343	0.40657
1	0.63212	0.36788
1.1	0.66713	0.33287
1.2	0.69881	0.30119
1.3	0.72747	0.27253
1.4	0.75340	0.24660
1.5	0.77687	0.22313
1.6	0.79810	0.20190
1.7	0.81732	0.18268
1.8	0.83470	0.16530
1.9	0.85043	0.14957
2	0.86466	0.13534
2.1	0.87754	0.12246
2.2	0.88920	0.11080
2.3	0.89974	0.10026
2.4	0.90928	0.09072
2.5	0.91792	0.08208
2.6	0.92573	0.07427

Continued

Exponential distribution *(continued)*

X	**Area to left of *X***	**Area to right of *X***
2.7	0.93279	0.06721
2.8	0.93919	0.06081
2.9	0.94498	0.05502
3	0.95021	0.04979
3.1	0.95495	0.04505
3.2	0.95924	0.04076
3.3	0.96312	0.03688
3.4	0.96663	0.03337
3.5	0.96980	0.03020
3.6	0.97268	0.02732
3.7	0.97528	0.02472
3.8	0.97763	0.02237
3.9	0.97976	0.02024
4	0.98168	0.01832
4.1	0.98343	0.01657
4.2	0.98500	0.01500
4.3	0.98643	0.01357
4.4	0.98772	0.01228
4.5	0.98889	0.01111
4.6	0.98995	0.01005
4.7	0.99090	0.00910
4.8	0.99177	0.00823
4.9	0.99255	0.00745
5	0.99326	0.00674
5.1	0.99390	0.00610
5.2	0.99448	0.00552
5.3	0.99501	0.00499
5.4	0.99548	0.00452
5.5	0.99591	0.00409
5.6	0.99630	0.00370
5.7	0.99665	0.00335
5.8	0.99697	0.00303
5.9	0.99726	0.00274
6	0.99752	0.00248

Appendix K

Poisson Distribution

Probability of x or fewer occurrences of an event

Poisson distribution

$\lambda\downarrow\ x\rightarrow$	0	1	2	3	4	5	6	7	8	9	10	11	12	13	14	15	16	17
0.005	0.995	1.000	1.000	1.000	1.000	1.000	1.000	1.000	1.000	1.000	1.000	1.000	1.000	1.000	1.000	1.000	1.000	1.000
0.01	0.990	1.000	1.000	1.000	1.000	1.000	1.000	1.000	1.000	1.000	1.000	1.000	1.000	1.000	1.000	1.000	1.000	1.000
0.02	0.980	1.000	1.000	1.000	1.000	1.000	1.000	1.000	1.000	1.000	1.000	1.000	1.000	1.000	1.000	1.000	1.000	1.000
0.03	0.970	1.000	1.000	1.000	1.000	1.000	1.000	1.000	1.000	1.000	1.000	1.000	1.000	1.000	1.000	1.000	1.000	1.000
0.04	0.961	0.999	1.000	1.000	1.000	1.000	1.000	1.000	1.000	1.000	1.000	1.000	1.000	1.000	1.000	1.000	1.000	1.000
0.05	0.951	0.999	1.000	1.000	1.000	1.000	1.000	1.000	1.000	1.000	1.000	1.000	1.000	1.000	1.000	1.000	1.000	1.000
0.06	0.942	0.998	1.000	1.000	1.000	1.000	1.000	1.000	1.000	1.000	1.000	1.000	1.000	1.000	1.000	1.000	1.000	1.000
0.07	0.932	0.998	1.000	1.000	1.000	1.000	1.000	1.000	1.000	1.000	1.000	1.000	1.000	1.000	1.000	1.000	1.000	1.000
0.08	0.923	0.997	1.000	1.000	1.000	1.000	1.000	1.000	1.000	1.000	1.000	1.000	1.000	1.000	1.000	1.000	1.000	1.000
0.09	0.914	0.996	1.000	1.000	1.000	1.000	1.000	1.000	1.000	1.000	1.000	1.000	1.000	1.000	1.000	1.000	1.000	1.000
0.1	0.905	0.995	1.000	1.000	1.000	1.000	1.000	1.000	1.000	1.000	1.000	1.000	1.000	1.000	1.000	1.000	1.000	1.000
0.15	0.861	0.990	0.999	1.000	1.000	1.000	1.000	1.000	1.000	1.000	1.000	1.000	1.000	1.000	1.000	1.000	1.000	1.000
0.2	0.819	0.982	0.999	1.000	1.000	1.000	1.000	1.000	1.000	1.000	1.000	1.000	1.000	1.000	1.000	1.000	1.000	1.000
0.25	0.779	0.974	0.998	1.000	1.000	1.000	1.000	1.000	1.000	1.000	1.000	1.000	1.000	1.000	1.000	1.000	1.000	1.000
0.3	0.741	0.963	0.996	1.000	1.000	1.000	1.000	1.000	1.000	1.000	1.000	1.000	1.000	1.000	1.000	1.000	1.000	1.000
0.35	0.705	0.951	0.994	1.000	1.000	1.000	1.000	1.000	1.000	1.000	1.000	1.000	1.000	1.000	1.000	1.000	1.000	1.000
0.4	0.670	0.938	0.992	0.999	1.000	1.000	1.000	1.000	1.000	1.000	1.000	1.000	1.000	1.000	1.000	1.000	1.000	1.000
0.5	0.607	0.910	0.986	0.998	1.000	1.000	1.000	1.000	1.000	1.000	1.000	1.000	1.000	1.000	1.000	1.000	1.000	1.000
0.6	0.549	0.878	0.977	0.997	1.000	1.000	1.000	1.000	1.000	1.000	1.000	1.000	1.000	1.000	1.000	1.000	1.000	1.000
0.7	0.497	0.844	0.966	0.994	0.999	1.000	1.000	1.000	1.000	1.000	1.000	1.000	1.000	1.000	1.000	1.000	1.000	1.000
0.8	0.449	0.809	0.953	0.991	0.999	1.000	1.000	1.000	1.000	1.000	1.000	1.000	1.000	1.000	1.000	1.000	1.000	1.000
0.9	0.407	0.772	0.937	0.987	0.998	1.000	1.000	1.000	1.000	1.000	1.000	1.000	1.000	1.000	1.000	1.000	1.000	1.000
1	0.368	0.736	0.920	0.981	0.996	0.999	1.000	1.000	1.000	1.000	1.000	1.000	1.000	1.000	1.000	1.000	1.000	1.000
1.2	0.301	0.663	0.879	0.966	0.992	0.998	1.000	1.000	1.000	1.000	1.000	1.000	1.000	1.000	1.000	1.000	1.000	1.000
1.4	0.247	0.592	0.833	0.946	0.986	0.997	0.999	1.000	1.000	1.000	1.000	1.000	1.000	1.000	1.000	1.000	1.000	1.000
1.6	0.202	0.525	0.783	0.921	0.976	0.994	0.999	1.000	1.000	1.000	1.000	1.000	1.000	1.000	1.000	1.000	1.000	1.000
1.8	0.165	0.463	0.731	0.891	0.964	0.990	0.997	0.999	1.000	1.000	1.000	1.000	1.000	1.000	1.000	1.000	1.000	1.000
2	0.135	0.406	0.677	0.857	0.947	0.983	0.995	0.999	1.000	1.000	1.000	1.000	1.000	1.000	1.000	1.000	1.000	1.000

Continued

Poisson distribution *(continued)*

λ↓ x→	0	1	2	3	4	5	6	7	8	9	10	11	12	13	14	15	16	17
2.2	0.111	0.355	0.623	0.819	0.928	0.975	0.993	0.998	1.000	1.000	1.000	1.000	1.000	1.000	1.000	1.000	1.000	1.000
2.4	0.091	0.308	0.570	0.779	0.904	0.964	0.988	0.997	0.999	1.000	1.000	1.000	1.000	1.000	1.000	1.000	1.000	1.000
2.6	0.074	0.267	0.518	0.736	0.877	0.951	0.983	0.995	0.999	1.000	1.000	1.000	1.000	1.000	1.000	1.000	1.000	1.000
2.8	0.061	0.231	0.469	0.692	0.848	0.935	0.976	0.992	0.998	0.999	1.000	1.000	1.000	1.000	1.000	1.000	1.000	1.000
3	0.050	0.199	0.423	0.647	0.815	0.916	0.966	0.988	0.996	0.999	1.000	1.000	1.000	1.000	1.000	1.000	1.000	1.000
3.2	0.041	0.171	0.380	0.603	0.781	0.895	0.955	0.983	0.994	0.998	1.000	1.000	1.000	1.000	1.000	1.000	1.000	1.000
3.4	0.033	0.147	0.340	0.558	0.744	0.871	0.942	0.977	0.992	0.997	0.999	1.000	1.000	1.000	1.000	1.000	1.000	1.000
3.6	0.027	0.126	0.303	0.515	0.706	0.844	0.927	0.969	0.988	0.996	0.999	1.000	1.000	1.000	1.000	1.000	1.000	1.000
3.8	0.022	0.107	0.269	0.473	0.668	0.816	0.909	0.960	0.984	0.994	0.998	0.999	1.000	1.000	1.000	1.000	1.000	1.000
4	0.018	0.092	0.238	0.433	0.629	0.785	0.889	0.949	0.979	0.992	0.997	0.999	1.000	1.000	1.000	1.000	1.000	1.000
4.5	0.011	0.061	0.174	0.342	0.532	0.703	0.831	0.913	0.960	0.983	0.993	0.998	0.999	1.000	1.000	1.000	1.000	1.000
5	0.007	0.040	0.125	0.265	0.440	0.616	0.762	0.867	0.932	0.968	0.986	0.995	0.998	0.999	1.000	1.000	1.000	1.000
5.5	0.004	0.027	0.088	0.202	0.358	0.529	0.686	0.809	0.894	0.946	0.975	0.989	0.996	0.998	0.999	1.000	1.000	1.000
6	0.002	0.017	0.062	0.151	0.285	0.446	0.606	0.744	0.847	0.916	0.957	0.980	0.991	0.996	0.999	0.999	1.000	1.000
6.5	0.002	0.011	0.043	0.112	0.224	0.369	0.527	0.673	0.792	0.877	0.933	0.966	0.984	0.993	0.997	0.999	1.000	1.000
7	0.001	0.007	0.030	0.082	0.173	0.301	0.450	0.599	0.729	0.830	0.901	0.947	0.973	0.987	0.994	0.998	0.999	1.000
7.5	0.001	0.005	0.020	0.059	0.132	0.241	0.378	0.525	0.662	0.776	0.862	0.921	0.957	0.978	0.990	0.995	0.998	0.999
8	0.000	0.003	0.014	0.042	0.100	0.191	0.313	0.453	0.593	0.717	0.816	0.888	0.936	0.966	0.983	0.992	0.996	0.998
8.5	0.000	0.002	0.009	0.030	0.074	0.150	0.256	0.386	0.523	0.653	0.763	0.849	0.909	0.949	0.973	0.986	0.993	0.997
9	0.000	0.001	0.006	0.021	0.055	0.116	0.207	0.324	0.456	0.587	0.706	0.803	0.876	0.926	0.959	0.978	0.989	0.995
9.5	0.000	0.001	0.004	0.015	0.040	0.089	0.165	0.269	0.392	0.522	0.645	0.752	0.836	0.898	0.940	0.967	0.982	0.991
10	0.000	0.000	0.003	0.010	0.029	0.067	0.130	0.220	0.333	0.458	0.583	0.697	0.792	0.864	0.917	0.951	0.973	0.986
10.5	0.000	0.000	0.002	0.007	0.021	0.050	0.102	0.179	0.279	0.397	0.521	0.639	0.742	0.825	0.888	0.932	0.960	0.978

Appendix L

Values of the t-Distribution

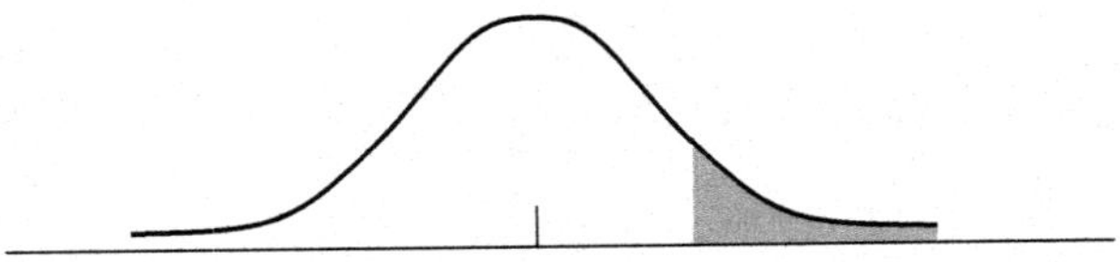

Values of t distribution

ν	$t_{0.100}$	$t_{0.050}$	$t_{0.025}$	$t_{0.010}$	$t_{0.005}$	ν
1	3.078	6.314	12.706	31.821	63.656	1
2	1.886	2.920	4.303	6.965	9.925	2
3	1.638	2.353	3.182	4.541	5.841	3
4	1.533	2.132	2.776	3.747	4.604	4
5	1.476	2.015	2.571	3.365	4.032	5
6	1.440	1.943	2.447	3.143	3.707	6
7	1.415	1.895	2.365	2.998	3.499	7
8	1.397	1.860	2.306	2.896	3.355	8
9	1.383	1.833	2.262	2.821	3.250	9
10	1.372	1.812	2.228	2.764	3.169	10
11	1.363	1.796	2.201	2.718	3.106	11
12	1.356	1.782	2.179	2.681	3.055	12
13	1.350	1.771	2.160	2.650	3.012	13
14	1.345	1.761	2.145	2.624	2.977	14
15	1.341	1.753	2.131	2.602	2.947	15
16	1.337	1.746	2.120	2.583	2.921	16
17	1.333	1.740	2.110	2.567	2.898	17
18	1.330	1.734	2.101	2.552	2.878	18
19	1.328	1.729	2.093	2.539	2.861	19
20	1.325	1.725	2.086	2.528	2.845	20
21	1.323	1.721	2.080	2.518	2.831	21
22	1.321	1.717	2.074	2.508	2.819	22
23	1.319	1.714	2.069	2.500	2.807	23
24	1.318	1.711	2.064	2.492	2.797	24
25	1.316	1.708	2.060	2.485	2.787	25
26	1.315	1.706	2.056	2.479	2.779	26
27	1.314	1.703	2.052	2.473	2.771	27
28	1.313	1.701	2.048	2.467	2.763	28

Continued

Values of *t* distribution *(continued)*

ν	$t_{0.10}$	$t_{0.05}$	$t_{0.025}$	$t_{0.01}$	$t_{0.005}$	ν
29	1.311	1.699	2.045	2.462	2.756	29
30	1.310	1.697	2.042	2.457	2.750	30
31	1.309	1.696	2.040	2.453	2.744	31
32	1.309	1.694	2.037	2.449	2.738	32
33	1.308	1.692	2.035	2.445	2.733	33
34	1.307	1.691	2.032	2.441	2.728	34
35	1.306	1.690	2.030	2.438	2.724	35
40	1.303	1.684	2.021	2.423	2.704	40
45	1.301	1.679	2.014	2.412	2.690	45
50	1.299	1.676	2.009	2.403	2.678	50
55	1.297	1.673	2.004	2.396	2.668	55
60	1.296	1.671	2.000	2.390	2.660	60
70	1.294	1.667	1.994	2.381	2.648	70
80	1.292	1.664	1.990	2.374	2.639	80
90	1.291	1.662	1.987	2.368	2.632	90
100	1.290	1.660	1.984	2.364	2.626	100
200	1.286	1.653	1.972	2.345	2.601	200
400	1.284	1.649	1.966	2.336	2.588	400
600	1.283	1.647	1.964	2.333	2.584	600
800	1.283	1.647	1.963	2.331	2.582	800
999	1.282	1.646	1.962	2.330	2.581	999

Appendix M

Acronym List

14 Points—Doctor Deming's 14 management practices

3C—cognition, comprehension, commitment

3D—dirty, dangerous, difficult

3P—people, planet, profit

3P—people, product, process

3P—production preparation process

4E—education, enforcement, engineering, environment

5F—find it, focus on it, fast, fix it, feedback

5M&P—materials, methods, machines, measurement, Mother Nature, and people

5P—Honda problem solving approach

5S—sort (seiri), straighten (seiton), shine (seiso), standardize (seiketsu), sustain (shitsuke)

5W1H—what, where, when, why, who, and how

6S—**5S** with *safety* added

7P—proper prior planning prevents piss poor performance

7S—**6S** with *oversight* added

8D—eight disciplines of problem solving

8M—man (people), machine (equipment), methods (operating procedures), materials, measurement, Mother Nature (environment), management, and money

A2LA—American Association for Laboratory Accreditation

A3—executive report on one page

ABET—ABET, Inc. (formerly the Accreditation Board of Education and Training)

AD—Anderson-Darling test

AIDR—aircraft inspection deficiency reports

AHP—analytic hierarchy process

AHT—average handling time

AIAG—Automotive Industry Action Group

AMA—American Management Association

ANAB—American National Accreditation Board

AND—activity network diagram

ANOM—analysis of means

ANOVA—analysis of variance

ANSI—American National Standards Institute

AOQ—average outgoing quality

AOQL—average outgoing quality limit

APQP—advanced product quality planning

AQAP—airspace quality advanced quality

AQL—acceptable quality level

AQP—advanced quality planning

AQP—Association for Quality and Participation (organization is now part of ASQ)

AQS—advanced quality system

AQT—acceptable quality test

ARL—average run length

AS—aerospace standards

ASA—American Statistical Association

ASCII—American standard code for information interchange

ASEE—American Society for Engineering Education

ASI—American Supplier Institute

ASME—American Society of Mechanical Engineers

ASN—average sample number

ASNT—American Society for Nondestructive Testing

ASQ—American Society for Quality

ASQC—American Society for Quality Control (ASQ name before 1997)

ASSE—American Society for Safety Engineers

ASTD—American Society for Training and Development

ASTM—ASTM International—formerly American Society for Testing and Materials

AV—appraiser variation

B/I—break in

B2C—business to customer

BAU—business-as-usual

BB—Black Belt

BBS—behavior based safety

BCI—Business Continuity Institute

BCM—business continuity management

BCMS—business continuity management system

BCP—business continuity plan

BCRRA—business continuity resource requirements analysis

BCSP—Board of Certified Safety Professionals

BIA—business impact analysis

BIB—balanced incomplete block design

BIC—best in class

BIC—business improvement coach

BIT—built-in test

BITE—built-in test equipment

BLT-bottom line on top

BOB—best of the best

BoK—body of knowledge

BOM—bill of materials

BOS—business operating system

BPO—best practice organization

BPR—business process reengineering

BRC—British Retail Consortium

BSI—British Standards Institute

BTW—by the way

C&E—cause and effect

C/N—change notice

C/O—changeover time

C/T—cycle time

CAA—Clean Air Act

CAA—compliance assurance agreement

CAD—computer-aided design

CADQAD—computer-aided development of quality assurance data

CAE—computer-aided engineering

CAFÉ—corporate average fuel economy

CAM—computer-aided manufacturing

CAMP—corrective action management process

CANDO—clean up, arranging, neatness, discipline, ongoing improvement

CAP—change acceleration process

CAP—corrective action plan

CAPA—corrective and preventive action

CAQ—computer-aided quality assurance

CAQ—condition adverse to quality

CAR—corrective action recommendation

CAR—corrective action report

CARE—customer acceptance review and evaluation

CASE—computer-aided software engineering

CASE—coordinated aerospace supplier evaluation

CBA—ASQ Certified Biomedical Auditor

CBP—customer benefits package

CBT—computer-based training

CC—critical characteristic

CCR—capacity constraint resource

CCR—critical customer requirement

CCT—ASQ Certified Calibration Technician

CE—cause and effect (for example, CE matrix)

CE—concurrent engineering

CEDAC—cause-and-effect diagram with additional of cards

CEO—chief executive officer

CEPT—Centre (for) Environmental Planning (and) Technology [India]

CERCLA—Comprehensive Environmental Response, Compensation, and Liability Act (Superfund)

CF—controlled forms

CFL—customer first leader

CFO—chief financial officer

CFR—USA Code of Federal Regulations

CGMP—current good manufacturing practice

CHA—ASQ Certified HACCP Auditor

CI—continual improvement

CIA—Certified Internal Auditor

CIM—change-in-mean-effect

CIO—chief information officer

CIP—clean in place

CIPO—continual improvement program owner

CIT—critical items list

CLCA—closed-loop corrective action

Cm—capability machine

CM—condition monitoring

CMI—ASQ Certified Mechanical Inspector

Cmk—machine capability index

CMM—capability maturity model for software (also known as SW-CMM)

CMM—coordinate measuring machine

CMP—crisis management plan

CMQ/OE—ASQ Certified Manager of Quality and Operational Excellence

CMQOE—ASQ Certified Manager of Quality Organizational Excellence

CNC—computer numerical control

COA—certificate of analysis

COB—chairman of board

COB—close of business

COC—certificate of conformance

COC—cost of conformance

COCQ—cost of current quality

CONC—cost of nonconformance

COO—chief operating officer

COP—code of practice

COP—customer oriented process

COPIS—customer, output, process, input, supplier

COPQ—cost of poor quality—measure of waste in operation

COQ—cost of quality (*see* COPQ)

COQC—certificate of quality compliance

CP—control plan

CPR—corrective preventive report

Cp—Process capability measurement—compares engineering specification divided by process six standard deviations

CPD—continuous professional development

Cpk—Process capability measurement—compares engineering specification to process mean divided by three standard deviations

CPM—critical path method

CPN—critical path network

CPU—cost per unit

CQA—ASQ Certified Quality Auditor

CQA—contract quality assurance

CQA—corporate quality assurance

CQE—ASQ Certified Quality Engineer

CQIA—ASQ Certified Quality Improvement Associate

CQM—Center for Quality of Management

CQMP—clinical quality management program

CQP—corporate quality policies

CQPA—ASQ Certified Quality Process Analyst

CQR—contract quality requirement

CQT—ASQ Certified Quality Technician

CR—conditionally required

Cr—ratio of process variation

CR/CR—concern report/change request

CRA—chemical risk assessment

CRE—ASQ Certified Reliability Engineer

CRM—certified reference material

CRM—corporate records management

CRM—customer relationship management

CS—customer satisfaction

CSA—compliance safety accountability

CSF—critical success factors

CSI—critical safety items

CSM—customer–supplier model

CSP—Certified Safety Professional

CSP—continuous sampling plan

CSQE—ASQ Certified Software Quality Engineer

CSSBB—ASQ Certified Six Sigma Black Belt

CSSGB—ASQ Certified Six Sigma Green Belt

CSSMBB—ASQ Certified Six Sigma Master Black Belt

CSSYB—ASQ Certified Six Sigma Yellow Belt

CT—controlled tables

CTC—critical to customer

CTQ—critical to quality

CTS—critical to satisfaction

CUSUM—cumulative sum control chart

CVEP—continuous value enhancement process

CWA—Clean Water Act

CWAP—clean water action plan

CWQC—company-wide quality control

D—detection

DAX—desire, attitude, execution

DBR—discounted cash flow

DCCDI—define–customer–concept–design–implement

DCF—discounted cash flow

DCMA—Defense Contract Management Agreement

DCOV—define–characterize–optimize–verify

DCP—dynamic control plan

DDW—drill deep and wide

DE—directed evolution

DER—designated engineering representative

df—degrees of freedom

DFA—design for assembly

DFD—design for disassembly

DFE—design for ergonomics

DFM—design for manufacturing

DFMA—design for manufacturing and assembly

DFMEA—design failure mode and effects analysis

DFSS—design for Six Sigma

DFX—design for X

DMADOV—define–measure–analyze–design–optimize–verify

DMADV—define–measure–analyze–design–verify

DMAIC—define, measure, analyze, improve, and control

DMEDI—define–measure–explore–develop–implement

DMR—discharge monitoring reports

DOE—design of experiment(s)

DOT—United States Department of Transportation

DPM—deficiencies (defects) per million units

DPM—downtime performance measurement

DPO—deficiencies (defects) per opportunity

DPU—deficiencies (defects) per unit

DQC—data quality control

DRBFM—design review based on failure mode (Toyota version of FMEA)

DRII—Disaster Recovery Institute International

DSL—digital subscriber line

DSU—digital service unit

DTD—dock to delivery

DV&PR—design verification and product reliability

DVP—design verification plan

DVP&PV—design verification, production and process validation

DVR—design verification report

DVT—design verification test

EARA—Environmental Auditors Registration Association

EC—European Community

ECC—estimated cost to complete

ECD—estimated completion date

ECDF—empirical cumulative distribution function

ECHO—enforcement and compliance history online

ECN—engineering change notice

ECO—engineer change order

ECR—engineering change request

EDA—exploratory data analysis

EDI—electronic data interchange

EI—employee involvement

EIO—engineering or installation caused outage

ELT—extract load transfer

EMI—electromagnetic interference

EMS—environmental management system

EOQ—economic order quantity

EP313—Emergency Planning and Community Right to Know Act, Section 313 (i.e., the Toxics Release Inventory (TRI) program)

EPSS—electronic performance support system

ER—engineering requirements

ERI—early return indicator

ERM—enterprise risk management

ERP—enterprise resource planning

ES—engineering specification

ESC—extreme service conditions

ESER—engineering sample evaluation report

ET—educational technology

ETA—event tree analysis

ETQ—excellence to quality

EU—European Union

EV—equipment variation

EVOP—evolutionary operation

EWMA—exponentially weighted moving average

FAHQMT—fully automatic high-quality machine translation

FAI—first article inspection

FAIR—first article inspection report

FAR—Federal Acquisition Regulation

FARA—field action risk assessment

FAST—function analysis system technique

FCE—frequently committed errors

FEA—finite element analysis

FEA—front-end analysis

FEFO—first expired, first out

FIFO—first in, first out

FIFRA—Federal Insecticide, Fungicide, and Rodenticide Act

FISH—first in still here

FMA—failure mode analysis

FMEA—failure mode and effects analysis

FMECA—failure mode effects and criticality analysis

FMEDA—failure modes, effects, and diagnostic analysis

FMEM—failure mode effects management

FPA—first party audit

FPS—Ford Production System

FQ&P—flight, quality, and performance

FQI—Federal Quality Institute (*see* OPM)

FR—field replaceable unit returns

FRT—fix response time

FSI—fatal and serious injury

FSL—flow synchronization leveling

FSP—function restoration plan

FSS—full service supplier

FST—food safety team

FTA—fault tree analysis

FTPM—Ford Total Productive Maintenance

FTQ—first time quality

FTT—first time through

G8D—global eight disciplines

GB—Green Belt

GD&T—geometric dimensioning and tolerancing

GE—General Electric Corporation

GHS—globally harmonized system

GLM—general linear model

GLP—good laboratory practice

GM—General Motors Corporation

GMP—good manufacturing practice

GPC—gage performance curve

GR&R—gage repeatability and reproducibility

GROW—goal, reality, options, way forward

GRPI—goals, roles, processes, interpersonal

GRR—gage repeatability and reproducibility

GQA—government quality assurance

GQAR—government quality assurance representative

GQTS—global quality tracking system

GSQA—government source quality assurance

GUM—*Guide to the Expression of Uncertainty of Measurement*

Ha—alternative hypothesis

HA—hazard analysis

HACCP—hazard analysis and critical control points

HALT—highly accelerated life test

HARM—high-availability, reliability, and maintainability

HASA—highly accelerated stress audits

HASS—highly accelerated stress screening

HAZOP—hazard and operability study

HCS—hazardous communication standard

HOQ—house of quality

HPT—human performance technology

HQS—high-quality screening

HR—human resources

HRM—human resources management

HSEQ—health safety environmental quality

HSPD—handling, storage, packaging, and delivery

HSSE—health safety security environment

HSSEQ—health safety security environment quality

IABLS—Institute of Advanced Business Learning Systems

IAQG—International Aerospace Quality Group

IATF—International Automotive Task Force

ICC—incident command centre

ICOFR—internal controls over financial reporting

ICOV—identify–characterize–optimize–validate

ICT—information communication technology

ID—interrelationship digraph

IDDOV—identify–define–develop–optimize–verify (and validate)

IDEA—identify–design–evaluate–affirm

IDOV—identify–design–optimize–verify (and validate)

IEC—International Electrotechnical Commission

IEEE—Institute of Electrical and Electronics Engineers

IID—independent identically distributed

IIE—Institute of Industrial Engineers

ILT—instructor lead training

IMDS—International Material Data System

IMP—incident management plan

IMR—individuals and moving range

IMS—incident management structure

INPO—Institute of Nuclear Power Operators

INT—interaction

IOBA—International Automotive Oversight Bureau

IOP—IOP The Packaging Society - Institute of Materials, Minerals, and Mining

IPIP—improving performance in practice

IPO—input–process–output

IPS—innovative problem solving

IQA—Institute for Quality Assurance

IQCS—in-service quality control system

IQF—International Quality Federation

IQR—interquartile range

IQUE—in-plant quality evaluation

IRC—issue review committee

IRCA—International Register of Certified Auditors

IRR—internal rate of return

IS—information security

ISD—instructional system design

ISIR—initial sample inspection report

ISO—International Organization for Standardization

ISPI—International Society for Performance Improvement

ISSSP—International Society of Six Sigma Practitioners

IT—industrial technology

IT—information technology (computers)

IT—instructional technology (education)

ITDR—information technology disaster recovery

ITP—inspection and test plan

ITU—International Telecommunication Union

JCAHO—Joint Commission on Accreditation of Healthcare Organizations

JDP—J. D. Power and Associates

JES—job element sheet (another name for work instructions)

JIS—Japan Industrial Standard

JIS—job instruction sheet (another name for work instructions)

JIT—just in time

JUSE—Union of Japanese Scientists and Engineers

KBC—knowledge based community

KBF—key business factors

KBI—key business issue

KBR—key business requirement

KC—key characteristic

KCC—key control characteristic

KISS—keep it simple and specific *or* keep it simple statistician

KLT—key life test

KPC—key product characteristic

KPI—key performance indicator

KPI—key process indicator

KPIV—key process input variable

KPOV—key process output variable

KSN—knowledge sharing network

LACL—lower acceptance control limit

LBC—level of business continuity

LCI—learner controlled instruction

LCL—lower control limit

LCR—linear responsibility chart (another name for RACI)

LEO—listen (observe and understand), enrich (explore and discover), and optimize (improve and perfect)

LIFO—last in, first out

LLL—lower lot limit

LMS—learning management system

LOB—line of business

LOTO—lock out tag out

LQ—limiting quality

LQIP—laboratory quality improvement program

LQL—limiting quality level

LRU—line replaceable unit

LSA—logistic support analysis

LSD—least significant difference

LSL—lower specification limit

LSS—Lean Six Sigma

LTI—lost time injury

LTPD—lot tolerance percentage defective

LTR—long-term return rate

LTS—long-term stability

m—mean

M&A—manufacturing and assembly

M&TE—measurement and test equipment

MAIC—measure, analyze, improve, and control

MAR—maximum allowable range

MBB—Master Black Belt

MBC—management of business continuity

MBO—management by objectives

MBNQA—Malcolm Baldrige National Quality Award

MBTI—Myers-Briggs Type Indicator

MBWA—management by walking around

MCF—mean cumulative function

MDR—medical device report

MEDIC—map + measure, explore + evaluate, define + describe, implement + improve, control + conform

MFMEA—machinery failure mode and effects analysis

MIL-STD—United States military standard

MIS—management information systems

MIS—months in service

MMBF—mean miles between failures

MODAPTS—modular arrangement of predetermined time standards

MOS—management operating system

MOT—moment of truth

MPRSA—Marine Protection, Research, and Sanctuaries Act

MPS—master production schedule

MQT—maintainability qualification test

MRA—mutual recognition arrangements

MRB—management review board

MRP—material requirements planning

MRSL—market required service level

MS—mean squares

MS (RES)—residual mean square

MS&R—mission solutions and readiness

MSA—measurement systems analysis

MSB—mean square between treatments

MSD—maximum standard deviation

MSDS—material data safety sheet (old phrase that should not be used any longer)

MSE—mean squared error

MSI—mean square for interaction

MSW—mean square within treatments

MT&E—measuring tools and equipment

MTBF—mean time between failures

MTC—manage the change

MTD—month to date

MTPD—maximum tolerable period of disruption

MTTF—mean time to failure

MTTN—mean time to notification

MTTR—mean time to recover

MTTR—mean time to repair

MWO—maintenance work order

NA—needs assessment

NA *or* **N/A**—not applicable

NACCB—National Accreditation Council for Certification Bodies

NADCAP—National Aerospace and Defense Contractors Accreditation Program

NATO—North Atlantic Treaty Organization

NCT—nonconformance ticket

ndc—number of distinct categories

NDE—nondestructive evaluation

NDT—nondestructive testing

NE *or* **N/E**—not evaluated

NGT—nominal group technique

NIH—not invented here

NIST—United States National Institute of Standards and Technology

NMI—near miss incident

NMQAO—Naval Material Quality Assessment Office

NPI—new product introduction

NPR—number of problem reports

NPV—net present value

NQCC—network quality control center

NTF—no trouble found

NTRM—NIST Traceable Reference Material

NVA—non-value-added

NVA-U—non-value-added, but unavoidable

NVH—noise, vibration, and harshness

O—occurrence

OASIS—Online Aerospace Supplier Information System

OBC—operator base care

OBS—observation

OC—operating characteristic

OCAP—out-of-control action plan

OCC—operating characteristic curve

OCM—operating committee meeting

OCM—organizational change management

OCS—operational control sheet (another name for control sheet)

OCT—operations cost target

OD—organization development

OE—organizational excellence

OEE—overall equipment effectiveness

OEM—original equipment manufacturer

OFI—opportunity for improvement

OFM—outage frequency measurement

OFR—overdue fix responsiveness

OHS—occupational health and safety

OI—operating instructions

OJT—on-the-job training

OLE—overall labor effectiveness

OOC—out of stock

ORT—ongoing reliability test

OSHA—United States Occupational Safety and Health Administration

OSS—operational support system

OTD—on-time delivery

OTED—one touch exchange of dies

OTI—on-time item delivery

OTIS—on-time installed system delivery

OTS—on-time service delivery

P&L—profit and loss

P&S—products and services

P/T—precision/tolerance

PaR—patients at risk

PAR—preventive action report

PART—program assessment rating tool

PAT—part average testing

PBC—process behavior charts

PBIB—partially balanced incomplete block design

PC—physical contradiction

PC—product complaint

PCD—process control document

PCR—product change request

PCS—production control system

PDA—personal data assistant

PDC—product development cycle

PDE—Proposal Development Engineer

PDCA—plan–do–check–act

PDM—precedence diagram method

PDPC—process decision program chart

PDQR—product quality deficiency reports

PDSA—plan–do–study–act

PE—professional engineer

PEL—permissable exposure limits

PERT—program evaluation review technique

PEST—political, economic, social and technological

PESTLE—political, economic, social, technology, legal, environmental

PFMEA—potential failure mode and effects analysis

PFQ—planning for quality

PI—principal inspector

PIPC—percent indices which are process capable

PISMOEA—part, instrument, standard, method, operator, environment, assumptions

PIST—percentage of inspection points satisfying tolerance

PIT—process improvement team

PM—preventive maintenance

PM—program management

PMA—premarket approval

PMA—president's management agenda

PMO—deficiencies (defects) per million opportunities

PMP—project management professional

PMS—planned maintenance system

PMTS—predetermined motion time system

PO—purchase order

POI—program of instruction

PONC—price of nonconformance

Pp—long-term process capability measurement

PP&B—prototype planning and build

PP&DC—product planning and design committee

PP&TC—product planning and technology committee

PPAP—production part approval process

PPCC—normal probability plot correlation coefficient

PPF—production process and product approval

Ppk—long-term process capability measurement

ppm—parts per million

PPPPP—prior planning prevents piss-poor performance

PPPPPP—proper planning prevents particularly poor performance

PPR—patients per run

PPS—production preparation schedule

PQ—process qualification

PQA—President's Quality Award

Pr—capability performance ration

PR—production release

PRAT—production reliability acceptance test

PRP—process recovery plan

PRR—problem reporting and resolution *or* product problem reporting

PSM—process safety management

PSO—process sign-off

PSP—problem-solving process

PSP—product support plan

PSW—part submission warrant

PTC—pass through characteristics

PTN—plant test number

PUMA—product usage measurements and applications

PV—part variation

PVP&R—production validation plan and report

PYR—pass yield rate

Q&R—quality and reliability

QA—quality assurance

QA—quick action

QAA—quality assurance analyst

QAA—quality assurance and assistance

QAA—quality assurance assessment

QAA—quality assurance audit

QAC—quality assurance checklist

QAC—quality assurance committee

QAD—quality assurance directorate

QAD—quality audit division

QADR—quality assurance discrepancy report

QAE—quality assurance engineer

QAE—quality assurance evaluation

QAE—quality assurance executive

QAER—quality acceptance equipment release

QAF—quality achievement factor

QAF—quality assurance fixture

QAF—quality assurance form

QAHB—Quality Assurance Program Handbook

QAI—quality assessment index

QAI—Quality Assurance Institute

QAI—quality assurance instruction

QALI—quality assurance letter of instruction

QAM—quality assurance manager

QAM—quality assurance monitoring

QAN—quality action notice

QAPI—quality assurance program index

QAPR—Quality Army Performance Review

QAR—quality acceptance report

QAR—quality assurance and reliability

QAR—quality assurance evaluator

QAR—quality assurance requirements

QAR—quality assurance review

QAR—quantitative analysis report

QAR—quarterly acceptance review

QARC—Quality Assurance Review Center

QAS—quality assurance, auditing, and security

QAS—quality assurance schedule

QAS—quality assurance screening program

QAS—quality assurance standard(s)

QAS—quality assurance study

QAS—quality assurance surveillance

QAS—quality assurance test system

QASP—quality assurance support plan

QATAP—quality assurance through attributes program

QATDP—quality assurance technical development program

QBP—quality and business planning

QC—quality center

QC—quality control

QCAI—quality control/assurance and inspection

QCCMM—quality control certified master model

QCE—quality control engineering

QCEM—quality control enforcement mechanism

QCI—Quality Circle Institute

QCI—quality control information

QCI—quality cost improve(ment)

QCI—Quality Council of India

QCI—Quality Council of Indiana

QCM—quality call monitoring

QCM—quality care monitoring

QCM—quality control manual

QCM—quality control master

QCP—quality commitment performance

QCP—quality control program

QCR—quality control reliability

QCR—quality control report

QCR—quality control representative

QCS—quality and customer satisfaction

QCS—quality customer service

QCT—quality, cost, timing

QCWF—quality, cost, weight, and function

QCWFT—quality, cost, weight, function, and timing attributes

QDR—quality, durability, reliability

QDR—quality deficiency report(s)

QEMS—quality and environmental management system

QEP—quality enhancement program

QEP—quality evaluation program

QF—quality form

QFD—quality function deployment

QFTF—quality function test fleet

QHC—quality in health care

QHNZ—Quality Health New Zealand

QHR—quality history records

QI—quality improvement

QI—quality increase

QIC—quality information using cycle time

QIES—quality improvement evaluation system

QIM—quality improvement meeting

QIP—quality improvement process

QIP—quality intervention plan

QIS—quality information system

QIT—quality in training

QIT—quality information and test

QITQM—*Quality Improvement Total Quality Management* (magazine)

QLA—quality level agreement

QLF—quality loss function

QLS—quality leadership system

QMAS—Quality Measurement Advisory Service

QMIS—quality management information system

QMMP—Quality Measurement and Management Project

QMP—quality, manufacturing, and purchasing

QMRP—Qantel manufacturing resource planning (MRP II) system

QMS—quality management system

QOS—quality of service

QOS—quality operating system

QP—quality plan

QP—quality procedure

QPC—quality and process control

QPC—quality performance consultant

QPI—quality performance indicator

QPIP—quality and productivity improvement program

QPM—quality and performance management

QPM—quality performance matrix

QPM—quality program manager

QPR—quality problem report

QPS—quality planning sheets

QPS—quality process sheets

QPS—quality process system

QPSS—quality process system sheets

QR—quality and reliability

QR—quality reject(s)

QR—quality report

QR—quantitative requirement

QR—quick response

QRA—quality and reliability assurance

QRA—quality reliability assurance

QRA—quick reaction assessment

QRA—quick readiness assessment

QRA—quick response audit

QRB—quality review board

QRC—quality record coordinator

QRC—quality risk and cost

QRD—quantitative risk management

QRM—quality risk management

QRO—quality review organization

QRS—quality review studies

QRT—quality responsible team

QS—quality systems

QS-9000—Quality System Requirements 9000

QSA—quality system analyst

QSC—quality strategy committee

QSDC—quality system document coordinator

QSF—quick service fix

QSHC—*Quality and Safety in Health Care* (magazine)

QSP—quality strategy and planning

QSR—quality system requirement(s)

QSRC—quality system record coordinator

QSS—quality support team

QSU—quality system update

QTS—quality tracking study

QUADS—quality document system

QUASAR—Quality and Safety Achievement Recognition

QUASAR—Quality Driven Software Architecture

QUEST—quality electrical systems test

QUEST—quality evaluation of settlement

QuEST—Quality Excellence for Suppliers of Telecommunications

QUGS—quality utilization generic screens

QUIP—quality assessment and improvement program

QUIP—quality assurance inspection procedure

QUIT—Quality in Training

QVI—quality verification inspection

QVP—quality vendor program

R—required

R^2—coefficient of determination

R2R—runs to reject

R&A—reliability and availability

R&D—research and development

R&M—reliability and maintainability

R&M—reliability and maintenance

R&MWG—reliability and maintainability working group

R&R—repeatability and reproducibility (*see also* GR&R)

RA—risk analysis

RA—risk assessment

RAB—registrar accreditation board

RABQSA—RABQSA International (formerly the Registrar Accreditation Board and the Quality Society of Australasia)

RACI—responsible, accountable, consulted, and informed

RADHAZ—radio and radar radiation hazards

RAM—reliability, availability, and maintainability

RAM—responsibility assignment matrix

RAMAS—reliability, availability, maintainability analysis system

RAMCAD—reliability and maintainability in computer-aided design

RAM-D—reliability, availability, maintainability, and durability

RAMDAS—reliability and maintainability data access system

RAMES—reliability, availability, maintainability, engineering system

RAMIS—reliability and maintainability information system

RAMS—range measurement system

RAMSH—reliability, availability, maintainability, safety, (and) human-factors (engineering)

RAMTIP—Reliability and Maintainability Technology Insertion Program

RAPID—rapid actions for process improvement deployment

RAS—reliability, availability, and serviceability

RBD—reliability block diagram

RBI—risk based inspection

RBM—risk based maintenance

RC—responsible care

RCA—root cause analysis

RCL—robustness checklist

RCM—reliability centered maintenance

RCPS—root cause problem solving

RCRA—Resource Conservation and Recovery Act

RD/GT—reliability development/growth test

RDCOV—recognize–define–characterize–optimize–verify

REG—regression

REM—reliability engineering model

RES—residual

RF—radio frequency

RF—remaining float

RFI—radio frequency interference

RFP—request for proposal

RFQ—request for quote

RFTA—reverse fault tree analysis

RII—required inspection item

RIW—reliability improvement warranty

RM—reference material

RM&A—reliability, maintainability, and availability

RM&S—reliability, maintainability, and supportability

RMA—reliability, maintainability, and availability

RMMP—reliability and maintainability management plan

RMS—root mean square

ROA—report of analysis

ROA—return on assets

ROE—return on equity

ROI—return on investment

RONA—return on net assets

RPL—rejectable process level

RPM—revolutions per minute

RPN—risk priority number

RPO—recovery point objective

RQL—rejectable quality level

RQMS—*Reliability and Quality Measurements for Telecommunications Systems*

RQL—rejectable quality level

RQT—reliability qualification test(ing)

RRA—residual risk assessment

RRM—risk review meeting

RSM—repair station manual

RSM—response surface methodology

RTO—recovery time objective

RtO—return to operation

RTOK—retest OK

RTY—rolled throughput yield

S—satisfactory

S—severity

S3—safety and suitability for service

SAE—Society of Automotive Engineers *or* SAE International

SAES—supplier audit evaluation summary

SB—service bulletin

SBP—strategic business plan

SC—significant characteristic

SCM—supply chain management

SCOT—strengths, challenge, opportunities, threats

SCP—service control point

SDCA—standardize–do–check–act

SDE—supplier development engineer

SDS—safety data sheet

SDWA—Safe Drinking Water Act

SDWT—self-directed work team

SE—simultaneous engineering

SE—standard error

SEP—safety and environmental procedure

SET—senior executive team

SF—secondary float

SHARP—Safety and Health Achievement Recognition Program

SIF—safety integrity analysis

SIPOC—supplier, input, process, output, and customer

SIT—systematic inventive thinking

SKSP—skip-lot sampling plan

SLACK—summary, learning objectives, application, context, knowledge base

SMART—specific, meaningful, agreed to, realistic, time-based

SMARTER—specific, measurable, acceptable, realistic, time-bound, evaluated, reviewed

SME—Society of Manufacturing Engineers

SME—small and medium enterprises

SME—subject matter expert

SMED—single-minute exchange of die

SMS—safety management system

SN—signal-to-noise ratio

SO—system outage measurement

SoA—statement of applicability

SOC2 Type 2—Service Organization Control audit related to ISO 27001

SOF—safety of flight

SOP—standard operating procedure

SoPK—System of Profound Knowledge (Dr. W. Edwards Deming)

SOQ—service-oriented architecture

SOR—sign-off report

SOS—standard operating sheets (another name for work instructions)

SOW—statement of work

SOW—Scope of work

SPA—second party audit

SPC—statistical process control

SPD—statistical process display

SPEAR—suppler performance and evaluation report

SPM—statistical process management

SPOF—single point of failure

SPOT—scope, purpose, overview, tangible benefits

SQC—statistical quality control

SQDCME—safety, quality, delivery, cost, moral, environment

SQA—supplier quality agreement

SQE—software quality evaluation

SQE—supplier quality engineer

SQI—supplier quality improvement

SQP—strategic quality plan

SQR—supplier quality representative

SQRTF—Supplier Quality Requirements Task Force

SREA—supplier request for engineering approval

SRG—statistical research group

SRM—supplier relationship management

SRMR—security risk management review

SRP—strategic regulatory plan

SS—Six Sigma

SS—sum of squares

SSB—between-treatments sum of squares

SSBB—Six Sigma Black Belt

SSBoK—Six Sigma Body of Knowledge

SSC—column sum of squares

SSE—error sum of squares

SSGB—Six Sigma Green Belt

SSI—interaction sum of squares

SSMBB—Six Sigma Master Black Belt

SSOS—Six Sigma operating system

SSR—residual sum of squares

SSR—row sum of squares

SSRA—system safety risk assessment

SST—total sum of squares

SSW—within-treatments sum of squares

SSYB—Six Sigma Yellow Belt

STA—supplier technical assistance

STD—standard deviation

STEL—short-term exposure limit

STF—Slip Trip & Fall

STOP—Safety Training Observation Program

STP—signaling transfer point

STS—synchronous transport signal

SUTR—set-up time reduction

SVA—source vulnerability assessment

SWAG—statistical wild ass guess

SWIPE—standard, workpiece, instrument, person and procedure, environment

SWL—safe working load

SWOT—strengths, weaknesses, opportunities, threats

T—target

T&D—test and diagnostic

T&D—training and development

T&E—test and evaluation

T&EO—training and evaluation outline

T&M—time and materials

T&O—test and operation

TACT—total average cycle time

TAT—turnaround time

TBD—to be determined

TBE—to be established

TC—technical contradiction

TDR—technical design review(s)

TE—tooling and equipment

TF—total float

TGR—things gone right

TGW—things gone wrong

TIE—technical information engineer

TMAP—thought process map

TNA—training needs assessment

TNA—training needs analysis

TOC—theory of constraints

TOPS—total operational performance system

TOU—terms of use

TPA—third-party audit

TPI—transaction process improvement

TPM—total productive maintenance

TPS—Toyota Production System

TQ—total quality

TQC—total quality control

TQHRM—total quality human resources management

TQM—total quality management

TRACE—total risk assessing cost estimate

TRACE—total risk assessing cost estimating

TRIZ—theory of inventive problem solving

TRLT—total replenishment lead time

TS—technical specification

TSCA—Toxic Substances Control Act

TSS—total sum of squares

TTI—targets to improve

TV—total variation

TVM—total value management

TWA—time weighted average

UACL—upper acceptable control limit

UCL—upper control limit

UKAS—United Kingdom Accreditation Service

ULL—upper lot limit

UP—unit price

UPC—uniform parts code

UQL—unacceptable quality level

UPS—uninterrupted power system

USL—upper specification limit

USPAP—Uniform Standards of Professional Appraisal Practice

V&V—verification and validation

VA—value-added

VA/VE—value analysis/value engineering

VC—virtual container

VDA—Verband der Automobilindustrie (German)

VIM—*International Vocabulary of Metrology—Basic and General Concepts and Associated Terms*

VIN—vehicle identification number

VIPER—verifiable integrated processor for enhanced reliability

VOB—voice of the business

VOC—voice of the customer

VOE—verification of effectiveness

VOE—voice of the employee

VOP—voice of the process

VPC—vendor product complaint

VPP—voluntary protection program

VQD—visual quality document

VSAS—vehicle situational awareness system

WAG—wild ass guess

WBS—work breakdown structure

WCP—world class process

WGD—worldwide guidance documents

WI—work instructions

WIIFM—what's in it for me

WIP—work in process

WOW—worst of the worst

WPS—Weld Procedure Specifications

WQP—worldwide quality procedures

WQS—worldwide quality standards

WUR—water use ratio

WYSIWYG—What you see is what you get

x—average

X—cause or process variable

Y—effect or process output

YRR—one-year return rate

ZD—zero defects

Appendix N

The Quality Profession, Historically Speaking

SELECTED NOTES OF EVENTS

1750's	Industrial Revolution begins in England
1760	Lloyd's Registrar starts auditing ship yards – charged ship builders – third party registration
January 22, 1901	Sir John Wolfe-Barry (the man who designed London's Tower Bridge) instigated the Council of the Institution of Civil Engineers to form a committee to consider standardizing iron and steel sections
1911	Frederick W. Taylor publishes "The Principles of Scientific Management"
December 1, 1913	Henry Ford installs the first moving assembly line for the mass production of an entire automobile
1919	The Technical Inspection Association (TIA) forms (forerunner of CQI)
1922	The TIA becomes the Institution of Engineering Inspection (IEI)
1924	Walter A. Shewhart, a statistician at Bell Laboratories, develops the control charts, and principles of statistical process control
1925	Sir Ronald Fisher publishes the book, Statistical Methods for Research Workers, and introduced the concept of ANOVA
1927-1933	Elton Mayo at Western Electric plant in Cicero IL – Harvard Business School – Hawthorne Studies
1930s	Statistical Quality Control (SQC) – application of statistics to quality control
1931	Shewhart publishes *Economic Control of Quality of Manufactured Products*
1937	Joseph Juran introduces the Pareto principle as a means of narrowing on the vital few
1940	The acceptance sampling plan is developed by Harold F. Dodge and Harry G. Roming
1943	Kaoru Ishikawa develops the cause and effect diagram (Ishikawa or fishbone diagram)

1943	Federated Societies forms: Society of Quality Control Engineers of Buffalos (Buffalo Section 0201); Michigan Society for Quality Control (Greater Detroit Section 1000); Quality Control Engineers of Rochester (Rochester Section 0204); The Society of Quality Control Engineers of Syracuse (Syracuse Section 0206)
1944	Industrial Quality Control Vol 1 No 1 July 1944 – Society of Quality Control Engineers with cooperation of the University of Buffalo
1944	Dodge and Roming publish *Sampling Inspection Tables Single and Double Sampling*
February 16, 1946	The American Society for Quality Control (ASQC) is formed
May 1946	The Union of Japanese Scientists and Engineering (JUSE) established
February 23, 1947	The International Organization for Standardization (ISO) founded Geneva, Switzerland
1947	Dr. W. Edwards Deming is sent to Japan to help Japanese rejuvenate their industries
1948	Shewhart Medal First Awarded by ASQC
November 9, 1949	USA Military Procedure MIL-P-1629, Procedures for Performing a Failure Mode, Effects and Criticality Analysis (FMECA) forerunner to Mil-Q-1629 Procedures for Performing a FMECA (later the FMEA) – 1974, 1977, MIL-STD 1629A:1980
1950	Genrich Altshuller develops the theory of inventive problem solving (TRIZ)
1950	JUSE publishes *Elementary Principles of The Statistical Control of Quality* by Deming
1951	JUSE institutes the Deming Prize
1951	Juran publishes the first edition of *Quality Control Handbook*
1951	Feigenbaum publishes the first edition of *Total Quality Control (QC Principles, Practices & Administration*
1954	JUSE invites Juran to Japan
@ 1955	England Training Film: Right First Time *The Deming Institute own copyright
1956	Western Electric Company publishes *Statistical Quality Control Handbook*
1956	Total Quality Control (TQC) – Feigenbaum book & articles
April 9, 1959	Mil-Q-9858:1959 – Military Specification: Quality Program Requirements
1959	Edwards Medal first awarded by ASQC
1959	Donald Kirkpatrick's Learning Evaluation Model
1959	ASQC and the American Statistical Association jointly create *Technometrics*
1960s	The first "quality control circles" are formed in Japan and simple statistical methods were used for quality improvement

1960s	Statistical Process Control (SPC) – use of control charts to monitor processes
1960s	The concept of Kaizen is developed
1960	ASQC Education and Training Institute Board is formed
1962	Japan Productivity Center opens in USA (founded 1955)
1961-1964	The concept of Poka Yoke is developed by Shigeo Shingo
1963	Mil-Q-9858A:1963 – Military Specification: Quality Program Requirements
1964	Ford Motor Company issues the first Q-101 for suppliers – Canada version 1965
1966	Dr. Yoji Akao introduces Quality Function Deployment (QFD) Methodology
1967	Grant Medal first awarded by ASQC
1968	ASQC Certified Quality Engineer develop – 1st professional certification in quality
1968	Company-Wide Quality Control (CWQC) – Japanese style total quality
1968	Kaoru Ishikawa publishes the Guide to Quality Control
1968	ASQC *Industrial Quality Control* is replaced by two new publications, *Quality Progress* magazine and *The Journal of Quality Technology*
1969	Dr. Shingo Shigeo, as part of JIT, pioneers the concept of Single Minute Exchange of Dies (SMED)
1969	Ishikawa emphasizes the use of Seven Quality Tools
1969	ASQC co-sponsors the first International Congress in Quality Control, hosted by the Union of Japanese Scientists and Engineers in Tokyo
1970s	Dr. Taguchi promotes the concept of Quality Loss Function
1973	ASQC publishes *Glossary & Tables for Statistical Quality Control*
1977	International Association for Quality Circles founded
1979	BS 5750 is issued. This was later replaced with ISO 9001:1987
1979	Philip Crosby publishes his book "Quality is Free."
1979	Deming Medal First Awarded by the ASQC Metropolitan Section and ASQC 1996
June 24, 1980	NBC airs the television documentary "If Japan Can, Why Can't We?" narrated by Lloyd Dobyns
1980s	Professor Noriaki Kano develops the Kano model which classifies customer preferences into five categories: Attractive, One-Dimensional, Must-Be, Indifferent, Reverse
1980	QCI International is founded by Donald Dewar
1981	Lancaster Medal first awarded by ASQC
1981	*Quality Digest* – first issue published under the title *Quality Circle Digest*

@ 1981	Eastern Michigan University begins developing first Master in Quality with Ford Motor Co & W. Edwards Deming
1982	*Out of the Crisis*, published in 1982, Deming offers a theory of management based on his famous 14 Points for Management
@ 1982	Transformation of American Industry – jointly developed by Ford Motor, Jackson Community College & W. Edwards Deming
1984	The Register of Lead Assessors is created and evolves to the International Register of Certificated Auditors (IRCA)
1984	October is designated National Quality Month by USA Congress
1985	NASA Excellence Award is established
1985	Total Quality Management – USA Department of Defense drive for organizational improvements
1986-1988	Six Sigma formulated by Bill Smith in Motorola
1986	Masaaki Imai establishes the Kaizen Institute to help Western companies introduce Kaizen concepts, systems, and tools
March 15, 1987	ISO issues the first version of the ISO 9000 series. (ISO 9001:1987)
1987	Malcolm Baldrige National Quality Award is established.
1987	*Quality Circle Digest* drops "Circle" from its name to become *Quality Digest*
1988	ASQ Certified Quality Auditor established
1988	ISO 9001:1988 Quality management systems – Requirements – 2nd 1994, 3rd 2000, 4th 2008, 5th 2015, 6th TBD
1988	Motorola becomes the first company to win Malcolm Baldrige National Quality Award
1988	Ford Motor Company creates Supplier Quality Improvement (SQI) – first pre-launch supplier program in the USA (tied to APQP) – merged with SQA in 1993
September 15, 1988	Presidents of 14 European companies come together to create the European Foundation for Quality Management
1989	IQA celebrates the first World Quality Day on 9 November 1989
1990	Registrar Accreditation Board (RAB) is founded to support ISO 9001
1990	AIAG Measurement System Analysis (MSA) – 2nd 1995, 3rd 2002, 4th 2010
1991	ASQC exceeds 100,000 members
@ 1992	General Electric officially adopts Six Sigma
1992	AIAG Statistical Quality Control (SPC) – 2nd 2005
1992	ISO 19011:1992 Guidelines for auditing management systems – 2nd 2010, 3rd 2018
1993	Ishikawa Medal first awarded by ASQC
1993	AIAG Production Part Approval Process – 2nd 1995, 3rd 1999, 4th 2006

1993	AIAG Failure Modes & Effects Analysis (FMEA) – 2nd 1995, 3rd 2001, 4th 2008
1993	*Quality Management Journal* debuts
1994	The First American Customer Satisfaction Index is released
1994	Eleven elementary schools participate in the Koalaty Kid training initiative
1994	AIAG USA Automotive Industry releases Advance Product Quality Planning & Control Plan (APQP) – 2nd Ed 2008
1994	Quality Systems Requirements 9000 (QS-9000) quality standard developed by a joint effort of the 'Big Three' automakers, GM, Chrysler, and Ford – 2nd 1996, 3rd 1998
1994	ISO issues the second version of the ISO 9000 series. (ISO 9001:1994)
1994	ISO 9001:1994 Quality management systems – Requirements
1994	SAE International releases J1739:1994 Surface Vehicle Recommended Practice FMEA – 2000, 2002, 2009, 2021
1995	ASQ Certified Quality Manager program is developed
1995	General Electric (GE) launches the Six Sigma initiative
1996	ASQ announces the Pioneers In Quality
1996	ISO 14001:1996 Environmental management systems – Requirements – 2004, 2015
1996	ISO 13485:1996 Medical devices – Quality management systems – Requirements for regulatory purposes released (1996, 2003, 2016)
1997	ASQC drops 'Control' from its name, becomes ASQ
1998	Ford Motor Company launches Six Sigma Black Belt training for 3,000 personnel
1998	ASQ is named administrator of the QuEST Forum, which develops the TL 9000 telecommunications standards
1998	The journal *Software Quality Professional* debuts
1998	ASQ rolls out Six Sigma black belt training for individuals and organizations
1999	ISO/TS 16949:1999 – 1st Edition is released with updates 2002, 2009
1999	British Standards (BS) OHSAS 18001:1999 – Occupational Health and Safety Assessment – updated 2007 – migrates to ISO 45001:2018
2000	ISO issues the third version of the ISO 9000 series. (ISO 9001:2000)
2000	Juran Medal first awarded by ASQ
2000	Feigenbaum Medal first awarded by ASQ
2000	ISO 9001:2000 Quality management systems – Requirements
2000	A "Hazard Analysis and Critical Control Point" (HACCP) add-on is developed for the ASQ Certified Quality Auditor program

2000	ASQ's WorldPartner Program is founded
2001	ASQ Certified Quality Improvement Associate (CQIA) program is introduced
2001	ASQ relocates to the ASQ Center, its current world headquarters, in the landmark Gimbel's building in downtown Milwaukee
2001	ASQ's Six Sigma Forum is launched
2001	*Six Sigma Forum Magazine* debuts
2001	*Gazeta Global*, an international newsletter, is founded
2001	Freund-Marquardt Medal First Awarded by ASQ
2001	Lean Six Sigma – blending of Lean & Six Sigma
2002	Distinguished Service Medal First Awarded by ASQ
2002	Crosby Medal First Awarded by ASQ
2002	Calibration Technician Certification is introduced
2003	The Association for Quality and Participation (AQP) merges with ASQ
2004	ASQ Costa Rica becomes the first International Member Unit (equivalent to a Section) outside of North America
2004	Shainin Medal first awarded by ASQ
2004	ISO 14001:2004 Environmental management systems – Requirements – 2nd 2015
2005	ASQ China subsidiary is established
2005	Annual Quality Congress (AQC) changes name to World Conference on Quality and Improvement (WCQI)
2005	First certification for Quality Process Analysts is offered
2006	ASQ launches initiative to promote the Economic Case for Quality
2006	ASQ is named secretariat for the US TAG to the ISO Committee on Social Responsibility
2006	The IQA is awarded Royal Charter status
2006	Quality Magazine launched *Quality Plant of the Year* & *Quality Professional of the Year* – Roderick Munro named first recipient of Professional
2007	The IQA reforms as the Chartered Quality Institute
2007	Ron Atkinson is the first Canadian to serve as ASQ president
2007	British Standard releases BSI 18001:2007 Occupational health and safety management systems – Requirements released
2007-2009	Robert Woods Johnson Foundation sponsors Improving Performance In Practice (IPIP) for the Medical Industry
2008	ISO issues the fourth version of the ISO 9000 series. (ISO 9001:2008)
2008	ISO 9001:2008 Quality management systems – Requirements
2009	Hutchens Medal first awarded by ASQ

2011	ISO 50001:2011 Energy management systems – Requirements – 2nd 2018
2015	ISO issued the ISO 9001:2015 & ISO 14001:2015 under Annex SL
2015	ISO 9001:2015 Quality management systems – Requirements
2016	International Automotive Task Force (IATF) – IATF 16949:2016
2015	ISO 14001:2015 Environmental management systems – Requirements
2017	The CQI launches the inaugural International Quality Awards
2017	Hromi Medal first awarded by ASQ
2018	ISO 45001:2018 Occupational Health and Safety management systems – Requirements – replaces OHSAS 18001
2019	ISO 22301:2019 Security and resilience – Business continuity management systems – Requirements released
2020	COVID-19 world wide pandemic – things become more virtual
2021	Subir Chowdhury named to ASQ Distinguished Service Award

Appendix O

Rice & Munro Training Evaluation Model

The Rice & Munro Training Evaluation Model was derived from and expanded upon the *Kirkpatrick Model of Training Evaluation*. In the Kirkpatrick Model, there are typically four levels of evaluation:

Level 1: Reaction to the learning event (sometime called the "Smiley Sheet") – conducted immediately on completion of the training event.

Leve 2: Learning Gain—how much new learning occurred during the event (typically evaluation with a pre/post test or just a skills test). This is conducted immediately upon completion of the training/learning event. Example: the driving test a new forklift driver has to show that they understand the machine.

Level 3: Learner usage over time and level of useful information to their respective jobs. Usually conducted about six months to one year after the learning event.

Level 4: Manager (supervisor) observations of applied skills. Usually conducted about one year to a year and a half after the learning event.

Optional – Level: Return on Investment (ROI) for sending personnel through the training with improved productivity back on the job (sometimes included in Level 4).

The practical application of the Kirkpatrick model noted that most of the time, the physical analysis of both the Level Three and Level Four was not completed as time was not available to conduct the analysis. This left many learning interventions in limbo as training had to rely on experience versus data in ways to improve the learning interventions being created by many organizations. The other challenge was that if the training was delivered by consultants or professional training organizations, they might not have access to the attendee or attendee managers in the timing needed to conduct the analysis.

The Rice & Munro Training Evaluation Model was designed to address the issues with lack of data for the Level Three and Level Four analysis. In what should be all cases, any training/learning intervention should have a stated list of learning objectives for that effort to help new learners. Given that the Human Resource (HR) function of any organization should have who took which courses (or at least the attendee's supervisor), then there should be some record of what was supposed to have been learned by the employee who took the training. The HR or Training

Coordinator can then review when employees took what training programs and determine when the Level 3 and Level 4 need to be conducted. From the learning goals, questions can then be developed on what information would be useful to understand if the training was useful to the employee, was there value in the learning at the time of delivery, what needs to be changed for practical application back on the job, what the supervisor sees as value to their personnel attending the learning, etc.

Armed with this information, but lacking the time to physically go out and track down the personnel who need to be questioned, the Training Coordinator should provide both the questions and who needs to be asked what questions, to the ISO Internal Audit Coordinator for inclusion by the internal auditors when they are auditing the departments with the targeted employees.

Given that all ISO Management System registrations (examples: ISO 9001, ISO 14001, ISO 45001, ISO 50001, ISO 22301, etc.) require ongoing internal audits to be conducted, this becomes a practical method for the Training Coordinators to have access to course attendees without having to go out and physically conduct the interviews themselves. The full course data may take a little longer to collect; however, functionally, courses will have the full data needed to make meaningful program adjustments as needed.

Endnotes

Chapter 1

1. S. A. Marash, P. Berman, and M. Flynn, *Fusion Management: Harnessing the Power of Six Sigma, Lean, ISO 9001:2000, Malcom Baldrige, TQM and other Quality Breakthroughs of the Past Century* (Fairfax, VA: QSU, 2003).
2. M. J. Harry and J. R. Lawson, *Six Sigma Producibility Analysis and Process Characterization* (Boston, MA: Addison-Wesley, 1992).
3. R. T. Westcott, ed. *The Certified Manager of Quality/Organizational Excellence Handbook* (Milwaukee: ASQ Quality Press, 2006).
4. W. E. Deming, *Out of the Crisis* (Cambridge, MA: MIT Press, 1982, 1986).
5. Westcott, *Certified Manager of Quality.*
6. J. M. Juran, *Managerial Breakthrough: A New Concept of the Manager's Job* (New York: McGraw-Hill, 1964).
7. R. Munro, *Six Sigma for the Shop Floor* (Milwaukee: ASQ Quality Press, 2001).

Chapter 2

1. J. P. Womack and D. T. Jones, *Lean Thinking: Banish Waste and Create Wealth in Your Corporation* (New York: Simon & Schuster, 1996).
2. J. M. Morgan and J. K. Liker, *The Toyota Product Development System: Integrating People, Process, and Technology* (New York: Productivity Press, 2006).
3. J. P. Womack and D. T. Jones, *Lean Thinking: Banish Waste and Create Wealth in Your Corporation* (New York: Productivity Press, 2003).
4. J. K. Liker and G. L. Convis, *The Toyota Way to Lean Leadership* (New York: McGraw-Hill, 2012).
5. J. Casey, "5S Shakeup: Three Secrets for Sustaining 5S Success," *Quality Progress* (October 2013): 18.
6. D. K. Sobek and A. Smalley, *Understanding A3 Thinking: A Critical Component of Toyota's PDCA Management System* (Boca Raton, FL: CRC Press, 2008).
7. E. M. Goldratt, *Critical Chain* (Great Barrington, MA: The North River Press, 1997).
8. M. Rother and J. Shook, *Learning to See: Value Stream Mapping to Add Value and Eliminate Muda* (Brookline, MA: The Lean Enterprise Institute, Inc., 1998).
9. Ibid.

Chapter 3

1. L. Mundy, *A Journey to Quality Leadership—Quality^3* (Milwaukee: ASQ Quality Press, 2011).

2. D. Woodford, Design for Six Sigma—IDOV Methodology. http://www.isixsigma.com/library/content/c020819a.asp. 2002.
3. F. Breyfogle, *Implementing Six Sigma*, 2nd ed. (New York: John Wiley & Sons, 2003).
4. Pritchard and Tate, *The Risk Management Memory Jogger* (Salem, NH: Goal/QPC, 2013).
5. First Edition, February 1993; Second Edition, February 1995; Third Edition, July 2001; Fourth Edition, June 2008. Copyright © 1993, © 1995, © 2001, © 2008. Chrysler LLC, Ford Motor Company, General Motors Corporation. ISBN: 978-1-60534-136-1.
6. C. S. Carlson, *Effective FMEAs: Achieving Safe, Reliable, and Economical Products and Processes Using Failure Mode and Effects Analysis* (Hoboken, NJ: John Wiley & Sons, 2012).

Chapter 4

1. E. J. Rice, "Automotive Excellence: Know the History of Training to Promote Learning," *ASQ Automotive Division* 3, no. 4 (Fall 1997): 12–15.

Chapter 5

1. W. E. Deming, *The New Economics for Industry, Government, Education* (Cambridge: MIT, Center for Advanced Engineering Study, 1993).
2. G. Taguchi, S. Chowdhury, and S. Taguchi, *Robust Engineering: Learn How to Boost Quality While Reducing Costs and Time to Market* (New York: McGraw Hill, 1999).

Chapter 6

1. P.M. Senge, *The Fifth Discipline: The Art and Practice of the Learning Organization* (New York: Doubleday/Currency, 1990).
2. S.R. Covey, *The Seven Habits of Highly Effective People: Restoring the Character Ethic* (New York: Simon and Schuster, 1989).
3. H. Kerzner, *Project Management: A Systems Approach to Planning, Schedule, and Controls*, 8th ed. (New York: John Wiley & Sons, 2003).

Chapter 7

1. D. W. Benbow and T. M. Kubiak, *The Certified Six Sigma Black Belt Handbook* (Milwaukee: ASQ Quality Press, 2005).
2. N. R. Tague, *The Quality Toolbox*, 2nd ed. (Milwaukee: ASQ Quality Press, 2004).
3. Ibid.
4. D. H. Stamatis, *Total Quality Service: Principles, Practices, and Implementation* (Boca Raton, FL: CRC Press, 1996).
5. J. M. Juran and A. B. Godfrey, *Juran's Quality Handbook*, 5th ed. (New York: McGraw-Hill, 1999).

Chapter 8

1. D. W. Benbow and T. M. Kubiak, *The Certified Six Sigma Black Belt Handbook* (Milwaukee: ASQ Quality Press, 2005).
2. Ibid.
3. Ibid.

Chapter 9

1. B. W. Tuckman, "Developmental Sequence in Small Groups," *Psychological Bulletin* 63, no. 6 (November–December, 1965): 384–99.

Chapter 13

1. www.andrews.edu/~calkins/math/webtexts/prod12.htm.

Chapter 14

1. Copyright © 2000–2005 Minitab Inc.
2. AIAG, *Measurement System Analysis*, 4th ed. (Chrysler Group LLC, Ford Motor Company, General Motors Corporation, 2010).

Chapter 15

1. D. C. Montgomery, *Introduction to Statistical Quality Control*, 6th ed. (Hoboken, NJ: Wiley, 2012).

Chapter 19

1. R. A. Munro, M. J. Maio, M. B. Nawaz, G. Ramu, and D. J. Zrymiak, *The Certified Six Sigma Green Belt Handbook*, 1st ed. (Milwaukee: ASQ Quality Press, 2007).
2. J. M. Juran and F. M. Gryna, *Juran's Quality Control Handbook*, 4th ed. (New York: McGraw-Hill, 1988).
3. J. M. Juran and J. A. De Feo, *Juran's Quality Handbook: The Complete Guide to Performance Excellence*, 6th ed. (New York: McGraw-Hill, 2010).
4. C. M. Borror, *The Certified Quality Engineer Handbook*, 3rd ed. (Milwaukee: ASQ Quality Press, 2009).

Chapter 21

1. R. Snee and R. Hoerl, "Integrating Lean and Six Sigma—A Holistic Approach," *ASQ Six Sigma Forum* (May 2007).
2. R. A. Munro, M. J. Maio, M. B. Nawaz, G. Ramu, and D. J. Zrymiak, *The Certified Six Sigma Green Belt Handbook*, 1st ed. (Milwaukee: ASQ Quality Press, 2007).
3. A. Manos and C. Vincent, eds., *The Lean Handbook: A Guide to the Bronze Certification Body of Knowledge* (Milwaukee: ASQ Quality Press, 2012), 124–30.
4. Ibid., 154–60.
5. Ibid., 167–71.
6. Ibid., 240–41.
7. Ibid., 243–46.
8. J. R. Meredith and S. M. Shafer, *Operations Management for MBAs*, 4th ed. (New York: John Wiley & Sons, 2010).
9. Manos and Vincent, *Lean Handbook*, 81–83.
10. Ibid., 235–40.
11. M. Imai, *Gemba Kaizen: A Commonsense Approach to a Continuous Improvement Strategy*, 2nd ed. (New York: McGraw-Hill, 2012).

Chapter 22

1. R. A. Munro, M. J. Maio, M. B. Nawaz, G. Ramu, and D. J. Zrymiak, *The Certified Six Sigma Green Belt Handbook*, 1st ed. (Milwaukee: ASQ Quality Press, 2007).
2. J. M. Juran and F. M. Gryna, *Juran's Quality Control Handbook*, 4th ed. (New York: McGraw-Hill, 1988).
3. J. M. Juran and A. B. Godfrey, *Juran's Quality Handbook*, 5th ed. (New York: McGraw-Hill, 1998).
4. J. M. Juran and J. A. Defeo, *Juran's Quality Handbook: The Complete Guide to Performance Excellence*, 6th ed. (New York: McGraw-Hill, 2010).
5. W. E. Deming, *Quality, Productivity, and Competitive Position* (Cambridge, MA: Massachusetts Institute of Technology, 1982).

Chapter 23

1. R. A. Munro, M. J. Maio, M. B. Nawaz, G. Ramu, and D. J. Zrymiak, *The Certified Six Sigma Green Belt Handbook*, 1st ed. (Milwaukee: ASQ Quality Press, 2007).
2. J. M. Juran and F. M. Gryna, *Juran's Quality Control Handbook*, 4th ed. (New York: McGraw-Hill, 1988).

Chapter 24

1. M. Imai, *Gemba Kaizen: A Commonsense Approach to a Continuous Improvement Strategy*, 2nd ed. (New York: McGraw-Hill, 2012).
2. J. R. Meredith and S. M. Shafer, *Operations Management for MBAs*, 4th ed. (New York: John Wiley & Sons, 2010).
3. A. Manos and C. Vincent, eds., *The Lean Handbook: A Guide to the Bronze Certification Body of Knowledge.* (Milwaukee: ASQ Quality Press, 2012), 116–23.
4. Ibid., 359–69.
5. Ibid., 61, 73.
6. Ibid., 86–108.
7. Ibid., 167–85.
8. G. L. Duffy, J. W. Moran, and W. J. Riley, *Quality Function Deployment and Lean-Six Sigma Applications in Public Health* (Milwaukee: ASQ Quality Press, 2010).

Glossary

A

acceptance number—The maximum number of defects or defectives allowable in a sampling lot for the lot to be acceptable.

acceptance quality limit (AQL)—In a continuing series of lots, a quality level that, for the purpose of sampling inspection, is the limit of a satisfactory process average.

acceptance sampling—Inspection of a sample from a lot to decide whether to accept that lot. There are two types: attributes sampling and variables sampling. In *attributes sampling,* the presence or absence of a characteristic is noted in each of the units inspected. In *variables sampling,* the numerical magnitude of a characteristic is measured and recorded for each inspected unit; this involves reference to a continuous scale of some kind.

acceptance sampling plan—A specific plan that indicates the sampling sizes and associated acceptance or nonacceptance criteria to be used. In attributes sampling, for example, there are single, double, multiple, sequential, chain, and skip-lot sampling plans. In variables sampling, there are single, double, and sequential sampling plans. For detailed descriptions of these plans, see the standard ANSI/ISO/ASQ A3534-2-1993: *Statistics—Vocabulary and symbols—Statistical quality control.*

accuracy—The closeness of agreement between a test result or measurement result and the accepted/true value.[2]

activity based costing—An accounting system that assigns costs to a product based on the amount of resources used to design, order, or make it.

activity network diagram—A diagram that links tasks with direct arrows showing the path through the task list. Tasks are linked when a task is dependent on a preceding task.[3] (AKA *arrow diagram.*)

Advanced Product Quality Planning (APQP)—High-level automotive process for product realization, from design through production part approval.

affinity diagram—A management tool for organizing information (usually gathered during a brainstorming activity).

American National Standards Institute (ANSI)—A private, nonprofit organization that administers and coordinates the U.S. voluntary standardization and conformity assessment system. It is the U.S. member body in the International Organization for Standardization, known as ISO.

American Society for Quality (ASQ)—A global community of people dedicated to quality who share the ideas and tools that make our world work better. With individual and organizational members around the world, ASQ has the reputation and reach to bring together the diverse quality champions who are transforming the world's corporations, organizations, and communities to meet tomorrow's critical challenges.

analysis of means (ANOM)—A statistical procedure for troubleshooting industrial processes and analyzing the results of experimental designs with factors at fixed levels. It provides a graphical display of data. Ellis R. Ott developed the procedure in 1967 because he observed that nonstatisticians had difficulty understanding analysis of variance. Analysis of means is easier for quality practitioners to use because it is an extension of the control chart. In 1973, Edward G. Schilling further extended the concept, enabling analysis of means to be used with nonnormal distributions and attributes data in which the normal approximation to the binomial distribution does not apply. This is referred to as *analysis of means for treatment effects.*

analysis of variance (ANOVA)—A basic statistical technique for determining the proportion of influence a factor or set of factors has on total variation. It subdivides the total variation of a data set into meaningful component parts associated with specific sources of variation to test a hypothesis on the parameters of the model or to estimate variance components. There are three models: fixed, random, and mixed.

analytical (inferential) studies—A set of techniques used to arrive at a conclusion about a population based upon the information contained in a sample taken from that population.[1]

arrow diagram—A planning tool used to diagram a sequence of events or activities (nodes) and their interconnectivity. It is used for scheduling and especially for determining the critical path through nodes. (AKA *activity network diagram.*)

assignable cause—A name for the source of variation in a process that is not due to chance and therefore can be identified and eliminated. Also called "special cause."

attributes (discrete) data—Go/no-go information. The control charts based on attributes data include percent chart, number of affected units chart, count chart, count per unit chart, quality score chart, and demerit chart.

attributes, method of—Method of measuring quality that consists of noting the presence (or absence) of some characteristic (attribute) in each of the units under consideration and counting how many units do (or do not) possess it. Example: go/no-go gauging of a dimension.

audit—The on-site verification activity, such as inspection or examination, of a product, process, or quality system, to ensure compliance to requirements. An audit can apply to an entire organization or might be specific to a product, function, process, or production step.

Automotive Industry Action Group (AIAG)—A global automotive trade association with about 1600 member companies that focuses on common business processes, implementation guidelines, education, and training.

average chart—A control chart in which the subgroup average, $\bar{x}$, is used to evaluate the stability of the process level.

average outgoing quality (AOQ)—The expected average quality level of an outgoing product for a given value of incoming product quality.

average outgoing quality limit (AOQL)—The maximum average outgoing quality over all possible levels of incoming quality for a given acceptance sampling plan and disposal specification.

average run length (ARL)—On a control chart, the number of subgroups expected to be inspected before a shift in magnitude takes place.

average sample number (ASN)—The average number of sample units inspected per lot when reaching decisions to accept or reject.

average total inspection (ATI)—The average number of units inspected per lot, including all units in rejected lots. Applicable when the procedure calls for 100 percent inspection of rejected lots.

B

balanced scorecard—A management system that provides feedback on both internal business processes and external outcomes to continuously improve strategic performance and results.

Baldrige Award—See *Malcolm Baldrige National Quality Award.*

baseline measurement—The beginning point, based on an evaluation of output over a period of time, used to determine the process parameters prior to any improvement effort; the basis against which change is measured.

batch and queue—Producing more than one piece and then moving the pieces to the next operation before they are needed.

Bayes's theorem—A formula to calculate conditional probabilities by relating the conditional and marginal probability distributions of random variables.

benchmarking—A technique in which a company measures its performance against that of best-in-class companies, determines how those companies achieved their performance levels, and uses the information to improve its own performance. Subjects that can be benchmarked include strategies, operations, and processes.

benefit–cost analysis—An examination of the relationship between the monetary cost of implementing an improvement and the monetary value of the benefits achieved by the improvement, both within the same time period.

bias—The influence in a sample of a factor that causes the data population or process being sampled to appear different from what it actually is, typically in a specific direction.[3]

binomial distribution—A discrete distribution that is applicable whenever an experiment consists of n independent Bernoulli trials and the probability of an outcome, say, success, is constant throughout the experiment.[1]

Black Belt (BB)—Full-time team leader responsible for implementing process improvement projects—define, measure, analyze, improve, and control (DMAIC) or define, measure, analyze, design, and verify (DMADV)—within a business to drive up customer satisfaction and productivity levels.

block diagram—A diagram that shows the operation, interrelationships, and interdependencies of components in a system. Boxes, or blocks (hence the name), represent the components; connecting lines between the blocks represent interfaces. There are two types of block diagrams: a *functional block diagram*, which shows a system's subsystems and lower-level products and their interrelationships and which interfaces with other systems; and a *reliability block diagram*, which is similar to the functional block diagram but is modified to emphasize those aspects influencing reliability.

brainstorming—A technique teams use to generate ideas on a particular subject. Each person on the team is asked to think creatively and write down as many ideas as possible. The ideas are not discussed or reviewed until after the brainstorming session.

breakthrough improvement—A dynamic, decisive movement to a new, higher level of performance.

business process reengineering (BPR)—The concentration on improving business processes to deliver outputs that will achieve results meeting the firm's objectives, priorities, and mission.

C

calibration—The comparison of a measurement instrument or system of unverified accuracy to a measurement instrument or system of known accuracy to detect any variation from the required performance specification.

capability—The total range of inherent variation in a stable process determined by using data from control charts.

causation—The relationship between two variables. The changes in variable x cause changes in y. For example, a change in outdoor temperature causes changes in natural gas consumption for heating. If we can change x, we can bring about a change in y.

cause—An identified reason for the presence of a defect, problem, or effect.

cause-and-effect diagram—A tool for analyzing process dispersion. It is also referred to as the "Ishikawa diagram," because Kaoru Ishikawa developed it, and the "fishbone diagram," because the completed diagram resembles a fish skeleton. The diagram illustrates the main causes and subcauses leading to an effect (symptom). The cause-and-effect diagram is one of the "seven tools of quality."

c-chart—See *count chart.*

centerline—A line on a graph that represents the overall average (mean) operating level of the process.

central limit theorem—A theorem that states that irrespective of the shape of the distribution of a population, the distribution of sample means is approximately normal when the sample size is large.[1]

central tendency—The tendency of data gathered from a process to cluster toward a middle value somewhere between the high and low values of measurement.

certification—The result of a person meeting the established criteria set by a certificate granting organization.

Certified Six Sigma Black Belt (CSSBB)—An ASQ certification.

Certified Six Sigma Green Belt (CSSGB)—An ASQ certification.

Certified Six Sigma Yellow Belt (CSSYB)—An ASQ certification.

chain reaction—A chain of events described by W. Edwards Deming: improve quality, decrease costs, improve productivity, increase market share with better quality and lower price, stay in business, provide jobs, and provide more jobs.

chain sampling plan—In acceptance sampling, a plan in which the criteria for acceptance and rejection apply to the cumulative sampling results for the current lot and one or more immediately preceding lots.

champion—A business leader or senior manager who ensures that resources are available for training and projects, and who is involved in periodic project reviews; also an executive who supports and addresses Six Sigma organizational issues.

change agent—An individual from within or outside an organization who facilitates change in the organization; might be the initiator of the change effort, but not necessarily.

changeover—A process in which a production device is assigned to perform a different operation or a machine is set up to make a different part—for example, a new plastic resin and new mold in an injection molding machine.

changeover time—The time required to modify a system or workstation, usually including both teardown time for the existing condition and setup time for the new condition.

characteristic—The factors, elements, or measures that define and differentiate a process, function, product, service, or other entity.

chart—A tool for organizing, summarizing, and depicting data in graphic form.

charter—A written commitment approved by management stating the scope of authority for an improvement project or team.

check sheet—A simple data recording device. The check sheet is custom-designed by the user, which allows him or her to readily interpret the results. The check sheet is one of the "seven tools of quality."

checklist—A tool for ensuring that all important steps or actions in an operation have been taken. Checklists contain items important or relevant to an issue or situation. Checklists are often confused with check sheets.

chi-square distribution—Probability distribution of sum of squares of n independent normal variables.[1]

classification of defects—The listing of possible defects of a unit, classified according to their seriousness. Note: Commonly used classifications: class A, class B, class C, class D; or critical, major, minor, and incidental; or critical, major, and minor. Definitions of these classifications require careful preparation and tailoring to the product(s) being sampled to ensure accurate assignment of a defect to the proper classification. A separate acceptance sampling plan is generally applied to each class of defects.

common causes—Causes of variation that are inherent in a process over time. They affect every outcome of the process and everyone working in the process. (AKA *chance causes*.) Also see *special causes*.

compliance—The state of an organization that meets prescribed specifications, contract terms, regulations, or standards.

conformance—An affirmative indication or judgment that a product or service has met the requirements of a relevant specification, contract, or regulation.

conformity assessment—All activities concerned with determining that relevant requirements in standards or regulations are fulfilled, including sampling, testing, inspection, certification, management system assessment and registration, accreditation of the competence of those activities, and recognition of an accreditation program's capability.

constraint—Anything that limits a system from achieving higher performance or throughput; also, the bottleneck that most severely limits the organization's ability to achieve higher performance relative to its purpose or goal.

consumer—The external customer to whom a product or service is ultimately delivered; also called end user.

continuous (variables) data—Data that vary with discontinuity across an interval. The values of continuous data are often represented by floating point numbers. In sampling, continuous data are often referred to as variables data.[3]

continuous flow production—A method in which items are produced and moved from one processing step to the next, one piece at a time. Each process makes only the one piece that the next process needs, and the transfer batch size is one. Also referred to as *one-piece flow* and *single-piece flow.*

continuous improvement (CI)—Sometimes called *continual improvement.* The ongoing improvement of products, services, or processes through incremental and breakthrough improvements.

continuous quality improvement (CQI)—A philosophy and attitude for analyzing capabilities and processes and improving them repeatedly to achieve customer satisfaction.

continuous sampling plan—In acceptance sampling, a plan, intended for application to a continuous flow of individual units of product, that involves acceptance and rejection on a unit-by-unit basis and employs alternate periods of 100 percent inspection and sampling. The relative amount of 100 percent inspection depends on the quality of submitted product. Continuous sampling plans usually require that each *t* period of 100 percent inspection be continued until a specified number *i* of consecutively inspected units is found clear of defects. Note: For single-level continuous sampling plans, a single *d* sampling rate (for example, inspect one unit in five or one unit in 10) is used during sampling. For multilevel continuous sampling plans, two or more sampling rates can be used. The rate at any given time depends on the quality of submitted product.

control chart—A chart with upper and lower control limits on which values of some statistical measure for a series of samples or subgroups are plotted. The chart frequently shows a central line to help detect a trend of plotted values toward either control limit.

control limits—The natural boundaries of a process within specified confidence levels, expressed as the upper control limit (UCL) and the lower control limit (LCL).

control plan (CP)—Written description of the systems for controlling part and process quality by addressing the key characteristics and engineering requirements.

corrective action—A solution meant to reduce or eliminate an identified problem.

corrective action recommendation (CAR)—The full cycle corrective action tool that offers ease and simplicity for employee involvement in the corrective action/process improvement cycle.

correlation (statistical)—A measure of the relationship between two data sets of variables.

cost of poor quality (COPQ)—The costs associated with providing poor-quality products or services. There are four categories: internal failure costs (costs associated with defects found before the customer receives the product or service), external failure costs (costs associated with defects found after the customer receives the product or service), appraisal costs (costs incurred to determine the degree of conformance to quality requirements), and prevention costs (costs incurred to keep failure and appraisal costs to a minimum).

cost of quality (COQ)—Another term for COPQ. It is considered by some to be synonymous with COPQ but is considered by others to be unique. While the two concepts emphasize the same ideas, some disagree as to which concept came first and which categories are included in each.

count chart—A control chart for evaluating the stability of a process in terms of the count of events of a given classification occurring in a sample; known as a "*c*-chart."

count per unit chart—A control chart for evaluating the stability of a process in terms of the average count of events of a given classification per unit occurring in a sample.

C_p—The ratio of tolerance to six sigma, or the upper specification limit (USL) minus the lower specification limit (LSL) divided by six sigma. It is sometimes referred to as the engineering tolerance divided by the natural tolerance and is only a measure of dispersion.

C_{pk} index—Equals the lesser of the USL minus the mean divided by three sigma (or the mean) minus the LSL divided by three sigma. The greater the C_{pk} value, the better.

C_{pm}—Used when a target value within the specification limits is more significant than overall centering.[3]

critical path method (CPM)—An activity-oriented project management technique that uses arrow-diagramming techniques to demonstrate both the time and the cost required to complete a project. It provides one time estimate: normal time.

critical to quality (CTQ)—A characteristic of a product or service that is essential to ensure customer satisfaction.[2]

cumulative sum control chart (CUSUM)—A control chart on which the plotted value is the cumulative sum of deviations of successive samples from a target value. The ordinate of each plotted point represents the algebraic sum of the previous ordinate and the most recent deviations from the target.

customer relationship management (CRM)—A strategy for learning more about customers' needs and behaviors to develop stronger relationships with them. It brings together information about customers, sales, marketing effectiveness, responsiveness, and market trends. It helps businesses use technology and human resources to gain insight into the behavior of customers and the value of those customers.

customer satisfaction—The result of delivering a product or service that meets customer requirements.

cycle time—The time required to complete one cycle of an operation. If cycle time for every operation in a complete process can be reduced to equal takt time, products can be made in single-piece flow. Also see *takt time.*

cyclical variation—Looks at the piece-to-piece changes in consecutive order. Patterns are identified in groups, batches, or lots of units.[3]

D

data—A set of collected facts. There are two basic kinds of numerical data: measured or variables data, such as "16 ounces," "4 miles," and "0.75 inches," and counted or attributes data, such as "go/no go" or "yes/no."

D-chart—See *demerit chart*.

decision matrix—A matrix teams use to evaluate problems or possible solutions. For example, a team might draw a matrix to evaluate possible solutions, listing them in the far left vertical column. Next, the team selects criteria to rate the possible solutions, writing them across the top row. Then, each possible solution is rated on a scale of 1 to 5 for each criterion, and the rating is recorded in the corresponding grid. Finally, the ratings of all the criteria for each possible solution are added to determine its total score. The total score is then used to help decide which solution deserves the most attention.

defect—A product's or service's nonfulfillment of an intended requirement or reasonable expectation for use, including safety considerations. There are four classes of defects: class 1, very serious, leads directly to severe injury or catastrophic economic loss; class 2, serious, leads directly to significant injury or significant economic loss; class 3, major, is related to major problems with respect to intended normal or reasonably foreseeable use; and class 4, minor, is related to minor problems with respect to intended normal or reasonably foreseeable use.

defective—A defective unit; a unit of product that contains one or more defects with respect to the quality characteristic(s) under consideration.

demerit chart—A control chart for evaluating a process in terms of a demerit (or quality score); in other words, a weighted sum of counts of various classified nonconformities.

Deming cycle—Another term for the plan–do–study–act cycle. Walter Shewhart created it (calling it the plan–do–check–act cycle), but W. Edwards Deming popularized it, calling it plan–do–study–act.

dependability—The degree to which a product is operable and capable of performing its required function at any randomly chosen time during its specified operating time, provided that the product is available at the start of that period. (Nonoperation related influences are not included.) Dependability can be expressed by the following ratio: time available divided by (time available + time required).

design for Six Sigma (DFSS)—Used for developing a new product or process, or for processes that need total overhaul. A process often used in DFSS is called DMADV: define, measure, analyze, design, verify.[4] See also *DMADV*.

design of experiments (DOE)—A branch of applied statistics dealing with planning, conducting, analyzing, and interpreting controlled tests to evaluate the factors that control the value of a parameter or group of parameters.

design record—Engineering requirements, typically contained in various formats; examples include engineering drawings, math data, and referenced specifications.

deviation—In numerical data sets, the difference or distance of an individual observation or data value from the center point (often the mean) of the set distribution.

dissatisfiers—The features or functions a customer expects that either are not present or are present but not adequate; also pertains to employees' expectations.

distribution (statistical)—The amount of potential variation in the outputs of a process, typically expressed by its shape, average, or standard deviation.

DMADV—A data-driven quality strategy for designing products and processes; it is an integral part of a Six Sigma quality initiative. It consists of five interconnected phases: define, measure, analyze, design, and verify.

DMAIC—A data-driven quality strategy for improving processes, and an integral part of a Six Sigma quality initiative. DMAIC is an acronym for define, measure, analyze, improve, and control.

Dodge-Romig sampling plans—Plans for acceptance sampling developed by Harold F. Dodge and Harry G. Romig. Four sets of tables were published in 1940: single sampling lot tolerance tables, double sampling lot tolerance tables, single sampling average outgoing quality limit tables, and double sampling average outgoing quality limit tables.

downtime—Lost production time during which a piece of equipment is not operating correctly due to breakdown, maintenance, power failures, or similar events.

E

effect—The result of an action being taken; the expected or predicted impact when an action is to be taken or is proposed.

effectiveness—The state of having produced a decided on or desired effect.

efficiency—The ratio of the output to the total input in a process.

efficient—A term describing a process that operates effectively while consuming minimal resources (such as labor and time).

eight wastes—Taiichi Ohno originally enumerated seven wastes (muda) and later added *underutilized people* as the eighth waste commonly found in physical production. The eight are (1) overproduction ahead of demand, (2) waiting for the next process, worker, material, or equipment, (3) unnecessary transport of materials (for example, between functional areas of facilities, or to or from a stockroom or warehouse), (4) overprocessing of parts due to poor tool and product design, (5) inventories more than the absolute minimum, (6) unnecessary movement by employees during the course of their work (such as to look for parts, tools, prints, or help), (7) production of defective parts, (8) underutilization of employees' brainpower, skills, experience, and talents.

eighty–twenty (80–20)—A term referring to the Pareto principle, which was first defined by J. M. Juran in 1950. The principle suggests that most effects come from relatively few causes; that is, 80 percent of the effects come from 20 percent of the possible causes. Also see *Pareto chart*.

enumerative (descriptive) studies—A group of methods used for organizing, summarizing, and representing data using tables, graphs, and summary statistics.[1]

error detection—A hybrid form of error-proofing. It means a bad part can be made but will be caught immediately, and corrective action will be taken to prevent another bad part from being produced. A device is used to detect and stop the process when a bad part is made. This is used when error-proofing is too expensive or not easily implemented.

error-proofing—Use of process or design features to prevent the acceptance or further processing of nonconforming products. Also known as *mistake-proofing*.

experimental design—A formal plan that details the specifics for conducting an experiment, such as which responses, factors, levels, blocks, treatments, and tools are to be used.

external customer—A person or organization that receives a product, service, or information but is not part of the organization supplying it. Also see *internal customer*.

external failure—Nonconformance identified by the external customers.

F

failure—The inability of an item, product, or service to perform required functions on demand due to one or more defects.

failure cost—The cost resulting from the occurrence of defects. One element of cost of quality or cost of poor quality.

failure mode analysis (FMA)—A procedure to determine which malfunction symptoms appear immediately before or after a failure of a critical parameter in a system. After all possible causes are listed for each symptom, the product is designed to eliminate the problems.

failure mode and effects analysis (FMEA)—A systematized group of activities to recognize and evaluate the potential failure of a product or process and its effects, identify actions that could eliminate or reduce the occurrence of the potential failure, and document the process.

***F*-distribution**—A continuous probability distribution of the ratio of two independent chi-square random variables.[1]

first in, first out (FIFO)—Use of material produced by one process in the same order by the next process. A FIFO queue is filled by the supplying process and emptied by the customer process. When a FIFO lane gets full, production is stopped until the next (internal) customer has used some of that inventory.

first-pass yield (FPY)—Also referred to as the *quality rate,* the percentage of units that completes a process and meets quality guidelines without being scrapped, rerun, retested, returned, or diverted into an offline repair area. FPY is calculated by dividing the units entering the process minus the defective units by the total number of units entering the process.

first-time quality (FTQ)—Calculation of the percentage of good parts at the beginning of a production run.

fishbone diagram—See *cause-and-effect diagram.*

fitness for use—A term used to indicate that a product or service fits the customer's defined purpose for that product or service.

five S (5S)—Five Japanese terms beginning with "s" used to create a workplace suited for visual control and lean production. *Seiri* means to separate needed tools, parts, and instructions from unneeded materials and to remove the unneeded ones. *Seiton* means to neatly arrange and identify parts and tools for ease of use. *Seiso* means to conduct a cleanup campaign. *Seiketsu* means to conduct seiri, seiton, and seiso daily to maintain a workplace in perfect condition. *Shitsuke* means to form the habit of always following the first four S's.

five whys—A technique for discovering the root causes of a problem and showing the relationship of causes by repeatedly asking the question, "Why?"

flow—The progressive achievement of tasks along the value stream so a product proceeds from design to launch, order to delivery, and raw to finished materials in the hands of the customer with no stoppages, scrap, or backflows.

flowchart—A graphical representation of the steps in a process. Flowcharts are drawn to better understand processes. One of the "seven tools of quality."

force-field analysis—A technique for analyzing what aids or hinders an organization in reaching an objective. An arrow pointing to an objective is drawn down the middle of a piece of paper. The factors that will aid the objective's achievement, called the driving forces, are listed on the left side of the arrow. The factors that will hinder its achievement, called the restraining forces, are listed on the right side of the arrow.

G

gage repeatability and reproducibility (GR&R)—The evaluation of a gauging instrument's accuracy by determining whether its measurements are repeatable (there is close agreement among a number of consecutive measurements of the output for the same value of the input under the same operating conditions) and reproducible (there is close agreement among repeated measurements of the output for the same value of input made under the same operating conditions over a period of time).

Gantt chart—A type of bar chart used in process planning and control to display planned and finished work in relation to time.

geometric dimensioning and tolerancing (GD&T)—A set of rules and standard symbols to define part features and relationships on an engineering drawing depicting the geometric relationship of part features and allowing the maximum tolerance that permits full function of the product.

go/no-go—State of a unit or product. Two parameters are possible: go (conforms to specifications) and no-go (does not conform to specifications).

Green Belt (GB)—An employee who has been trained in the Six Sigma improvement method at a Green Belt level and will lead a process improvement or quality improvement team as part of his or her full-time job.

H

Hawthorne effect—The concept that every change results (initially, at least) in increased productivity.

heijunka—A method of leveling production, usually at the final assembly line, that makes just-in-time production possible. It involves averaging both the volume and sequence of different model types on a mixed-model production line. Using this method avoids excessive batching of different types of product and volume fluctuations in the same product.

histogram—A graphic summary of variation in a set of data. The pictorial nature of a histogram lets people see patterns that are difficult to detect in a simple table of numbers. One of the "seven tools of quality."

hoshin kanri—The selection of goals, projects to achieve the goals, designation of people and resources for project completion, and establishment of project metrics.

hoshin planning—Breakthrough planning. A Japanese strategic planning process in which a company develops up to four vision statements that indicate where the company should be in the next five years. Company goals and work plans are developed based on the vision statements. Periodic submitted audits are then conducted to monitor progress. Also see *value stream*.

house of quality—A product planning matrix, somewhat resembling a house, that is developed during quality function deployment and shows the relationship of customer requirements to the means of achieving these requirements.

I

in-control process—A process in which the statistical measure being evaluated is in a state of statistical control; in other words, the variations among the observed sampling results can be attributed to a constant system of chance causes (common causes). Also see *out-of-control process*.

incremental improvement—Improvement implemented on a continual basis.

indicators—Established measures to determine how well an organization is meeting its customers' needs and other operational and financial performance expectations.

inputs—The products, services, and material obtained from suppliers to produce the outputs delivered to customers.

inspection—Measuring, examining, testing, and gauging one or more characteristics of a product or service and comparing the results with specified requirements to determine whether conformity is achieved for each characteristic.

inspection, normal—Inspection used in accordance with a sampling plan under ordinary circumstances.

inspection, 100 percent—Inspection of all the units in the lot or batch.

inspection cost—The cost associated with inspecting a product to ensure that it meets the internal or external customer's needs and requirements; an appraisal cost.

inspection lot—A collection of similar units or a specific quantity of similar material offered for inspection and acceptance at one time.

internal customer—The recipient (person or department) within an organization of another person's or department's output (product, service, or information). Also see *external customer.*

internal failure—A product failure that occurs before the product is delivered to external customers.

International Organization for Standardization—A network of national standards institutes from 157 countries working in partnership with international organizations, governments, industry, business, and consumer representatives to develop and publish international standards; acts as a bridge between public and private sectors.

interrelationship diagram—A management tool that depicts the relationship among factors in a complex situation; also called a *relations diagram.*

Ishikawa diagram—See *cause-and-effect diagram.*

J

jidoka—The deliberate effort to automate a process with a human touch. It means that when a problem occurs on a production line, a worker or machine is able to stop the process and prevent defective goods from being produced.

just-in-time (JIT) manufacturing—An optimal material requirement planning system for a manufacturing process in which there is little or no manufacturing material inventory on hand at the manufacturing site and little or no incoming inspection.

K

kaizen—A Japanese term that means gradual unending improvement by doing little things better and setting and achieving increasingly higher standards. Masaaki Imai made the term famous in his book *Kaizen: The Key to Japan's Competitive Success.*

kanban—A Japanese term for one of the primary tools of a just-in-time system. It maintains an orderly and efficient flow of materials throughout the entire manufacturing process. It is usually a printed card that contains specific information such as part name, description, and quantity.

key performance indicator (KPI)—A statistical measure of how well an organization is doing in a particular area. A KPI could measure a company's financial performance or how it is holding up against customer requirements.

key process characteristic—A process parameter that can affect safety or compliance with regulations, fit, function, performance, or subsequent processing of product.

key product characteristic—A product characteristic that can affect safety or compliance with regulations, fit, function, performance, or subsequent processing of product.

L

leadership—An essential part of a quality improvement effort. Organization leaders must establish a vision, communicate that vision to those in the organization, and provide the tools and knowledge necessary to accomplish the vision.

lean—Producing the maximum sellable products or services at the lowest operational cost while optimizing inventory levels and eliminating waste.

lean enterprise—A manufacturing company organized to eliminate all unproductive effort and unnecessary investment, both on the shop floor and in office functions.

lean manufacturing/production—An initiative focused on eliminating all waste in manufacturing processes. Principles of lean manufacturing include zero waiting time, zero inventory, scheduling (internal customer pull instead of push system), batch to flow (cut batch sizes), line balancing, and cutting actual process times. The production systems are characterized by optimum automation, just-in-time supplier delivery disciplines, quick changeover times, high levels of quality, and continuous improvement.

lean migration—The journey from traditional manufacturing methods to one in which all forms of waste are systematically eliminated.

linearity—Refers to measurements being statistically different from one end of the measurement space to the other. For example, a measurement process may be very capable of measuring small parts but much less accurate measuring large parts, or one end of a long part can be measured more accurately than the other.[3]

lot—A defined quantity of product accumulated under conditions considered uniform for sampling purposes.

lot, batch—A definite quantity of some product manufactured under conditions of production that are considered uniform.

lot quality—The value of percentage defective or of defects per hundred units in a lot.

lot size (also referred to as *N*)—The number of units in a lot.

lower control limit (LCL)—Control limit for points below the central line in a control chart.

M

maintainability—The probability that a given maintenance action for an item under given usage conditions can be performed within a stated time interval when the maintenance is performed under stated conditions using stated procedures and resources.

Malcolm Baldrige National Quality Award (MBNQA)—An award established by the U.S. Congress in 1987 to raise awareness of quality management and recognize U.S. companies that have implemented successful quality management systems. Awards can be given annually in six categories: manufacturing, service, small business, education, healthcare, and nonprofit. The award is named after the late Secretary of Commerce Malcolm Baldrige, a proponent of quality management. The U.S. Commerce Department's National Institute of Standards and Technology manages the award, and ASQ administers it.

Master Black Belt (MBB)—Six Sigma or quality expert responsible for strategic implementations in an organization. An MBB is qualified to teach other Six Sigma facilitators the methods, tools, and applications in all functions and levels of the company, and is a resource for using statistical process control in processes.

matrix diagram—A planning tool for displaying the relationships among various data sets.

mean—A measure of central tendency; the arithmetic average of all measurements in a data set.

mean time between failures (MTBF)—The average time interval between failures for repairable product for a defined unit of measure; for example, operating hours, cycles, and miles.

measure—The criteria, metric, or means to which a comparison is made with output.

measurement—The act or process of quantitatively comparing results with requirements.

median—The middle number or center value of a set of data in which all the data are arranged in sequence.

metric—A standard for measurement.

MIL-STD-105E—A military standard that describes the sampling procedures and tables for inspection by attributes.

mistake-proofing—Use of production or design features to prevent the manufacture or passing downstream of a nonconforming product; also known as *error-proofing*.

mode—The value occurring most frequently in a data set.

muda—Japanese for *waste;* any activity that consumes resources but creates no value for the customer.

multivariate control chart—A control chart for evaluating the stability of a process in terms of the levels of two or more variables or characteristics.

multivoting—Typically used after brainstorming, multivoting narrows a large list of possibilities to a smaller list of the top priorities (or to a final selection) by allowing items to be ranked in importance by participants. Multivoting is preferable to straight voting because it allows an item that is favored by all, but not the top choice of any, to rise to the top.[4]

N

n—The number of units in a sample.

N—The number of units in a population.

nominal group technique (NGT)—A technique, similar to brainstorming, used to generate ideas on a particular subject. Team members are asked to silently write down as many ideas as possible. Each member is then asked to share one idea, which is recorded. After all the ideas are recorded, they are discussed and prioritized by the group.

nonconformity—The nonfulfillment of a specified requirement.

nondestructive testing and evaluation (NDT, NDE)—Testing and evaluation methods that do not damage or destroy the product being tested.

nonlinear parameter estimation—A method whereby the arduous and labor-intensive task of multiparameter model calibration can be carried out automatically under the control of a computer.

nonparametric tests—All tests involving ranked data (data that can be put in order). Nonparametric tests are often used in place of their parametric counterparts when certain assumptions about the underlying population are questionable. For example, when comparing two independent samples, the Wilcoxon Mann-Whitney test (see entry) does not assume that the difference between the samples is normally distributed, whereas its parametric counterpart, the two-sample *t*-test, does. Nonparametric tests can be, and often are, more powerful in detecting population differences when certain assumptions are not satisfied.

non-value-added—A term that describes a process step or function that is not required for the direct achievement of process output. This step or function is identified and examined for potential elimination. Also see *value-added.*

normal distribution (statistical)—The charting of a data set in which most of the data points are concentrated around the average (mean), thus forming a bell-shaped curve.

O

operating characteristic curve (OC curve)—A graph to determine the probability of accepting lots as a function of the lots' or processes' quality level when using various sampling plans. There are three types: type A curves, which give the probability of acceptance for an individual lot coming from finite production (will not continue in the future); type B curves, which give the probability of acceptance for lots coming from a continuous process; and type C curves, which (for a continuous sampling plan) give the long-run percentage of product accepted during the sampling phase.

operations—Work or steps to transform raw materials to finished product.

out of spec—A term that indicates a unit does not meet a given requirement or specification.

out-of-control process—A process in which the statistical measure being evaluated is not in a state of statistical control. In other words, the variations among the observed sampling results cannot be attributed to a constant system of chance causes. Also see *in-control process.*

outputs—Products, materials, services, or information provided to customers (internal or external), from a process.

P

paired-comparison tests—Examples are two-mean, equal variance *t*-test; two-mean, unequal variance *t*-test; paired *t*-test; and *F*-test.

Pareto chart—A graphical tool for ranking causes from most significant to least significant. It is based on the Pareto principle, which was first defined by Joseph M. Juran in 1950. The principle, named after 19th-century economist Vilfredo Pareto, suggests that most effects come from relatively few causes; that is, 80 percent of the effects come from 20 percent of the possible causes. One of the "seven tools of quality."

parts per million (ppm)—A method of stating the performance of a process in terms of actual nonconforming material, which can include rejected, returned, or suspect material in the calculation.

***p*-chart**—See *percent chart.*

percent chart—A control chart for evaluating the stability of a process in terms of the percentage of the total number of units in a sample in which an event of a given classification occurs. Also referred to as a *proportion chart.*

plan–do–check–act (PDCA) cycle—A four-step process for quality improvement. In the first step (plan), a way to effect improvement is developed. In the second step (do), the plan is carried out, preferably on a small scale. In the third step (check), a study takes place comparing what was predicted and what was observed in the previous step. In the last step (act), action is taken on the causal system to effect the desired change. The plan–do–check–act cycle is sometimes referred to as the Shewhart cycle, because Walter A. Shewhart discussed the concept in his book *Statistical Method from the Viewpoint of Quality Control*, and as the Deming cycle, because W. Edwards Deming introduced the concept in Japan. The Japanese subsequently called it the Deming cycle. Also called the *plan–do–study–act (PDSA) cycle*.

point of use—A technique that ensures people have exactly what they need to do their jobs—work instructions, parts, tools, and equipment—where and when they need them.

Poisson distribution—A discrete probability distribution that expresses the probability of a number of events occurring in a fixed time period if these events occur with a known average rate and are independent of the time since the last event.

poka-yoke—Japanese term that means mistake-proofing. A poka-yoke device is one that prevents incorrect parts from being made or assembled, or easily identifies a flaw or error.

positional variation—Type of variation frequently within-piece, but can also include machine-to-machine variation, line-to-line or plant-to-plant variation, within-batch variation, and test positioning variation.[3]

P_p (process performance index)—An index describing process performance in relation to specified tolerance.[2]

P_{pk} (minimum process performance index)—The smaller of upper process performance index and lower process performance index.[2]

practical significance—At least as important as the question of statistical significance, practical or economic significance determines whether an observed sample difference is large enough to be of practical interest.

precision—The aspect of measurement that addresses repeatability or consistency when an identical item is measured several times.

prevention cost—The cost incurred by actions taken to prevent a nonconformance from occurring; one element of cost of quality or cost of poor quality.

preventive action—Action taken to remove or improve a process to prevent potential future occurrences of a nonconformance.

prioritization matrix—An L-shaped matrix that uses pairwise comparisons of a list of options to a set of criteria in order to choose the best option(s). First, the importance of each criterion is decided. Then, each criterion is considered separately, with each option rated for how well it meets the criterion. Finally, all the ratings are combined for a final ranking of options. Numerical calculations ensure a balance between the relative importance of the criteria and the relative merits of the options.[4]

probability (statistical)—The likelihood of occurrence of an event, action, or item.

procedure—The steps in a process and how these steps are to be performed for the process to fulfill a customer's requirements; usually documented.

process—A set of interrelated work activities characterized by a set of specific inputs and value-added tasks that make up a procedure for a set of specific outputs.

process average quality—Expected or average value of process quality.

process capability—A statistical measure of the inherent process variability of a given characteristic. The most widely accepted formula for process capability is Six Sigma.

process capability index—The value of the inherent tolerance specified for the characteristic divided by the process capability. The several types of process capability indices include the widely used C_{pk} and C_p.

process control—The method for keeping a process within boundaries; the act of minimizing the variation of a process.

process decision program charts (PDPC)—A variant of tree diagrams, a PDPC can be used as a simple alternative to FMEA.[3]

process flow diagram—A depiction of the flow of materials through a process, including any rework or repair operations; also called a *process flow chart*.

process improvement—The application of the plan–do–check–act cycle (see entry) to processes to produce positive improvement and better meet the needs and expectations of customers.

process management—The pertinent techniques and tools applied to a process to implement and improve process effectiveness, hold the gains, and ensure process integrity in fulfilling customer requirements.

process map—A type of flowchart depicting the steps in a process and identifying responsibility for each step and key measures.

process owner—The person who coordinates the various functions and work activities at all levels of a process, has the authority or ability to make changes in the process as required, and manages the entire process cycle to ensure performance effectiveness.

process performance management—The overseeing of process instances to ensure their quality and timeliness; can also include proactive and reactive actions to ensure a good result.

process quality—The value of percentage defective or of defects per hundred units in product from a given process. Note: The symbols "*p*" and "*c*" are commonly used to represent the true process average in fraction defective or defects per unit, and "100p" and "100c" the true process average in percentage defective or in defects per hundred units.

production part approval process (PPAP)—A "Big Three" automotive process that defines the generic requirements for approval of production parts, including production and bulk materials. Its purpose is to determine during an actual production run at the quoted production rates whether all customer engineering design record and specification requirements are properly understood by the supplier and that the process has the potential to produce product consistently meeting these requirements.

program evaluation and review technique (PERT) charts—Developed during the Nautilus submarine program in the 1950s, a PERT chart resembles an activity network diagram in that it shows task dependencies. It calculates best, average, and worst expected completion times.[3]

project management—The application of knowledge, skills, tools, and techniques to a broad range of activities to meet the requirements of a particular project.

project team—Manages the work of a project. The work typically involves balancing competing demands for project scope, time, cost, risk, and quality, satisfying stakeholders with differing needs and expectations, and meeting identified requirements.

proportion chart—See *percent chart*.

pull system—An alternative to scheduling individual processes in which the customer process withdraws the items it needs as at a supermarket, and the supplying process produces to replenish what was withdrawn; used to avoid push. Also see *kanban*.

Q

quality—A subjective term for which each person or sector has its own definition. In technical usage, quality can have two meanings: 1. the characteristics of a product or service that bear on its ability to satisfy stated or implied needs; 2. a product or service free of deficiencies. According to Joseph M. Juran, quality means "fitness for use"; according to Philip Crosby, it means "conformance to requirements."

quality assurance/quality control (QA/QC)—Two terms that have many interpretations because of the multiple definitions for the words "assurance" and "control." For example, "assurance" can mean the act of giving confidence, the state of being certain, or the act of making certain; "control" can mean an evaluation to indicate needed corrective responses, the act of guiding, or the state of a process in which the variability is attributable to a constant system of chance causes. (For a detailed discussion on the multiple definitions, see ANSI/ISO/ASQ A3534-2, *Statistics—Vocabulary and symbols—Statistical quality control.*) One definition of quality assurance is: all the planned and systematic activities implemented within the quality system that can be demonstrated to provide confidence that a product or service will fulfill requirements for quality. One definition for quality control is: the operational techniques and activities used to fulfill requirements for quality. Often, however, "quality assurance" and "quality control" are used interchangeably, referring to the actions performed to ensure the quality of a product, service, or process.

quality audit—A systematic, independent examination and review to determine whether quality activities and related results comply with plans and whether these plans are implemented effectively and are suitable to achieve the objectives.

quality costs—See *cost of poor quality*.

quality function deployment (QFD)—A structured method in which customer requirements are translated into appropriate technical requirements for each stage of product development and production. The QFD process is often referred to as listening to the voice of the customer.

quality loss function—A parabolic approximation of the quality loss that occurs when a quality characteristic deviates from its target value. The quality loss function is expressed in monetary units: the cost of deviating from the target increases quadratically the farther the quality characteristic moves from the target. The formula used to compute the quality loss function depends on the type of quality characteristic being used. The quality loss function was first introduced in this form by Genichi Taguchi.

quality management (QM)—The application of a quality management system in managing a process to achieve maximum customer satisfaction at the lowest overall cost to the organization while continuing to improve the process.

quality management system (QMS)—A formalized system that documents the structure, responsibilities, and procedures required to achieve effective quality management.

queue time—The time a product spends in a line awaiting the next design, order processing, or fabrication step.

quick changeover—The ability to change tooling and fixtures rapidly (usually within minutes) so multiple products can be run on the same machine.

R

random cause—A cause of variation due to chance and not assignable to any factor.

random sampling—A commonly used sampling technique in which sample units are selected so all combinations of n units under consideration have an equal chance of being selected as the sample.

range (statistical)—The measure of dispersion in a data set (the difference between the highest and lowest values).

range chart (R chart)—A control chart in which the subgroup range R evaluates the stability of the variability within a process.

rational subgrouping—Subgrouping wherein the variation is presumed to be only from random causes.[2]

regression analysis—A statistical technique for determining the best mathematical expression describing the functional relationship between one response variable and one or more independent variables.

relations diagram—See *interrelationship diagram.*

reliability—The probability of a product's performing its intended function under stated conditions without failure for a given period of time.

repeatability—The variation in measurements obtained when one measurement device is used several times by the same person to measure the same characteristic on the same product.

reproducibility—The variation in measurements made by different people using the same measuring device to measure the same characteristic on the same product.

requirements—The ability of an item to perform a required function under stated conditions for a stated period of time.

risk management—Using managerial resources to integrate risk identification, risk assessment, risk prioritization, development of risk handling strategies, and mitigation of risk to acceptable levels.

risk priority number (RPN)—The product of the severity, occurrence, and detection values determined in FMEA. The higher the RPN, the more significant the failure mode.

robustness—The condition of a product or process design that remains relatively stable, with a minimum of variation, even though factors that influence operations or usage, such as environment and wear, are constantly changing.

root cause—A factor that caused a nonconformance and should be permanently eliminated through process improvement.

run chart—A chart showing a line connecting numerous data points collected from a process running over time.

S

sample—In acceptance sampling, one or more units of product (or a quantity of material) drawn from a lot for purposes of inspection to reach a decision regarding acceptance of the lot.

sample size (*n*)—The number of units in a sample.

sample standard deviation chart (*s*-chart)—A control chart in which the subgroup standard deviation s is used to evaluate the stability of the variability within a process.

scatter diagram—A graphical technique to analyze the relationship between two variables. Two sets of data are plotted on a graph, with the y-axis being used for the variable to be predicted and the *x*-axis being used for the variable to make the prediction. The graph will show possible relationships (although two variables might appear to be related, they might not be; those who know most about the variables must make that evaluation). One of the "seven tools of quality."

seven tools of quality—Tools that help organizations understand their processes to improve them. The tools are the cause-and-effect diagram, check sheet, control chart, flowchart, histogram, Pareto chart, and scatter diagram.

seven wastes—See *eight wastes.*

Shewhart cycle—See *plan–do–check–act cycle.*

sigma—One standard deviation in a normally distributed process.

single-piece flow—A process in which products proceed one complete product at a time, through various operations in design, order taking, and production without interruptions, backflows, or scrap.

SIPOC diagram—A tool used by Six Sigma process improvement teams to identify all relevant elements (suppliers, inputs, process, outputs, customers) of a process improvement project before work begins.

Six Sigma—A method that provides organizations tools to improve the capability of their business processes. This increase in performance and decrease in process variation lead to defect reduction and improvement in profits, employee morale, and quality of products or services. Six Sigma quality is a term generally used to indicate that a process is well controlled ($\pm 6\sigma$ from the centerline in a control chart).

six sigma quality—A term generally used to indicate process capability in terms of process spread measured by standard deviations in a normally distributed process.

special causes—Causes of variation that arise because of special circumstances. They are not an inherent part of a process. Special causes are also referred to as *assignable causes.* Also see *common causes.*

specification—A document that states the requirements to which a given product or service must conform.

stages of team growth—Four stages that teams move through as they develop maturity: forming, storming, norming, and performing.

standard deviation (statistical)—A computed measure of variability indicating the spread of the data set around the mean.

standard work—A precise description of each work activity, specifying cycle time, takt time, the work sequence of specific tasks, and the minimum inventory of parts on hand needed to conduct the activity. All jobs are organized around human motion to create an efficient sequence without waste. Work organized in such a way is called standard(ized) work. The three elements that make up standard work are takt time, working sequence, and standard in-process stock.

standard work instructions—A lean manufacturing tool that enables operators to observe a production process with an understanding of how assembly tasks are to be performed. It ensures that the quality level is understood and serves as an excellent training aid, enabling replacement or temporary individuals to easily adapt and perform the assembly operation.

statistical process control (SPC)—The application of statistical techniques to control a process; often used interchangeably with the term *statistical quality control.*

statistical quality control (SQC)—The application of statistical techniques to control quality. Often used interchangeably with the term *statistical process control,* although statistical quality control includes acceptance sampling, which statistical process control does not.

statistical significance—Level of accuracy expected of an analysis of data. Most frequently it is expressed as either a "95 percent level of significance" or "five percent confidence level."[5]

strengths, weaknesses, opportunities, threats (SWOT) analysis—A strategic technique used to assess an organization's competitive position.

Student's *t*-distribution—A continuous distribution of the ratio of two independent random variables—a standard normal and a chi-square.[1]

supplier—A source of materials, service, or information input provided to a process.

supplier quality assurance—Confidence that a supplier's product or service will fulfill its customers' needs. This confidence is achieved by creating a relationship between the customer and supplier that ensures that the product will be fit for use with minimal corrective action and inspection. According to Joseph M. Juran, nine primary activities are needed: (1) define product and program quality requirements, (2) evaluate alternative suppliers, (3) select suppliers, (4) conduct joint quality planning, (5) cooperate with the supplier during the execution of the contract, (6) obtain proof of conformance to requirements, (7) certify qualified suppliers, (8) conduct quality improvement programs as required, (9) create and use supplier quality ratings.

supply chain—The series of suppliers to a given process.

system—A group of interdependent processes and people that together perform a common mission.

T

Taguchi methods—The American Supplier Institute's trademarked term for the quality engineering methodology developed by Genichi Taguchi. In this engineering approach to quality control, Taguchi calls for off-line quality control, online quality control, and a system of experimental design to improve quality and reduce costs.

takt time—The rate of customer demand, takt time is calculated by dividing production time by the quantity of product the customer requires in that time. Takt is the heartbeat of a lean manufacturing system. Also see *cycle time.*

team—A group of individuals organized to work together to accomplish a specific objective. Also see *stages of team growth.*

temporal variation—The time-to-time or shift-to-shift variation—that is, variation across time.[3]

theory of constraints (TOC)—A lean management philosophy that stresses removal of constraints to increase throughput while decreasing inventory and operating expenses. TOC's set of tools examines the entire system for continuous improvement. The current reality tree, conflict resolution diagram, future reality tree, prerequisite tree, and transition tree are the five tools used in TOC's ongoing improvement process. Also called *constraints management.*

throughput—The rate at which the system generates money through sales, or the conversion rate of inventory into shipped product.

tolerance—The maximum and minimum limit values a product can have and still meet customer requirements.

total productive maintenance (TPM)—A series of methods, originally pioneered by Nippondenso (a member of the Toyota group), to ensure that every machine in a production process is always able to perform its required tasks so production is never interrupted.

total quality management (TQM)—A term coined by the Naval Air Systems Command to describe its Japanese-style management approach to quality improvement. Since then, TQM has taken on many meanings. Simply put, it is a management approach to long-term success through customer satisfaction. TQM is based on all members of an organization participating in improving processes, products, services, and the culture in which they work. The methods for implementing this approach are found in the teachings of such quality leaders as Philip B. Crosby, W. Edwards Deming, Armand V. Feigenbaum, Kaoru Ishikawa, and Joseph M. Juran.

Toyota Production System (TPS)—The production system developed by Toyota Motor Corp. to provide best quality, lowest cost, and shortest lead time through eliminating waste. TPS is based on two pillars: just-in-time and jidoka. TPS is maintained and improved through iterations of standardized work and kaizen.

tree diagram—A management tool that depicts the hierarchy of tasks and subtasks needed to complete an objective. The finished diagram bears a resemblance to a tree.

trend—The graphical representation of a variable's tendency, over time, to increase, decrease, or remain unchanged.

trend control chart—A control chart in which the deviation of the subgroup average, $\bar{x}$, from an expected trend in the process level is used to evaluate the stability of a process.

TRIZ—A Russian acronym for a theory of innovative problem solving.

***t*-test**—A method to assess whether the means of two groups are statistically different from each other.

type I error—An incorrect decision to reject something (such as a statistical hypothesis or a lot of products) when it is acceptable.

type II error—An incorrect decision to accept something when it is unacceptable.

U

***u*-chart**—Count-per-unit chart.

unit—An object for which a measurement or observation can be made; commonly used in the sense of a "unit of product," the entity of product inspected to determine whether it is defective or nondefective.

upper control limit (UCL)—Control limit for points above the central line in a control chart.

V

validation—The act of confirming that a product or service meets the requirements for which it was intended.

validity—The ability of a feedback instrument to measure what it was intended to measure; also, the degree to which inferences derived from measurements are meaningful.

value stream—All activities, both value-added and non-value-added, required to bring a product from raw material state into the hands of the customer, bring a customer requirement from order to delivery, and bring a design from concept to launch. Also see *hoshin planning*.

value stream mapping—A pencil and paper tool used in two stages. First, follow a product's production path from beginning to end and draw a visual representation of every process in the material and information flows. Second, draw a future state map of how value should flow. The most important map is the future state map.

value-added—A term used to describe activities that transform input into a customer (internal or external)–usable output.

variables (attributes) data—Measurement information. Control charts based on variables data include average ($\bar{x}$) chart, range (R) chart, and sample standard deviation (s) chart.

variation—A change in data, characteristic, or function caused by one of four factors: special causes, common causes, tampering, or structural variation.

verification—The act of determining whether products and services conform to specific requirements.

voice of the customer—The expressed requirements and expectations of customers relative to products or services, as documented and disseminated to the providing organization's members.

W

waste—Any activity that consumes resources and produces no added value to the product or service a customer receives. Also known as *muda.*

Wilcoxon Mann-Whitney test—Used to test the null hypothesis that two populations have identical distribution functions against the alternative hypothesis that the two distribution functions differ only with respect to location (median), if at all. It does not require the assumption that the differences between the two samples are normally distributed. In many applications, it is used in place of the two-sample *t*-test when the normality assumption is questionable. This test can also be applied when the observations in a sample of data are ranks, that is, ordinal data, rather than direct measurements.

X

x-bar ($\bar{x}$) chart—Average chart.

Z

zero defects—A performance standard and method Philip B. Crosby developed, which states that if people commit themselves to watching details and avoiding errors, they can move closer to the goal of zero defects.

Endnotes

Source: Except where noted, definitions reproduced with permission of ASQ, http://www.asq.org/glossary/index.html. The glossary was compiled by *Quality Progress* magazine editorial staff members Dave Nelsen, Assistant Editor, and Susan E. Daniels, Editor at Large. Volunteers James Bossert, R. Dan Reid, and James Rooney reviewed the content.

1. Reproduced by permission of Bhisham C. Gupta and H. Fred Walker, *Applied Statistics for the Six Sigma Green Belt* (Milwaukee: ASQ Quality Press, 2005).
2. Reproduced by permission of the ASQ Statistics Division, *Glossary and Tables for Statistical Quality Control,* 4th ed. (Milwaukee: ASQ Quality Press, 2004).
3. Reproduced by permission of Kim H. Pries, *Six Sigma for the Next Millennium* (Milwaukee: ASQ Quality Press, 2006).
4. Reproduced by permission of Nancy R. Tague, *The Quality Toolbox,* 2nd ed. (Milwaukee: ASQ Quality Press, 2005).
5. Reproduced by permission of Donald L. Siebels, *The Quality Improvement Glossary.* (Milwaukee: ASQ Quality Press, 2004).

References

AIAG (Automotive Industry Action Group). *Measurement System Analysis Reference Manual*, 3rd ed. DaimlerChrysler Corporation, Ford Motor Company, General Motors Corporation, 2001.

———. *Statistical Process Control (SPC)*, 2nd ed. DaimlerChrysler Corporation, Ford Motor Company, General Motors Corporation, 2005.

———. *Advance Product Quality Planning*, 2nd ed. DaimlerChrysler Corporation, Ford Motor Company, General Motors Corporation, 2008.

———. *Potential Failure Mode and Effects Analysis (FMEA) Reference Manual*, 4th ed. DaimlerChrysler Corporation, Ford Motor Company, General Motors Corporation, 2008.

———. *Measurement Systems Analysis (MSA)*, 4th ed. DaimlerChrysler Corporation, Ford Motor Company, General Motors Corporation, 2010.

AT&T. *Statistical Quality Control Handbook*. Milwaukee: ASQC Quality Press, 1985.

Barrentine, L. *Concepts for R&R Studies*, 2nd ed. Milwaukee: ASQ Quality Press, 2003.

Benbow, D. W., and T. M. Kubiak. *The Certified Six Sigma Black Belt Handbook*. Milwaukee: ASQ Quality Press, 2005.

Berger, R., D. Benbow, A. Elshennawy, and H. F. Walker. *The Certified Quality Engineer Handbook*, 2nd ed. Milwaukee: ASQ Quality Press, 2006.

Bodek, N. *Kaikaku: The Power and Magic of Lean*. Vancouver, WA: PCS Press, 2004.

Borror, C. M. *The Certified Quality Engineer Handbook*, 3rd ed. Milwaukee: ASQ Quality Press, 2009.

Brassard, M., and D. Ritter. *The Memory Jogger 2: Tools for Continuous Improvement and Effective Planning*. Salem, NH: GOAL/QPC, 2010.

Breyfogle, F. W. *Implementing Six Sigma*. New York: John Wiley & Sons, 1999.

Chowdhury, S. *Design for Six Sigma: The Revolutionary Process for Achieving Extraordinary Profit*. Chicago: Dearborn Trade, 2002.

———. *The Ice Cream Maker*. New York: Doubleday, 2005.

———. *The Power of LEO: The Revolutionary Process for Achieving Extraordinary Results*. New York: McGraw-Hill, 2012.

Coleman, L.B. *The ASQ Certified Quality Auditor Handbook*, 5th ed. Milwaukee: ASQ Quality Press, 2020.

Day, R. G. *Quality Function Deployment*. Milwaukee: ASQC Quality Press, 1993.

Deming, W. E. *Out of the Crisis*. Cambridge, MA: MIT Press, 2000.

Duffy, G. L., J. W. Moran, and W. J. Riley. *Quality Function Deployment and Lean Six Sigma Applications in Public Health*. Milwaukee: ASQ Quality Press, 2010.

eStats: Engineering Statistics Handbook by NIST. www.itl.nist.gov/div898/handbook/index.htm.

Galloway, D. *Mapping Work Processes*. Milwaukee: ASQC Quality Press, 1994.

Gould, L. "Building Better Vehicles via Axiomatic Design." *Automotive Design and Production* (June 2000).

Gryna, F., R. Chua, and J. Defeo. *Juran's Quality Planning and Analysis for Enterprise Quality,* 5th ed. New York: McGraw-Hill, 2005.

Imai, M. *Gemba Kaizen: A Commonsense Approach to a Continuous Improvement Strategy,* 2nd ed. New York: McGraw-Hill, 2012.

ISO (International Organization for Standardization). ISO 9000 *Quality management systems—Fundamentals and vocabulary.* Geneva: ISO, 2015.

———. ISO 9001 *Quality management systems—Requirements.* Geneva: ISO, 2015.

———. ISO 5807:1985 *Information processing—Documentation symbols and conventions for data, program and system flowcharts, program network charts and system resources charts.* Geneva: ISO, 1985.

———. ISO 19011:2018 *Guidelines for Auditing Management Systems.* Geneva: ISO, 2018.

Juran, J. M. *Juran's Quality Control Handbook,* 5th ed. New York: McGraw Hill, 1999.

Juran, J. M., and J. A. Defeo. *Juran's Quality Handbook: The Complete Guide to Performance Excellence,* 6th ed. New York: McGraw-Hill, 2010.

Juran, J. M., and F. M. Gryna. *Juran's Quality Control Handbook,* 4th ed. New York: McGraw-Hill, 1988.

Keys, T and A.R. Zeff. *How to Be a Successful Technical Trainer: Core Skills for Instructor Certification.* New York: McGraw-Hill, 2000.

Kirkpatrick, J. D., and W. K. Kirkpatrick, *Kirkpatrick's Four Levels of Training Evalution.* Alexandria, VA: Association for Talent Development, 2016.

Kubiak, T. M. "An Integrated Approach System." *Quality Progress* (July 2003).

———. *The Certified Six Sigma Master Black Belt Handbook.* Milwaukee: ASQ Quality Press, 2012.

Lane, B and J.-D. Phyper, *Systems Integration, ISO 9001 & ISO 14001: Convergence of the Standards.* Toronto: Templegate Information Services, 1998.

Liker, J. K., and D. P. Meier. *Toyota Talent: Developing Your People the Toyota Way.* New York: McGraw-Hill, 2007.

Manos, A., and C. Vincent, eds. *The Lean Handbook: A Guide to the Bronze Certification Body of Knowledge.* Milwaukee: ASQ Quality Press, 2012.

McNair, C. J., and K. H. J. Leibfried. *Benchmarking: A Tool for Continuous Improvement.* Esssex Junction, VT: Oliver Wight, 1992.

Meredith, J. R., and S. M. Shafer. *Operations Management for MBAs,* 4th ed. New York: John Wiley & Sons, 2010.

Montgomery, D. C. *Design and Analysis of Experiments,* 8th ed. Hoboken, NJ: John Wiley & Sons, 2008.

———. *Introduction to Statistical Quality Control,* 7th ed. Hoboken, NJ: John Wiley & Sons, 2012.

Munro, R. A. *Six Sigma for the Office.* Milwaukee: ASQ Quality Press, 2003.

———. *Six Sigma for the Shop Floor.* Milwaukee: ASQ Quality Press, 2001.

———. "Training In Quality." *Quality Digest* (August 2006).

———. "Green Belts: A Road Map for Success." *Quality Digest* (December 2005).

Munro, R. A., M. J. Maio, M. B. Nawaz, G. Ramu, and D. J. Zrymiak. *The Certified Six Sigma Green Belt Handbook,* 1st ed. Milwaukee: ASQ Quality Press, 2007.

Munro, R.A., and E.J. Rice-Munro. "Continual Improvement of Training." *Quality Digest* (August 2004).

———. "Learning Styles, Teaching Approaches, and Technology." *ASQ Journal of Quality and Participation* (Spring 2004).

Ramu, Govind. "Expert Answers" *Quality Progress* (November 2014).

———. "Expert Answers" *Quality Progress* (October 2018).

———. "Expert Answers" *Quality Progress* (May 2019).

Rantanen, K., and E. Domb. *Simplified TRIZ: New Problem-Solving Applications for Engineers and Manufacturing Professionals*. Boca Raton, FL: St. Lucie Press, 2002.

Snee, R., and R. Hoerl. "Integrating Lean and Six Sigma—A Holistic Approach." *ASQ Six Sigma Forum* (May 2007).

Tague, N. R. *The Quality Toolbox*. Milwaukee: ASQ Quality Press, 2004.

Wader, M. *Lean Tools—Pocket Guide to Implementing Lean Practices*. Surfside, FL: QSU, 2003.

Westcott, R. *Certified Manager of Quality/Organizational Excellence Handbook*. Milwaukee: ASQ Quality Press, 2005.

Wheeler, D. *Understanding Statistical Process Control*, 2nd ed. Knoxville, TN: SPC Press, 1992.

Womack, J. P., and D. T. Jones. *Lean Thinking*. New York: Free Press, 2003.

Selected Websites/Organizations

Deming Prize. https://juse.or.jp/deming_en/
GOAL/QPC. www.goalqpc.com/
ASQ Quality Press. www.asq.org/quality-press
Juran Institute. www.juran.com/
Leader to Leader Institute (Drucker). www.pfdf.org/
Malcolm Baldrige Award Criteria. http://www.nist.gov/baldrige/
Minitab. www.minitab.com/
NIST. www.nist.gov/
PQ Systems. https://www.pqsystems.com/
Shainin Red X. https://shainin.com/
Shingo Prize. www.shingoprize.org/
Six Sigma Forum Magazine. http://asq.org/sixsigma/
The Gemba Academy. https://www.gembaacademy.com/
The W. Edwards Deming Institute. www.deming.org/
www.animatedsoftware.com/statglos/sgcltheo.htm
www.qualitytrainingportal.com/resources/fmea/index.htm
www.todaysengineer.org/2005/Mar/team_dynamics.asp

Index

A

B

C

D

E

H

I

J

K

N

O

P

Q

R

S

T

U

V